Binomial Theorem

$$(a + b)^n = a^n + na^{n-1}b + \frac{n(n-1)}{2}a^{n-2}b^2 + \frac{n(n-1)(n-2)}{3 \cdot 2}a^{n-3}b^3 + \cdots + b^n.$$

The rth term is $\dfrac{n(n-1)(n-2)\cdots(n-r+2)}{(r-1)(r-2)\cdots 2}a^{n-(r-1)}b^{r-1}$.

Progressions

Arithmetic

$$a_n = a_1 + (n-1)d$$

$$\sum_{i=1}^{n} a_i = n\left(\frac{a_1 + a_n}{2}\right)$$

Geometric

$$a_n = r^{(n-1)}a_1$$

$$\sum_{i=1}^{n} a_i = a_1\left(\frac{1 - r^n}{1 - r}\right)$$

DISCARDED

College Algebra

College Algebra

Chris Vancil

University of Kentucky

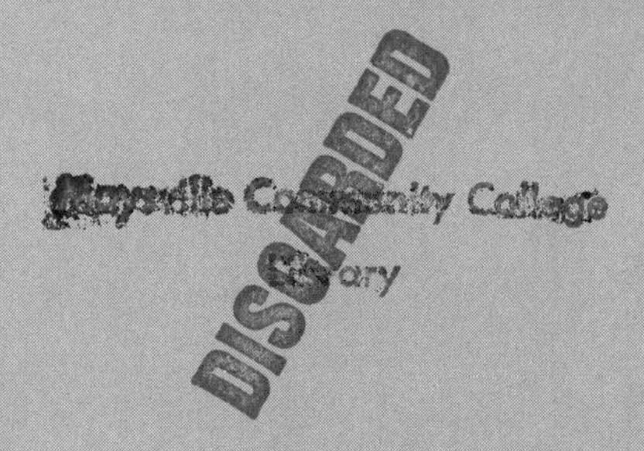

Macmillan Publishing Co., Inc.
New York
Collier Macmillan Publishers
London

Macmillan Publishing Co., Inc.
866 Third Avenue, New York, New York 10022

Collier Macmillan Canada, Inc.

Library of Congress Cataloging in Publication Data

Vancil, Chris N.
College algebra.

Includes index.
1. Algebra. I. Title.
QA154.2.V367 1983 512.9 82-18653
ISBN 0-02-422420-0

Printing: 12345678 Year: 34567890

ISBN 0-02-422420-0

To My Parents
with Love

Preface

This book is a presentation of the fundamentals of algebra. One objective has been to write a text that students find understandable and that instructors find easy to use. Because proper organization facilitates student mastery of the material, another objective has been to develop the topics into a natural, logical progression. Toward achieving these goals, the book contains the following features.

- The presentation is precise, yet the writing style is informal. The format is basically that of theorem-proof-examples, although not every proof appears. Proofs are included only when they are especially interesting or when they provide insight.
- Each section contains exercises (with answers) integrated into the exposition. These problems are similar to preceding examples and allow students to test their understanding before proceeding. They may also be used by the instructor to provide additional classroom examples.
- Each section has a set of carefully written problems that provide practice in the mechanical and conceptual aspects of the subject. The end of each chapter contains a problem set that spans the entire chapter. Problems within each set are carefully arranged, becoming progressively more difficult, conceptually or manipulatively. For the most part, the problems occur in matching pairs, with an even-numbered problem very similar to the problem immediately preceding it.
- Useful, realistic applications appear throughout the text.
- Excluding proofs, answers to all odd-numbered problems of the problem sets and to all problems of the chapter reviews are included in the Appendix.
- Proofs and answers to the remaining problems appear in the Instructor's Resource Manual, which is available to instructors upon request. This supplement also provides a test bank with answers. It contains three exams per chapter and three comprehensive exams. The manual is perforated to provide easy duplication.
- Chapter 0 serves as a refresher of the fundamentals of algebra. Because the instructor may wish to skip this chapter or make it a reading assignment, answers to all of the problems in this chapter are included in the Appendix.
- The conic sections occur in Chapter 2, "Relations." This early presence helps sharpen a student's manipulative skills before reaching the more conceptual topics and allows a wider range of functions to be studied in Chapter 3.
- The chapter on polynomial and rational functions, Chapter 4, follows the chapter on functions. It is thus integrated into the mainstream of the text.

Further, the topics of synthetic division and zeros of polynomials appear in this chapter, where the presentations are linked to the graphing of polynomial and rational functions.

Comments About Calculators

With the availability of inexpensive, electronic calculators, I have assumed that each student possesses one that performs the arithmetic operations and takes square roots. However, there are no specially marked problems designed for calculators. Because my goal is to teach concepts, develop manipulative skills, and promote understanding, most problems may be solved about as easily without a calculator as with one.

When necessary, tables are introduced, and calculations are based on the tables. However, because many students have scientific calculators, the use of tables is not overemphasized. There is no discussion of linear interpolation. And although a common logarithms table and its explanation are included in the Appendix, only two problems require its use. The use of calculators should be encouraged where appropriate. For those with scientific calculators, the principal advantages are efficiency and increased accuracy. (Students must be reminded, however, that calculations in the text use tables and may differ slightly from those obtained with a calculator.)

Acknowledgments

I wish to thank those who have listened to, and responded to, my ideas and questions. This includes Mike Mears, Marylynn Blackburn, Don Bennett, Betty Vanderpool, Brauch Fugate, Ken Spackman, Tom Moss, and Mike Seiler. In this regard, a special thanks goes to my good friend Cliff Swauger.

I also wish to thank the following individuals who reviewed the manuscript and offered helpful suggestions: Professors M. Michael Awad, Southwest Missouri State University; Samuel Councilman, California State University—Long Beach; Larry S. Dilley, Central Missouri State University; Marilyn Gilchrist, North Lake College; Arthur M. Hobbs, Texas A&M University; Kendall Hyde, Weber State College; Joseph E. Quinn, University of North Carolina—Charlotte; Jerry Reed, Mississippi State University; Richard Semmler, Northern Virginia Community College; Carroll G. Wells, Western Kentucky University; Thomas J. Woods, Central Connecticut State University; and Michael Ziegler, Marquette University.

I am grateful to Beth Kaufman, Elsa Nadler, and Virginia Gandy for typing the preliminary notes for use in the classroom, and to Sue Carl and Angie Charlton for typing the manuscript.

Finally, I wish to thank Paula and Judy for their encouragement and the numerous unmentioned students and teachers who have used the preliminary editions. Their comments, suggestions, and support are sincerely appreciated.

C.V.

Contents

3 Functions 155

4 Polynomial and Rational Functions 187

5 Exponential and Logarithmic Functions 231

6 Systems of Equations 257

College Algebra

Topics for Review

0

The study of college algebra requires basic manipulative skills from intermediate algebra. This chapter is a review of topics that are fundamental to the rest of the text.

0.1 Sets

A **set** is a collection of objects. When we refer to a set of dishes or to a set of tires, we are talking about a collection of these items. The objects belonging to a set are called **elements** or **members** of the set.

One method of specifying a set is to list the elements between braces. For example, the set of the three smallest natural numbers may be written

$$\{1, 2, 3\}.$$

We shall frequently use symbols, often capital letters, to denote sets. Further, we use the symbol "$\in$" to indicate that an object belongs to a particular set. Hence, if we let A denote the set above, then the statement

$$1 \in A$$

means that 1 is an element of A. On the other hand, the statement

$$4 \notin A$$

tells us that 4 is not an element of A.

Although most of the sets discussed in this text will consist of real numbers, the elements may be anything. For example, we could discuss a set of letters like

$$\{a, b, c\},$$

or a set of names like

{John, Mary, Judy, Cliff},

or even a set of sets like

$$\{\{1, 2\}, \{a, b, c\}\}.$$

When there are numerous objects belonging to a set, we may sometimes use an abbreviated notation for the listing. For example, the set of all positive odd integers less than 100 may be written

$$\{1, 3, 5, \ldots, 99\}.$$

The ". . ." means "and so on." That is, continue in the same pattern until we reach 99. The disadvantage of this notation is that we must discover the pattern. Hence, there are other methods of *defining* a set.

To **define a set** means to provide criteria which enable one to determine which objects belong to the set. In addition to simply listing all the elements, we may define a set by describing its elements. For example, the preceding set may be defined as

$$\{x | x \text{ is a positive odd integer less than } 100\},$$

which is read "the set of all elements x such that x is a positive odd integer less than 100."

The letter x used in defining this set is an example of a *variable*. A **variable** is a symbol which represents each of *two or more elements*. A symbol which represents *one element only*, such as 2 or π, is called a **constant.**

As mentioned earlier, most sets considered in this text will be sets of real numbers. Some commonly used sets are the set of all real numbers, which will be denoted R, the set of all integers, which will be denoted Z, and the set of all positive integers (the natural numbers), which will be denoted $\mathbf{Z}^+$.

In a given discussion, the specified set to which all elements under consideration belong is called the **universal set** (and is usually denoted U). That special set which contains *no* elements is called the **empty set** or **null set.** It may be written { } but is usually denoted $\varnothing$. Note that the set

$$\{\varnothing\}$$

is *not* the empty set. It is a set having one element, which happens to be another set, the empty set.

Let A and B be nonempty sets. Then A and B are said to be **equal,** denoted $A = B$, provided that they have the same elements. Consequently, a given set remains the same when the elements are rearranged or repeated. Thus we see that

$$\{1, 2, 3\} = \{3, 2, 1\}$$

and

$$\{1, 2, 3\} = \{1, 1, 2, 3\}.$$

Rearrangement or repetition of the elements of a set does not produce a new set. In order for two sets A and B to be unequal, denoted $A \neq B$, there must exist an element in one that is not in the other.

Let A and B be sets. The **union** of A and B, denoted $A \cup B$, is the set of all elements which belong to A or to B (or to both). Thus

$$A \cup B = \{x | x \in A \text{ or } x \in B\}.$$

Such an operation may be depicted in terms of a **Venn diagram** (a plane figure illustrating the relationships between sets). The shaded portion of each diagram in Figure 0.1 corresponds to the union of A and B since that region is the one which contains all points belonging to region A or to region B. For example, if

$$A = \{1, 2, 3\} \quad \text{and} \quad B = \{3, 4, 5\},$$

then

$$A \cup B = \{1, 2, 3, 4, 5\}.$$

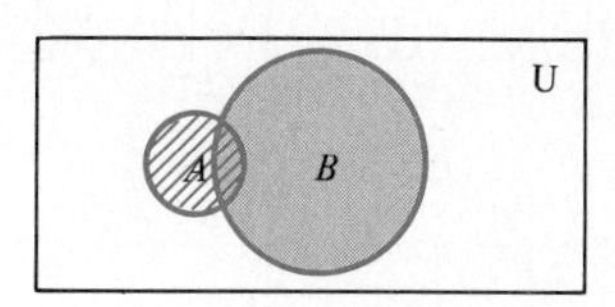

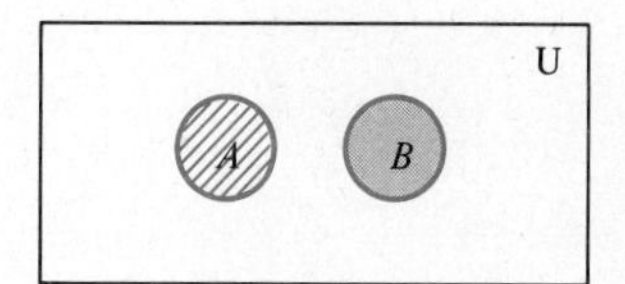

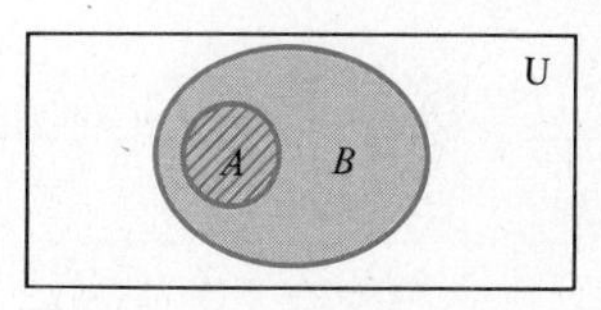

Figure 0.1 $A \cup B$

From the definition of union of two sets we see that

$$A \cup B = B \cup A.$$

Furthermore, we may write $A \cup B \cup C$ without confusion because

$$(A \cup B) \cup C = A \cup (B \cup C).$$

Let A and B be sets. The **intersection** of A and B, denoted $A \cap B$, is the set of all elements which belong to both A and B. Thus

$$A \cap B = \{x | x \in A \text{ and } x \in B\}.$$

This operation may also be depicted with Venn diagrams. As seen in Figure 0.2, the shaded portion of each diagram corresponds to the intersection of A and B since that region contains all points which belong to both of

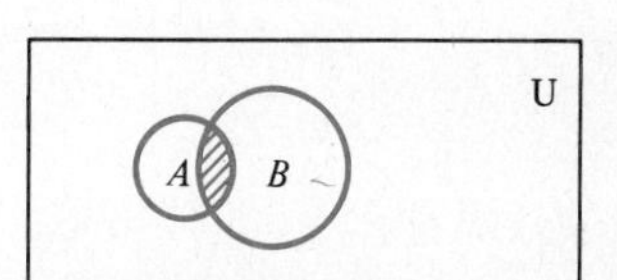

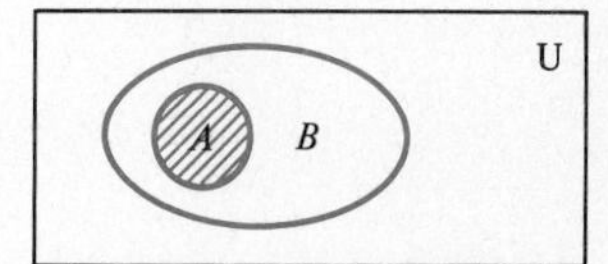

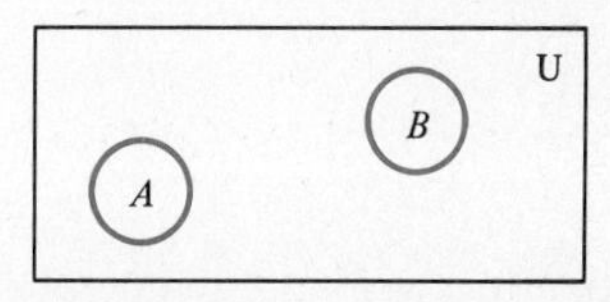

Figure 0.2 $A \cap B$

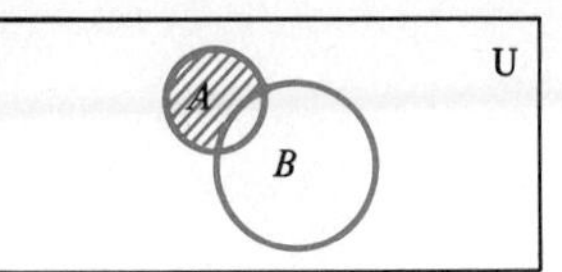

the regions A and B. Note that there is no shaded area in the last diagram because there are no common points. Two sets A and B are called **disjoint** provided that $A \cap B = \varnothing$. (Thus the empty set and every other set are disjoint.)

For example, if $A = \{1, 2, 3\}$ and $B = \{3, 4, 5\}$, then

$$A \cap B = \{3\}.$$

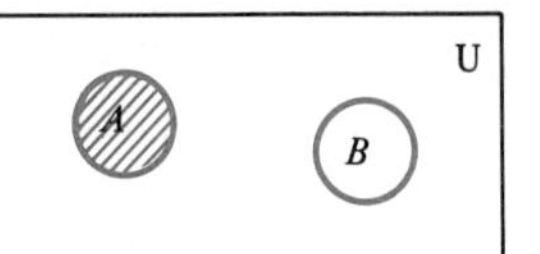

From the definition of intersection we see that

$$A \cap B = B \cap A.$$

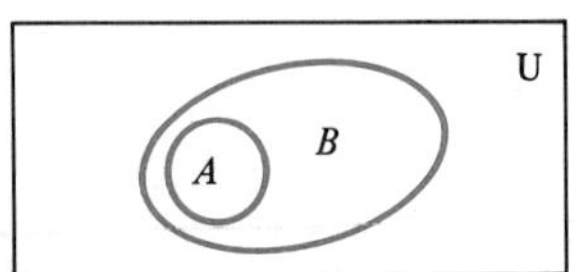

Furthermore, we may write $A \cap B \cap C$ without confusion because

$$(A \cap B) \cap C = A \cap (B \cap C).$$

Let A and B be sets. The **difference** of A and B, denoted $A - B$, is the set of all elements which belong to A but not to B. Thus

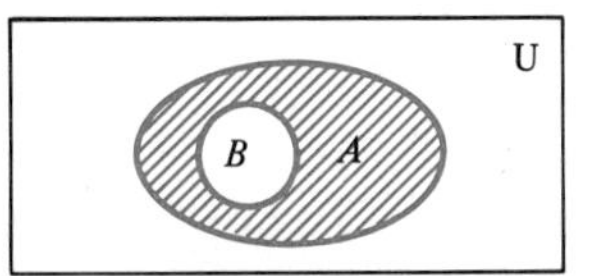

Figure 0.3 $A - B$

$$A - B = \{x | x \in A \text{ and } x \notin B\}.$$

In a Venn diagram, $A - B$ corresponds to the region consisting of all points contained in A that are *not* contained in B. The shaded areas in Figure 0.3 depict $A - B$. For example, if

$$A = \{1, 2, 3\} \quad \text{and} \quad B = \{3, 4, 5\},$$

then

$$A - B = \{1, 2\} \quad \text{and} \quad B - A = \{4, 5\}.$$

We see immediately that, in general, $A - B \neq B - A$.

Exercise

Let $A = \{0, 1, 4\}$ and $B = \{2, 3, 4\}$. Determine each of the following.
(a) $A \cup B$ (b) $A \cap B$ (c) $A - B$ (d) $B - A$

Answers. (a) $\{0, 1, 2, 3, 4\}$ (b) $\{4\}$ (c) $\{0, 1\}$ (d) $\{2, 3\}$ ■

Problem Set 0.1

In Problems 1–6, determine whether each statement is true or false.

1. $5 \in \{1, 2, 3\}$

2. $0 \notin \{3, 4, 5\}$

3. $\{0, 1, -1\} \neq \{1, -1, 0\}$

4. $\{a, a\} = \{a\}$

5. $\{0\}$ is the empty set.

6. $\{\varnothing\}$ is the empty set.

In Problems 7–12, determine the specified set, where $A = \{1, 3, 5\}$ *and* $B = \{2, 4, 5\}$.

7. $A \cup B$ **8.** $A \cap B$ **9.** $A - B$
10. $B - A$ **11.** $B \cap (B - A)$ **12.** $A \cup (A - B)$

The **complement** of a set A is $\mathsf{U} - A$, the difference of the universal set and A, and is denoted $\overline{A}$.

In Problems 13–18, determine the specified set, where $\mathsf{U} = \{1, 2, 3, 4, 5, 6\}$, $A = \{1, 3, 5\}$, *and* $B = \{2, 4, 5\}$.

13. $A \cup \overline{B}$ **14.** $\overline{A} \cup \overline{B}$ **15.** $\overline{A} \cap \overline{B}$
16. $\overline{A} \cap B$ **17.** $\overline{A \cup B} \cup B$ **18.** $A \cap \overline{A \cap B}$

A set A is said to be a **subset** of a set B provided that A is the empty set or every element of A is an element of B. This relationship is denoted $A \subseteq B$ and may be illustrated with a Venn diagram.

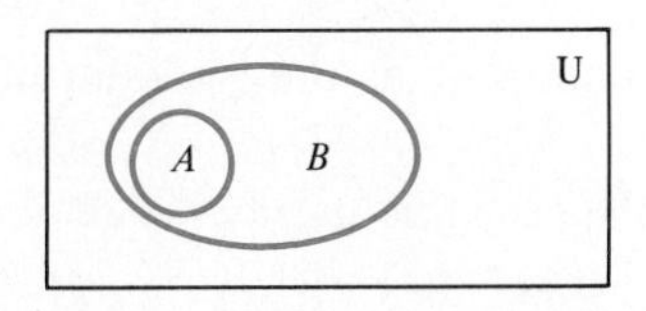

19. Use Venn diagrams to determine which of the following statements are true for all subsets A and B of U.
(a) $A \cup B \subseteq \mathsf{U}$ **(b)** $A \cap B \subseteq B$
(c) $\overline{A \cup B} = \overline{A} \cup \overline{B}$ **(d)** $\overline{\overline{A}} = A$
(e) $A - B = A \cap \overline{B}$ **(f)** $\overline{A \cap B} = \overline{A} \cup \overline{B}$

0.2 Real Numbers

The counting numbers: 1, 2, 3, 4, 5, . . . , were the first numbers each of us learned. These numbers are called the positive integers or natural numbers, and the set of positive integers is denoted $\mathbf{Z}^+$. Next we encountered the negative integers: $-1, -2, -3, \ldots$, and then 0. The set of all integers (positive, negative, and zero) is denoted Z.

A **rational number** is a number which may be expressed as the quotient of two integers. Thus any number which may be written as p/q, where p and q are integers, $q \neq 0$, is a rational number. (Division by zero is not defined.) This includes all fractions, like $\frac{2}{3}$, $\frac{1}{2}$, $-\frac{3}{5}$, as well as all integers (because every integer may be written as a fraction with a denominator of 1). The set of all rational numbers is frequently denoted Q.

By using ordinary division, every rational number may be written as a **terminating decimal**, like

$$\frac{1}{2} = 0.5 \qquad \text{and} \qquad \frac{9}{8} = 1.125,$$

or as a **periodic decimal,** a decimal in which a block of one or more digits in the decimal repeats itself indefinitely. For example,

$$\frac{2}{3} = 0.66666\ .\ .\ . = 0.\overline{6}$$

and

$$\frac{15}{11} = 1.363636\ .\ .\ . = 1.\overline{36}$$

are periodic decimals. The bar above the digit or group of digits indicates which are repeated.

It is also true that every terminating or periodic decimal represents a rational number. Examples 1 and 2 show how to write a periodic decimal as a fraction.

Example 1
Express the rational number $0.\overline{45}$ as the quotient of two integers.

Solution. We'll first represent this number algebraically by letting

$$k = 0.454545.\ .\ .\ .$$

Then

$$100k = 45.454545.\ .\ .\ .$$

Subtracting the first equation from the second, we get

$$\begin{array}{rrl} & 100k = & 45.454545\ .\ .\ . \\ & k = & 0.454545\ .\ .\ . \\ \hline \text{(subtract)} & 99k = & 45 \end{array}$$

Thus $k = 45/99 = 5/11$. ■

Note that to get an integer upon subtraction, the two numbers must have identical decimal parts. That's why we multiplied k by 100 in Example 1. As seen in Example 2, we must sometimes use two multiples of k (instead of k and one multiple) to get two numbers with identical decimal parts.

Example 2
Express the rational number $0.4\overline{9}$ as the quotient of two integers.

Solution. Let

$$k = 0.4\overline{9}.$$

Then

$$10k = 4.\overline{9}$$

and

$$100k = 49.\overline{9}.$$

Subtracting the first equation from the second, we have

$$\begin{array}{rrl} & 100k = & 49.\overline{9} \\ & 10k = & 4.\overline{9} \\ \hline \text{(subtract)} & 90k = & 45 \end{array}$$

Thus $k = \frac{1}{2}$. ■

From Example 2 we see that the decimal representation of a rational number is not unique. The number $\frac{1}{2}$ may be written as 0.5 or $0.4\overline{9}$.

Exercise
Express each of the following rational numbers as the quotient of two integers.
(a) $0.\overline{27}$ (b) $0.5\overline{6}$

Answers. (a) $\dfrac{3}{11}$ (b) $\dfrac{51}{90}$ ■

There do exist quantities, however, which cannot be represented as rational numbers. For example, the length of the hypotenuse of a right triangle whose legs have length 1 is a number whose square is 2. (The positive number whose square is 2 is denoted $\sqrt{2}$, read "radical two.") And $\sqrt{2}$ cannot be expressed as the ratio of two integers. (For a proof of this statement, see the Appendix.) Such numbers are called irrational numbers. An **irrational number** is a real number which cannot be expressed as the ratio of two integers. Besides $\sqrt{2}$, another well-known example is π (the ratio of the circumference of a circle to its diameter). The decimal representation of irrational numbers is not periodic. And any number whose decimal representation is not periodic, like 0.1010010001 . . . , formed by adding an extra zero between successive 1's, is an irrational number.

Exercise
Which of the numbers -3, $\frac{3}{4}$, $\pi/2$, 0, $3\sqrt{2}$, 0.02, $0.\overline{13}$, and 0.1919919991 . . . are rational numbers?

Answer. -3, $\frac{3}{4}$, 0, 0.02, and $0.\overline{13}$ ■

The irrational numbers together with the rational numbers form the set of real numbers. The relationships between the types of numbers discussed in this section are shown in Figure 0.4.

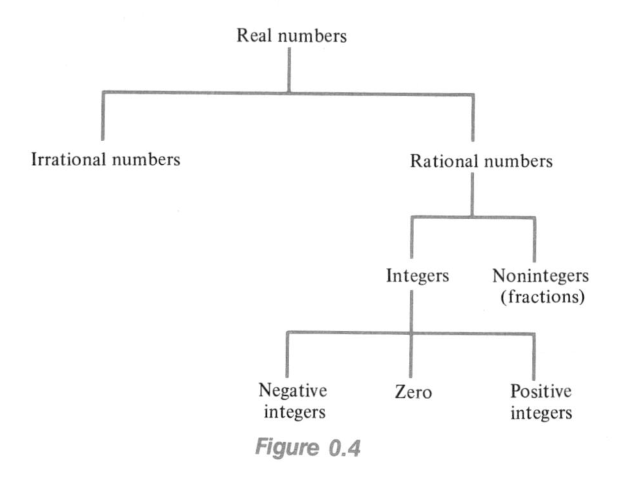

Figure 0.4

It is possible to associate the set of real numbers with the points on a straight line. Given a line, we indicate one direction as positive with an arrowhead. Next we select an arbitrary point P_0 called the **origin,** and associate that point with the number 0. Then we choose another point P_1, in the positive direction from the origin, and associate this point with the number 1. The length of the line segment from P_0 to P_1 is referred to as the unit length. From P_1 we locate the point P_2, in the positive direction, such that the line segment from P_1 to P_2 has unit length. The number 2 is associated with P_2. Similarly, we locate $P_3, P_4, P_5, \ldots$, which will be associated with 3, 4, 5, . . . , respectively. (*Note:* We shall use a bar to denote line segments. Thus $\overline{P_1P_2}$ denotes the line segment connecting P_1 and P_2.) Then from the origin we locate the point P_{-1}, in the negative direction, such that $\overline{P_0P_{-1}}$ has unit length. The point P_{-1} will be associated with the number -1. Similarly, we locate the points $P_{-2}, P_{-3}, P_{-4}, \ldots$, which correspond to the numbers $-2, -3, -4, \ldots$, respectively. This construction is pictured in Figure 0.5. It is conventional that the number line be horizontal with the positive direction to the right.

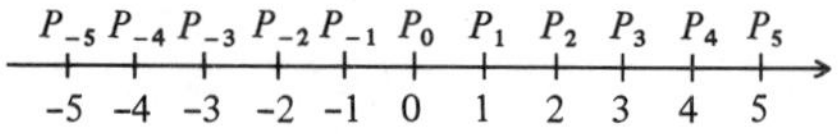

P_{-5} P_{-4} P_{-3} P_{-2} P_{-1} P_0 P_1 P_2 P_3 P_4 P_5

-5 -4 -3 -2 -1 0 1 2 3 4 5

Figure 0.5

With this graphical representation, each real number is associated with a unique point on the line and each point on the line is associated with a unique real number, called the **coordinate** of the point. Because of this one-to-one correspondence, we refer to a coordinate line as a **number line,** and frequently refer to a number as a point, and vice versa. Some points and their coordinates are noted in Figure 0.6.

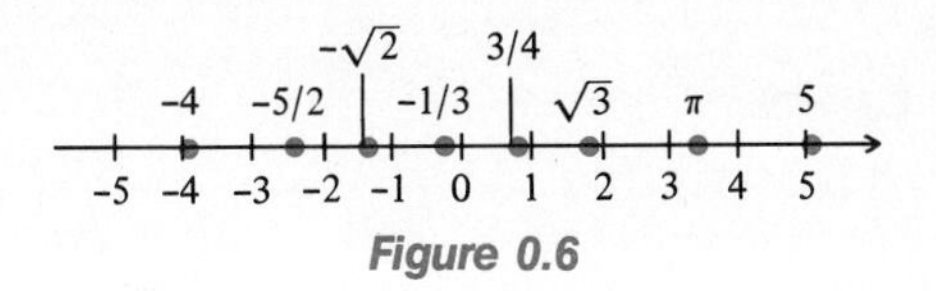

Figure 0.6

Exercise
Locate the numbers $-\frac{3}{2}$ and $\frac{5}{2}$ on the number line.

Answer

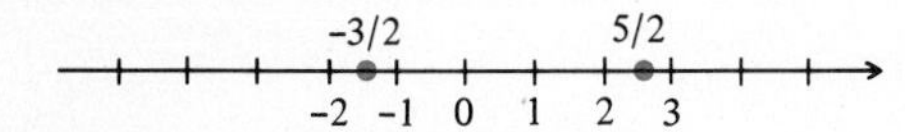

■

Problem Set 0.2

1. Let $S = \{-5, 2\pi, \frac{22}{7}, -\sqrt{5}, 7, -\frac{1}{3}, 0.87, 0, 0.1\overline{3}, 0.121221222\ .\ .\ .\}$. Determine each of the following sets.
   (a) $S \cap \mathsf{Z}^+$ (b) $S \cap \mathsf{Z}$ (c) $S \cap \mathsf{Q}$
2. Determine the repeating decimal expansions of $\frac{2}{9}$, $\frac{3}{11}$, and $\frac{5}{12}$.
3. Express $0.2\overline{9}$, $0.\overline{18}$, and $0.0\overline{6}$ as the quotient of two integers.
4. Can a rational number have two different decimal expansions?
5. Is the sum of two rational numbers necessarily a rational number?
6. Is the sum of two irrational numbers necessarily an irrational number?
7. Is the product of two rational numbers necessarily a rational number?
8. Is the product of two irrational numbers necessarily an irrational number?
9. Give the real numbers associated with the points A, B, and C.

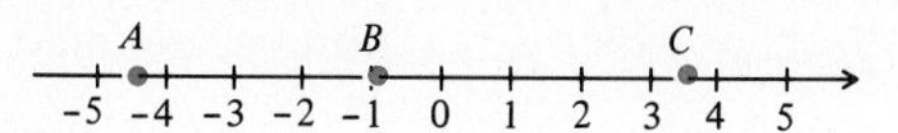

10. Draw a real number line and plot the following points.

    (a) $\sqrt{5}$ (b) $-\pi$ (c) 4 (d) $-\dfrac{7}{2}$

0.3 Properties of Real Numbers

In this section we review some of the properties, notation, and terminology of the set of real numbers with the operations of addition (denoted by "+") and multiplication (denoted by "·"). For every $a, b \in \mathsf{R}$, there exists a unique real number $a + b$ called the sum of a and b and a unique real number $a \cdot b$ (frequently written ab) called the product of a and b. If a, b, $c \in \mathsf{R}$, we have the following properties.

Commutative Laws

$$a + b = b + a \qquad \text{and} \qquad ab = ba$$

Associative Laws

$$(a + b) + c = a + (b + c) \qquad \text{and} \qquad (ab)c = a(bc)$$

Distributive Laws

$$a(b + c) = ab + ac \qquad \text{and} \qquad (a + b)c = ac + bc$$

Identity Elements

There exist two special real numbers, denoted 0 and 1, such that

$$a + 0 = 0 + a = a \qquad \text{and} \qquad a \cdot 1 = 1 \cdot a = a$$

for every real number a.

Inverse Elements

For every real number a, there exists a real number denoted $-a$ such that

$$a + (-a) = (-a) + a = 0,$$

and for every $a \neq 0$, there exists a real number denoted $1/a$ such that

$$a\left(\frac{1}{a}\right) = \left(\frac{1}{a}\right) a = 1.$$

Because of the associative laws, we may write $a + b + c$ without confusion since we may add a and b and then c, or we may add b and c and then a. In fact, we may write the sum of any number of real numbers without regard to the order in which they are to be added. An analogous situation exists for multiplication. Instead of $(ab)c$ or $a(bc)$, we simply write abc. And the product of any number of real numbers can be written without regard to the order.

Unless altered by parentheses, multiplication occurs before addition. Hence, to evaluate $a \cdot b + c$, we multiply a by b and then add this result to c. On the other hand, to evaluate $a \cdot (b + c)$, we add b to c and then multiply this result by a.

The number $-a$ is called the **negative** (or **additive inverse**) of a, and the number $1/a$ is called the **reciprocal** (or **multiplicative inverse**) of a $(a \neq 0)$. The number 0 has no reciprocal.

Other properties can be proven from the ones listed. It can be shown that $a \cdot 0 = 0 \cdot a = 0$ for every real number a, and that if $ab = 0$, then $a = 0$ or $b = 0$. Some theorems concerning negatives are $-(-a) = a$, $(-a)b = -(ab)$, $a(-b) = -(ab)$, $(-a)(-b) = ab$, and $(-1)a = -a$. And we have the **cancellation laws,** which state that if $a + c = b + c$, then $a = b$, and if $ac = bc$, where $c \neq 0$, then $a = b$.

Also, if $a = b$, then $a + c = b + c$ and $ac = bc$. This fact enables us to substitute a for b or b for a in an equation without changing the truth value.

The operations of subtraction (denoted by "$-$") and division (denoted by "$\div$") are defined in terms of addition and multiplication, respectively. If $a, b \in R$, then $a - b = a + (-b)$ and $a \div b = a(1/b)$. Now $a \div b$ is frequently written a/b, the fraction a over b. In this form the numbers a and b are called, respectively, the **numerator** and the **denominator.** Because 0 has no reciprocal, a/b is not defined if $b = 0$. That is, *division by zero is not defined.* Thus, when we write any fraction, we are assuming that the denominator is not zero.

For any fractions a/b and c/d, we shall write

$$\frac{a}{b} = \frac{c}{d}$$

provided that $ad = bc$. With this definition we see that $\frac{4}{10} = \frac{2}{5}$ because $4 \cdot 5 = 10 \cdot 2$, and we immediately get the following theorem.

Theorem 0.1

If $c \neq 0$, then for any fraction a/b,

$$\frac{ac}{bc} = \frac{a}{b}.$$

This theorem says that if both the numerator and denominator are multiplied (or divided) by the same nonzero expression, the resulting fraction is the same.

An integer which forms part of a product is called a **factor.** Thus factors that are common to both numerator and denominator may be canceled. Hence

$$\frac{6}{9} = \frac{3 \cdot 2}{3 \cdot 3} = \frac{2}{3}.$$

In essence, we have canceled the like factor of 3.

Fractions in which there are no factors (other than 1 and -1) common to both numerator and denominator are said to be in **reduced** (or **simplest**) form.

A fraction may be reduced by factoring the numerator and denominator and then eliminating common factors.

We may think of every fraction as having three signs: one for the fraction, one for the numerator, and one for the denominator. Based on the definition of equality of, we get the following:

Theorem 0.2

For any fraction a/b, we have

(i) $\frac{a}{b} = \frac{-a}{-b} = -\frac{-a}{b} = -\frac{a}{-b}$

(ii) $-\frac{a}{b} = \frac{-a}{b} = \frac{a}{-b} = -\frac{-a}{-b}$

This theorem reveals that we may simultaneously change any two of the three signs of a fraction and still have an equivalent fraction. With arithmetic fractions it is most common to have the denominator positive. Thus, in practice, we shall usually write the negative of $\frac{1}{5}$ as $\frac{-1}{5}$ or $-\frac{1}{5}$ instead of $\frac{1}{-5}$.

For any fractions a/b and c/d, we define multiplication by

$$\frac{a}{b} \cdot \frac{c}{d} = \frac{ac}{bd}.$$

That is, we multiply fractions by multiplying numerator by numerator and denominator by denominator.

For any fractions a/b and c/d (with $c/d \neq 0$), we define division by

$$\frac{a}{b} \div \frac{c}{d} = \frac{a}{b} \cdot \frac{d}{c}.$$

Thus we must simply invert the divisor and multiply.

Exercise

Perform the indicated operation.

(a) $\frac{3}{4} \cdot \frac{8}{15}$ (b) $\frac{8}{9} \div \frac{4}{3}$

Answers. (a) $\frac{2}{5}$ (b) $\frac{2}{3}$ ■

Addition of fractions is defined for fractions which have the same denominator.

$$\frac{a}{b} + \frac{c}{b} = \frac{a+c}{b}$$

Thus

$$\frac{2}{5} + \frac{1}{5} = \frac{2 + 1}{5} = \frac{3}{5}.$$

When the denominators are different, we may get equivalent fractions with a common denominator by multiplying the numerator and denominator of each fraction by the denominator of the other. Hence

$$\frac{a}{b} + \frac{c}{d} = \frac{ad}{bd} + \frac{bc}{bd} = \frac{ad + bc}{bd},$$

so

$$\frac{3}{4} + \frac{1}{6} = \frac{3}{4} \cdot \frac{6}{6} + \frac{1}{6} \cdot \frac{4}{4} = \frac{18}{24} + \frac{4}{24} = \frac{22}{24} = \frac{11}{12}.$$

Of course, any common denominator will do, and the product of the two given denominators may be larger than necessary. To find the least common denominator we need to discuss the concept of least common multiple. A common multiple of two positive integers is a positive integer into which each of the given integers can be divided with a remainder of zero. Thus some common multiples of *4* and *6* are *24, 48,* and *12*. Because *12* is the smallest of these common multiples, it is called the least common multiple.

To find the least common multiple of a group of positive integers, we write each integer as a product of primes. (A prime number is a positive integer p, $p \neq 1$, which has only two positive integral factors, 1 and p.) Then the least common multiple will be the product formed by using all prime factors in the list with each factor used the greatest number of times that it appears in any single integer. Thus, to calculate the least common multiple of 8, 12, and 15, we write

$$\begin{aligned} 8 &= 2 \cdot 2 \cdot 2 \\ 12 &= 2 \cdot 2 \cdot 3 \\ 15 &= 3 \cdot 5. \end{aligned}$$

The least common multiple will be the product formed by using 2, 3, and 5, with the 2 being used three times. We get

$$2 \cdot 2 \cdot 2 \cdot 3 \cdot 5 = 120.$$

Exercise
Find the least common multiple of 12, 15, and 20.

Answer. 60 ■

The least common multiple of denominators is called the least common denominator. Thus, to add $\frac{1}{4}$ and $\frac{3}{10}$, we may use the least common denominator 20 and get

$$\frac{1}{4} + \frac{3}{10} = \frac{5}{20} + \frac{6}{20} = \frac{11}{20}.$$

Subtraction of fractions is defined

$$\frac{a}{c} - \frac{b}{c} = \frac{a - b}{c},$$

and the least common denominator again plays a role. Thus

$$\frac{7}{12} - \frac{8}{9} = \frac{21}{36} - \frac{32}{36} = \frac{21 - 32}{36} = \frac{-11}{36}.$$

Exercise

Perform the indicated operation.

(a) $\frac{3}{10} + \frac{4}{15}$ (b) $\frac{5}{12} - \frac{7}{16}$

Answers. (a) $\frac{17}{30}$ (b) $\frac{-1}{48}$ ■

Problem Set 0.3

1. State the negative of each of the following.

(a) 7 **(b)** -5 **(c)** 0

(d) $-(-\pi)$ **(e)** $-\frac{2}{5}$

2. State the reciprocal of each of the following.

(a) -3 **(b)** 6 **(c)** 0

(d) $\frac{1}{3}$ **(e)** $\frac{-1}{\pi}$

3. Show that the operation of subtraction is not commutative.

4. Show that the operation of division is not commutative.

5. Show that the operation of subtraction is not associative.

6. Show that the operation of division is not associative.

7. Name the property being illustrated by each of the following:

(a) $5 + 0 = 5$ **(b)** $7 \cdot \frac{1}{7} = 1$

(c) $3 + (-3) = 0$ **(d)** $2(a + 5) = 2a + 2 \cdot 5$

(e) $5(ab) = (5a)b$ (f) $3 + a = a + 3$
(g) $6 \cdot 1 = 6$

8. Determine whether each of the following statements is true or false.
   (a) $0 \cdot a = 0$ for all real numbers a
   (b) $-(-10) = 10$
   (c) $2 \cdot 3 + 1 = 8$
   (d) If $a = 10$, then $a + 5 = 15$.
   (e) If $ac = bc$, then $a = b$ for all real numbers a, b, and c.
   (f) If $a = -3$, then $2a = 6$.
   (g) If $a + c = b + c$, then $a = b$ for all real numbers a, b, and c.
   (h) If $ab = 0$, then $a = 0$ or $b = 0$.
   (i) If $a = 5$, then $a + 2 = 7$.

In Problems 9–14, evaluate.

9. $2 \cdot 3 + 5$ 10. $2 \cdot (3 + 5)$ 11. $-3(2 - 5) + 4$
12. $-5 - 4(6 - 1)$ 13. $-3(2(3 - 4) + 1) - 6$ 14. $5(3(4 + 1) - 2) + 7$

In Problems 15–18, find the least common multiple.

15. 12 and 15 16. 10 and 14 17. 12, 18, and 30 18. 12, 15, and 25

In Problems 19 and 20, express as fractions with the least common denominator.

19. $\frac{5}{12}$ and $\frac{7}{8}$ 20. $\frac{3}{10}, \frac{1}{4}$, and $\frac{2}{3}$

In Problems 21–26, perform the indicated operation.

21. $\frac{3}{4} \cdot \frac{2}{9}$ 22. $\frac{2}{3} \div \frac{8}{9}$ 23. $\frac{2}{9} + \frac{5}{18}$

24. $\frac{3}{14} - \frac{8}{21}$ 25. $\frac{5}{48} - \frac{3}{16} + \frac{1}{9}$ 26. $\frac{5}{12} + \frac{1}{15} - \frac{7}{20}$

0.4 Order Relations and Absolute Value

Let a and b be real numbers. Then we shall say that a is **greater than** b, and write $a > b$ (or say that b is **less than** a and write $b < a$), provided that $a - b$ is positive. For example, $6 > 5$ (or equivalently $5 < 6$) because $6 - 5$ is positive. With respect to the number line (in standard position), $a < b$ if and only if the point with coordinate a lies to the left of the point with coordinate b.

For any two real numbers a and b, either $a - b$ is positive, negative, or zero. Hence it follows that either

$$a > b, \qquad a < b, \qquad \text{or} \qquad a = b.$$

Because $a - 0 = a$, $a > 0$ if and only if a is positive, and similarly, $a < 0$ if and only if a is negative.

Theorem 0.3

Let $a, b, c \in \mathsf{R}$.

(i) *If $a > b$ and $b > c$, then $a > c$.*
(ii) *If $a > b$, then $a + c > b + c$.*
(iii) *If $a > b$ and $c > 0$, then $ac > bc$.*
(iv) *If $a > b$ and $c < 0$, then $ac < bc$.*

The proof of this theorem is straightforward, and these important results will be used in the next chapter. Stating this theorem somewhat differently by using the symbol "<," we have the following:

(i) *If $a < b$ and $b < c$, then $a < c$.*
(ii) *If $a < b$, then $a + c < b + c$.*
(iii) *If $a < b$ and $c > 0$, then $ac < bc$.*
(iv) *If $a < b$ and $c < 0$, then $ac > bc$.*

Item (iv) states an important fact. When both sides of an inequality are multiplied by a negative number, the inequality symbol "reverses its direction." For example, consider the statement $5 > 3$. Upon multiplying both sides by -2, we have $-10 < -6$. Similarly, after multiplying both sides of $-3 < -2$ by -3, we have $9 > 6$.

Theorem 0.4

Let $a, b \in \mathsf{R}$. Then

(i) *$ab > 0$ if and only if $a > 0$ and $b > 0$, or $a < 0$ and $b < 0$.*
(ii) *$ab < 0$ if and only if $a < 0$ and $b > 0$, or $a > 0$ and $b < 0$.*

This theorem says that the product of two numbers is positive if and only if both are positive or both are negative. And the product of two numbers is negative if and only if one of the numbers is positive and the other is negative.

Two other symbols must be defined. We write $a \geq b$ (and say "a is greater than or equal to b") provided that $a > b$ or $a = b$. And we write $a \leq b$ (and say "a is less than or equal to b") provided that $a < b$ or $a = b$. Hence we see that $5 \geq 3$, $\pi \geq \pi$, $-11 \leq -10$, and $-2 \leq -2$. The symbols, $<$, $>$, $\leq$, and $\geq$ are called the **inequality symbols.** And Theorems 0.3 and 0.4 may be naturally extended for the latter two inequality symbols.

A statement like $a < b < c$ is simply a compound statement: $a < b$ *and* $b < c$. If true, it means that a is less than c, and b is in between.

Let $a \in \mathsf{R}$. The **absolute value** of a, denoted $|a|$, is defined as follows:

$$|a| = a \qquad \text{if } a \geq 0$$
$$|a| = -a \qquad \text{if } a < 0.$$

We see immediately that $|0| = 0$ and that the absolute value of any nonzero number is positive. If $a > 0$, then $|a| = a > 0$, and if $a < 0$, then $|a| = -a > 0$. For example,

$$|3| = 3,$$
$$|-2| = -(-2) = 2.$$

What about $|\sqrt{2} - 2|$? Since $\sqrt{2} - 2$ is negative,

$$|\sqrt{2} - 2| = -(\sqrt{2} - 2) = 2 - \sqrt{2}.$$

Exercise

Find the value.

(a) $|-6|$ (b) $|5|$ (c) $|3 - \sqrt{10}|$

Answers. (a) 6 (b) 5 (c) $\sqrt{10} - 3$ ■

Theorem 0.5

Let $a, b \in \mathsf{R}$. Then

(i) $|-a| = |a|$

(ii) $|ab| = |a| \cdot |b|$

(iii) $\left|\dfrac{a}{b}\right| = \dfrac{|a|}{|b|} \qquad (b \neq 0)$

(iv) $|a + b| \leq |a| + |b|$

Part (iv) of this theorem is called the **triangle inequality.** Two of the consequences of the triangle inequality are that for any real numbers a and b,

$$|a - b| \leq |a| + |b|$$

and

$$|a| - |b| \leq |a - b|.$$

Let A and B be points on the number line with coordinates a and b, respectively. Then the distance between A and B, denoted AB, is

$$AB = |b - a|.$$

We see that the distance from A to B, being an absolute value, is always nonnegative and is the length of the line segment $\overline{AB}$. Furthermore, since $|b - a| = |a - b|$, we have $AB = BA$. For any point C on the number line, we have $AC \leq AB + BC$. This is because

$$\begin{aligned} AC = |c - a| &= |(c - b) + (b - a)| \leq |c - b| + |b - a| \\ &= BC + AB = AB + BC, \end{aligned}$$

where c is the coordinate of C. The relationship $AC \leq AB + BC$ is also referred to as the triangle inequality.

Note that the distance from A to the origin is $|a - 0|$, which is $|a|$. Thus the absolute value of a number is its distance from the origin on the number line.

Problem Set 0.4

1. Replace the $*$ by a $<$ or a $>$ to make a true statement.
   (a) $5 * 7$
   (b) $-1 * -5$
   (c) $0 * -2$
   (d) $-2 * 6$
2. Using the following figure, replace the $*$ by a $<$ or a $>$ to make a true statement.

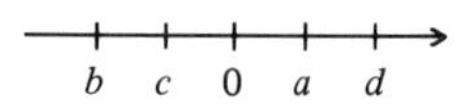

   (a) $a * b$
   (b) $0 * a$
   (c) $b * c$
   (d) $d * a$
3. Write each of the following statements using inequality symbols.
   (a) c is not greater than 4.
   (b) a is greater than b.
   (c) a is nonnegative.
   (d) b is greater than 3 and less than or equal to 5.
   (e) b is less than a.
   (f) a is between -3 and -2.
   (g) a is not less than -5.
4. Determine whether each of the following is true or false.
   (a) If $a > b$ and $ac > bc$, then c is positive.
   (b) If $a \leq b$, then $a - c \leq b - c$.
   (c) If $a \geq b$, then $ac \geq bc$.
   (d) If $a < b \leq c$, then $a - b \leq c$.
   (e) If $a > b$ and $b > a/2$, then $a - b > a/2$.
   (f) If $a < b$ and $ac \geq bc$, then c is negative.
5. Is there a smallest positive rational number?
6. For each of the following pairs of numbers, determine which lies leftmost on the number line (in standard position).
   (a) $-5, 0$
   (b) $3, \frac{9}{2}$
   (c) $4, \pi$
   (d) $-7, -8$

In Problems 7–18, find the value.

7. $|7|$

8. $|-3|$

9. $\left|\frac{-2}{3}\right|$

10. $|-3(4)|$

11. $-|3|$

12. $-|-5|$

13. $|2 - 3|$

14. $|2| - |3|$

15. $|\pi - 4|$

16. $|\pi - 3|$

17. $|a - 2b| - |2b - a|$

18. $\frac{-|ab|}{|a| \cdot |-b|}$

19. Must $|a + b| > |a|$?

20. Must $-|-a| = -|a|$?

21. Can $|a| > |-a| + |b|$?

22. Can $|-a| \leq |a| + |b|$?

23. Under what conditions is $|a + b| = |a| + |b|$?

0.5 Exponents

Let $a \in \mathsf{R}$, and let n be any positive integer. Then we define

1. $a^1 = a$
2. $a^n = \underbrace{a \cdot a \cdots a}_{n \text{ times}}$ if $n > 1$

Hence, if $n > 1$, a^n means that n factors of a are multiplied together. Thus $a^2 = a \cdot a$, $a^3 = a \cdot a \cdot a$, and so on. For example, $4^1 = 4$, $4^2 = 16$, and $4^3 = 64$. The number a^n is called the nth power of a, a is called the **base,** and n is called the **exponent** (or **power**). If $a \neq 0$, we define

$$a^0 = 1.$$

The expression 0^0 is *not* defined.

Theorem 0.6

Let $a, b \in \mathsf{R}$, and let m and n be nonnegative integers. Then

(i) $a^m \cdot a^n = a^{m+n}$

(ii) $(a^m)^n = a^{mn}$

(iii) $(ab)^m = a^m b^m$

(iv) $\left(\frac{a}{b}\right)^m = \frac{a^m}{b^m} \quad (b \neq 0)$

(v) $\frac{a^m}{a^n} = a^{m-n}$ *if* $m \geq n \quad (a \neq 0)$

(vi) $\frac{a^m}{a^n} = \frac{1}{a^{n-m}}$ *if* $m < n \quad (a \neq 0)$

Part (i) says that when multiplying two numbers with the same base, we may *add* the exponents. For example, $2^2 \cdot 2^3 = 2^5$. Part (ii) states that if a number is raised to a power, and then that result is raised to another power, we may *multiply* exponents. Thus $(2^2)^3 = 2^6$. Part (iii) says that if the product of two numbers is raised to a power, we may raise each factor to that power and then multiply. For example, $(2 \cdot 3)^4 = 2^4 \cdot 3^4$. According to part (iv), a quotient raised to a power is equal to the numerator to that power divided by the denominator to that power. Thus $(2/3)^2 = 2^2/3^2$. In dividing two numbers with the same base, we may subtract exponents. According to (v), $3^3/3^2 = 3^{3-2} = 3^1$, and $2^2/2^2 = 2^{2-2} = 2^0 = 1$. By (vi) we see that $3^2/3^4 = 1/3^2$.

Let $a \in \mathrm{R}$ such that $a \neq 0$, and let n be any positive integer. We define

$$a^{-n} = \frac{1}{a^n}.$$

For example, $3^{-2} = 1/3^2 = 1/9$. With this definition, Theorem 0.6 is true even when m and n are negative. In fact, parts (v) and (vi) may be combined.

$$\frac{a^m}{a^n} = a^{m-n} = \frac{1}{a^{n-m}}$$

The important thing to remember is that if the exponential form is to be put in the numerator, we must subtract the exponent in the denominator from the exponent in the numerator. On the other hand, if the exponential form is to be put in the denominator, we must subtract the exponent in the numerator from the exponent in the denominator.

Furthermore, with negative exponents thus defined, we may move any factor from the numerator to the denominator, or vice versa, provided that we change the sign of the exponent. For example,

$$\frac{a^2b^{-3}}{c} = \frac{a^2}{b^3c}, \qquad \frac{a^2}{bc^{-2}} = \frac{a^2c^2}{b}, \qquad \text{and}$$

$$\frac{1+a}{b} = b^{-1}(1+a)$$

Exercise
Evaluate.

(a) $\left(\frac{1}{2}\right)^3$ (b) $(-3)^4$ (c) $\left(\frac{2}{3}\right)^{-2}$ (d) 4^{-1} (e) π^0 (f) -3^2

Answers. (a) $\frac{1}{8}$ (b) 81 (c) $\frac{9}{4}$ (d) $\frac{1}{4}$ (e) 1 (f) -9 ■

Exercise

Use Theorem 0.6 to simplify each of the following.

(a) $x^2 \cdot x^3$ (b) $(2a)^3$ (c) $\dfrac{a^2}{a^3}$

(d) $(y^2)^3$ (e) $\left(\dfrac{x}{y}\right)^3$ (f) $\dfrac{y^3}{2y^2}$

Answers. (a) x^5 (b) $8a^3$ (c) $\dfrac{1}{a}$ (d) y^6 (e) $\dfrac{x^3}{y^3}$ (f) $\dfrac{y}{2}$ ■

Let $a \in \mathrm{R}$, and let n be a positive integer. We shall say that b is an **nth root** of a provided that $b^n = a$. For example, a fourth root of 16 is 2 because $2^4 = 16$. And since $(-2)^4 = 16$, -2 is also a fourth root of 16. When $n = 2$ we shall say "square root," and when $n = 3$ we shall say "cube root." Thus 2 and -2 are square roots of 4 because $2^2 = 4$ and $(-2)^2 = 4$. And -3 is the cube root of -27 because $(-3)^3 = -27$. The **principal nth root** of a number is the nth root which agrees with the sign of the number. Thus 2 is the principal square root of 4, and -3 is the principal cube root of -27. When n is even and a is positive, there are two real nth roots, one positive and one negative. The positive one is the principal one. (When n is even and a is negative, the nth roots are *not* real numbers. For example, the fourth root of -16 is not real.) When n is odd, there is only one real nth root, so it is automatically the principal one. The notation which refers to the principal nth root of a is

$$\sqrt[n]{a}.$$

The number n is called the **index**, and a is called the **radicand.** When the index is omitted, it is assumed to be 2, and the notation refers to the principal square root. Thus

$$\sqrt{4} = 2 \quad \text{and} \quad \sqrt[3]{-27} = -3.$$

Note that the radicand and the result agree in sign.

Theorem 0.7

Let n be an integer >1, and let a and b be real numbers. If $\sqrt[n]{a}$ and $\sqrt[n]{b}$ are real numbers, then

(i) $\sqrt[n]{ab} = \sqrt[n]{a}\sqrt[n]{b}$

(ii) $\sqrt[n]{\dfrac{a}{b}} = \dfrac{\sqrt[n]{a}}{\sqrt[n]{b}} \quad (b \neq 0)$

For example, $\sqrt{8} = \sqrt{4 \cdot 2} = \sqrt{4}\,\sqrt{2} = 2\sqrt{2}$ and $\sqrt[3]{2/27} = \sqrt[3]{2}/\sqrt[3]{27} = \sqrt[3]{2}/3$.

Exercise
Simplify.

(a) $\sqrt{9}$ (b) $\sqrt[5]{-32}$ (c) $\sqrt{52}$ (d) $\sqrt{\frac{5}{9}}$ (e) $\sqrt{a^2}$

Answers. (a) 3 (b) -2 (c) $2\sqrt{13}$ (d) $\frac{\sqrt{5}}{3}$ (e) $|a|$ ■

Pay particular attention to (e). In general, $\sqrt{a^2} \neq a$. If $a < 0$, then $\sqrt{a^2} = -a$. Thus

$$\sqrt{a^2} = |a|.$$

Another method of simplifying radicals is used when the denominator of a fraction contains a radical. The method is called **rationalizing the denominator** and involves multiplying both numerator and denominator by the same appropriate constant. For example, to rationalize $1/\sqrt{2}$, we multiply both numerator and denominator by $\sqrt{2}$. We have

$$\frac{1}{\sqrt{2}} = \frac{1}{\sqrt{2}} \cdot \frac{\sqrt{2}}{\sqrt{2}} = \frac{\sqrt{2}}{\sqrt{2^2}} = \frac{\sqrt{2}}{2}.$$

Another technique of rationalizing uses the fact that

$$(a + b)(a - b) = a^2 - ab + ba - b^2 = a^2 - b^2.$$

To rationalize $1/(\sqrt{5} - \sqrt{2})$, we multiply both the numerator and denominator by $\sqrt{5} + \sqrt{2}$. We get

$$\frac{1}{\sqrt{5} - \sqrt{2}} = \frac{1}{\sqrt{5} - \sqrt{2}} \cdot \frac{\sqrt{5} + \sqrt{2}}{\sqrt{5} + \sqrt{2}} = \frac{\sqrt{5} + \sqrt{2}}{(\sqrt{5})^2 - (\sqrt{2})^2}$$
$$= \frac{\sqrt{5} + \sqrt{2}}{5 - 2} = \frac{\sqrt{5} + \sqrt{2}}{3}.$$

Exercise
Rationalize each of the following.

(a) $\frac{1}{\sqrt{3}}$ (b) $\frac{1}{\sqrt{7} + \sqrt{3}}$

Answers. (a) $\frac{\sqrt{3}}{3}$ (b) $\frac{\sqrt{7} - \sqrt{3}}{4}$ ■

Let a be any real number and let n be a positive integer. Then we shall define

$$a^{1/n} = \sqrt[n]{a}.$$

For example, $4^{1/2} = \sqrt{4} = 2$ and $(-27)^{1/3} = \sqrt[3]{-27} = -3$.

Let m and n be positive integers and let a be a real number. Then we define

$$a^{m/n} = (a^{1/n})^m = (\sqrt[n]{a})^m.$$

For example,

$$8^{2/3} = (\sqrt[3]{8})^2 = 2^2 = 4$$

and

$$81^{3/4} = (\sqrt[4]{81})^3 = 3^3 = 27.$$

If $\sqrt[n]{a}$ is real, then $a^{m/n} = (a^m)^{1/n} = \sqrt[n]{a^m}$. The two previous examples may be calculated as

$$8^{2/3} = \sqrt[3]{8^2} = \sqrt[3]{64} = 4$$

and

$$81^{3/4} = \sqrt[4]{81^3} = \sqrt[4]{531{,}441} = 27.$$

Let m and n be positive integers, and let a be a real number. Define

$$a^{-m/n} = \frac{1}{a^{m/n}}.$$

Thus

$$8^{-2/3} = \frac{1}{8^{2/3}} = \frac{1}{(\sqrt[3]{8})^2} = \frac{1}{4}$$

and

$$27^{-1/3} = \frac{1}{27^{1/3}} = \frac{1}{3}.$$

In general, Theorem 0.6 is true when m and n are rational numbers. The only exception is part (ii), which states that $(a^m)^n = a^{mn}$. By this rule we would get

$$((-2)^2)^{1/2} = (-2)^1 = -2,$$

which is false. We really have

$$((-2)^2)^{1/2} = 4^{1/2} = \sqrt{4} = 2.$$

The exception to part (ii) occurs when a is negative, m is a nonzero even integer, and n is the reciprocal of a nonzero even integer. In such a case we actually get

$$(a^m)^n = |a|^{mn}.$$

Exercise

Evaluate.

(a) $16^{1/2}$ (b) $27^{-2/3}$ (c) $32^{2/5}$ (d) $8^{-1/3}$

Answers. (a) 4 (b) $\frac{1}{9}$ (c) 4 (d) $\frac{1}{2}$ ■

Problem Set 0.5

In Problems 1–12, evaluate.

1. 9^2 **2.** $(-5)^2$ **3.** 5^{-1} **4.** 2^{-3}

5. -5^2 **6.** $(-4)^0$ **7.** $\left(\frac{1}{6}\right)^2$ **8.** $\left(\frac{2}{3}\right)^{-1}$

9. $\left(\frac{2}{5}\right)^{-2}$ **10.** $\frac{5^{-2}}{3^{-3}}$ **11.** $3^2 + 3^{-2}$ **12.** $5 + 5^{-1}$

In Problems 13–22, use the laws of exponents to simplify, expressing your answer with positive exponents.

13. a^2a^{-5} **14.** $3x^3x^{-2}$ **15.** $(x^2)^5$ **16.** $(a^{-3})^2$

17. $\frac{a^3}{a^{-4}}$ **18.** $\frac{2x^{-1}}{x^4}$ **19.** $(x^2y^{-1})^{-3}$ **20.** $(3a)^2$

21. $\left(\frac{3y^2}{x^{-3}}\right)^2$ **22.** $\left(\frac{x^{-1}}{xy^{-1}}\right)^{-3}$

In Problems 23–30, simplify.

23. $-\sqrt{16}$ **24.** $\sqrt[3]{-27}$ **25.** $\sqrt[3]{16}$ **26.** $\sqrt{12}$

27. $\sqrt[3]{\frac{5}{8}}$ **28.** $\sqrt{\frac{4}{3}}$ **29.** $\sqrt{4x^2}$ **30.** $-\sqrt[3]{x^6}$

In Problems 31–34, rationalize.

31. $\frac{-2}{\sqrt{5}}$ **32.** $\frac{6}{\sqrt{3}}$ **33.** $\frac{-1}{\sqrt{2} - \sqrt{6}}$ **34.** $\frac{8}{\sqrt{5} + 1}$

In Problems 35–42, evaluate.

35. $16^{3/4}$ **36.** $-25^{3/2}$ **37.** $(-32)^{2/5}$ **38.** $27^{-2/3}$

39. $(-27)^{-1/3}$ **40.** $4^{-1/2}$ **41.** $\left(\frac{1}{8}\right)^{1/3}$ **42.** $\left(\frac{1}{4}\right)^{-3/2}$

In Problems 43–48, use the laws of exponents to simplify, expressing your answer with positive exponents.

43. $x^{3/4} \cdot x^{1/4}$ **44.** $(a^3)^{1/6}$ **45.** $\dfrac{a^{4/3}}{a^{2/3}}$

46. $\dfrac{x^{-1/2}}{x^{3/2}}$ **47.** $(4a^{2/3})^{-1/2}$ **48.** $\dfrac{x^2y^{-1/2}}{x^{1/2}y}$

In Problems 49–52, rewrite each radical by using fractional exponents.

49. $\sqrt{x^3}$ **50.** $\sqrt[4]{a^3}$ **51.** $\sqrt{\sqrt{a}}$ **52.** $\sqrt[3]{\sqrt{x}}$

0.6 Algebraic Expressions

An algebraic expression is a constant, a variable, or any combination of constants and variables made by addition, subtraction, multiplication, division, or extraction of roots. For example,

$$x^2 - 5, \quad \frac{x^2 - y^2}{5xy}, \quad y, \quad \text{and} \quad \frac{x^2\sqrt{2 - x}}{xy - y^2}$$

are algebraic expressions. When an algebraic expression is written as the sum of other expressions, each expression forming the sum is called a **term.** Thus the terms of the expression

$$\frac{x}{y} + x^2 - 5y + 6$$

are x/y, x^2, $-5y$, and 6. When the variable factors of two terms are exactly alike, the terms are said to be **like terms.** Hence $2x^2$ and $-3x^2$ are like terms, whereas $5y$ and $3y^2$ are not.

When each variable of an expression is replaced by a number it represents, the expression equals a constant. This number which results because of the substitution is called the value of the expression.

Example 1

Find the value of $\dfrac{x^2 + xy + 3}{x - 2y}$ when $x = 2$, $y = -1$.

Solution. When $x = 2$ and $y = -1$,

$$\frac{x^2 + xy + 3}{x - 2y} = \frac{2^2 + 2(-1) + 3}{2 - 2(-1)} = \frac{4 - 2 + 3}{2 + 2} = \frac{5}{4}. \qquad \blacksquare$$

Exercise

Find the value of $\dfrac{2x - y}{x^2 - 2xy + 5}$ when $x = -2$, $y = 3$.

Answer. $\dfrac{-1}{3}$

A **monomial** is an algebraic expression which can be written as the product of a real number and nonnegative integral powers of variables. Hence

$$2x^3, \quad -5xy, \quad 3xy^2, \quad \text{and} \quad -2x^2yz$$

are monomials. The constant of a monomial is called the **numerical coefficient.** For example, 6 is the numerical coefficient of $6x^2$, -2 is the numerical coefficient of $-2xy$, and 1 is the numerical coefficient of y^3.

A **polynomial** is an algebraic expression which may be written as the sum of monomials. Thus

$$x^2 + 5x + 6, \quad 2x^2 - 5xy, \quad \text{and} \quad 5$$

are polynomials. A polynomial which is the sum of two monomials is called a binomial, and one which is the sum of three monomials is called a trinomial. The largest exponent for a given variable is called the **degree** of the polynomial in that variable. Thus $2x^3 - x^2y + y^2$ is a polynomial of degree 3 in x and of degree 2 in y.

Since the variables of an algebraic expression represent real numbers, the properties of real numbers previously discussed apply to algebraic expressions. Using these properties, we see that when adding or subtracting polynomials, we can combine like terms by adding or subtracting the numerical coefficients.

Example 2

Add $2x^3 + 5x^2 - x + 5$ and $3x^3 - 2x^2 + x + 6$.

Solution. Grouping and combining like terms, we have

$$\begin{aligned}(2x^3 + 5x^2 - x + 5) + (3x^3 - 2x^2 + x + 6)& \\ &= (2x^3 + 3x^3) + (5x^2 - 2x^2) + (-x + x) + (5 + 6) \\ &= 5x^3 + 3x^2 + 11. \qquad \blacksquare\end{aligned}$$

Example 3
Subtract $2x^2 - 3xy + y^2 - 1$ from $5x^2 - 3y^2 + 2$.

Solution. We may proceed as in Example 2 or use a vertical array. Aligning like terms and subtracting the numerical coefficients, we have

$$\begin{array}{rrrr} 5x^2 & & -\ 3y^2 & +\ 2 \\ 2x^2 & -\ 3xy & +\ y^2 & -\ 1 \\ \hline 3x^2 & +\ 3xy & -\ 4y^2 & +\ 3. \end{array}$$

■

Exercise
(a) Add $7x^3 - 5x^2y + y^2 + 2$ and $2x^3 + 3x^2y - y^2 + 3$.
(b) Subtract $7x^3 - 5x^2 + x + 2$ from $2x^3 - x + 3$.

Answers. (a) $9x^3 - 2x^2y + 5$ (b) $-5x^3 + 5x^2 - 2x + 1$ ■

To find the product of two polynomials, we use the distributive laws, together with the laws of exponents, and then combine like terms.

Example 4

Multiply $3x + 5$ by $-2x^2 + x - 1$.

Solution
$$\begin{aligned}(3x + 5)(-2x^2 + x - 1) &= -6x^3 + 3x^2 - 3x - 10x^2 + 5x - 5 \\ &= -6x^3 - 7x^2 + 2x - 5\end{aligned}$$

Using a vertical array, we could have written

$$\begin{array}{rrrrl} -2x^2 & +\ x & -\ 1 & & \\ & 3x & +\ 5 & & \\ \hline -6x^3 & +\ 3x^2 & -\ 3x & & [3x(-2x^2 + x - 1)] \\ & -\ 10x^2 & +\ 5x & -\ 5 & [5(-2x^2 + x - 1)] \\ \hline -6x^3 & -\ 7x^2 & +\ 2x & -\ 5. & [\text{sum}] \end{array}$$

■

Exercise
Multiply $2x^2 - x$ by $3x^2 + x - 5$.

Answer. $6x^4 - x^3 - 11x^2 + 5x$ ■

Following are some special products that occur frequently. Their validity can be established by actual multiplication.

$$\begin{aligned}(x + y)^2 &= x^2 + 2xy + y^2 \\ (x - y)^2 &= x^2 - 2xy + y^2 \\ (x + y)(x - y) &= x^2 - y^2\end{aligned}$$

$$(x + a)(x + b) = x^2 + (a + b)x + ab$$
$$(x + y)^3 = x^3 + 3x^2y + 3xy^2 + y^3$$
$$(x - y)^3 = x^3 - 3x^2y + 3xy^2 - y^3$$

Example 5

Use the rules for special products to perform the indicated operation.
(a) $(3x + 2)^2$ (b) $(x + 2y)(x - 2y)$ (c) $(x + 5)(x - 4)$ (d) $(2x - y)^3$

Solution

(a) $(3x + 2)^2 = (3x)^2 + 2(3x)(2) + 2^2 = 9x^2 + 12x + 4$
(b) $(x + 2y)(x - 2y) = x^2 - 4y^2$
(c) $(x + 5)(x - 4) = x^2 + (5 - 4)x + 5(-4) = x^2 + x - 20$
(d) $(2x - y)^3 = (2x)^3 - 3(2x)^2y + 3(2x)y^2 - y^3$
$= 8x^3 - 12x^2y + 6xy^2 - y^3$ ■

Exercise

Use the rules for special products to perform the indicated operation.
(a) $(2x + 3)^2$ (b) $(3x + y)(3x - y)$ (c) $(x + 4)(x - 3)$ (d) $(x - 2y)^3$

Answers. (a) $4x^2 + 12x + 9$ (b) $9x^2 - y^2$ (c) $x^2 + x - 12$
(d) $x^3 - 6x^2y + 12xy^2 - 8y^3$ ■

The technique for dividing a polynomial by a monomial is seen in the next example.

Example 6

Divide $10x^2y^2 - 6xy^2 + 5x$ by $2xy$.

Solution

$$\frac{10x^2y^2 - 6xy^2 + 5x}{2xy} = \frac{10x^2y^2}{2xy} - \frac{6xy^2}{2xy} + \frac{5x}{2xy}$$
$$= 5xy - 3y + \frac{5}{2y}$$ ■

Exercise

Divide $10x^3y - 5x^2y^2 + 7y$ by $5xy$.

Answer. $2x^2 - xy + \dfrac{7}{5x}$ ■

When the divisor is not a monomial, we may use the **long-division algorithm** to determine the quotient. This technique is illustrated in Examples 7 and 8.

Example 7
Divide $8x^2 - 6x + 3$ by $2x - 1$.

Solution. We begin by writing

$$2x - 1\,\overline{)\,8x^2 - 6x + 3}$$

The first step is to divide $2x$ into $8x^2$. This quotient is $4x$ and we have

$$\begin{array}{r@{\,}l} & \;4x \\ 2x - 1 & \overline{)\,8x^2 - 6x + 3} \end{array}$$

Upon multiplying the divisor $2x - 1$ by $4x$, we have

$$\begin{array}{r@{\,}l} & \;4x \\ 2x - 1 & \overline{)\,8x^2 - 6x + 3} \\ & \;\;8x^2 - 4x \end{array}$$

Next we subtract $8x^2 - 4x$ from $8x^2 - 6x$ and bring down the next term. We get

$$\begin{array}{r@{\,}l} & \;4x \\ 2x - 1 & \overline{)\,8x^2 - 6x + 3} \\ & \;\;\underline{8x^2 - 4x} \\ & \qquad\;\; -2x + 3 \end{array}$$

Now we repeat the entire process by dividing $2x$ into $-2x$. This leads to

$$\begin{array}{r@{\,}l} & \;4x - 1 \\ 2x - 1 & \overline{)\,8x^2 - 6x + 3} \\ & \;\;\underline{8x^2 - 4x} \\ & \qquad\;\; -2x + 3 \\ & \qquad\;\; \underline{-2x + 1} \\ & \qquad\qquad\quad\;\; 2 \end{array}$$

The process terminates when the subtraction results in a polynomial of lower degree than the divisor. The 2 is thus a remainder and we may write

$$\frac{8x^2 - 6x + 3}{2x - 1} = \text{quotient} + \frac{\text{remainder}}{\text{divisor}} = 4x - 1 + \frac{2}{2x - 1}. \qquad \blacksquare$$

Example 8
Divide $2x^4 + 5 - 2x^2 + 3x$ by $2x^2 - x$.

Solution. Note that we must rearrange the terms of the dividend (the exponents must descend) and leave a space for the missing x^3 term. We get

$$\begin{array}{rl}
 & x^2 + \tfrac{1}{2}x - \tfrac{3}{4} \\
2x^2 - x & \overline{)\,2x^4 \qquad\qquad - 2x^2 + 3x + 5} \\
 & \underline{2x^4 - x^3} \\
 & \qquad\quad x^3 - 2x^2 \\
 & \qquad\quad \underline{x^3 - \tfrac{1}{2}x^2} \\
 & \qquad\qquad - \tfrac{3}{2}x^2 + 3x \\
 & \qquad\qquad \underline{- \tfrac{3}{2}x^2 + \tfrac{3}{4}x} \\
 & \qquad\qquad\qquad\quad \tfrac{9}{4}x + 5
\end{array}$$

Therefore,

$$\frac{2x^4 - 2x^2 + 3x + 5}{2x^2 - x} = x^2 + \frac{1}{2}x - \frac{3}{4} + \frac{\frac{9}{4}x + 5}{2x^2 - x}.$$ ■

Exercise

(a) Divide $6x^2 - 4x + 2$ by $3x + 1$.
(b) Divide $2x^4 + 3 - 5x - 4x^2$ by $2x^2 + x$.

Answers. (a) $\dfrac{6x^2 - 4x + 2}{3x + 1} = 2x - 2 + \dfrac{4}{3x + 1}$

(b) $\dfrac{2x^4 - 4x^2 - 5x + 3}{2x^2 + x} = x^2 - \dfrac{1}{2}x - \dfrac{7}{4} + \dfrac{-\frac{13}{4}x + 3}{2x^2 + x}$

Problem Set 0.6

In Problems 1–4, find the value of the algebraic expression when $x = -2$ and $y = 3$.

1. $x^2 - 3xy$ **2.** $(2x + y)(x - 3y)$ **3.** $\dfrac{xy - 5}{3x^2 - y^2}$ **4.** $\dfrac{x^2y + y^2 + 3}{xy^2 + y}$

In Problems 5–16, perform the indicated operation.

5. $(2x^2 + x) + (3x^2 - 2x)$
6. $(x^2 - 2xy + 2y^2) + (3x^2 - y^2)$
7. $(y^2 - 5y) - (2y^2 - 3)$
8. $(5x^2 + 6x - 2) - (x^2 + x - 5)$
9. $x(2x - 3y)$
10. $(y^2 - xy + 5x)2xy^2$
11. $(2x - 5)(3x + 2)$
12. $(x - 2y)(3x + 5y)$
13. $(2x - y)(x^2 + xy - 3y^2)$
14. $(x^2 + 3x - 1)(2x^2 - 2x + 5)$
15. $(9x^2y - 6xy + 3x^2y^2) \div 3xy$
16. $(4x^2y - 2x^2y^3 + 4x^4y^2) \div 2x^2y^2$
17. Divide $4x^2 + 6x - 1$ by $2x - 5$.
18. Divide $x^3 - 5x + 7 + 6x^4$ by $3x^2 - 2x$.

In Problems 19–30, use the rules for special products to perform the indicated operation.

19. $(2x + 5)^2$
20. $(3x^2y + y^2)^2$
21. $(3y - 2)^2$
22. $(5xy^2 - x^2)^2$
23. $(2x + y)(2x - y)$
24. $(3y^2 + 4xy)(3y^2 - 4xy)$
25. $(x + 3)(x + 5)$
26. $(x - 4)(x + 3)$
27. $(x + 5)(x - 2)$
28. $(x - 4)(x - 6)$
29. $(2x + y)^3$
30. $(x - 3y)^3$

0.7 Factoring

When a polynomial is written as the product of algebraic expressions, it is said to be factored. In this section we study various techniques for factoring polynomials.

A given polynomial may be expressed as a product in more than one way. For example, $x^2 - 1$ may be written as

$$(x + 1)(x - 1) \quad \text{or} \quad 4\left(\frac{x + 1}{2}\right)\left(\frac{x - 1}{2}\right) \quad \text{or} \quad x^{-2}(x^4 - x^2).$$

However, in general, we shall want the factors to be polynomials with integer coefficients. Thus, of the preceding three expressions, the first is the desired factored form.

A monomial that is a factor of each term of the polynomial to be factored is called a **common factor.** For example, x is a common factor of $x^2 + xy$, and 3 is a common factor of $3y - 9$. By using the distributive property, expressions which have common factors may be factored. Thus

$$x^2 + xy = x(x + y)$$

and

$$3y - 9 = 3(y - 3).$$

Exercise
Factor.
(a) $xy + y^3$ (b) $6x - 2$

Answers. (a) $y(x + y^2)$ (b) $2(3x - 1)$ ■

Example 1
Factor $6x^2y + 4xy^2 - 2x^2y^2$.

Solution. This polynomial has several common factors: 2, x, and y. Thus we write the expression as

$$2xy(3x + 2y - xy).$$ ■

Example 2
Factor $y(x - 2) + z(x - 2) - x + 2$.

Solution. This expression, consisting of four terms, has no common factors. However, we note that the binomial $x - 2$ is a factor of the first two terms and is the negative of the last two terms. Because of this, we may view the given expression as

$$y(x - 2) + z(x - 2) - (x - 2).$$

Consisting of three terms with a common factor of $x - 2$, this expression is factored as

$$(x - 2)(y + z - 1).$$ ■

Exercise
Factor.
(a) $6x^2yz + 3y^2z - 3yz^2$ (b) $x(y + 1) + y(y + 1) - y - 1$

Answers. (a) $3yz(2x^2 + y - z)$ (b) $(y + 1)(x + y - 1)$ ■

Example 3
Factor $2x^3 + x^2y - 2xy - y^2$.

Solution. Upon examining these four terms, we see that there are no common factors. However, it sometimes happens that by grouping terms and factoring each group, a common factor emerges. The given polynomial may be viewed as

$$(2x^3 + x^2y) + (-2xy - y^2).$$

The first group may be factored as $x^2(2x + y)$, and the second group as $y(-2x - y)$. Of course, in order for this technique to be of benefit, each factored group must have a common factor. We see that if we factor out $-y$ instead of y from the second group, this will be so. We get

$$x^2(2x + y) - y(2x + y)$$

and finally,

$$(2x + y)(x^2 - y).$$ ■

Exercise
Factor $xy - 3x^3 - 2y^2 + 6x^2y$.

Answer. $(y - 3x^2)(x - 2y)$ ■

The rule for factoring the difference of two squares is expressed as follows:

$$x^2 - y^2 = (x + y)(x - y).$$

This says that the difference of two squares is factored as the product of the sum and difference of the expressions being squared.

Example 4
Factor $x^2 - 4$.

Solution. Recognizing this as the difference of two squares, we get

$$x^2 - 4 = (x + 2)(x - 2).$$ ■

Example 5
Factor $9x^2 - 25$.

Solution. We must first recognize that this is indeed the difference of two squares. The given expression may be written as

$$(3x)^2 - 5^2.$$

Thus it factors as

$$(3x + 5)(3x - 5).$$ ■

Example 6
Factor $(2x + y)^2 - (x - 2y)^2$.

Solution. We get

$$[(2x + y) + (x - 2y)][(2x + y) - (x - 2y)].$$

By combining like terms, this simplifies to

$$(3x - y)(x - y).$$ ■

Example 7
Factor $x^4 - y^4$.

Solution. We have

$$x^4 - y^4 = (x^2)^2 - (y^2)^2 = (x^2 + y^2)(x^2 - y^2).$$

Note that the last factor is again the difference of two squares. By factoring it, we get

$$x^4 - y^4 = (x^2 + y^2)(x + y)(x - y).$$

Can we factor $x^2 + y^2$? The answer is "no." In general, we cannot factor the *sum* of two squares. ■

Next we consider factoring trinomials, recalling that $(x + a)(x + b) = x^2 + (a + b)x + ab$. Thus, given a polynomial like the right side of this equation, we know that if it factors, the factors will be binomials with first terms of x and second terms constant. We merely have to find the two numbers whose product equals the constant of the trinomial to be factored, and whose sum equals the coefficient of x of the trinomial. For example, let's factor $x^2 + 5x + 6$. If it factors, it looks like $(x + a)(x + b)$, where a and b are constants whose product is 6 and whose sum is 5. Let's list the pairs of numbers whose product is 6:

a	−1	−2	−3	−6	1	2	3	6
b	−6	−3	−2	−1	6	3	2	1

Checking the possibilities, we find that the desired numbers are 2 and 3 (or 3 and 2, changing the order of the factors). Therefore,

$$x^2 + 5x + 6 = (x + 2)(x + 3).$$

Although there are eight pairs of numbers in our list, a couple of observations limit the "real" possibilities to only two. First, four of the pairs are simply a change in order of the other four. For instance, the pair 6 and 1 is just a reversal of 1 and 6. Obviously, switching the numbers around doesn't change their product or sum. Hence our list could have been simplified.

a	−1	−2	1	2
b	−6	−3	6	3

Now a word about the signs. If the sign of the constant of the trinomial to be factored is plus (as it is here), then the two constants we are seeking are either both positive or both negative. The sign of the coefficient of x in the trinomial determines which case occurs. If positive (indicating that the sum must be positive), then both constants are positive. If the coefficient of x is negative, then both numbers are negative. So our list can be reduced even more.

a	1	2
b	6	3

Note also that if the constant of the trinomial is negative, then one of the numbers is positive and the other is negative. The sign of the coefficient of x matches the sign of the number largest in absolute value.

Example 8
Factor.
(a) $x^2 - 5x + 6$ (b) $x^2 - x - 6$ (c) $x^2 + x - 6$

Solution. (a) Since the constant is positive and the coefficient of x is negative, we are looking for two negative numbers whose product is 6. The only possibilities are -1 and -6 or -2 and -3. Since the latter pair has the proper sum,

$$x^2 - 5x + 6 = (x - 2)(x - 3).$$

(b) Since the constant is negative, we are looking for a positive and a negative number whose sum is -1. And since this sum is negative, we know that the negative number must be larger in absolute value than the positive number. The only possibilities are -6 and 1, and -3 and 2. The latter pair has the proper sum, thus

$$x^2 - x - 6 = (x - 3)(x + 2).$$

(c) We are looking for a positive and a negative number whose sum is 1. Thus we know that the positive number is larger in absolute value. We get

$$x^2 + x - 6 = (x + 3)(x - 2).$$ ■

Exercise
Factor.
(a) $x^2 + 7x + 12$ (b) $x^2 - 8x + 12$ (c) $x^2 - 11x - 12$
(d) $x^2 + 4x - 12$

Answers. (a) $(x + 3)(x + 4)$ (b) $(x - 2)(x - 6)$ (c) $(x - 12)(x + 1)$
(d) $(x + 6)(x - 2)$ ■

To factor expressions like $2x^2 + 7x - 4$, where the coefficient of x^2 is not 1, we cannot simply look for numbers whose product equals the constant and whose sum equals the coefficient of x. Instead, we must play a much more complicated guessing game. Let's factor $2x^2 + 7x - 4$. Now the first terms of the two binomials must have a product of $2x^2$. Thus they must be x and $2x$ (or of course $2x$ and x). And since the product of the last terms of the two binomials must be -4, the factors must be of the form

$$(x + a)(2x - b) \quad \text{or} \quad (x - a)(2x + b) \tag{0-1}$$

for positive constants a and b whose product is 4. The only way to determine which numbers work is by trial and error. We'll simply choose factors of 4 and calculate the middle term of the product until we get $7x$. Which form of line (0-1) should we use? Let's not concern ourselves with the signs and use the first. If our middle term turns out to be a $-7x$, we'll just switch the $+$ and $-$. Using the pairs 1 and 4, 2 and 2, 4 and 1, we get the following:

$$\begin{array}{ll} (x + 1)(2x - 4) & \text{middle term is } -2x \\ (x + 2)(2x - 2) & \text{middle term is } +2x \\ (x + 4)(2x - 1) & \text{middle term is } +7x. \end{array}$$

Thus we have

$$2x^2 + 7x - 4 = (x + 4)(2x - 1).$$

Note that, unlike the case when factoring trinomials whose coefficient of x^2 is 1, the numbers 1 and 4 give us a different set of factors than 4 and 1. This is because the first terms of the binomial are not equal.

Example 9
Factor $6x^2 - x - 12$.

Solution. Now the first terms of the binomials are x and $6x$ or $2x$ and $3x$, and one of the signs is plus, the other negative. Let's begin by trying numbers whose product is 12 in the expression

$$(x + \quad)(6x - \quad).$$

We get

$$\begin{array}{ll} (x + 1)(6x - 12) & \text{middle term is } -6x \\ (x + 2)(6x - 6) & \text{middle term is } 6x \\ (x + 3)(6x - 4) & \text{middle term is } 14x \\ (x + 4)(6x - 3) & \text{middle term is } 21x \\ (x + 6)(6x - 2) & \text{middle term is } 34x \\ (x + 12)(6x - 1) & \text{middle term is } 71x. \end{array}$$

We didn't get a middle term of $-x$. Thus, if the given trinomial is to be factored, the first terms of the binomials must be $2x$ and $3x$. Let's try numbers for

$$(2x + \quad)(3x - \quad).$$

We get

$$\begin{array}{ll} (2x + 1)(3x - 12) & \text{middle term is } -21x \\ (2x + 2)(3x - 6) & \text{middle term is } -6x \\ (2x + 3)(3x - 4) & \text{middle term is } x. \end{array}$$

We need a middle term of $-x$. By changing the signs of our last guess, we'll have it. Thus

$$6x^2 - x - 12 = (2x - 3)(3x + 4).$$ ■

In Example 9 we see that several guesses were needed in order to get the result. In general, there is no shortcut to this guessing and calculating the corresponding middle term. However, if we note that there are no common factors in the trinomial, we can eliminate some of the middle-term calculation. Consider the second guess made in Example 9:

$$(x + 2)(6x - 6).$$

Without even calculating the middle term, we can observe that these aren't the proper factors. If these were the correct factors, then 6 would be a common factor of the trinomial, *because* 6 *is a common factor of the second binomial.* But the trinomial has no common factors. Similarly, we may reject the third, fourth, fifth, seventh, and eighth guesses without calculating the middle term.

Exercise
Factor.
(a) $3x^2 - 7x - 6$ (b) $4x^2 - 5x - 6$ (c) $5x^2 + 27x + 10$
(d) $10x^2 - 19x + 6$

Answers. (a) $(x - 3)(3x + 2)$ (b) $(x - 2)(4x + 3)$
(c) $(5x + 2)(x + 5)$ (d) $(5x - 2)(2x - 3)$ ■

Problem Set 0.7

Factor.

1. $2x + 8$
2. $x^2y + x$
3. $x^2y - 3xy^2 + x^2y^2$
4. $9x^3y^2 + 3xy^2 - 6x^2y^3$
5. $x(y + 1) + y(y + 1)$
6. $2x^2(2y - 3) + 3 - 2y$
7. $x^2 + 3x - 2xy - 6y$
8. $2y^2 - 6y + 3x^2 - x^2y$
9. $4y^2 - 9$
10. $x^4 - 16$
11. $x^2 + 7x + 10$
12. $x^2 + 9x + 20$
13. $x^2 + x - 12$
14. $x^2 - 2x - 15$
15. $x^2 - 7x + 12$
16. $x^2 - 9x + 14$
17. $2x^2 + 11x + 5$
18. $3x^2 - 13x + 12$

19. $3x^2 - 5x - 8$

20. $5x^2 - 7x - 6$

21. $12x^2 + 19x + 4$

22. $6x^2 - 17x + 10$

23. $10x^2 - 7x - 12$

24. $12x^2 + 23x - 2$

25. $2x^3 + 4x^2y + 2xy^2$

26. $6x^4 - 7x^3y - 3x^2y^2$

27. $2x^3 + 7x^2y - 4xy^2$

28. $2x^3 - 8x - 3x^2 + 12$

0.8 Rational Expressions

The quotient of two polynomials is called a rational expression. Some examples are

$$\frac{x}{2x+1}, \quad \frac{x^2+2xy}{5x+y}, \quad \text{and} \quad \frac{1}{y}.$$

From Theorem 0.1 we see that a rational expression may be simplified by canceling common factors, as seen in Example 1.

Example 1

Reduce the fraction $\dfrac{x^3 - 3x^2 - 10x}{x^4 - 4x^2}$ to simplest form.

Solution. Upon factoring both the numerator and denominator, we get

$$\frac{x(x-5)(x+2)}{x^2(x+2)(x-2)}.$$

After canceling the common factors of x and $x + 2$, we have

$$\frac{x-5}{x(x-2)}.$$

■

Exercise

Reduce the fraction $\dfrac{2x^4 - 2x^2}{3x^3 - 2x^2 - x}$ to simplest form.

Answer. $\dfrac{2x(x+1)}{3x+1}$ ■

We multiply rational expressions the same way we multiply arithmetic fractions. That is, we multiply numerator by numerator and denominator by denominator.

Example 2

Evaluate $\dfrac{4x(x-4)}{9(x+4)} \cdot \dfrac{x^2-16}{2x^2}$.

Solution. First we apply the definition, and then factor numerator and denominator to see if the result may be reduced. We get

$$\begin{aligned}\frac{4x(x-4)}{9(x+4)} \cdot \frac{x^2-16}{2x^2} &= \frac{4x(x-4)(x^2-16)}{18x^2(x+4)} \\ &= \frac{4x(x-4)(x+4)(x-4)}{18x^2(x+4)} \\ &= \frac{2(x-4)^2}{9x}.\end{aligned}$$

■

Exercise

Evaluate $\dfrac{x^2+4x+3}{2x^2} \cdot \dfrac{6x^4}{x^2-1}$.

Answer. $\dfrac{3x^2(x+3)}{x-1}$ ■

When dividing by a rational expression, we invert and multiply, as seen in Example 3.

Example 3

Evaluate $\dfrac{x^2-9}{4x} \div \dfrac{x^2-x-6}{2x^2}$.

Solution. We invert the divisor and multiply. Then we see if the result may be reduced. We get

$$\begin{aligned}\frac{x^2-9}{4x} \div \frac{x^2-x-6}{2x^2} &= \frac{x^2-9}{4x} \cdot \frac{2x^2}{x^2-x-6} \\ &= \frac{2x^2(x^2-9)}{4x(x^2-x-6)} \\ &= \frac{2x^2(x+3)(x-3)}{4x(x-3)(x+2)} \\ &= \frac{x(x+3)}{2(x+2)}.\end{aligned}$$

■

Exercise

Evaluate $\dfrac{3x^2}{2x^2 - 5x - 3} \div \dfrac{9x^5}{4x^2 - 1}$.

Answer. $\dfrac{2x - 1}{3x^3(x - 3)}$ ■

To add and subtract rational expressions, we must find a common denominator, write the given expressions as equivalent ones with the common denominator, and then combine numerators, as seen in Examples 4 and 5.

Example 4

Evaluate $\dfrac{2}{x - 3} + \dfrac{x}{2x + 1}$.

Solution. The least common denominator for these two fractions is $(x - 3)(2x + 1)$. Therefore, we have

$$\begin{aligned}\frac{2}{x-3} + \frac{x}{2x+1} &= \frac{2(2x+1)}{(x-3)(2x+1)} + \frac{x(x-3)}{(x-3)(2x+1)}\\ &= \frac{2(2x+1) + x(x-3)}{(x-3)(2x+1)}\\ &= \frac{4x + 2 + x^2 - 3x}{(x-3)(2x+1)}\\ &= \frac{x^2 + x + 2}{(x-3)(2x+1)}.\end{aligned}$$

Since the numerator cannot be factored, we know that the last result is in lowest terms. ■

Example 5

Evaluate $\dfrac{2x}{x + 3} - \dfrac{1}{x^2 + 3x}$.

Solution. Our first task is to find the least common denominator. To this end, it is best to factor the denominators. We get

$$\frac{2x}{x+3} - \frac{1}{x^2 + 3x} = \frac{2x}{x+3} - \frac{1}{x(x+3)}.$$

Thus we see that the least common denominator is $x(x + 3)$. Proceeding, we get

$$\frac{2x^2}{x(x+3)} - \frac{1}{x(x+3)} = \frac{2x^2-1}{x(x+3)}$$ ■

Exercise

Evaluate each of the following.

(a) $\dfrac{2}{x^2-1} + \dfrac{2x+1}{x^2-x}$ (b) $\dfrac{x}{x^2+2x-3} - \dfrac{x+1}{2x^2+5x-3}$

Answers. (a) $\dfrac{2x^2+5x+1}{x(x+1)(x-1)}$ (b) $\dfrac{x^2-x+1}{(x+3)(x-1)(2x-1)}$ ■

Problem Set 0.8

In Problems 1–8, reduce the given fraction to simplest form.

1. $\dfrac{15x^3}{-3x^2}$

2. $\dfrac{2xy}{6x^2y^3}$

3. $\dfrac{x^2-1}{x-x^2}$

4. $\dfrac{3x-2}{3x^2+x-2}$

5. $\dfrac{2x^2+7x-4}{4x^2-1}$

6. $\dfrac{x^3+4x^2+3x}{3x^2-6x-9}$

7. $\dfrac{x^2+2x-2y-xy}{x^2-y^2}$

8. $\dfrac{2x^4+3x^2-5}{x^4-1}$

In Problems 9–24, evaluate, expressing the result in reduced form.

9. $\dfrac{x}{x+2} \cdot \dfrac{x}{x-2}$

10. $\dfrac{x+y}{2y} \cdot \dfrac{x-y}{2x}$

11. $\dfrac{y}{x^2-xy} \cdot \dfrac{xy-y^2}{x}$

12. $\dfrac{2x^2+x}{2x-4} \cdot \dfrac{x^2-4}{2xy+y}$

13. $\dfrac{x+5}{2x+5} \div \dfrac{x^2+10x+25}{4x^2+20x+25}$

14. $\dfrac{x^2+4x+4}{x^2+4x+3} \div \dfrac{x+2}{x+1}$

15. $\dfrac{2x^2-3x-5}{2x^3+10x^2+12x} \div \dfrac{x^2-1}{2x}$

16. $\dfrac{x^2-y^2}{3x^2y-5xy^2-2y^3} \div \dfrac{x^2+xy}{y}$

17. $\dfrac{x-1}{7x} + \dfrac{x+2}{7x}$

18. $\dfrac{x+1}{x} - \dfrac{x-1}{x}$

19. $\dfrac{x}{x^2-4x-5} + \dfrac{1}{x^2-4x-5}$

20. $\dfrac{2}{2x^2+x-6} + \dfrac{2x-1}{2x^2+x-6}$

21. $\dfrac{x+1}{x-2} + \dfrac{4}{x+1}$

22. $\dfrac{x-1}{x+5} - \dfrac{2}{x-1}$

23. $\dfrac{x}{2x+2y} + \dfrac{x^2}{x^2-y^2}$

24. $\dfrac{4}{x^2-4} + \dfrac{4}{10-5x} + \dfrac{1}{x+2}$

Chapter 0 Review

1. Let $A = \{0, 1, 4\}$ and $B = \{2, 3, 4\}$. Determine each of the following.
   (a) $A \cup B$ **(b)** $A \cap B$ **(c)** $A - B$ **(d)** $B - A$
2. Let $A = \{1, 2, 3\}$ and determine whether each of the following is true or false.
   (a) $3 \in A$ **(b)** $1 \notin A$
   (c) $A \neq \{1, 2, 3, 3\}$ **(d)** $A = \{3, 2, 1\}$
   (e) $A - A$ is the empty set. **(f)** $\{\emptyset\}$ is the empty set.
3. Express each of the following statements in terms of inequalities.
   (a) a is not positive. **(b)** a is between -8 and -10.
   (c) a is not less than 1.

In Problems 4–6, simplify.

4. $\sqrt{\frac{5}{18}}$ **5.** $\sqrt{45}$ **6.** $\sqrt[3]{-16}$

In Problems 7–9, rationalize.

7. $\dfrac{9}{2\sqrt{3}}$ **8.** $\dfrac{1}{2 - \sqrt{3}}$ **9.** $\dfrac{5}{3\sqrt{2} - \sqrt{3}}$

In Problems 10–13, evaluate.

10. $64^{5/6}$ **11.** $\left(\dfrac{1}{9}\right)^{-1/2}$ **12.** $\left(\dfrac{1}{16}\right)^{1/4}$ **13.** $-4^{3/2}$

In Problems 14–17, factor.

14. $12x^2 + 32x + 5$ **15.** $a^6b^2 - b^6a^2$
16. $2x^3 - x^2 - 2x + 1$ **17.** $ac + bd - ad - bc$

In Problems 18–20, simplify.

18. $\dfrac{a^4 + 6a^2 + 9}{9 - a^4}$ **19.** $\dfrac{3x + 6x^2}{3x}$ **20.** $\dfrac{a^2 - b^2}{a^2b + ab^2}$

In Problem 21–31, write each expression as a single fraction with positive exponents.

21. $(x + y)^{-1}$ **22.** $(x - y)^{-1}$ **23.** $x^{-1} - y^{-1}$

24. $(x^{-1} + y^{-1})^{-1}$ **25.** $(x - y)(x^{-1} + y^{-1})$ **26.** $\dfrac{(xy)^{-1}}{x^{-1} + y^{-1}}$

27. $\dfrac{1}{x^{1/2}} + \dfrac{x^{-1/2}}{y^{1/2}}$ **28.** $\dfrac{1}{x^{1/3}} - \dfrac{1}{y^{1/3}}$ **29.** $(x^{1/2} + y^{1/2})^2$

30. $(x^{1/2} - y^{1/2})^2$ **31.** $\left[\left(\dfrac{x^6(x + 3)^3}{27}\right)^{1/3}\right]^{-1}$

In Problems 32–37, evaluate, expressing the result in reduced form.

32. $\dfrac{x^2 + 2xy}{4x - 8y} \cdot \dfrac{2xy - 4y^2}{x^4 + 2x^3y}$

33. $\dfrac{x(x + 8) - 9}{x(x + 9) + 20} \cdot \dfrac{x^2 - 2x - 24}{x^2 + 8x - 9}$

34. $\dfrac{x^2 - y^2}{x - 2y} \div \dfrac{x + y}{x^2 - 2xy}$

35. $\dfrac{x + y}{x - y} + \dfrac{2y^2}{x^2 - 3xy + 2y^2}$

36. $\dfrac{1}{2x^2} + \dfrac{2x^3 + x - 5}{x^3 - 5x^2} - \dfrac{2x}{x - 5}$

37. $\dfrac{x + 3}{x + 4} + \dfrac{x - 6}{x^2 + 3x - 4} - \dfrac{x - 2}{x - 1}$

38. Divide $x^4 + 3x^3 + 5x - 2$ by $x^2 - 1$.

Equations and Inequalities in One Variable

1

This chapter is a study of techniques for solving various types of equations and inequalities.

1.1 Linear Equations

Recall that a **variable** is a symbol which represents each element of a specified set containing more than one element. The symbol is frequently a letter from the end of the alphabet, usually x, y, or z. The specified set is called the **universe** of the variable. When the universe is a subset of R (the set of real numbers), the variable is called a real variable. Unless noted otherwise, the term "variable" will mean "real variable."

Note that a variable represents *two or more elements*. A symbol which represents *one element only* is called a **constant.** When letters of the alphabet are used as constants, we generally use letters from the front of the alphabet, frequently a, b, or c. In this text each constant, unless noted otherwise, will represent a single *real* number.

An **equation in the variable** x is a statement of the form

$$p_x = q_x,$$

where p_x and q_x are algebraic expressions in x. Thus an equation in the variable x equates two algebraic expressions in x. For example,

$$2x + 3 = 5, \qquad x^2 = x + 2, \qquad \text{and} \qquad \frac{2}{x - 1} = 3$$

are equations in x.

Let x be a variable with universe $\mathsf{U} = \{1, 2, 3\}$, and consider the equation $x + 5 = 7$. Because x represents 1, 2, and 3, the equation $x + 5 = 7$ actually stands for three statements: $1 + 5 = 7$, $2 + 5 = 7$, and $3 + 5 = 7$. Hence it is inappropriate to ask whether the equation $x + 5 = 7$ is true or false. Instead, one should ask: "For which element(s) in the universe is the equation true when the variable is replaced?" In this instance it is clear that the question is true only when 2 is substituted for x.

A real **root** of an equation in one variable is a real number which makes the equation true when the variable is replaced by that real number. A real **solution** of an equation is a real root which belongs to the universe of the variable. Because most of the variables discussed in this text will be real variables, the adjective "real" will frequently be omitted when we speak of real roots and real solutions.

The set of all solutions is called the **solution set.** Therefore, for the equation just considered, 2 is a solution, and the solution set is $\{2\}$. Two equations are called **equivalent** provided that they have the same solution set. For example, $2x = 10$ and $x + 2 = 7$ are equivalent.

Unfortunately, the universe in subsequent problems will not be as simple as that of our example. Frequently, the universe of the variable is not explicitly stated. In such cases we shall assume that the universe is the set of all real numbers for which all terms of the equation are defined and real upon replacement. Given the equation $1/x = \sqrt{x}$, we see that the left side is undefined if x is replaced by 0, and the right side is not real if x is negative. Therefore, the universe is the set of all *positive* real numbers.

Exercise

Determine the universe of x given the equation $\sqrt{x - 2} = \dfrac{1}{x - 3}$.

Answer. All reals ≥ 2, except 3. ■

Usually, the equation will be too complicated to find solutions by observation. In such instances we shall find an equivalent equation whose solution set can be easily noted. We do this by using the following theorem.

Theorem 1.1

Let p_x, q_x, and r_x be algebraic expressions in the variable x. Then the equation

$$p_x = q_x$$

is equivalent to each of the following:

(i) $p_x + r_x = q_x + r_x$
(ii) $p_x \cdot r_x = q_x \cdot r_x$ *when* $r_x \neq 0$

This theorem says that an algebraic expression (including one consisting of a number only) may be added to both sides of an equation, and the resulting equation is equivalent to the beginning equation. And because $p_x - r_x = p_x + (-r_x)$, an algebraic expression may be *subtracted* from both sides of an equation without changing the solution set. Further, the theorem says that both sides of an equation may be multiplied by a *nonzero* algebraic expression, and the resulting equation is equivalent to the initial equation. And since division is the same as multiplication by a reciprocal, both sides of an equation may be *divided* by a *nonzero* algebraic expression without changing the solution set. We shall use these facts in a moment to prove Theorem 1.2.

The first general type of equation that we will study is the linear equation. An equation in a variable x is called **linear** in x provided that it may be expressed in the form

$$ax + b = 0,$$

where a and b are constants with $a \neq 0$. So linear equations are those which, when like terms are combined, the variable is raised to the first power only. For example,

$$2x + 3 = 0, \qquad 2x - 1 = x + 5, \qquad \text{and} \qquad x^2 + x = x^2 - 2x + 5$$

are linear equations. But

$$\sqrt{x} + 1 = 4, \qquad x^3 - 8 = 0, \qquad \text{and} \qquad x^2 + 5x = x - 3$$

are *not* linear equations. Theorem 1.2 tells us how to solve linear equations.

Theorem 1.2

The linear equation $ax + b = 0$, *where* a *and* b *are constants with* $a \neq 0$, *has the unique root* $-b/a$.

Proof. Using Theorem 1.1, we subtract b from both sides of the given equation (which is the same as adding $-b$ to both sides) to obtain the equivalent equation

$$ax = -b. \tag{1-1}$$

Using Theorem 1.1 again, we divide both sides of equation (1-1) by a and get

$$x = \frac{-b}{a}. \tag{1-2}$$

(Note that this last step is the same as multiplying both sides by $1/a$.) Equation (1-2) has the unique root $-b/a$. ■

In general, then, to solve a linear equation, we get the x terms on one side of the equation and the constants on the other. Then we combine like terms and divide each side by the coefficient of x. This technique is illustrated in Examples 1, 2, and 3.

Example 1
Find the solution set of $3x + 5 = 0$.

Solution. Subtracting 5 from each side of the given equation, we get

$$3x = -5.$$

Upon dividing both sides by 3, we have

$$x = \frac{-5}{3}.$$

Thus the solution set is $\{-5/3\}$. ■

Example 2
Find the solution set of $2(x - 1) = 3(2x - 1) + 5$.

Solution. The given equation is equivalent to each of the following:

$$\begin{aligned} 2x - 2 &= 6x - 3 + 5 \\ 2x - 6x &= -3 + 5 + 2 \\ -4x &= 4 \\ x &= -1. \end{aligned}$$

The solution set is $\{-1\}$. ■

Example 3

Find the solution set of $3 = \dfrac{2}{x - 1}$.

Solution. After multiplying both sides by $x - 1$, we have

$$3(x - 1) = 2.$$

This is equivalent to each of the following:

$$\begin{aligned} 3x - 3 &= 2 \\ 3x &= 5 \\ x &= \frac{5}{3} \end{aligned}$$

The solution set is {5/3}. ■

Exercise
Find the solution set.

(a) $2x + 3 = 0$ (b) $3(x + 1) - 2 = -2(2x - 3)$ (c) $2 = \dfrac{1}{x + 1}$

Answers. (a) $\left\{\dfrac{-3}{2}\right\}$ (b) $\left\{\dfrac{5}{7}\right\}$ (c) $\left\{\dfrac{-1}{2}\right\}$ ■

In Example 3 we multiplied both sides of the given equation by $x - 1$, an expression containing the variable. One must be careful when this is done. We must note the universe of the variable or must check the root of the final equation in the original. In Example 3 the universe of x is $\mathsf{R} - \{1\}$. (Recall that when the universe is not explicitly stated, it is assumed to be the set of all real numbers for which all terms of the equation are defined and real. If x is replaced by 1, we get a zero in a denominator.) By observing that $\frac{5}{3}$ is in this universe or by noting that $2/(x - 1)$ is defined when x is replaced by $\frac{5}{3}$, we can safely state that $\frac{5}{3}$ is a solution.

But consider the problem of finding the solution set of

$$\frac{2x}{x - 1} = 1 + \frac{2}{x - 1}.$$

Upon multiplying both sides by $x - 1$, we get

$$2x = x - 1 + 2. \tag{1-3}$$

Solving for x, we have

$$x = 1. \tag{1-4}$$

Now 1 is a *root* of equation (1-4), but it is *not* a *solution* of that equation. The *initial* equation tells us that the universe of x is $\mathsf{R} - \{1\}$. Since a root is *any* real number which makes the equation true when it is substituted for the variable, 1 is a root of equation (1-4). But because this root is not in the universe, it is not a solution of equation (1-4). Consequently, equation (1-4) has no solutions. And since the given equation is equivalent to this one, the solution set is $\varnothing$. [Note how important it is to know that the initial equation determines the universe. If equation (1-3) had been the beginning equation, the universe would have been R, and the solution set would have been $\{1\}$.]

If we had not noted the universe of x but had replaced x by 1 in the initial equation, we would have seen two undefined terms. This also would have indicated that 1 is not a solution. Thus, in multiplying both sides of an equation by an expression containing the variable, we must either note the universe or check the roots of our final equation in the original. This situation is encountered again in Example 4.

Example 4

Find the solution set of $\dfrac{x}{x+1} + 2 = \dfrac{-1}{x+1}$.

Solution. After multiplying both sides by $x + 1$, we have

$$x + 2(x + 1) = -1.$$

This is equivalent to each of the following:

$$\begin{aligned} x + 2x + 2 &= -1 \\ x + 2x &= -1 - 2 \\ 3x &= -3 \\ x &= -1. \end{aligned}$$

By noting that -1 is not in the universe $R - \{-1\}$, or by substituting -1 for x in the original equation and obtaining undefined terms, we conclude that the solution set is $\varnothing$. ■

Example 5

Find the solution set of $\dfrac{1}{x-1} - \dfrac{2}{x+1} = \dfrac{5x+3}{(x+1)(x-1)}$.

Solution. Upon multiplying both sides by $(x - 1)(x + 1)$, we have

$$x + 1 - 2(x - 1) = 5x + 3,$$

which is equivalent to each of the following:

$$\begin{aligned} x + 1 - 2x + 2 &= 5x + 3 \\ x - 2x - 5x &= 3 - 2 - 1 \\ -6x &= 0 \\ x &= 0. \end{aligned}$$

Since 0 is in the universe of x (none of the denominators equal zero if x is replaced by 0), the solution set is $\{0\}$. ■

Exercise
Find the solution set of each of the following.

(a) $\frac{2}{x-2} = 1 + \frac{x}{x-2}$ (b) $\frac{2}{x+1} - \frac{2x+3}{(x-1)(x+1)} = \frac{-5}{x-1}$

Answers. (a) $\varnothing$ (b) $\{0\}$ ■

Example 6
Solve $5xy - 3 = 2x - y$ for y, noting any restrictions on x.

Solution. When an equation contains other variables in addition to the one for which we are asked to solve, these additional variables may be viewed as constants. Thus, in solving the given equation for y, we shall treat the variable x as a constant. Viewed in this fashion, we see that the equation is linear in y. Hence we want to get the y terms on one side and the remaining terms on the other. We get

$$y + 5xy = 2x + 3.$$

In order to determine the coefficient of y, we need to factor the left side. We have

$$y(1 + 5x) = 2x + 3.$$

Dividing both sides by the coefficient of y, we get

$$y = \frac{2x+3}{1+5x}.$$

But this equation makes sense only if the denominator, $1 + 5x$, is not zero. Thus we must have $x \neq -\frac{1}{5}$. (Note that if $x = -\frac{1}{5}$, the given equation reduces to $-3 = -\frac{2}{5}$, a false statement.) ■

Exercise
Solve $3x - 2xy = 5 - 3y$ for y, noting any restrictions on x.

Answer. $y = \frac{5-3x}{3-2x}, x \neq \frac{3}{2}$ ■

Problem Set 1.1

In Problems 1–12, determine the universe of x.

1. $3x^2 - x = \frac{1}{2x}$ **2.** $x^2 - 5x = \frac{1}{x-2}$ **3.** $\frac{1}{2x-4} = \frac{x}{x+5}$

4. $\dfrac{x}{2x+6} = \dfrac{5}{x-4}$

5. $2x - \sqrt{x-1} = 5$

6. $x^2 + \sqrt{3x} = 1$

7. $2x^2 + \sqrt{-5x} = 2$

8. $3x - \sqrt[3]{2-x} = x^2$

9. $\dfrac{1}{\sqrt{x}} + x^2 = 3$

10. $x - \dfrac{2}{\sqrt{-x}} = 3x^2$

11. $\dfrac{1}{x} = \sqrt{x+2}$

12. $\sqrt{3-x} + x^2 = \dfrac{1}{x-1}$

In Problems 13–42, find the solution set.

13. $3x + 10 = 0$

14. $4x - 7 = 0$

15. $4x - 5 = 2x + 3$

16. $6x + 7 = 3x - 5$

17. $4 - 2x = 5x - 2$

18. $2x + 1 = 5x - 6$

19. $3(2-x) = 2(x-5)$

20. $4(x-3) = 3(1-x)$

21. $3(2x-5) = 2(4-3x) + 1$

22. $5(2x-1) - 3 = 2(3-2x)$

23. $x(x+1) = x^2 + 3x + 5$

24. $2x^2 + 2x - 1 = x(2x-3)$

25. $\dfrac{3x+1}{2x-3} = \dfrac{1}{8}$

26. $\dfrac{2x-1}{x+1} = \dfrac{-2}{3}$

27. $\dfrac{5x+4}{2-3x} + \dfrac{4}{3} = 0$

28. $\dfrac{2-x}{2x+3} - \dfrac{1}{2} = 0$

29. $\dfrac{3}{2x-1} = \dfrac{-5}{x+3}$

30. $\dfrac{2}{x-3} = \dfrac{3}{x+2}$

31. $\dfrac{5}{x+1} - \dfrac{16}{6-x} = 0$

32. $\dfrac{1}{x+6} + \dfrac{5}{x-3} = 0$

33. $\dfrac{x+1}{x-2} - \dfrac{2x+1}{2x-2} = 0$

34. $\dfrac{x+1}{x} + \dfrac{x+2}{5-x} = 0$

35. $\dfrac{6}{2x-1} + \dfrac{x+1}{2x-1} = 3$

36. $\dfrac{1}{x+1} - \dfrac{x}{x+1} = -2$

37. $\dfrac{x+1}{x-2} = 3 + \dfrac{3}{x-2}$

38. $\dfrac{4x}{2x-3} = \dfrac{6}{2x-3} + 2$

39. $\dfrac{x}{x+3} = 1 - \dfrac{5}{2x-1}$

40. $\dfrac{2x}{2x+1} - 1 = \dfrac{3}{x-2}$

41. $\dfrac{2}{x} = \dfrac{1+\sqrt{5}}{x-2}$

42. $\dfrac{\sqrt{3}x}{x+1} = \dfrac{2}{\sqrt{3}+1}$

In Problem 43–54, solve for the indicated variable, noting any restrictions on other variables.

43. $\dfrac{x}{3} + a = \dfrac{x-a}{2}$, for x

44. $\dfrac{3}{2x-c} = \dfrac{5}{x+3}$, for x

45. $d = rt$, for r

46. $v = gt + k$, for t

47. $F = \dfrac{9}{5}C + 32$, for C

48. $A = \dfrac{h}{2}(b+c)$, for c

49. $\dfrac{y - y_1}{x - x_1} = 6$, for y

50. $\dfrac{y - y_1}{x - x_1} = 2$, for x

51. $6x + 5y = 3xy$, for y

52. $2xy - 5 = 3x + y$, for y

53. $2xy' - 3y' + x^2 = 0$, for y'

54. $x^2y' - 3x - 2y^3y' = 1$, for y'

55. Find the solution set of $\dfrac{x}{a} = \dfrac{x + 2}{b}$ if a and b are constants such that $a \neq b$. What is the solution set if $a = b$?

56. For what values of a and b will $ax + b = 0$ have a solution set of $\{\frac{2}{3}\}$?

57. Determine k so that $2x + 3 = k$ is equivalent to $3x - 5 = 0$.

58. Determine k so that $3x + 2 = k$ is equivalent to $2x - 3 = 0$.

59. Determine k so that the solution set of $\dfrac{3}{x - 2} = \dfrac{k}{x + 1}$ is $\{1\}$.

60. Determine k so that the solution set of $\dfrac{2y + k}{y - 5} + \dfrac{3}{4} = 8$ is $\{9\}$.

1.2 Quadratic Equations

An equation in a variable x is called **quadratic** in x provided that it can be expressed in the form

$$ax^2 + bx + c = 0,$$

where a, b, and c are constants with $a \neq 0$. So quadratic equations are those which, when like terms are combined, the variable is raised to the second and first powers only or to the second power only. For example,

$$2x^2 + x + 1 = 0 \quad \text{and} \quad x^2 - 1 = 0$$

are quadratic, whereas

$$3x + 2 = 0, \quad x^3 - 2x^2 = 0, \quad \text{and} \quad \sqrt{x} = 4$$

are not.

The method for solving linear equations does not extend to quadratic equations. But one of several other procedures may be used. One technique, called the **method of factoring,** follows from Theorem 1.3.

Theorem 1.3

Let p_x and q_x be algebraic expressions in the variable x. Then the equation

$$p_x \cdot q_x = 0$$

is equivalent to the compound statement

$$p_x = 0 \text{ or } q_x = 0.$$

This theorem is a direct result of the fact that if the product of two numbers is zero, then one (or both) of the numbers must be zero.

This theorem states that an equation in which the product of two expressions equals zero is equivalent to an "or" statement in which each expression equals zero. Because the solution set of an "or" statement is the union of the solution sets of each part, we find the solution set of $p_x = 0$ and union it with the solution set of $q_x = 0$. This procedure is illustrated in Examples 1, 2, and 3.

Example 1

Find the solution set of $2x^2 + 5x - 3 = 0$.

Solution. Factoring the left side, we get

$$(2x - 1)(x + 3) = 0.$$

By Theorem 1.3 this equation is equivalent to the compound statement

$$2x - 1 = 0 \qquad \text{or} \qquad x + 3 = 0.$$

Upon solving each of these linear equations for x, we have

$$x = \frac{1}{2} \qquad \text{or} \qquad x = -3.$$

Hence the solution set is $\{1/2, -3\}$. ■

Example 2

Find the solution set of $(x + 1)(x - 2) = 4$.

Solution. Although the left side of this equation is already factored, the right side is not zero. After subtracting 4 from both sides, multiplying the binomials, and combining like terms, we have

$$x^2 - x - 6 = 0.$$

Factoring the left side, we get

$$(x - 3)(x + 2) = 0.$$

This is equivalent to

$$x - 3 = 0 \qquad \text{or} \qquad x + 2 = 0.$$

Thus the solution set is $\{3, -2\}$. ■

Example 3

Find the solution set of $2x - 5 = \frac{12}{x}$.

Solution. After multiplying each side by x, we have

$$2x^2 - 5x = 12.$$

Now we get one side to zero and factor, obtaining

$$2x^2 - 5x - 12 = 0$$
$$(2x + 3)(x - 4) = 0,$$

which is equivalent to

$$2x + 3 = 0 \qquad \text{or} \qquad x - 4 = 0$$
$$x = \frac{-3}{2} \qquad\qquad x = 4.$$

Since each root is in the universe $\mathsf{R} - \{0\}$, the solution set is $\{-3/2, 4\}$. ■

Exercise
Find the solution set.

(a) $6x^2 + x - 2 = 0$ (b) $(x + 2)(x - 3) = 6$ (c) $3x + 14 = \frac{5}{x}$

Answers. (a) $\left\{\frac{-2}{3}, \frac{1}{2}\right\}$ (b) $\{4, -3\}$ (c) $\left\{\frac{1}{3}, -5\right\}$ ■

Problem Set 1.2

In Problem 1–26, find the solution set by factoring.

1. $x^2 - 3x = 0$
2. $2x^2 + 5x = 0$
3. $x^2 - 5x + 6 = 0$
4. $x^2 + 3x - 10 = 0$
5. $2x^2 - x - 21 = 0$
6. $3x^2 + 10x + 3 = 0$
7. $10x^2 - 3x - 1 = 0$
8. $6x^2 - 23x - 4 = 0$
9. $x^2 + x = 20$
10. $x^2 + 3x = 5x + 15$
11. $x(2x - 1) = 6$
12. $x(3x + 5) = 2$
13. $4x(x + 4) = x + 4$
14. $2x(3x + 1) = 15 + x$
15. $(x - 5)(x + 2) = 8$
16. $(2x + 1)(x + 2) = 5$
17. $(6x + 1)(2x + 3) = 3(x - 1)$
18. $(3x - 5)^2 = x^2 + 5x + 13$
19. $5 = \frac{6}{x^2} - \frac{7}{x}$
20. $6 + \frac{5}{x} = \frac{-1}{x^2}$

21. $x + 2 = \dfrac{8}{x}$

22. $x + \dfrac{1}{x} = 2$

23. $x + \dfrac{4}{x - 2} = \dfrac{2x}{x - 2} - 3$

24. $3x - \dfrac{6x}{2x + 1} = \dfrac{14}{2x + 1} - 2$

25. $\dfrac{x - 2}{2x - 5} = \dfrac{x - 2}{3x - 2}$

26. $3 = \dfrac{2}{x + 5} - \dfrac{2}{x + 2}$

27. Determine k so that the solution set of $3x - k^2 - 2 = 2x + k$ is $\{14\}$.

28. Determine k so that $3x + 3k + 1 = k^2 + 2x$ is equivalent to $5x + 2k^2 - 4 = 3x + 4k$.

29. Determine k so that the solution set of $x^2 + 7kx + 10k^2 = 0$ is $\{4, 10\}$.

1.3 Quadratic Equations Continued

Another method sometimes employed to solve quadratic equations is called the **method of extraction** and is based on this theorem.

Theorem 1.4

Let p_x be an algebraic expression in the variable x, and let c be a constant. Then the equation

$$(p_x)^2 = c$$

is equivalent to the compound statement

$$p_x = \sqrt{c} \quad \text{or} \quad p_x = -\sqrt{c}.$$

The proof of this theorem follows from the definition of square root. Examples 1 and 2 demonstrate how this theorem is used.

Example 1
Find the solution set of $x^2 = 4$ by the method of extraction.

Solution. By Theorem 1.4 the given equation is equivalent to the compound statement

$$x = \sqrt{4} \quad \text{or} \quad x = -\sqrt{4}.$$

Simplified, this is

$$x = 2 \quad \text{or} \quad x = -2.$$

Thus the solution set is $\{2, -2\}$. ■

Example 2
Find the solution set of $(x + 2)^2 = 6$ by the method of extraction.

Solution. The given equation is equivalent to

$$x + 2 = \sqrt{6} \quad \text{or} \quad x + 2 = -\sqrt{6}. \tag{1-5}$$

Solving each of these linear equations for x, we get

$$x = \sqrt{6} - 2 \quad \text{or} \quad x = -\sqrt{6} - 2. \tag{1-6}$$

The solution set is $\{\sqrt{6} - 2, -\sqrt{6} - 2\}$. ■

Sometimes an "or" statement is abbreviated by using the symbol $\pm$. For example, statement (1-5) in Example 2 could have been written

$$x + 2 = \pm\sqrt{6},$$

and statement (1-6) could have been written

$$x = \pm\sqrt{6} - 2.$$

If the constant of Theorem 1.4 is negative, then the solution set will be empty. (This is because no *real* number squared is negative. Remember, we're studying *real* variables.) If we failed to observe this and used the method of extraction, we would get a negative under the radical sign. Since such a number is not real, we would still conclude that there are no solutions.

Exercise

Find the solution set by extraction.

(a) $(x - 2)^2 = 5$ (b) $2x^2 = 16$ (c) $(x + 1)^2 = -4$

Answers. (a) $\{2 + \sqrt{5}, 2 - \sqrt{5}\}$ (b) $\{2\sqrt{2}, -2\sqrt{2}\}$ (c) $\varnothing$ ■

Sometimes neither the method of extraction nor the method of factoring can be used to find the solution set of a quadratic equation. In such cases it will be necessary to apply the following theorem.

Theorem 1.5

The equation

$$ax^2 + bx + c = 0,$$

where a, b, and c are constants with $a \neq 0$, is equivalent to the compound statement

$$x = \frac{-b \pm \sqrt{b^2 - 4ac}}{2a}. \tag{1-7}$$

Proof. Subtracting c from both sides of the given equation and dividing by a, we get

$$x^2 + \frac{b}{a}x = \frac{-c}{a}. \tag{1-8}$$

Now we want to produce an equivalent equation on which we can apply the method of extraction. So we want to add a constant to both sides of equation (1-8) so that the resulting left side, which will be a trinomial, will be a perfect square (i.e., the square of a binomial). The appropriate constant is $b^2/(4a^2)$. (We'll see how we get this constant in a moment.) Thus we get

$$x^2 + \frac{b}{a}x + \frac{b^2}{4a^2} = \frac{b^2}{4a^2} - \frac{c}{a}.$$

Upon writing the left side as a perfect square and simplifying the right side, we have

$$\left(x + \frac{b}{2a}\right)^2 = \frac{b^2 - 4ac}{4a^2}.$$

By the method of extraction we get

$$x + \frac{b}{2a} = \pm\sqrt{\frac{b^2 - 4ac}{4a^2}}.$$

Since $\sqrt{a^2} = |a|$, $\sqrt{4a^2} = 2|a|$, and we have

$$x + \frac{b}{2a} = \pm\frac{\sqrt{b^2 - 4ac}}{2|a|}. \tag{1-9}$$

If $a > 0$, then $|a| = a$ and the right side of (1-9) becomes

$$\pm\frac{\sqrt{b^2 - 4ac}}{2a}.$$

If $a < 0$, then $|a| = -a$, and the right side of (1-9) becomes

$$\pm\frac{\sqrt{b^2 - 4ac}}{-2a},$$

which may be written

$$\mp\frac{\sqrt{b^2 - 4ac}}{2a}.$$

Consequently, whether a is positive or negative, we may write (1-9) as

$$x + \frac{b}{2a} = \pm \frac{\sqrt{b^2 - 4ac}}{2a}.$$

Solving for x, we get

$$x = \frac{-b \pm \sqrt{b^2 - 4ac}}{2a}. \qquad \blacksquare$$

Perhaps the most important issue regarding the proof of this theorem is how we came up with the constant $b^2/(4a^2)$ that was added to both sides of equation (1-8). Its determination depends on the fact that for every constant k,

$$(x + k)^2 = x^2 + 2kx + k^2.$$

Looking at the right side of this equation, we see that k, the constant being squared, is $\frac{1}{2}$ of the coefficient of x. In the left side of equation (1-8), the coefficient of x is b/a. Thus $b/(2a)$ is the constant to be squared and added.

This method of adding a constant to make the resulting trinomial a perfect square is called **completing the square.** We will use this procedure again in later sections.

Equation (1-7) is referred to as the **quadratic formula** and can be used to find the solution set of *any* quadratic equation. It is used in Examples 3 and 4.

Example 3
Find the solution set of $12x^2 - 5x - 2 = 0$ by using the quadratic formula.

Solution. In this problem we have $a = 12$, $b = -5$, and $c = -2$. Substituting these values into the quadratic formula, we get

$$\begin{aligned} x &= \frac{-(-5) \pm \sqrt{(-5)^2 - 4(12)(-2)}}{2(12)} \\ &= \frac{5 \pm \sqrt{121}}{24} \\ &= \frac{5 \pm 11}{24} \end{aligned}$$

$$x = \frac{5 + 11}{24} \quad \text{or} \quad x = \frac{5 - 11}{24}.$$

Thus the solution set is $\{2/3, -1/4\}$. $\blacksquare$

Example 4
Find the solution set of $4x^2 = 8x + 1$ by using the quadratic formula.

Solution. Before attempting to apply the quadratic formula, let's put the equation in the form

$$4x^2 - 8x - 1 = 0.$$

Hence we have $a = 4$, $b = -8$, and $c = -1$. After substituting these values into the quadratic formula, we have

$$\begin{aligned} x &= \frac{-(-8) \pm \sqrt{(-8)^2 - 4(4)(-1)}}{2(4)} \\ &= \frac{8 \pm \sqrt{80}}{8} \\ &= \frac{8 \pm 4\sqrt{5}}{8} \\ &= \frac{2 \pm \sqrt{5}}{2}. \end{aligned}$$

Thus the solution set is $\left\{\frac{2 + \sqrt{5}}{2}, \frac{2 - \sqrt{5}}{2}\right\}$. ■

Note that the method of factoring can be used in Example 3 but will not work in Example 4.

The number $b^2 - 4ac$ which appears under the radical sign in the quadratic formula is called the **discriminant** of the quadratic equation. Its value determines the number of (real) roots, as follows:

1. If the discriminant is positive, then there are two roots.
2. If the discriminant is zero, then there is one root.
3. If the discriminant is negative, then there are no roots.

Example 5
Find the solution set of $x^2 - 2x + 5 = 0$ by using the quadratic formula.

Solution. We have $a = 1$, $b = -2$, and $c = 5$. Substituting, we get

$$x = \frac{2 \pm \sqrt{4 - 20}}{2}.$$

We need not simplify. Since the discriminant is negative, there are no real solutions. (The square root of a negative number isn't real.) The solution set is $\varnothing$. ■

Exercise
Find the solution set of each of the following by using the quadratic formula.
(a) $3x^2 + 5x - 2 = 0$ (b) $4x^2 + 2x = 3$ (c) $2x^2 - x + 1 = 0$

Answers. (a) $\left\{\frac{1}{3}, -2\right\}$ (b) $\left\{\frac{-1 + \sqrt{13}}{4}, \frac{-1 - \sqrt{13}}{4}\right\}$ (c) $\varnothing$ ■

Problem Set 1.3

In Problems 1–10, find the solution set by extraction.

1. $x^2 = 25$
2. $2x^2 = 32$
3. $(2x - 3)^2 = 16$
4. $(3x + 5)^2 = 36$
5. $(x - 2)^2 = 7$
6. $(2x + 1)^2 = 8$
7. $x^2 = -9$
8. $(x - 1)^2 = -1$
9. $\left(\frac{x}{2x + 1}\right)^2 = \frac{1}{25}$
10. $\left(\frac{2x}{x - 5}\right)^2 = 12$

In Problems 11–24, find the solution set by using the quadratic formula.

11. $2x^2 - 3x - 2 = 0$
12. $3x^2 + 7x = 0$
13. $x^2 - 6x + 6 = 0$
14. $\frac{x^2}{24} + \frac{x}{2} + 1 = 0$
15. $2x^2 - x = 5$
16. $x^2 + 3x = 5x + 7$
17. $x(3x - 2) = 4$
18. $(2x - 1)(x + 3) = -4$
19. $\frac{x}{x^2 + 4} = \frac{1}{3}$
20. $\frac{2x}{x - 1} = \frac{3x - 1}{x + 1}$
21. $\frac{5}{x + 2} + \frac{3}{x - 2} = 4 + \frac{1}{x^2 - 4}$
22. $\frac{1}{x^2 - 1} - \frac{x}{x + 1} = \frac{3}{x - 1}$
23. $x^2 - \sqrt{2}x - 12 = 0$
24. $x^2 + 2\sqrt{2}x - 1 = 0$
25. Determine k so that $kx^2 - 20x + 25 = 0$ has exactly one solution.
26. Determine a quadratic equation with integral coefficients whose solution set is $\left\{\frac{-3 + \sqrt{21}}{2}, \frac{-3 - \sqrt{21}}{2}\right\}$.

1.4 Radical Equations

Equations in which the variable appears within a radical sign are called **radical equations.** For example,

$$\sqrt{x + 1} = 5, \qquad \sqrt[3]{2x - 1} = 10, \qquad \text{and} \qquad \sqrt{x^2 - 1} = \sqrt{2x + 5}$$

are radical equations. Note that an equation like $2x + \sqrt{3} = 5x - 2$ is *not* a radical equation because the variable is not within the radical sign! (This equation is linear.) To solve radical equations, we need the following theorem.

Theorem 1.6

Let p_x and q_x be algebraic expressions in the variable x, and let n by any positive integer. Every solution of the equation

$$p_x = q_x \tag{1-10}$$

is also a solution of

$$(p_x)^n = (q_x)^n. \tag{1-11}$$

This theorem says that if we raise each side of a given equation to the nth power, where n is a positive integer, then all solutions of the given equation are solutions of the derived equation. It does *not* say that the two equations are equivalent. There may be some solutions of equation (1-11) which are *not* solutions of equation (1-10). Such "extra" solutions of the derived equation are called **extraneous solutions.** We can see this by considering a very simple equation like $x = 1$. Upon squaring both sides we have $x^2 = 1$. Now the first equation obviously has only one solution, but the second has two (1 and -1). The -1 is extraneous. Because of the possibility of extraneous solutions when raising both sides of an equation to a positive integral power, it is *essential* to check solutions of the final equation by substituting in the original.

In solving radical equations we should isolate a radical term on one side of the equation and then raise both sides to the appropriate power. This is demonstrated in Examples 1 and 2. As seen in Example 2, this procedure may be needed more than once.

Example 1
Find the solution set of $x = \sqrt{x^2 - x - 2} + 1$.

Solution. We must first isolate the radical term. (Get it on one side of the equation by itself.) Subtracting 1 from both sides, we get

$$x - 1 = \sqrt{x^2 - x - 2}.$$

Now we square each side and get the following:

$$\begin{aligned} (x - 1)^2 &= (\sqrt{x^2 - x - 2})^2 \\ x^2 - 2x + 1 &= x^2 - x - 2 \\ -2x + x &= -2 - 1 \\ x &= 3. \end{aligned}$$

Because 3 is the only root of the last equation, 3 is the only *possible* solution of the original equation. We *must* check to find out. Since $3 = \sqrt{3^2 - 3 - 2} + 1$ is a true statement, the solution set is $\{3\}$. ■

In general, before writing the solution set of an equation, we must note whether the roots of the final equation are in the universe of the variable or we must substitute the roots in the original equation. But because we *must* check the roots in solving radical equations, we need not concern ourselves with the universe of the variable.

Example 2
Find the solution set of $\sqrt{5x - 11} + \sqrt{x - 3} = 4$.

Solution. When there are two radical terms, we shall still isolate one of them. Subtracting $\sqrt{x - 3}$ from both sides, we get

$$\sqrt{5x - 11} = 4 - \sqrt{x - 3}.$$

Upon squaring both sides, we have the following:

$$\begin{aligned} (\sqrt{5x - 11})^2 &= (4 - \sqrt{x - 3})^2 \\ 5x - 11 &= 16 - 8\sqrt{x - 3} + x - 3. \end{aligned}$$

Because we still have a radical equation, we must again isolate the radical term. We have

$$4x - 24 = -8\sqrt{x - 3}.$$

Upon dividing both sides by 4 (to keep the numbers smaller) and squaring, we have the following:

$$\begin{aligned} x - 6 &= -2\sqrt{x - 3} \\ (x - 6)^2 &= (-2\sqrt{x - 3})^2 \\ x^2 - 12x + 36 &= 4(x - 3) \\ x^2 - 12x + 36 &= 4x - 12 \\ x^2 - 16x + 48 &= 0 \\ (x - 4)(x - 12) &= 0 \\ x = 4 \quad \text{or} \quad x &= 12 \end{aligned}$$

The only *possible* solutions are 4 and 12. Because $\sqrt{5 \cdot 4 - 11} + \sqrt{4 - 3} = 4$ is true but $\sqrt{5 \cdot 12 - 11} + \sqrt{12 - 3} = 4$ is false, the solution set is $\{4\}$. ■

Exercise
Find the solution set.

(a) $x = \sqrt{x^2 - x - 5} - 3$ (b) $\sqrt{7 - 2x} + \sqrt{3 - x} = 1$

Answers. (a) $\{-2\}$ (b) $\{3\}$ ■

Problem Set 1.4

Find the solution set.

X 1. $\sqrt{3x - 2} = \sqrt{x + 8}$

2. $\sqrt{4x + 2} = \sqrt{5x - 6}$

3. $\sqrt{x + 14} - \sqrt{3x - 7} = 0$

4. $\sqrt{x + 4} - \sqrt{4x + 5} = 0$

X 5. $\sqrt[3]{2 - 3x} = \sqrt[3]{3x - 5}$

6. $\sqrt[3]{15 - 2x} - \sqrt[3]{5 + 3x} = 0$

7. $\sqrt{1 - 2x} = \sqrt{3x - 9}$

8. $\sqrt{1 - 5x} = -2$

X 9. $\sqrt{\dfrac{x + 5}{5}} = \sqrt{\dfrac{3x + 1}{2}}$

10. $\sqrt{\dfrac{3}{2x - 5}} - \sqrt{\dfrac{2}{x + 6}} = 0$

11. $\sqrt{x^2 - 3x - 3} = \sqrt{3x + 4}$

12. $\sqrt{x^2 - 5x - 11} - \sqrt{2x + 7} = 0$

X 13. $x - 1 = \sqrt{x^2 - x - 2}$

14. $\sqrt{4x^2 - 3x + 6} = 2x - 8$

15. $2x - 3 = \sqrt{x^2 - 2x + 6}$

16. $\sqrt{2x^2 - 7x} = x + 4$

X 17. $3 + \sqrt{3x + 1} = x$

18. $2x = 1 + \sqrt{2x^2 - x + 3}$

19. $\sqrt[3]{x^2 + 3x} = 2$

20. $\sqrt[4]{x^2 + 2x} = 1$

X 21. $\sqrt{2(8x^2 - 2x - 3)} + 3 = 2x$

22. $\sqrt{12x^2 + x - 1} - 2x = 3$

X 23. $\sqrt{2x - 7} - 1 = \sqrt{x - 4}$

24. $\sqrt{4x + 5} + 1 = \sqrt{6x + 6}$

25. $\sqrt{2x + 7} - \sqrt{x - 5} = 3$

26. $\sqrt{5x + 1} - \sqrt{4x - 3} = 1$

27. $\sqrt{3x + 3} - \sqrt{7x + 2} = 1$

28. $\sqrt{5x + 6} = 3 + \sqrt{x + 3}$

29. $\sqrt{x^2 + x + 2} - \sqrt{x^2 + x - 1} = 1$

30. $\sqrt{x^2 + x + 4} - \sqrt{x^2 - 2x + 1} = 2$

31. $2 - x = \sqrt{3 - x}$

32. $2x = 1 + \sqrt{2x^2 - 3x + 3}$

33. $x = \sqrt{3x^2 + x + 6} - 2$

34. $(3x - 2) - (x + 1) = \sqrt{6}$

1.5 Other Types of Equations

Equations of types other than the ones studied so far are usually difficult to solve. However, certain favorable circumstances enable us to solve some of these. In the first two examples a method used to solve quadratic equations enables us to find the solution set.

Example 1
Find the solution set of $x^5 - 16x = 0$.

Solution. We want to factor the left side as completely as possible. We get the following:

$$\begin{aligned} x^5 - 16x &= 0 \\ x(x^4 - 16) &= 0 \\ x(x^2 - 4)(x^2 + 4) &= 0 \\ x(x - 2)(x + 2)(x^2 + 4) &= 0. \end{aligned}$$

Because this last equation is true only when at least one of the factors is zero, it is equivalent to

$$x = 0 \quad \text{or} \quad x - 2 = 0 \quad \text{or} \quad x + 2 = 0 \quad \text{or} \quad x^2 + 4 = 0.$$

There are no solutions of the fourth equation of this compound statement, but the first three determine the solution set: $\{0, 2, -2\}$. ■

Example 2
Find the solution set of $x^3 - 2x^2 = x - 2$.

Solution. We first need to get one side to zero. Subtracting $x - 2$ from both sides, we have

$$x^3 - 2x^2 - x + 2 = 0.$$

Factoring the left side, we have the following:

$$\begin{aligned} x^2(x - 2) - (x - 2) &= 0 \\ (x - 2)(x^2 - 1) &= 0 \\ (x - 2)(x - 1)(x + 1) &= 0. \end{aligned}$$

Hence the solution set is $\{2, 1, -1\}$. ■

Exercise
Find the solution set.
(a) $x^7 - 4x^3 = 0$ (b) $2x^3 + x^2 = 8x + 4$

Answers. (a) $\{0, \sqrt{2}, -\sqrt{2}\}$ (b) $\left\{-\frac{1}{2}, 2, -2\right\}$ ■

An equation of the form

$$au^2 + bu + c = 0,$$

where a, b, and c are constants, with $a \neq 0$, and where u is an algebraic expression in x, is said to be of **quadratic type** in x. For example,

$$(x^2)^2 + 5x^2 + 6 = 0 \qquad \text{and} \qquad (x^2 - 1)^2 - 2(x^2 - 1) - 3 = 0$$

are equations of quadratic type.

Such equations can be solved by an appropriate substitution and by using the methods for solving quadratic equations, as demonstrated in Examples 3 and 4.

Example 3
Find the solution set of $x^4 - 5x^2 - 14 = 0$.

Solution. If we let $u = x^2$, then the equation becomes

$$u^2 - 5u - 14 = 0,$$

a quadratic equation in u. (We say that the given equation is quadratic in x^2.) Factoring we get

$$(u + 2)(u - 7) = 0$$
$$u = -2 \qquad \text{or} \qquad u = 7.$$

After substituting x^2 for u, we have

$$x^2 = -2 \qquad \text{or} \qquad x^2 = 7.$$

There are no solutions to the first equation of the preceding statement, but the second equation has two. The solution set is $\{\sqrt{7}, -\sqrt{7}\}$. ■

Example 4
Find the solution set of $(x^2 - x)^2 - 3(x^2 - x) + 2 = 0$.

Solution. Letting $u = x^2 - x$, our equation becomes

$$u^2 - 3u + 2 = 0.$$

After factoring, we have

$$(u - 2)(u - 1) = 0$$
$$u = 2 \qquad \text{or} \qquad u = 1.$$

Replacing u by $x^2 - x$, we get

$$x^2 - x = 2 \qquad \text{or} \qquad x^2 - x = 1$$
$$x^2 - x - 2 = 0 \qquad \text{or} \qquad x^2 - x - 1 = 0$$

$$(x - 2)(x + 1) = 0 \qquad \text{or} \qquad x = \frac{1 \pm \sqrt{1 + 4}}{2}$$

$$x = 2 \quad \text{or} \quad x = -1 \qquad \text{or} \qquad x = \frac{1 \pm \sqrt{5}}{2}.$$

So the solution set is $\left\{2, -1, \frac{1 + \sqrt{5}}{2}, \frac{1 - \sqrt{5}}{2}\right\}$. ■

Exercise

Find the solution set of each of the following.
(a) $x^4 - x^2 - 2 = 0$ (b) $(x^2 + 3x)^2 - (x^2 + 3x) - 6 = 0$

Answers. (a) $\{\sqrt{2}, -\sqrt{2}\}$ (b) $\left\{-1, -2, \frac{-3 + \sqrt{21}}{2}, \frac{-3 - \sqrt{21}}{2}\right\}$ ■

The following theorem will be used in solving equations in which the variable appears within an absolute value symbol.

Theorem 1.7

Let a be a nonnegative constant, and let p_x be an algebraic expression in x. Then the equation

$$|p_x| = a$$

is equivalent to the compound statement

$$p_x = a \quad \text{or} \quad p_x = -a.$$

The proof of this theorem depends on the fact that $|a| = |-a|$. Its use is illustrated in Example 5.

Example 5

Find the solution set of $|2x + 3| = 5$.

Solution. By Theorem 1.7, the given equation is equivalent to

$$2x + 3 = 5 \qquad \text{or} \qquad 2x + 3 = -5.$$

Solving each of these equations for x, we get

$$x = 1 \qquad \text{or} \qquad x = -4.$$

Thus the solution set is $\{1, -4\}$. ■

Because the absolute value of any number is nonnegative, the solution set of $|p_x| = a$, where a is negative, is $\varnothing$.

Example 6
Find the solution set of $|5x^2 - 7x| = |7x + 3|$.

Solution. Because the right side of this equation is not a constant, Theorem 1.7 does not apply. But since the absolute value of each expression equals either plus or minus the expression, the given equation is equivalent to the four-part compound statement

$$5x^2 - 7x = 7x + 3 \qquad \text{or} \qquad 5x^2 - 7x = -(7x + 3)$$

or

$$-(5x^2 - 7x) = -7x + 3 \qquad \text{or} \qquad -(5x^2 - 7x) = 7x + 3.$$

But note that the third equation of this statement is equivalent to the first (it's just the first multiplied by -1), and similarly, the fourth equation is equivalent to the second. Thus the compound statement may be simplified to

$$5x^2 - 7x = 7x + 3 \qquad \text{or} \qquad 5x^2 - 7x = -(7x + 3).$$

Solving the former equation for x, we have

$$5x^2 - 14x - 3 = 0$$
$$(5x + 1)(x - 3) = 0$$
$$x = \frac{-1}{5} \qquad \text{or} \qquad x = 3.$$

Solving the other equation for x, we get

$$5x^2 = -3$$
$$x^2 = \frac{-3}{5}.$$

Since there are no solutions to this last equation, the solution set is $\{-1/5, 3\}$. ■

Exercise
Find the solution set of each of the following.
(a) $|3x + 2| = 1$ (b) $|x^2 - x| = |x + 3|$

Answers. (a) $\left\{-1, \frac{-1}{3}\right\}$ (b) $\{3, -1\}$ ■

Problem Set 1.5

Find the solution set.

1. $x^3 - 4x = 0$

2. $x^4 - 9x^2 = 0$

3. $x^4 = 5x^2$

4. $x^3 = 8x$

5. $x^4 + 4x^2 = 0$

6. $x^6 + 6x^4 = 0$

7. $x^3 = 8$

8. $x^3 + 8 = 0$

9. $2x^3 - 8x = x^2 - 4$

10. $3x^3 = 3x - 2x^2 + 2$

11. $2x^3 + x^2 + 8x + 4 = 0$

12. $3x^3 - 2x^2 + 27x = 18$

13. $5x^3 + 10x^2 = 2x^4 + 25x$

14. $5x^4 - x^3 - 5x + 1 = 0$

15. $2x^4 - 3x^3 - x^2 = 0$

16. $x^3 = 2x^2 - 3x$

17. $(3x^2 - 4)^2 = 25$

18. $(2x^2 + 5)^2 = 49$

19. $x^4 - 13x^2 + 36 = 0$

20. $x^4 - 5x^2 + 4 = 0$

21. $x^4 - 2x^2 - 3 = 0$

22. $x^4 - 3x^2 - 10 = 0$

23. $12x^4 + 28x^2 = -15$

24. $5x^4 + 5x^2 + 1 = 0$

25. $2x^4 + 5x^2 - 1 = 0$

26. $x^4 + 2\sqrt{2}x^2 - 6 = 0$

27. $(x^2 + 2x)^2 - 2(x^2 + 2x) - 3 = 0$

28. $(x^2 - 3x)^2 - 2(x^2 - 3x) - 8 = 0$

29. $(x^2 + x)^2 - 2(x^2 + x) - 3 = 0$

30. $(x^2 - x)^2 - (x^2 - x) - 6 = 0$

31. $\left(\dfrac{x + 2}{3x + 1}\right)^2 - 5\left(\dfrac{x + 2}{3x + 1}\right) + 6 = 0$

32. $\left(\dfrac{x^2 - 3}{x}\right)^2 - 6\left(\dfrac{x^2 - 3}{x}\right) + 8 = 0$

33. $x^{-4} - 5x^{-2} + 4 = 0$

34. $x^{-4} - 3x^{-2} - 4 = 0$

35. $|2x - 3| = 3$

36. $|3x - 2| = 5$

37. $|x^2 - 2x| = 1$

38. $|x^2 - 2x| = 15$

39. $\left|\dfrac{x^2}{3x + 1}\right| = 2$

40. $\left|\dfrac{2}{x + 1} - \dfrac{x}{x - 1}\right| = \dfrac{3}{2}$

41. $|2x + 1| = -5$

42. $|(x + 1)^2| = 6$

43. $2x + 3 = |-5|$

44. $2x^2 + 2x - |5| = 0$

45. $|x - 2| = |3x - 4|$

46. $|5x + 7| = |2 - 3x|$

47. $|3x^2 + x| = |x - 5|$

48. $|x^2 - 3x + 5| = |7x^2 - 4x|$

49. $|x^2 + 1| = 0$

1.6 Word Problems

In many applications of algebra, problems are stated in words, and one must produce a correct equation as well as solve it. In this section we study word problems that produce linear and quadratic equations. Using the following steps will help solve word problems.

1. Represent the quantity asked for with an algebraic symbol.
2. Find a statement of equality.
3. If necessary, represent other unknown quantities in terms of the variable already introduced.
4. Translate the statement of equality.

5. Solve the equation and state your answer.
6. Make certain that your answer satisfies the conditions stated in the problem.

It is essential that you *understand the problem*. Read the problem as many times as necessary to comprehend the information being given and the information being requested. Then proceed to step 1. And if the variable represents a measurable quantity such as distance, length, or time, identify the units of measurement.

Step 2 is the most difficult. The statement of equality may be explicit or implied. And some problems must be identified as being a certain type to determine the statement of equality.

The statement of equality may tell us that there are other unknown quantities, besides the one asked for, which need to be represented. Other information in the problem must be used to represent such quantities.

In step 4 we convert the statement of equality into an equation. If correctly translated, we should be able to identify what each term (or group of terms) represents in the problem. Writing the statement of equality in English before attempting to write the equation will help produce a correct equation.

Examples 1 through 5 illustrate the procedure to be used in solving word problems. Note how each of the recommended steps is applied.

Example 1
The sum of a number and twice its reciprocal is $\frac{17}{6}$. What is the number?

Solution. The quantity asked for is the number. Let

$$x = \text{the number.}$$

The first sentence is the statement of equality. That is, "the number plus twice its reciprocal is $\frac{17}{6}$." Translated into an equation, it becomes

$$x + 2\left(\frac{1}{x}\right) = \frac{17}{6}.$$

Solving this equation for x, we get

$$\begin{aligned} 6x^2 + 12 &= 17x \\ 6x^2 - 17x + 12 &= 0 \\ (2x - 3)(3x - 4) &= 0 \end{aligned}$$

$$x = \frac{3}{2} \quad \text{or} \quad x = \frac{4}{3}.$$

Thus the number is $\frac{3}{2}$ or $\frac{4}{3}$. To check we note that

$$\frac{3}{2} + 2 \cdot \frac{2}{3} = \frac{17}{6}$$

and

$$\frac{4}{3} + 2 \cdot \frac{3}{4} = \frac{17}{6}. \qquad \blacksquare$$

Example 2
A rancher made a trip of 657 miles by traveling part of the way in a car that averaged 40 miles per hour and the remainder of the way on a train that averaged 60 miles per hour. If the entire trip took 12 hours and there were no delays, find the distance traveled in the car.

Solution. The quantity asked for is the distance traveled in the car. Let

$$x = \text{distance traveled in car (in miles).}$$

The statement of equality is ". . . the entire trip took 12 hours. . .," that is,

$$\text{(time in car)} + \text{(time on train)} = 12.$$

To calculate these times, we use the formula $d = rt$, where d represents distance, r represents rate (average speed), and t represents time. Solving this equation for t, we find that $t = d/r$. Therefore,

$$\text{time in car} = \frac{x}{40}.$$

To figure the time on the train, we need to know the distance. Since the sum of the two distances is 657,

$$657 - x = \text{distance traveled on the train.}$$

Hence

$$\text{time on train} = \frac{657 - x}{60}.$$

Thus the correct translation of the statement of equality is

$$\frac{x}{40} + \frac{657 - x}{60} = 12.$$

Solving this equation for x, we get

$$\begin{aligned} 3x + 2(657 - x) &= 12(120) \\ x &= 126. \end{aligned}$$

Our answer is 126 miles. To check, we need to note that 3.15 hours ($126/40 = 3.15$) were spent in the car, 8.85 hours ($531/60 = 8.85$) were spent on the train, and the sum of these is 12. ■

Example 3
How many milliliters of a 60% acid solution must be added to 500 ml of a 5% acid solution to obtain a 10% acid solution?

Solution. Let

x = volume of 60% acid solution to be added (in milliliters).

Now the statement of equality in this problem is not as evident as those in the previous two examples because it is not explicit. In mixture problems of this type, it is, in general,

(amount in first ingredient) + (amount in second ingredient)
= (amount in final mixture).

In particular, the statement of equality in this example is

(amount of acid in 60% solution) + (amount of acid in 5% solution)
= (amount of acid in 10% solution).

To find each of these amounts, we must multiply the per cent acid by the volume of the solution. The volume of the 60% solution is x, the volume of the 5% solution is 500, and the volume of the 10% solution is the sum $500 + x$. Hence the equation is

$$0.60x + 0.05(500) = 0.10(500 + x).$$

Upon solving this equation for x, we have

$$x = 50.$$

After checking, we conclude that the answer is 50 ml. ■

Example 4
If Jim can paint a room in 4 hours and his brother can paint the same room in 6 hours, how long would it take the two working together to paint the room?

Solution. Let

x = number of hours it would take them to paint the room together.

As in the previous example, the statement of equality is not apparent. It is

(that part of job done by Jim) + (that part of job done by his brother) = 1.

Jim does $\frac{1}{4}$ of the job in 1 hour, and his brother does $\frac{1}{6}$ of the job in 1 hour. And since it takes x hours to do the job together, that part of the job done by Jim is $x/4$, and that part done by his brother is $x/6$. Therefore, our equation is

$$\frac{x}{4} + \frac{x}{6} = 1.$$

Solving for x, we get

$$x = \frac{12}{5}.$$

Thus our answer is 2.4 hours. We check this by noting that in 2.4 hours, Jim can paint 0.6 of the room ($2.4/4 = 0.6$), and his brother can paint 0.4 of the room ($2.4/6 = 0.4$). Hence together they can paint 1 room ($0.6 + 0.4 = 1$) in 2.4 hours. ■

Example 5
A page contains 48 square inches of printed area and has a margin of 2 inches at the top and bottom and a margin of 1 inch at the sides. What is the length of the page if its width is two-thirds of its length?

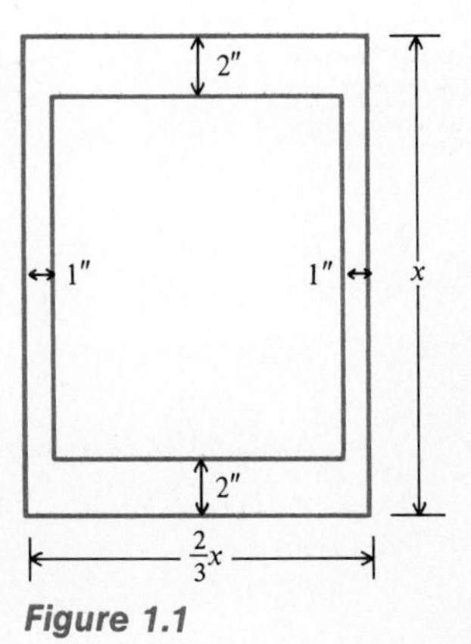

Figure 1.1

Solution. Let

x = length (in inches).

The statement of equality is "a page contains 48 square inches of printed area . . . ." However, because of the margins, this area is not the product of the length and width of the page. As seen in Figure 1.1, the length of the printed region is $x - 4$, and the width is $\frac{2}{3}x - 2$. Therefore, the equation is

$$(x - 4)\left(\frac{2}{3}x - 2\right) = 48.$$

Solving this quadratic equation, we get the following equivalent equations:

$$\frac{2}{3}x^2 - \frac{14}{3}x + 8 = 48$$

$$2x^2 - 14x + 24 = 144$$
$$x^2 - 7x - 60 = 0$$
$$(x - 12)(x + 5) = 0$$
$$x = 12 \quad \text{or} \quad x = -5.$$

Because x represents a length, we reject the value -5. Thus our answer is that the length is 12 inches. To check, we note that such a page with the margins described has a printed region 8 inches long and 6 inches wide—48 square inches of printed area. ■

Problem Set 1.6

1. If a student's midterm grades are 85, 62, and 73, what score is needed on the final to average 75?
2. (Continuation of Problem 1.) What if the final carries double weight?
3. A homeowner wishes to make \$47,000 from the sale of his house. At what price must it be sold if the real estate broker's commission is 6%?
4. During the first half of the game, a basketball team shot 45% from the field on 40 shots. If the team took 35 shots in the second half and finished the game with a shooting percent of 52%, how many shots did they hit in the second half?
5. How long is a rectangular plot whose length is 9 feet greater than its width and whose perimeter is 94 feet?
6. The width of a rectangle is $\frac{1}{3}$ its length. What is its length if the perimeter is 32 feet?
7. Find the altitude of a triangle whose area is 8 in.^{2} and whose base is twice the altitude.
8. Kevin agreed to work 3 months on his uncle's farm for \$600 and a used car. After 2 months, Kevin had to quit, and his uncle gave him the car plus \$100. What was the value of the car?
9. A passenger train travels 150 miles in the same time that a freight train travels 100 miles. If the freight train travels 20 mph slower than the passenger train, what is the speed of the passenger train?
10. A freight train leaves town A for town B, traveling at an average rate of 50 mph. Three hours later a passenger train also leaves town A for town B, on a parallel track, traveling at an average rate of 80 mph. If the two towns are 500 miles apart, how far from town A does the passenger train catch up to the freight train?
11. A motorist traveled from city A to city B at an average speed of 50 mph. On the return trip, he left at 3:00 P.M. and drove at an average rate of 60 mph, making the trip in 33 minutes less time. What is the distance from city A to city B?
12. An airplane flew at cruising speed with the wind for 50 minutes and covered the same distance against the wind in 1 hour. If the cruising speed of the plane was 330 mph, what was the velocity of the wind?
13. A pistol is fired at a target, and the sound of the bullet's impact is heard 1.35 seconds after its firing. If the speed of the bullet is 5500 ft/sec and the speed of sound is 1100 ft/sec, how far away is the target?

14. A boy ran to his friend's house at the rate of 10 mph, rested for 15 minutes, and then walked back home at the rate of 4 mph. How far apart are the two houses if the entire trip took 2 hours?
15. Paducah is 180 miles west of Elizabethtown and 260 miles west of Lexington. A woman drove from Paducah to Lexington. If the trip from Paducah to Elizabethtown took 2 hours more than the trip from Elizabethtown to Lexington, what was the average speed? (Assume that the average speeds on each trip are equal.)
16. Three airports, A, B, and C, are located on an east-west line. A is 450 miles west of B and 855 miles west of C. A pilot flew from A to B, waited 2 hours, and then flew to C. During the first part of the trip the wind was blowing from the west at 25 mph, but during the delay it quit. If each flight required the same amount of time, what was the average airspeed? (Assume that the average airspeeds on each flight are equal.)
17. How many pounds of an alloy containing 40% silver must be melted with 30 pounds of an alloy containing 68% silver to obtain an alloy containing 61% silver?
18. Kentucky bluegrass seed sells for $2.40 per pound, and ryegrass seed sells for $2.00 per pound. If a seed dealer wants to make a mixture weighing 40 pounds to sell for $94.00, how many pounds of bluegrass must he use?
19. How much water should be added to 6 gallons of pure acid to obtain a 30% acid solution?
20. A nurse is told to dilute a 50-ml solution which is 80% medicine. If the solution is to be 20% medicine, how much water must be added?
21. A chemist has 12 liters of acid solution which is 25% acid. How much should be removed and replaced with a 50% solution in order to obtain a solution which is 30% acid?
22. A radiator contains 8 quarts of a mixture of water and antifreeze. If 40% of the mixture is antifreeze, how much of the mixture should be drained and replaced by pure antifreeze to obtain a new mixture which is 60% antifreeze?
23. Suppose that a professor can grade a set of papers in 5 hours, and his grader can grade a set in 7 hours. How long will it take to grade a set if both work together?
24. Suppose that one pump can fill a certain tank in 2 hours, and a second pump can fill the tank in 3 hours. How long will it take them together to fill the tank?
25. A man and his son working together can paint their house in 3 days. The man can do the job alone in $2\frac{1}{2}$ days less than the son can do it. How long would it take the man to paint the house alone?
26. Working together, Kurt and his younger brother can mow the yard in 50 minutes. Working alone, the young brother takes twice as long as Kurt. How long would it take Kurt to do the job alone?
27. A farmer can plow a certain field in 6 hours. Using a smaller tractor his hired hand can plow the same field in 10 hours. If the hired hand works 2 hours before being joined by the farmer, how long will the farmer have to work to finish the job?
28. If a chicken and a half can lay an egg and a half in a day and a half, how long does it take one chicken to lay one egg?
29. Suppose that one pump can fill a certain tank in 12 hours and a second pump can empty the tank (from full) in 8 hours. If the tank is half full when both pumps are started, how long will it take to empty the tank?
30. A company offered two proposals for the accumulation and use of sick leave. Under system A, an employee would earn sick leave at a rate of 90 sick hours per 1860 non-sick hours reported, and sick leave would not be earned on sick-time turned in. Under system B, an employee would earn sick leave at a rate of 90 sick hours per 1905 hours reported, and sick leave would be earned on sick-time

turned in. Show that these two systems are equivalent. (*Hint:* Show that if an employee has 0 hours of sick leave under one system, he has 0 hours under the other.)

1.7 Inequalities

In the remainder of this chapter we shall study inequalities. An **inequality in the variable** x is a statement of the form

$$p_x < q_x, \quad p_x \leq q_x, \quad p_x \geq q_x, \quad \text{or} \quad p_x > q_x,$$

where p_x and q_x are algebraic expressions in x. For example,

$$2x + 1 > x - 5, \qquad x^2 \leq x + 1, \qquad \text{and} \qquad \frac{x - 1}{x^2} \geq 2$$

are inequalities in x.

Let x be a variable with universe U = {1, 2, 3}, and consider the inequality $x \leq 2$. Because x represents 1, 2, and 3, the inequality $x \leq 2$ actually stands for three inequalities: $1 \leq 2$, $2 \leq 2$, and $3 \leq 2$. Hence it is inappropriate to ask whether the inequality $x \leq 2$ is true or false. Instead, one should ask: "For which element(s) in the universe is the inequality true when the variable is replaced?" In this case it is clear that the inequality is true only when 1 and 2 are substituted for x.

A real **root** of an inequality in one variable is a real number which makes the inequality true when the variable is replaced by that number.

A real **solution** of an inequality in one variable is a real root which belongs to the universe of the variable. The set of all solutions is called the **solution set.** Therefore, for the inequality just considered, 1 and 2 are solutions, and the solution set is {1, 2}.

Two inequalities are called **equivalent** provided that they have the same solution set. For example, $2x < 6$ and $x + 1 < 4$ are equivalent.

As was the case with equations, neither the universe nor the inequality in subsequent problems will be as simple as those of our example. Just as with equations, when the universe is not explicitly stated, we shall assume that it is the set of all real numbers for which all terms of the inequality are defined and real upon replacement. Given the inequality $1/x \leq \sqrt{x}$, we see that the left side is undefined if x is replaced by 0, and the right side is not real if x is negative. Therefore, the universe is the set of all *positive* real numbers.

An inequality may be too complicated to find solutions by observation. In such cases we shall find an equivalent inequality whose solution set may be easily noted. We do this by using the following theorem.

Theorem 1.8

Let p_x, q_x, and r_x be algebraic expressions in the variable x. Then the inequality

$$p_x < q_x$$

is equivalent to each of the following:

(i) $p_x + r_x < q_x + r_x$
(ii) $p_x \cdot r_x < q_x \cdot r_x$ *when* $r_x > 0$
(iii) $p_x \cdot r_x > q_x \cdot r_x$ *when* $r_x < 0$

This theorem is a direct result of Theorem 0.3. This theorem remains true when $<$ and $>$ are interchanged throughout (excluding the qualifiers "when $r_x > 0$" and "when $r_x < 0$") and is true if $<$ and $>$ are replaced by $\leq$ and $\geq$, respectively, or $\geq$ and $\leq$, respectively (again excluding the qualifiers). It says that an algebraic expression (including one consisting of a number only) may be added to or subtracted from both sides of an inequality, and the resulting inequality is equivalent to the beginning inequality. Further, the theorem says that both sides of an inequality may be multiplied by or divided by a *nonzero* algebraic expression, and the resulting inequality is equivalent to the initial inequality, provided that the inequality symbol is reversed when the multiplier or divisor is negative. Hence the methods for obtaining equivalent inequalities are the same as those for obtaining equivalent equations, with one exception: *with inequalities, a negative multiplier or divisor reverses the inequality symbol.* We shall use these facts in Example 1.

Example 1
Find the solution set of $x - 1 < 3x + 5$.

Solution. Subtracting $3x$ and adding 1 to both sides, we get

$$x - 3x < 5 + 1.$$

Upon combining like terms, we have

$$-2x < 6.$$

Because we have merely added and subtracted to both sides, we have not been concerned with the inequality symbol (it stays the same). It is when we multiply or divide that we must be careful. Upon dividing both sides by -2, and reversing the inequality symbol since the divisor is negative, we have

$$x > -3.$$

Thus the solution set consists of all real numbers greater than -3 and may be written $\{x|x > -3\}$. ■

The **graph** of an inequality in a single variable is the set of all points on a number line whose coordinates are solutions of the inequality. In sketching the graph it is customary to darken that portion of the number line which corresponds to the solution set of the inequality. For example, the graph of the inequality in Example 1 is

-3

The symbol ○ at -3 indicates that the point with coordinate -3 is not part of the graph (because -3 is not in the solution set). If -3 were a solution, we would have used the symbol ● at -3.

An inequality in a variable x is called **linear** in x provided that it is equivalent to

$$ax + b < 0, \quad ax + b \leq 0, \quad ax + b > 0, \quad \text{or} \quad ax + b \geq 0,$$

where a and b are constants with $a \neq 0$.

Finding the solution set of a linear inequality is very much like finding the solution set of a linear equation. In general, we get the x terms on one side of the inequality and the constants on the other. Then we combine like terms and divide both sides by the coefficient of x (reversing the inequality symbol if the divisor is negative). This technique was used in Example 1 and is again illustrated in Example 2.

Example 2

Determine and graph the solution set of $2(x + 1) \geq 5(x - 2) + 1$.

Solution. The given inequality is equivalent to each of the following:

$$\begin{aligned} 2x + 2 &\geq 5x - 10 + 1 \\ 2x - 5x &\geq -10 + 1 - 2 \\ -3x &\geq -11 \\ x &\leq \frac{11}{3}. \end{aligned}$$

The solution set is $\{x | x \leq \frac{11}{3}\}$, and its graph is

$\frac{11}{3}$

■

Exercise

Determine and graph the solution set of each of the following.
(a) $x + 1 > 2x + 3$ (b) $2(x - 1) \leq 4(x + 2) - 1$

Answers. (a) $\{x | x < -2\}$ -2

(b) $\left\{x | x \geq -\frac{9}{2}\right\}$ $\frac{-9}{2}$

■

An interval is a set of real numbers with the property that for any two numbers in the set, all numbers between those two are also in the set. Hence the graph of an interval has no "gaps."

In addition to R, the set of all real numbers, there are eight other types of intervals. They will be introduced with interval notation, together with their graphs. The symbols a and b are constants, with $a < b$. The symbols ∞ **(infinity)** and $-\infty$ **(minus infinity)** are *not* numbers. At present, they have meaning only when used with other symbols in denoting intervals.

Type	*Definition*	*Notation*	*Graph*
open interval from a to b	$\{x \mid a < x < b\}$	(a, b)	
closed interval from a to b	$\{x \mid a \le x \le b\}$	$[a, b]$	
half-open (or half-closed) interval from a to b	$\{x \mid a < x \le b\}$	$(a, b]$	
	$\{x \mid a \le x < b\}$	$[a, b)$	
open ray from $-\infty$ to a	$\{x \mid x < a\}$	$(-\infty, a)$	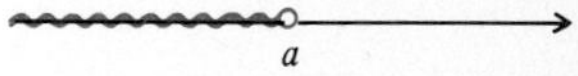
closed ray from $-\infty$ to a	$\{x \mid x \le a\}$	$(-\infty, a]$	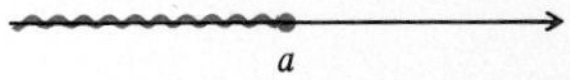
open ray from a to ∞	$\{x \mid a < x\}$	(a, ∞)	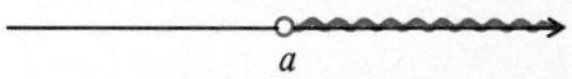
closed ray from a to ∞	$\{x \mid a \le x\}$	$[a, \infty)$	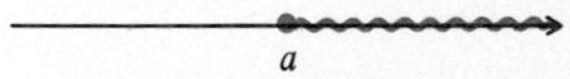

From now on we shall use interval notation instead of the more cumbersome set notation. For instance, in Example 1, instead of writing the solution set as $\{x \mid x > -3\}$, we will write $(-3, \infty)$.

Problem Set 1.7

In Problems 1–8, determine and graph the solution set. (Use interval notation.)

1. $3x < 5$

2. $2x > -5$

3. $5x \ge -10$

4. $2x \le 7$

5. $-3x > 6$

6. $-5x \le -15$

7. $2x + 1 < 3x + 2$

8. $5x + 3 \ge 7x$

In Problems 9–20, find the solution set.

9. $2(3x + 5) \ge 5(3x + 2)$

10. $5(2x - 1) \le -3(2x + 3)$

11. $2(3x - 1) + 5 < -2(x + 1)$

12. $5 - 3(2x + 1) > 3(x - 2)$

13. $x(x + 1) < x^2 + 3x + 4$

14. $2x^2 + 2x - 1 \ge x(2x - 3)$

15. $\dfrac{3x - 2}{-3} \le 0$

16. $\dfrac{6x - 5}{-2} > 0$

17. $\dfrac{2x-1}{2} \le \dfrac{-2}{3}$

18. $\dfrac{5x+4}{3} - 3 > -2$

19. $\dfrac{2x-3}{6} \le \dfrac{2-5x}{8}$

20. $\dfrac{6-5x}{5} > \dfrac{3x-1}{6}$

In Problems 21–26, determine the universe of x.

21. $\sqrt{1-2x} = 3$

22. $\sqrt{5x-3} = 4$

23. $\sqrt{2x+7} - \sqrt{x-5} = 3$

24. $\sqrt{3-2x} = 2 + \sqrt{4-x}$

25. $\sqrt{9-3x} = 1 + \sqrt{2x-3}$

26. $\sqrt{1-2x} = \sqrt{3x-5}$

27. Determine the values of k such that the solution set of $x^2 - 3x + k - 1 = 0$ is $\varnothing$.

28. Determine the values of k such that $3x^2 + 2x + 5 - 2k = 0$ has two distinct roots.

1.8 Inequalities Continued

In this section we learn to determine the solution set of some nonlinear inequalities.

Example 1
Determine and graph the solution set of $3x^2 + 8x > 3$.

Solution. After getting one side to zero, we factor and obtain

$$(3x-1)(x+3) > 0. \tag{1-12}$$

Since the product of two numbers is positive if and only if both are positive or both are negative, the solution set of inequality (1-12) is S ∪ T, where S is the solution set of the compound statement

$$3x - 1 > 0 \quad \text{and} \quad x + 3 > 0, \tag{1-13}$$

and T is the solution set of

$$3x - 1 < 0 \quad \text{and} \quad x + 3 < 0. \tag{1-14}$$

Statement (1-13) is equivalent to

$$x > \frac{1}{3} \quad \text{and} \quad x > -3.$$

Since both of these inequalities must be true for the statement to be true, we see that the solution set is $S = (\frac{1}{3}, \infty)$. (Note that this is simply the intersection of the solution sets of each inequality.) Statement (1-14) is equivalent to

$$x < \frac{1}{3} \quad \text{and} \quad x < -3.$$

The solution set of this statement is T $= (-\infty, -3)$. Therefore, the solution set of the given inequality is $(-\infty, -3) \cup (\frac{1}{3}, \infty)$. Its graph:

-3 $\frac{1}{3}$

■

Example 2

Determine and graph the solution set of $\dfrac{x-2}{2x+1} \leq 0$.

Solution. The fraction $\dfrac{x-2}{2x+1}$ equals zero if and only if $x = 2$. Therefore, 2 is in the solution set. It remains to find the solution set of

$$\frac{x-2}{2x+1} < 0. \tag{1-15}$$

Since the quotient of two numbers is negative if and only if one is positive and the other is negative, the solution set of inequality (1-15) is S ∪ T, where S is the solution set of the compound statement

$$x - 2 > 0 \quad \text{and} \quad 2x + 1 < 0, \tag{1-16}$$

and T is the solution set of

$$x - 2 < 0 \quad \text{and} \quad 2x + 1 > 0. \tag{1-17}$$

Statement (1-16) is equivalent to

$$x > 2 \quad \text{and} \quad x < -\frac{1}{2}.$$

The solution set to this statement is S $= \varnothing$. Statement (1-17) is equivalent to

$$x < 2 \quad \text{and} \quad x > -\frac{1}{2}.$$

The solution set of this statement is T $= (-\frac{1}{2}, 2)$. Therefore, the solution set of the given inequality is

$$\varnothing \cup \left(-\frac{1}{2}, 2\right) \cup \{2\},$$

which may be written $(-\frac{1}{2}, 2]$. Its graph:

$-\frac{1}{2}$ 2

■

Let's take another look at the inequality of the preceding example.

$$\frac{x-2}{2x+1} \leq 0$$

Why can't we simply multiply both sides by $2x + 1$ and get

$$x - 2 \leq 0$$
$$x \leq 2?$$

Thus the solution set would be $(-\infty, 2] - \{-\frac{1}{2}\}$ ($-\frac{1}{2}$ is not in the universe of the variable), which may be written $(-\infty, -\frac{1}{2}) \cup (-\frac{1}{2}, 2]$. But this differs from the solution set we previously obtained. That's because, upon multiplying by $2x + 1$, we neglected to consider the direction of the inequality symbol. We wrote

$$x - 2 \leq 0.$$

This is fine when $2x + 1$ is positive. But $2x + 1$ can be negative, too. And for those values we should reverse the inequality symbol and write

$$x - 2 \geq 0.$$

Thus, without considering the sign of $2x + 1$, we don't know what to do with the inequality symbol. So there is nothing to be gained (and a lot to be lost if we're careless) by eliminating fractions in an inequality when the denominator contains the variable and can be both positive and negative.

Another method of solving inequalities, sometimes preferred by students, is called the **critical number method.** Given an inequality of the form

$$\frac{p_x}{q_x} > 0, \quad \frac{p_x}{q_x} < 0, \quad \frac{p_x}{q_x} \geq 0, \quad \text{or} \quad \frac{p_x}{q_x} \leq 0,$$

where p_x and q_x are algebraic expressions in x, a *critical number* is a real number for which p_x or q_x is zero.

Let's rework Example 2 and use the critical number method. To determine the critical numbers of

$$\frac{x-2}{2x+1} \leq 0, \tag{1-18}$$

we must find the roots of $x - 2 = 0$ and $2x + 1 = 0$. We see that the critical numbers are 2 and $-\frac{1}{2}$. These two numbers determine the following three subsets of R:

$-\frac{1}{2}$ 2

$(-\infty, -\frac{1}{2})$ $(-\frac{1}{2}, 2)$ $(2, \infty)$

When the subsets are determined in this fashion, *the sign of a factor in the numerator or denominator does not change throughout a subset.* This means that a specific factor is either positive for every number in the interval or negative for every number in the interval. To find out which, it suffices to choose *one* number in the interval and substitute. To illustrate, consider the interval $(-\infty, -\frac{1}{2})$ and choose the number -1. When $x = -1$, the factor $x - 2$ is negative, and hence $x - 2$ is negative throughout $(-\infty, -\frac{1}{2})$. When $x = -1$, the factor $2x + 1$ is also negative, and thus $2x + 1$ is negative throughout $(-\infty, -\frac{1}{2})$. Therefore, when x is replaced by any number in the interval $(-\infty, -\frac{1}{2})$, the quotient of $x - 2$ and $2x + 1$ is positive. We summarize this information in the following table:

Critical numbers on the number line: $-\frac{1}{2}$ and 2.

	$(-\infty, -\frac{1}{2})$	$(-\frac{1}{2}, 2)$	$(2, \infty)$
$x - 2$	$-$		
$2x + 1$	$-$		
$\frac{x-2}{2x+1}$	$+$		

Continue in the same fashion for the other two intervals. Remember, any number in the interval will do, so choose a "nice" one. Also, note that you are not looking for the exact value upon substitution. You merely need to know whether the factors, and thus the quotient, are positive or negative. The completed table should be as follows.

Critical numbers on the number line: $-\frac{1}{2}$ and 2.

	$(-\infty, -\frac{1}{2})$	$(-\frac{1}{2}, 2)$	$(2, \infty)$
$x - 2$	$-$	$-$	$+$
$2x + 1$	$-$	$+$	$+$
$\frac{x-2}{2x+1}$	$+$	$-$	$+$

From inequality (1-18) we note that we are searching for the numbers which make the quotient of the two factors negative or zero. The table tells us that $(-\frac{1}{2}, 2)$ is included in the solution set. What about the critical numbers? Certainly, 2 is to be included, but $-\frac{1}{2}$ is *not* in the universe of the variable. (It makes the denominator zero.) Therefore, the solution set is $(-\frac{1}{2}, 2]$.

This method may also be used for an inequality like that of Example 3. (With respect to finding the critical numbers, it is viewed as a quotient p_x/q_x where $q_x = 1$.)

Example 3

Determine and graph the solution set of $5x^2 + 7x < 6$.

Solution. First, we get one side to zero. We have

$$5x^2 + 7x - 6 < 0. \tag{1-19}$$

Next we want to determine the critical numbers. We can do this by finding the roots of $5x^2 + 7x - 6 = 0$. If the left side of equation (1-19) is factored,

$$(5x - 3)(x + 2) < 0, \tag{1-20}$$

we see that the critical numbers are $\frac{3}{5}$ and -2. Our table is as follows.

	$(-\infty, -2)$	$(-2, \frac{3}{5})$	$(\frac{3}{5}, \infty)$
$5x - 3$	−	−	+
$x + 2$	−	+	+
$(5x - 3)(x + 2)$	+	−	+

From inequality (1-20) we note that we are searching for the numbers which make the product of two factors negative (less than zero). The table tells us that $(-2, \frac{3}{5})$ is included in the solution set. Because the inequality symbol does not include equality, the critical numbers are not solutions. Therefore, the solution set is $(-2, \frac{3}{5})$. ■

Although the critical number method may seem more complicated than the one used in Examples 1 and 2, it simplifies matters with an inequality like that in Example 4.

Example 4

Determine and graph the solution set of $\dfrac{x(x + 2)}{x - 3} \leq 0$.

Solution. The critical numbers are 0, −2, and 3. Our table is as follows:

	$(-\infty, -2)$	$(-2, 0)$	$(0, 3)$	$(3, \infty)$
x	−	−	+	+
$x + 2$	−	+	+	+
$x - 3$	−	−	−	+
$\dfrac{x(x+2)}{x-3}$	−	+	−	+

Since we are looking for numbers that make the quotient negative or zero, we see from the table that $(-\infty, 2)$ and $(0, 3)$ are included in the solution set. What about the critical numbers? Certainly, 0 and −2 are to be included, but 3 is *not* in the universe of the variable. (It makes the denominator zero.) Therefore, the solution set is $(-\infty, -2] \cup [0, 3)$. The graph:

■

Exercise

Determine and graph the solution set of each of the following.

(a) $2x^2 + 5x < 3$ (b) $\dfrac{x(x - 1)}{x + 2} \leq 0$

Answers. (a) $\left(-3, \frac{1}{2}\right)$

(b) $(-\infty, -2) \cup [0, 1]$ ■

Problem Set 1.8

In Problems 1–8, determine and graph the solution set.

1. $(x - 1)(x + 3) < 0$
2. $(x + 2)(x - 3) > 0$
3. $(2x - 1)(x + 2) \geq 0$
4. $(3x - 2)(x + 5) \leq 0$
5. $x^2 + 5x + 6 \leq 0$
6. $x^2 - 3x - 10 \geq 0$
7. $2x^2 - x - 3 > 0$
8. $6x^2 - 7x - 3 < 0$

In Problems 9–40, find the solution set.

9. $10x^2 - 2x \leq 1 - 5x$
10. $6x^2 - 35x > 6$
11. $4 - 7x \geq 2x^2$
12. $x + 15 < 6x^2$
13. $x(x - 2) < 0$
14. $x^2 \geq 5x$
15. $x^2 - 2 < 0$
16. $(x - 3)^2 > 5$
17. $x^2 + 2x \geq 5$
18. $x^2 + 5x \leq 2$
19. $(x + 1)^2 \leq 0$
20. $\dfrac{1}{(2x + 1)^2} > 0$
21. $2x^2 + 5x + 4 \geq 0$
22. $3x^2 + 3x + 1 < 0$
23. $2x^3 > 2x + 9 - 9x^2$
24. $x^3 - 4x \leq 5x^2 - 20$
25. $\dfrac{x + 1}{x - 2} < 0$
26. $\dfrac{2x - 1}{x + 5} > 0$
27. $\dfrac{x + 5}{x^2 + 5x + 6} \geq 0$
28. $\dfrac{x - 1}{2x^2 - 5x - 3} \leq 0$
29. $\dfrac{1}{x} < 1$
30. $2 - \dfrac{1}{x} > x$
31. $\dfrac{x}{x - 2} \geq 3$
32. $\dfrac{x - 2}{x} \leq 5$
33. $\dfrac{1}{x - 1} > \dfrac{2}{x + 2}$
34. $\dfrac{2}{x + 1} < \dfrac{3}{x + 5}$
35. $\dfrac{1}{x - 1} \leq \dfrac{x}{x + 3}$
36. $\dfrac{x}{x + 5} < \dfrac{1}{x - 3}$
37. $\dfrac{x^3 - 3x^2}{x^2 + 7x + 10} > 0$
38. $\dfrac{x^2 + 3x - 10}{x^3 + 2x^2} \leq 0$
39. $\dfrac{x + 3}{(x - 1)^2} > 0$
40. $\dfrac{2x^2 - x + 5}{(x + 1)^2} \geq 0$

41. Determine the values of k such that $2x^2 - 3kx + 6k - 6 = 0$ has two distinct roots.
42. Determine the values of k such that the solution set of $x^2 + (2k - 1)x + 1 = 0$ is $\varnothing$.

1.9 Other Types of Inequalities

In this section we study compound inequalities and inequalities with an absolute value symbol. Examples 1 and 2 deal with the first topic.

Example 1
Determine and graph the solution set of $-2 \leq 3x + 1 < 5$.

Solution. This compound inequality is simply shorthand notation for the following compound statement:

$$-2 \leq 3x + 1 \qquad \text{and} \qquad 3x + 1 < 5.$$

As with all "and" statements, both parts must be true for the entire statement to be true. Thus the solution set is the intersection of the solution set of the first inequality with the solution set of the second. The first inequality simplifies to

$$-1 \leq x,$$

and the second reduces to

$$x < \frac{4}{3}.$$

Hence the solution set of the compound inequality is $[-1, \infty) \cap (-\infty, \frac{4}{3})$, which is $[-1, \frac{4}{3})$. Its graph:

-1 $\frac{4}{3}$

■

Example 2
Determine and graph the solution set of $0 < x^2 - 2x \leq 8$.

Solution. This compound inequality is equivalent to the statement

$$0 < x^2 - 2x \qquad \text{and} \qquad x^2 - 2x \leq 8.$$

We first find the solution set of

$$0 < x^2 - 2x.$$

The critical number method tells us that the solution set is $(-\infty, 0) \cup (2, \infty)$. Next we find the solution set of

$$x^2 - 2x \le 8.$$

Here we find that the solution set is $[-2, 4]$. So the solution set of the given inequality is $((-\infty, 0) \cup (2, \infty)) \cap [-2, 4]$, which is $[-2, 0) \cup (2, 4]$. Its graph:

-2 0 2 4 ■

There is an abbreviated version to finding the solution set of the inequality in Example 1:

$$-2 \le 3x + 1 < 5.$$

Whenever the expression in the middle is linear and there is a constant on each side, we may simplify the expression between the inequality symbols to x by operating on all three parts of the compound inequality simultaneously. To this end we first subtract 1 from the left, right, and middle to get

$$-3 \le 3x < 4.$$

Now we divide everything by 3 to obtain

$$-1 \le x < \frac{4}{3},$$

and the solution set to this is $[-1, \frac{4}{3})$.

But this method cannot be applied to the inequality of Example 2 because the middle expression isn't linear.

Exercise

Determine and graph the solution set of each of the following.

(a) $4 < 2x + 5 \le 11$ (b) $-6 \le x^2 + 5x < 0$.

Answers. (a) $\left(-\frac{1}{2}, 3\right]$ $-\frac{1}{2}$ 3

(b) $(-5, -3] \cup [-2, 0)$ -5 -3 -2 0 ■

To solve inequalities in which the variable appears within an absolute value symbol, we will use the following important theorem.

Theorem 1.9

Let a be a nonnegative constant, and let p_x be an algebraic expression in x. Then

$$|p_x| \le a$$

is equivalent to

$$-a \le p_x \le a,$$

and

$$|p_x| \ge a$$

is equivalent to the compound statement

$$p_x \ge a \qquad or \qquad p_x \le -a.$$

An analogous theorem is obtained when $\le$ and $\ge$ are replaced by $<$ and $>$, respectively. Applications of this theorem appear in Examples 3 and 4.

Example 3
Determine and graph the solution set of $|2x - 3| \le 6$.

Solution. By Theorem 1.9 this inequality is equivalent to

$$-6 \le 2x - 3 \le 6.$$

Solving this compound inequality for x, we get

$$\frac{-3}{2} \le x \le \frac{9}{2}.$$

Hence the solution set is $[-\frac{3}{2}, \frac{9}{2}]$. The graph:

Example 4
Determine and graph the solution set of $|3x + 2| > 4$.

Solution. This inequality is equivalent to the compound statement

$$3x + 2 > 4 \qquad \text{or} \qquad 3x + 2 < -4.$$

Solving each of the inequalities for x, we get

$$x > \frac{2}{3} \qquad \text{or} \qquad x < -2.$$

Thus the solution set is $(\frac{2}{3}, \infty) \cup (-\infty, -2)$. Its graph:

-2 $\qquad$ $\frac{2}{3}$

■

Exercise
Determine the solution set and graph.
(a) $|2x + 1| < 3$ (b) $|4x - 3| \geq 2$

Answers. (a) $(-2, 1)$

(b) $\left(-\infty, \frac{1}{4}\right] \cup \left[\frac{5}{4}, \infty\right)$ ■

Example 5
Find the solution set of $|x - 6| < 3x - 2$.

Solution. Because the right side of this inequality is not a constant, Theorem 1.9 does not apply. We must therefore appeal to the definition of absolute value. When $x - 6 \geq 0$, $|x - 6| = x - 6$. Thus, when $x \geq 6$, the given inequality is equivalent to the following:

$$\begin{aligned} x - 6 &< 3x - 2 \\ -2x &< 4 \\ x &> -2 \end{aligned} \tag{1-21}$$

Now the set of roots to inequality (1-21) is $(-2, \infty)$. But (1-21) is equivalent to the given inequality when $x \geq 6$. Taking this restriction into account, we get a solution set of $[6, \infty)$. And we still have one case to consider. When $x - 6 < 0$, $|x - 6| = -x + 6$. Thus, when $x < 6$, the given inequality is equivalent to the following:

$$\begin{aligned} -x + 6 &< 3x - 2 \\ -4x &< -8 \\ x &> 2. \end{aligned} \tag{1-22}$$

Although the set of roots to inequality (1-22) is $(2, \infty)$, this inequality is equivalent to the original when $x < 6$. Considering this condition, we have a solution set of $(2, 6)$. Hence the solution set of the given inequality is

$$(2, 6) \cup [6, \infty) = (2, \infty).$$ ■

Exercise
Find the solution set of $|x - 3| < 4x - 7$.

Answer. $(2, \infty)$ ■

Problem Set 1.9

In Problems 1–8, determine and graph the solution set.

1. $-2 < 2x - 1 \leq 5$ **2.** $5 \leq 3x + 2 < 10$ **3.** $3 < 4 - 5x < 11$

4. $-2 < 2 - 3x < 5$

5. $0 \le \dfrac{3x - 2}{-5} < 1$

6. $-6 \le \dfrac{2 - 3x}{-2} \le -3$

7. $0 \le x^2 - x < 20$

8. $-2 < 2x^2 + 5x \le 12$

In Problems 9–34, find the solution set.

9. $2 < x^2 - x < 6$
10. $-1 \le 3x^2 + 4x < 15$
11. $2x^2 - 10 \le 2x^2 + 5x < 5 - 4x$
12. $5x - 1 < 3x \le x^2 - 4$
13. $2x - 1 < 3 \le 2x + 1$
14. $-3x + 1 \le 2x < 5x - 2$
15. $|2x| < 5$
16. $|4x| \le 6$
17. $|3x| \ge 7$
18. $1 - 5x| > 10|$
19. $|2x + 1| \ge 5$
20. $|3x - 1| > 8$
21. $|3x + 5| < 6$
22. $12x - 5| \le 1|$
23. $|3 - 2x| \ge 2$
24. $|5 - 2x| > 7$
25. $\left|\dfrac{3 + 2x}{5}\right| < 2$
26. $\left|\dfrac{3 - 2x}{3}\right| \ge 1$
27. $|5x - 4| \le -2$
28. $|4x + 1| > -1$
29. $|2x^2 + 5x| < 3$
30. $\left|\dfrac{2 - 3x}{x}\right| \ge 1$
31. $|x - 5| < 2x + 4$
32. $|2x + 5| \ge x - 3$
33. $x|2x| - x < 6$
34. $x|3x| - x + 4 \le 0$

Chapter 1 Review

In Problems 1–16, find the solution set.

1. $3(x - 1) = 2(x + 3) - 1$
2. $x^2 - 1 = (x + 1)(x - 2)$
3. $\dfrac{x}{x + 3} - \dfrac{5}{3} = 1$
4. $x(2x + 3) = -1$
5. $(2x + 3)^2 = 5$
6. $12x^2 - 10x = 12$
7. $\dfrac{x}{x^2 + 1} = \dfrac{1}{3x + 1}$
8. $3x^2 + 2x = 2$
9. $-2x^2 + \sqrt{3}x + 3 = 0$
10. $x^2 + 1 = \dfrac{-3x}{2}$
11. $2x + \sqrt{x^2 + 7x + 5} = 1$
12. $2 + \sqrt{2x - 3} = \sqrt{x + 7}$
13. $3x^3 + 6x = 10 + 5x^2$
14. $2x^4 - x^2 - 3 = 0$
15. $\left|\dfrac{x^2}{2x - 1}\right| = 2$
16. $\left|\dfrac{2x - 3}{5}\right| = 4$

In Problems 17–26, determine and graph the solution set.

17. $2x - 1 \le -5$

18. $|3x - 2| < 5$

19. $x^2 - 3x - 10 > 0$

20. $\dfrac{x}{x - 1} \ge 2$

21. $\dfrac{1}{|x - 2|} > -1$

22. $\left|\dfrac{3 - x}{x}\right| < 2$

23. $-3 < 1 - 2x \leq 4$

24. $0 \leq x^2 + 3x < 4$

25. $|x^2 + 5x| < 4$

26. $|x + 6| \leq 2x - 3$

In Problems 27–28, determine the universe of x.

27. $\sqrt{-x} = \dfrac{7}{(x + 1)(x + 7)}$

28. $\sqrt{x^2 + 5x + 6} = 9$

In Problems 29–32, solve for the indicated variable, noting any restrictions on other variables.

29. $\dfrac{x}{x + 3} = 1 - \dfrac{3}{2x - a}$, for x

30. $2(x + y)(1 + y') = 2(x - y + 1)(1 - y')$, for y'

31. $3x^2 - 8xy - 3y^2 = 0$, for x

32. $2y^2 + xy - 3x = 0$, for y

33. During his first two games, a St. Louis pitcher pitched 18 innings and had a 3.00 earned run average. In his third game he pitched only 6 innings. If his ERA after the third game was 3.75, how many runs did he allow in the third game? (ERA, earned run average, is the averaged number of earned runs allowed per 9 innings.)

34. On a trip to Florida, John noted that his car averaged 26 miles per gallon without the air conditioning and 23 miles per gallon with the air conditioning. If he used 30 gallons of gas to travel 720 miles, on how many of these miles did he use the air conditioning?

35. Sue drove 30 miles to work each day and picked up a friend along the way. If she averaged 40 mph on each part of the trip and drove 21 minutes longer with the friend than without him, how far did she live from the friend's house?

36. How many gallons of a 40% sulfuric acid solution should be mixed with 9 gallons of a 20% sulfuric acid solution to obtain a 25% solution?

37. A secretary can do a certain typing job in 6 hours while another secretary takes 9 hours to do the same job. If both secretaries work together, in how many hours can the job be done?

Relations

2

The subject of this chapter is analytic geometry, a way of representing equations with geometry and describing geometric figures with equations.

2.1 The Cartesian Plane

The symbolism (a, b) is used to designate the ordered pair of numbers a and b, where a is the first coordinate and b is the second coordinate. The order in which the numbers are listed is important. The ordered pair $(1, 2)$ is different from the ordered pair $(2, 1)$. Thus two ordered pairs are said to be equal provided that the first coordinates and the second coordinates are the same.

Of course, the notation (a, b), where $a < b$, has also been used to denote the open interval from a to b. But it will be apparent from the context of the problem whether the symbolism refers to the set of numbers between a and b or to the ordered pair.

Let A and B be nonempty sets. The **Cartesian product** of A and B, denoted $A \times B$, is the set

$$A \times B = \{(a, b) | \, a \in A, \, b \in B\}.$$

Thus $A \times B$, read "A cross B," is the set of all possible ordered pairs in which the first coordinate is from A and the second coordinate is from B. If $A = \{1, 2, 3\}$ and $B = \{2, 4\}$, then

$$A \times B = \{(1, 2), (1, 4), (2, 2), (2, 4), (3, 2), (3, 4)\}$$

and

$$B \times A = \{(2, 1), (2, 2), (2, 3), (4, 1), (4, 2), (4, 3)\}.$$

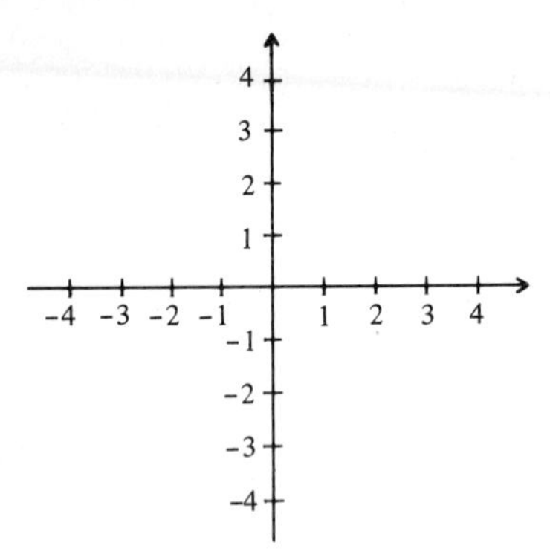

Figure 2.1

Let us now consider two number lines intersecting at their origins. If the lines are perpendicular, if the positive directions are up and right, and if the lines have equal unit lengths, then we would have the system shown in Figure 2.1. Such a system is called the **Cartesian coordinate system.** Although it is not necessary, it is conventional to have the number lines vertical and horizontal with the positive directions up and right. And it is convenient to have the unit lengths of each line the same, but this is not always practical.

In this system each number line is called an **axis.** The horizontal line is called the x-axis, and the vertical line is called the y-axis. The point of intersection is called the **origin.** The two axes divide the plane into four distinct regions called **quadrants,** numbered I through IV, as indicated in Figure 2.2. The points on the axes are not part of any quadrant.

With this system each point in the plane is associated with an ordered pair, and each ordered pair corresponds to a point in the plane. Given a point P, the ordered pair associated with it can be found as follows. Construct the line passing through P perpendicular to the x-axis. The x-axis coordinate of the point of intersection of the x-axis with the constructed line is the first coordinate of the ordered pair which corresponds to P. Construct the line passing through P perpendicular to the y-axis. The y-axis coordinate of the point of intersection of the y-axis with the constructed line is the second coordinate of the ordered pair corresponding to P. In Figure 2.3 we see that the point P is associated with (a, b).

Every point on the x-axis has 0 as a second coordinate, and the first coordinate of every point on the y-axis is 0. The ordered pair associated with the origin is $(0, 0)$.

If a and b are positive real numbers, then the point corresponding to (a, b) is the one that lies a units to the right of the y-axis and b units above the x-axis (in quadrant I). The point associated with $(-a, -b)$ is a units to the left of the y-axis and b units below the x-axis (in quadrant III). Similarly, $(a, -b)$ and $(-a, b)$ identify points in quadrants IV and II, respectively. Some points and associated ordered pairs are indicated in Figure 2.4.

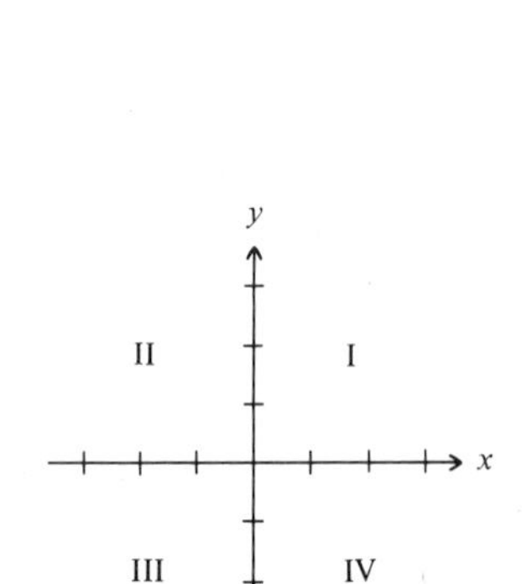

Figure 2.2

Whenever the axis coordinates do not appear, as in Figure 2.4, we shall

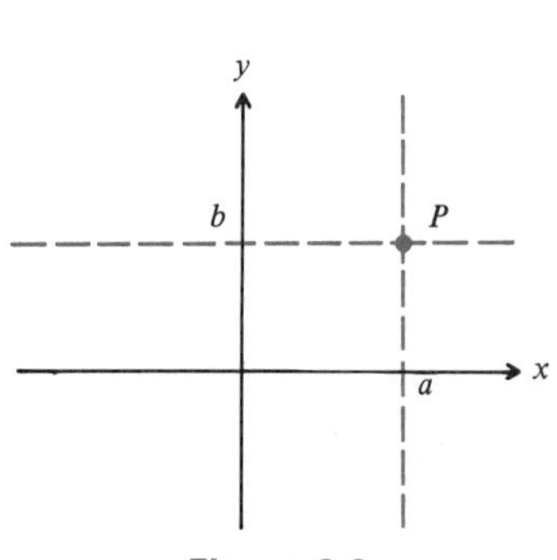

Figure 2.3

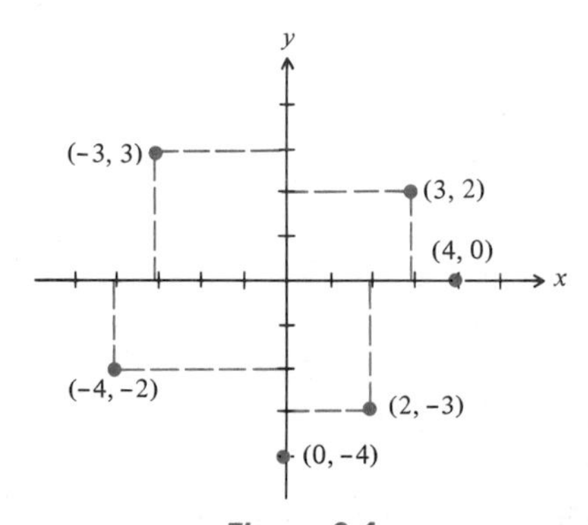

Figure 2.4

assume that the length represented by consecutive hash marks is one unit. If the axes are not labeled, we shall assume that the horizontal axis is the x-axis and that the vertical axis is the y-axis.

Because of the one-to-one relationship which exists between points in the plane and ordered pairs, we shall frequently interchange the terms. And when we refer to the point $P(a, b)$, we mean that the point is labeled P, and its corresponding ordered pair is (a, b). Furthermore, we shall often refer to the first coordinate as the **x-coordinate** and to the second coordinate as the **y-coordinate.**

With the Cartesian coordinate system the distance between any two points in the plane is calculated with the **distance formula,** equation (2-1), which is proven in the following theorem.

Theorem 2.1

The distance between the points $P_1(x_1, y_1)$ and $P_2(x_2, y_2)$, denoted P_1P_2, is

$$P_1P_2 = \sqrt{(x_2 - x_1)^2 + (y_2 - y_1)^2}. \tag{2-1}$$

Proof. If P_1 and P_2 determine a nonhorizontal and nonvertical line, then we may construct a triangle, as in Figure 2.5.

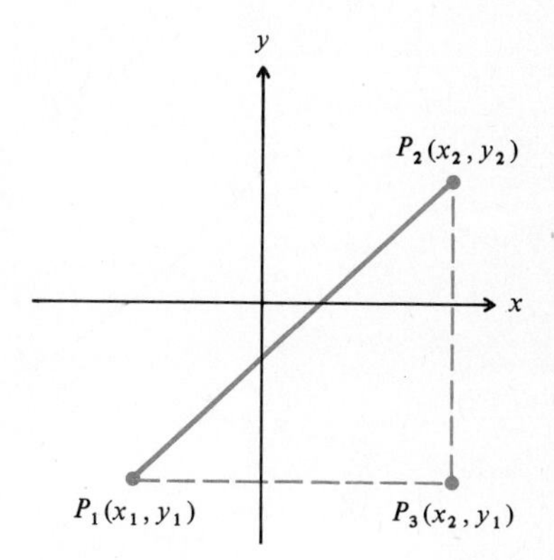

Figure 2.5

The point P_3 is the intersection of two perpendicular lines: the line passing through P_1 that is parallel to the x-axis and the line passing through P_2 that is parallel to the y-axis. Thus the y-coordinate of P_3 must be the same as the y-coordinate of P_1, and the x-coordinate of P_3 must be the same as the x-coordinate of P_2.

Because P_1 and P_3 lie along the same horizontal line,

$$P_1P_3 = |x_2 - x_1|,$$

and since P_2 and P_3 lie along the same vertical line,

$$P_2P_3 = |y_2 - y_1|.$$

By the Pythagorean Theorem, we have

$$(P_1P_2)^2 = (P_1P_3)^2 + (P_2P_3)^2.$$

Substituting, we get

$$(P_1P_2)^2 = |x_2 - x_1|^2 + |y_2 - y_1|^2 = (x_2 - x_1)^2 + (y_2 - y_1)^2.$$

Therefore,

$$P_1P_2 = \sqrt{(x_2 - x_1)^2 + (y_2 - y_1)^2}$$

since distance is always nonnegative.

If P_1 and P_2 determine a horizontal line (one perpendicular to the y-axis), then $y_1 = y_2$. And the distance between P_1 and P_2 is $x_2 - x_1$ if P_2 is to the right of P_1, and is $x_1 - x_2$ if P_1 is on the right. Either way we have

$$P_1P_2 = |x_2 - x_1| = \sqrt{(x_2 - x_1)^2} = \sqrt{(x_2 - x_1)^2 + (y_2 - y_1)^2}.$$

A similar analysis may be made if P_1 and P_2 determine a vertical line. ■

Because $(x_2 - x_1)^2 = (x_1 - x_2)^2$ and $(y_2 - y_1)^2 = (y_1 - y_2)^2$, the order in which the coordinates are subtracted is of no consequence. The distance formula is used in Examples 1 and 2.

Example 1
Determine the distance between $P_1(2, -3)$ and $P_2(-1, 3)$.

Solution. Using the distance formula, we have

$$P_1P_2 = \sqrt{(2 - (-1))^2 + (-3 - 3)^2} = \sqrt{3^2 + (-6)^2} = \sqrt{45} = 3\sqrt{5}. \quad ■$$

Example 2
Find the point on the x-axis which is equidistant from $P_1(2, 3)$ and $P_2(-1, -2)$.

Solution. Let $P(x, 0)$ be such a point. Since $PP_1 = PP_2$, we have

$$\sqrt{(x - 2)^2 + (0 - 3)^2} = \sqrt{(x + 1)^2 + (0 + 2)^2}. \tag{2-2}$$

Squaring both sides gives

$$(x - 2)^2 + 9 = (x + 1)^2 + 4.$$

Simplifying, we get

$$\begin{aligned} x^2 - 4x + 13 &= x^2 + 2x + 5 \\ -6x &= -8 \\ x &= \frac{4}{3}. \end{aligned}$$

Because we squared both sides of equation (2-2) to get this result, $\frac{4}{3}$ must be checked. After doing so, we conclude that the desired point is $(\frac{4}{3}, 0)$. ■

Exercise
(a) Determine the distance between $P_1(-3, 4)$ and $P_2(1, -2)$.
(b) Find the point on the x-axis which is equidistant from $P_1(-1, 2)$ and $P_2(3, -4)$.

Answers. (a) $2\sqrt{13}$ (b) $\left(\frac{5}{2}, 0\right)$ ■

In the next theorem, we show how to find the coordinates of the midpoint of a line segment.

Theorem 2.2

The midpoint of the line segment joining the points $P_1(x_1, y_1)$ and $P_2(x_2, y_2)$ is

$$\left(\frac{x_1 + x_2}{2}, \frac{y_1 + y_2}{2}\right).$$

Proof. Let $M(a, b)$ denote the midpoint of $\overline{P_1P_2}$. If $\overline{P_1P_2}$ is neither horizontal nor vertical, then we may construct two triangles, as seen in Figure 2.6. (Without loss of generality we may assume that the P_1 is above and to the left of P_2.)

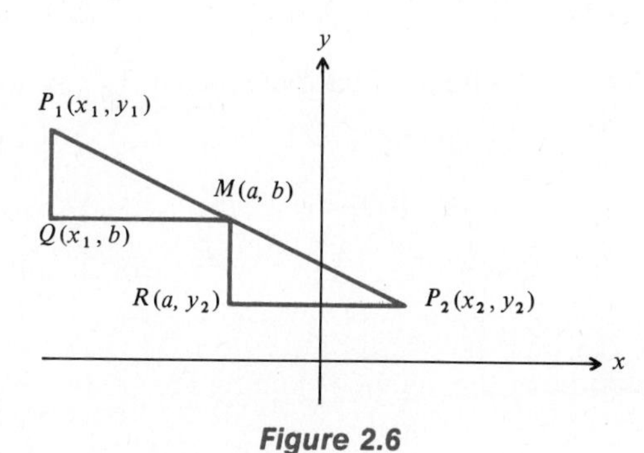

Figure 2.6

The point $Q(x_1, b)$ is the intersection of two perpendicular lines: the vertical line passing through $P_1(x_1, y_1)$ and the horizontal line passing through $M(a, b)$. Similarly, $R(a, y_2)$ is the intersection of two perpendicular lines.

Now triangles P_1MQ and MP_2R are congruent (identical) by angle–side–angle. Therefore, $QM = RP_2$. That is,

$$|x_1 - a| = |a - x_2|.$$

Since $x_1 < a < x_2$, we have

$$a - x_1 = x_2 - a.$$

Solving this equation for a, we get

$$a = \frac{x_1 + x_2}{2}.$$

Similarly, $P_1Q = MR$ and we have

$$|y_1 - b| = |b - y_2|.$$

Since $y_1 > b > y_2$, we have

$$y_1 - b = b - y_2.$$

Solving for b, we get

$$b = \frac{y_1 + y_2}{2}.$$

If $\overline{P_1P_2}$ is vertical, then x_1 equals x_2 and we still have $a = \frac{x_1 + x_2}{2}$. Because $P_1M = MP_2$, $|y_1 - b| = |b - y_2|$. Thus $b = \frac{y_1 + y_2}{2}$, regardless of relationship between y_1 and y_2. A similar situation exists if $\overline{P_1P_2}$ is horizontal. Therefore, the midpoint is

$$\left(\frac{x_1 + x_2}{2}, \frac{y_1 + y_2}{2}\right).$$

■

Example 3
Calculate the midpoint of the segment joining (2, 4) and (−3, 8).

Solution. Using the midpoint formula, we see that the x-coordinate of the midpoint is

$$\frac{2 + (-3)}{2} = \frac{-1}{2},$$

and the y-coordinate is

$$\frac{4 + 8}{2} = 6.$$

Thus the midpoint is $(-\frac{1}{2}, 6)$. ■

Exercise
Calculate the midpoint of the segment joining (3, −5) and (7, 2).

Answer. $\left(5, -\frac{3}{2}\right)$ ■

Problem Set 2.1

In Problems 1–6, compute $A \times B$.

1. $A = \{1, 2\}, B = \{3, 4\}$
2. $A = \{5, 6\}, B = \{7, 8\}$
3. $A = \{3, 4\}, B = \{3, 4\}$
4. $A = \{\pi\}, B = \{\pi, 0, 1\}$
5. $A = \{a, b\}, B = \{b, c\}$
6. $A = \{a, 1\}, B = \{1, b\}$

In Problem 7–9, describe $A \times B$.

7. $A = \{0\}, B = \mathsf{R}$
8. $A = \mathsf{Z}^+, B = \{0\}$
9. $A = \mathsf{Z}, B = \mathsf{R}$

10. Let $P(x, y)$ be a point in the plane. In which quadrant(s) is P if we have the following?
    (a) $x > 0$
    (b) $y < 0$
    (c) $x > 0$ and $y < 0$
    (d) $x > 0$ or $y < 0$
    (e) $xy > 0$
    (f) $\dfrac{x}{y} < 0$

11. List the ordered pair which corresponds to each of the points in the following plane.

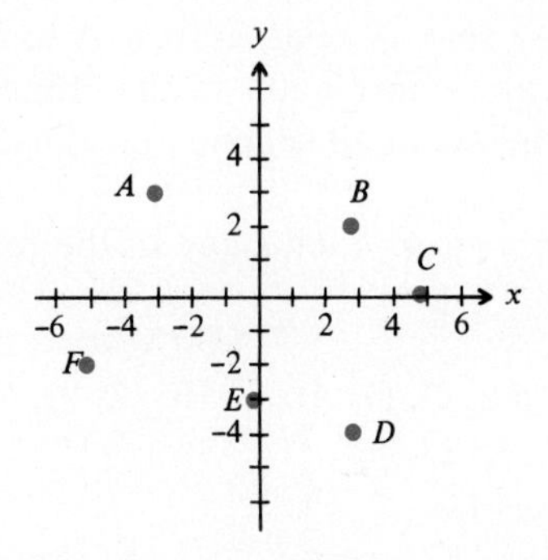

12. Using a Cartesian coordinate system, plot the following points:

$$(-2, 4), (0, 2), \left(1, \frac{5}{2}\right) (2, -4), (-3, 0), \left(\frac{-3}{2}, -1\right).$$

In Problems 13–26, determine the distance between the given points.

13. $(1, 2), (4, 6)$
14. $(2, 5), (7, 17)$
15. $(-2, -3), (10, -8)$
16. $(-2, 3), (1, -1)$
17. $(-7, -5), (-6, -2)$
18. $(-3, 6), (2, 2)$
19. $(1, -1), (3, 3)$
20. $(-5, 2), (1, 6)$
21. $(3\sqrt{5}, -1), (\sqrt{5}, 3)$
22. $(0, -2), (\sqrt{5}, 3)$
23. $(2\sqrt{6}, \sqrt{5}), (-\sqrt{6}, 3\sqrt{5})$
24. $(0, -\sqrt{3}), (-\sqrt{13}, \sqrt{3})$
25. $\left(\frac{3}{5}, -1\right), \left(\frac{7}{5}, \frac{1}{2}\right)$
26. $\left(\frac{-1}{4}, \frac{1}{6}\right), \left(\frac{3}{2}, \frac{2}{3}\right)$

In Problems 27–34, calculate the midpoint of the segment connecting the given points.

27. $(3, 6), (1, 4)$ **28.** $(-3, 5), (-1, 7)$

29. $(-7, 6), (6, 11)$ **30.** $(-1, -2), (-6, -1)$

31. $(\sqrt{3}, 2\sqrt{5}), (3\sqrt{3}, \sqrt{5})$ **32.** $(-\sqrt{7}, \sqrt{6}), (2\sqrt{7}, -5\sqrt{6})$

33. $\left(\frac{3}{5}, -1\right), \left(\frac{-1}{4}, \frac{1}{6}\right)$ **34.** $\left(\frac{7}{5}, \frac{1}{2}\right), \left(\frac{3}{2}, \frac{-2}{3}\right)$

35. Find the point on the x-axis which is equidistant from $(-3, -5)$ and $(2, 7)$.

36. Find the point on the y-axis which is equidistant from $(2, 3)$ and $(-3, 7)$.

37. The midpoint of (a, b) and $(-5, 3)$ is $(3, 2)$. Find a and b.

38. Find the distance between $(-2, 5)$ and the midpoint of the segment connecting $(-3, 1)$ and $(-4, 2)$.

2.2 Graphs of Relations

Let A and B be nonempty sets. A **relation** from A to B is a nonempty subset of $A \times B$. The set of all first coordinates of the ordered pairs of a relation is called the **domain,** and the set of all second coordinates of the ordered pairs is called the **range.**

If $A = \{1, 2, 3\}$ and $B = \{2, 4\}$, then *some* of the relations from A to B are

$$\begin{aligned} r_1 &= A \times B = \{(1, 2), (1, 4), (2, 2), (2, 4), (3, 2), (3, 4)\}, \\ r_2 &= \{(1, 4), (2, 4), (3, 4)\}, \\ r_3 &= \{(1, 2), (1, 4)\}, \\ r_4 &= \{(2, 2)\}, \\ r_5 &= \{(1, 2), (2, 4), (3, 2)\}. \end{aligned}$$

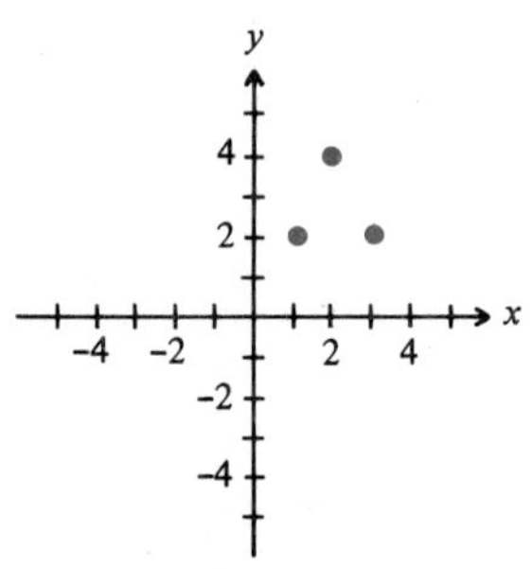

Figure 2.7 Graph of $\{(1, 2), (2, 4), (3, 2)\}$

The domain of r_1, r_2, and r_5 is A. The domain of r_3 is $\{1\}$, and the domain of r_4 is $\{2\}$. The range of r_1, r_3, and r_5 is B. The range of r_2 is $\{4\}$, and the range of r_4 is $\{2\}$.

In order to study relations, it is valuable to have a pictorial representation. Thus the **graph of a relation** from R to R is the set of all points in the Cartesian plane which correspond to the ordered pairs of the relation. The graph of r_5 is shown in Figure 2.7.

Exercise

Let $r = \{(-1, 3), (-2, 4), (-3, 4)\}$. Determine the domain and range and graph.

Answer. The domain is $\{-1, -2, -3\}$, and the range is $\{3, 4\}$. The graph:

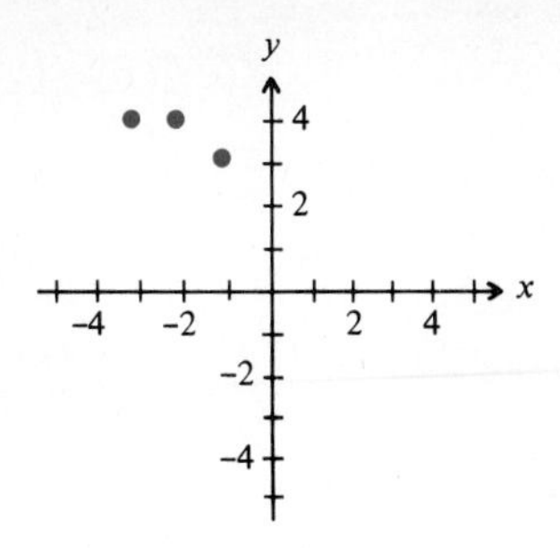

■

Let p_{xy} and q_{xy} be algebraic expressions in the variables x and y. Then a statement of the form

$$p_{xy} = q_{xy}$$

is called an **equation in x and y.** Examples include

$$2x - y = x + y - 3, \qquad y = x^2, \qquad \text{and} \qquad x^2 + y^2 = 9.$$

A real **solution** of an equation in x and y is an ordered pair (a, b) which makes the equation true when x is replaced by a and y is replaced by b. Hence $(2, 1)$ is a solution of $2x + y = 5$, but $(3, 1)$ is not. The set of all solutions is called the **solution set.** Equations which have the same solution set are called **equivalent.**

Many equations have infinitely many solutions. Hence the graph of the solution set cannot be constructed by plotting all solutions. We will plot a few selected points and connect them with a smooth curve.

Example 1
Graph the solution set of $y = x + 1$.

Solution. We begin by constructing a table which lists some of the infinitely many solutions of this equation. We will choose some values for x and find the corresponding value for y. For example, when $x = -3$, $y = -3 + 1 = -2$. When $x = -2$, $y = -2 + 1 = -1$, and so on. We get the following table:

x	−3	−2	−1	0	1	2	3
y	−2	−1	0	1	2	3	4

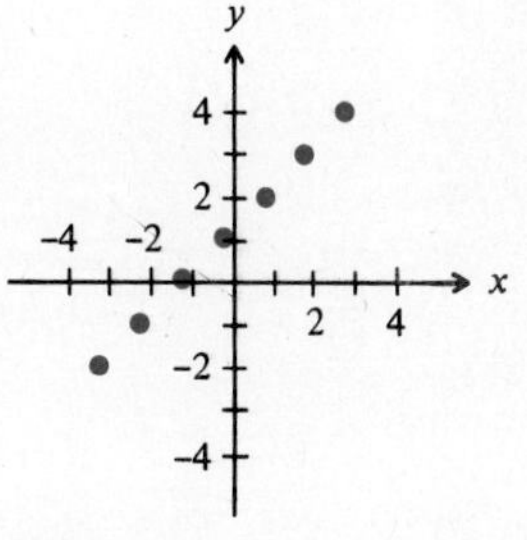

Figure 2.8

This table is simply another way of representing the ordered pairs. Each column refers to an ordered pair, with the entry in the top row being the x-coordinate and the entry in the bottom row being the y-coordinate. The graph of these ordered pairs is shown in Figure 2.8. Since the solution set is infinite, and these solutions appear to lie along a straight line, we suspect

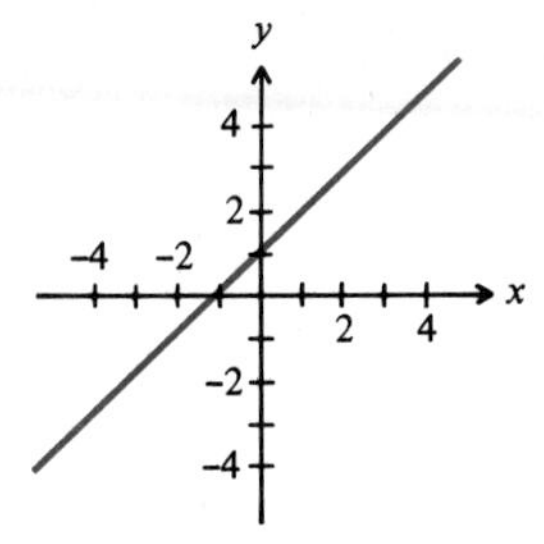

Figure 2.9 $y = x + 1$

that the graph of the solution set of $y = x + 1$ is the line determined by these points. See Figure 2.9. ■

Example 2
Graph the solution set of $x = y^2$.

Solution. We will construct a table which lists some of the solutions by selecting some values for y and finding the corresponding values for x. We get

x	4	1	0	1	4
y	−2	−1	0	1	2

The graph of these ordered pairs is shown in Figure 2.10. Although these points do not lie along a straight line, we can still make a guess about the graph. See Figure 2.11. ■

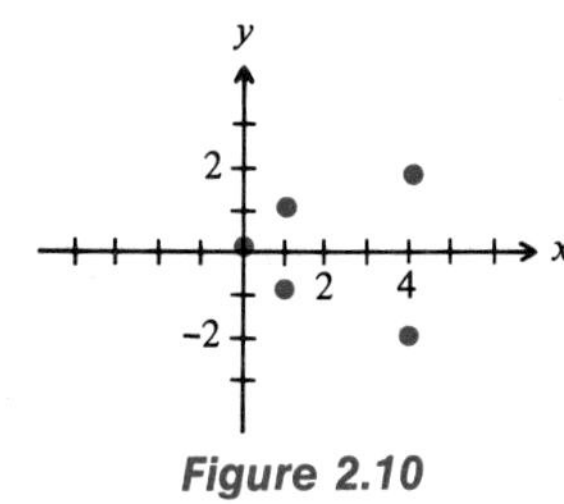

Figure 2.10

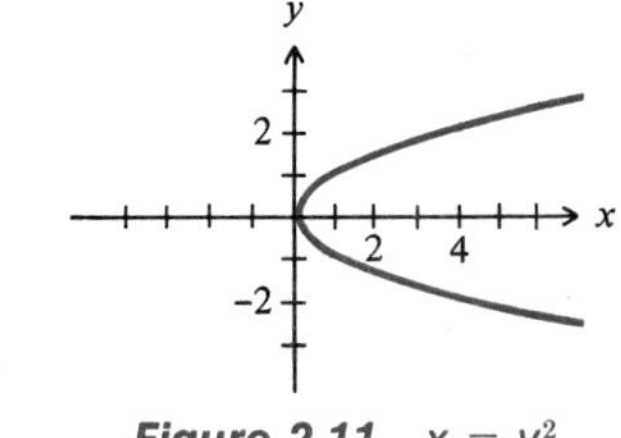

Figure 2.11 $x = y^2$

When graphing equations by plotting points, how can we be certain that we have selected representative points and connected them properly to give us an accurate sketch? The fact is that we cannot unless we have a general idea of what the graph looks like before we begin. Much of the remainder of this text is devoted to obtaining a general idea of the graph by examining the equation. Until such techniques are developed, suggested values for plotting points will be included with the problems.

Note that the solution set of an equation in x and y is a relation from R to R. Because of this we say that an equation in x and y *defines* a relation. And when we speak of the relation defined by an equation, we are referring to the solution set of the equation. Also, when we speak of the graph of an equation in x and y, we are referring to the graph of its solution set.

Exercise
Graph the relation defined by the given equation after plotting points with the given values:

(a) $y = -x + 1$, x-values: -2, -1, 0, 1, and 2
(b) $x = -y^2$, y-values: -2, -1, 0, 1, and 2 [recall that $-y^2 = -(y^2)$]

Answers

(a)

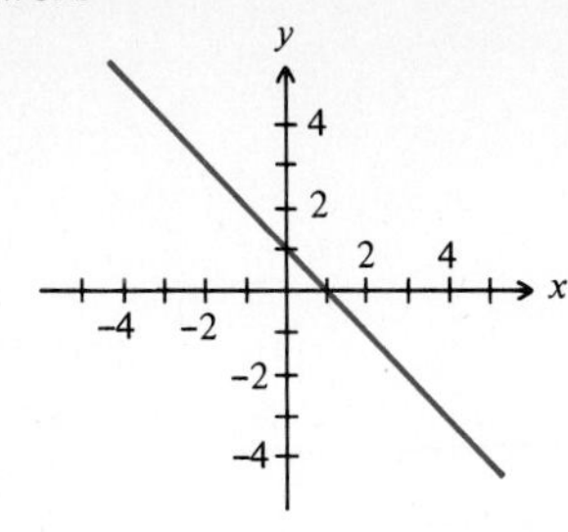

(b)

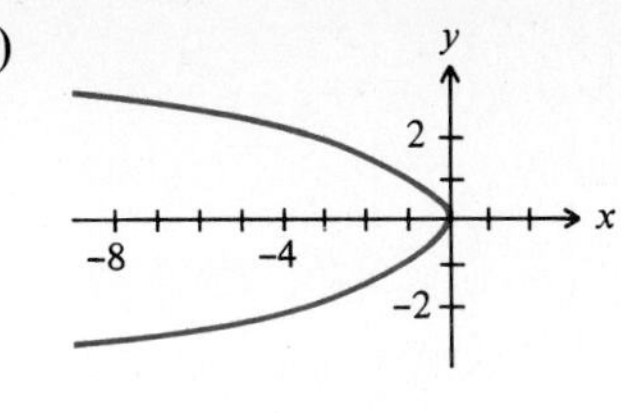

■

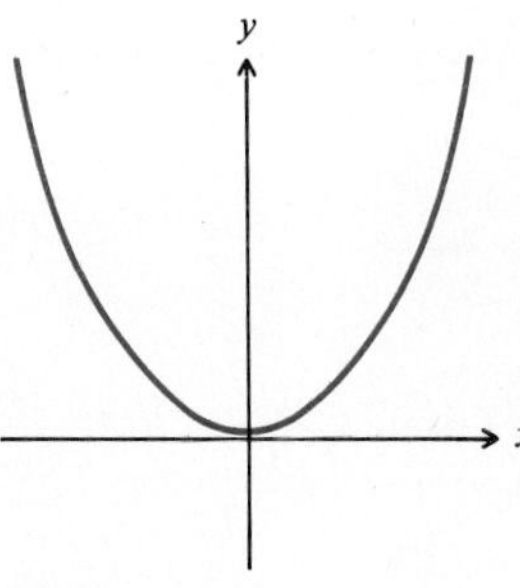

Figure 2.12

Now the domain of a relation from R to R is the set of all x-coordinates of the points on its graph, and the range is the set of all y-coordinates. Thus if we project the graph of a relation onto the x-axis, the domain will consist of that part of the x-axis which has a point projected onto it. And if we project the graph onto the y-axis, the range of the relation will be that part of the y-axis which has a point projected onto it. This technique of finding the domain and range of a relation is illustrated in Example 3.

Example 3
Determine the domain and range of the relation graphed in Figure 2.12.

Solution. Projecting the graph onto the x-axis, as shown in Figure 2.13, we see that the domain consists of all real numbers. Thus the domain is R. Projecting the graph onto the y-axis, we see in Figure 2.14 that the range includes all reals greater than or equal to zero. Hence the range is $[0, \infty)$. ■

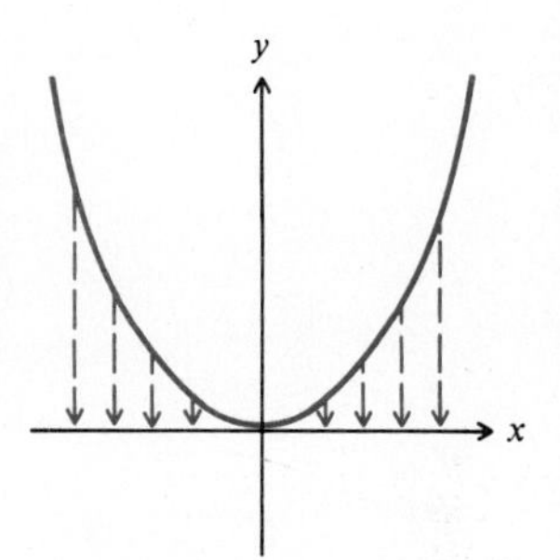

Figure 2.13

Exercise
Determine the domain and range of the relation graphed.

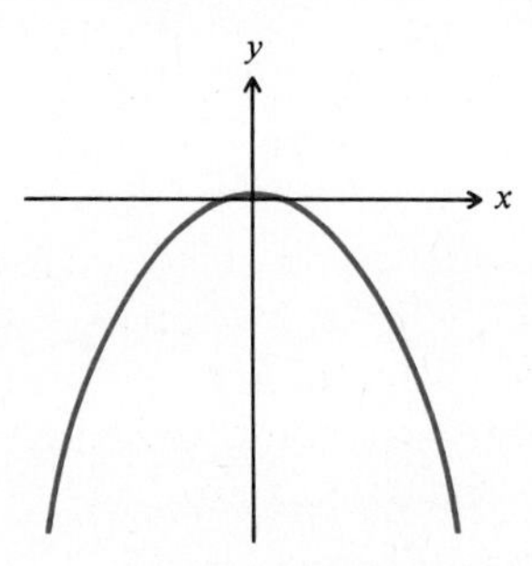

Answer. The domain is R, and the range is $(-\infty, 0]$. ■

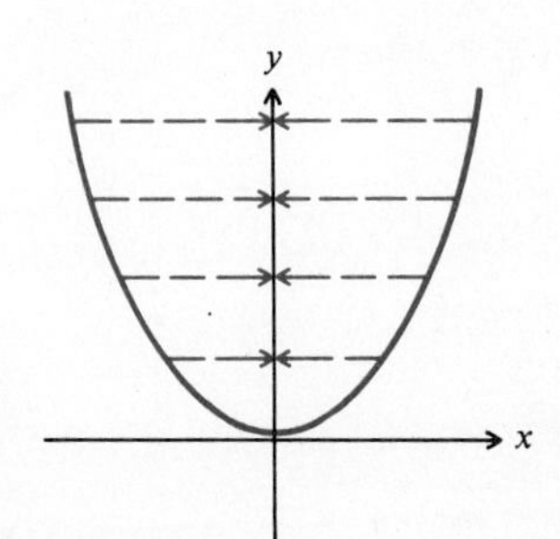

Figure 2.14

Problem Set 2.2

In Problems 1–6, determine the domain and range of r.

1. $r = \{(5, 7), (-3, 2), (6, 4)$

2. $r = \{(-3, 2), (6, 1), (-10, 0)\}$

3. $r = \{(2, 0), (3, 0), (-1, 0)\}$

4. $r = \{(-1, 6), (-1, 2), (-1, 0)\}$

5. $r = \{(0, \pi), (\pi, 0)\}$

6. $r = \{(a, b), (b, c)\}$

In Problems 7–10, list three ordered pairs which belong to the given relation.

7. $\{(x, y) | y = \sqrt{x + 2}\}$

8. $\{(x, y) | y = \sqrt{x - 3}\}$

9. $\{(x, y) | x^2 + y^2 = 1\}$

10. $\{(x, y) | x = y^3\}$

In Problems 11–18, sketch the graph of the equation after plotting points with the given values.

11. $y = x + 3$, x-values: $-4, -2, 0, 2, 4$
12. $y = 3 - 2x$, x-values: $-4, -2, 0, 2, 4$
13. $y = -x^2$, x-values: $-2, -1, 0, 1, 2$
14. $y = x^2$, x-values: $-2, -1, 0, 1, 2$
15. $x = |y|$, y-values: $-2, -1, 0, 1, 2$
16. $y = x^3$, x-values: $-2, -1, 0, 1, 2$
17. $x^2 + y^2 = 4$, x-values: $-2, -1, 0, 1, 2$

18. $y = \sqrt{x}$, x-values: 0, 1, 4, 9, 16

In Problems 19–28, determine the domain and range of the relation graphed.

19.

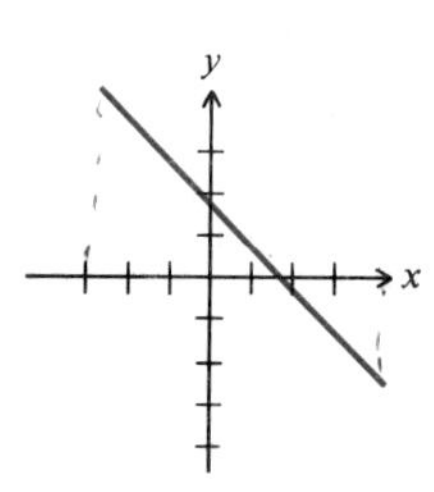

20.

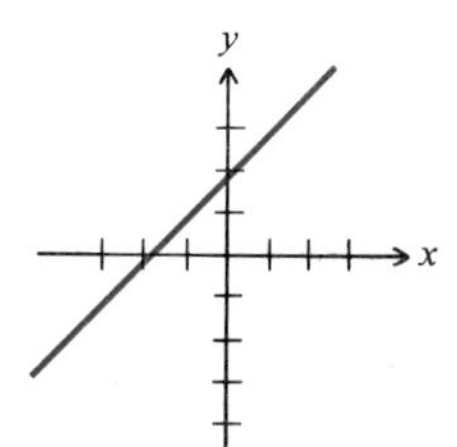

21.

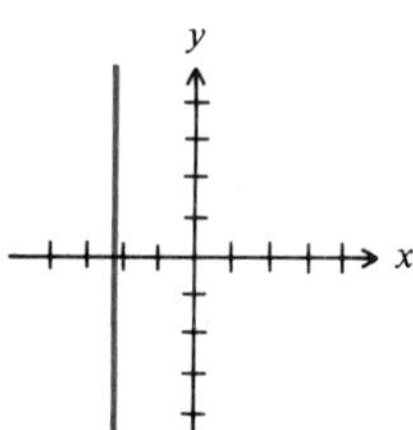

22.

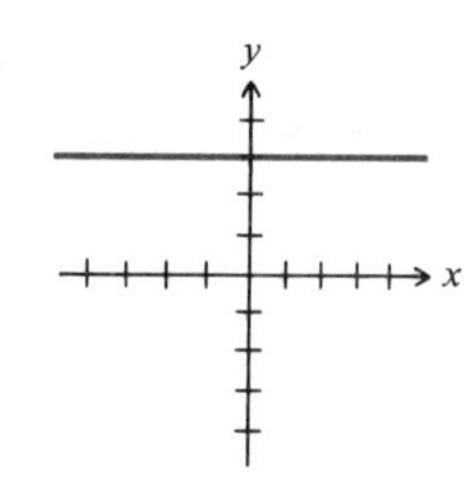

23.

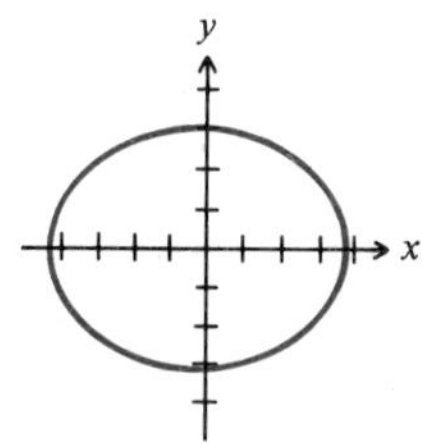

24.

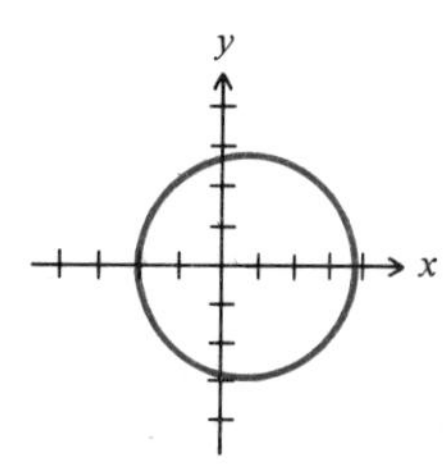

25.

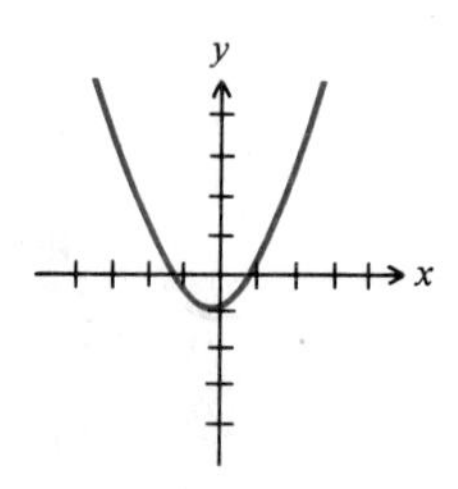

26.

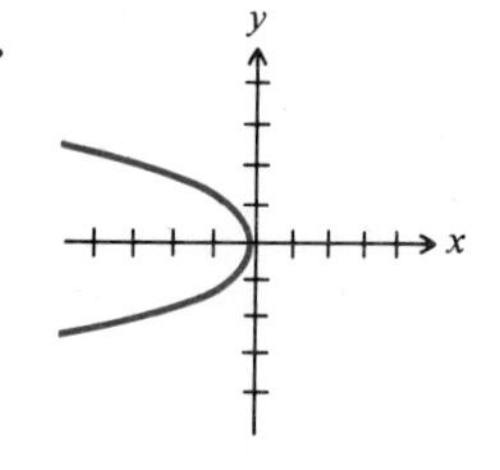

27.

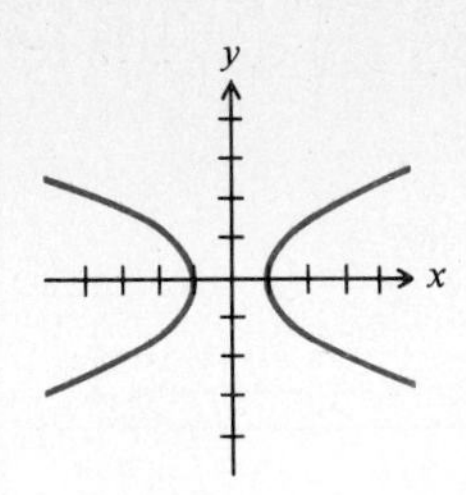

28.

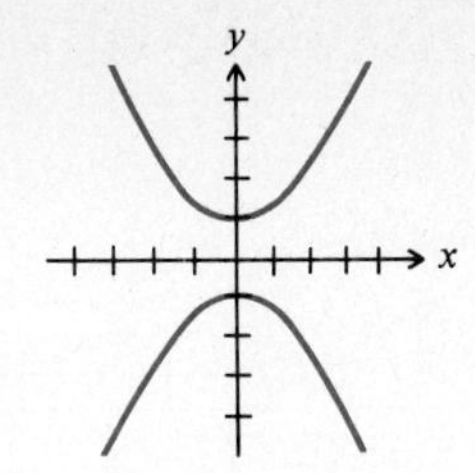

2.3 Lines

An equation in variables x and y is called **linear** in x and y provided that it can be expressed in the form

$$Ax + By + C = 0,$$

where A, B, and C are constants, with A and B not both zero. Some examples of linear equations are

$$x + y + 2 = 0, \quad y = 2x - 1, \quad y = -2, \quad \text{and} \quad x = 5.$$

On the other hand,

$$y = x^2, \quad x^2 + y^2 = 4, \quad \text{and} \quad y + \sqrt{x} = 0$$

are *not* linear.

As its name suggests, the graph of a linear equation is a straight line. Although we won't prove this for the general case, let's examine the proof for $a, b > 0$ and $c = 0$. Then the general linear equation becomes

$$ax + by = 0,$$

and $(0, 0)$ and $(1, -a/b)$ are solutions. Let $\mathscr{L}$ be the line passing through these two points. See Figure 2.15.

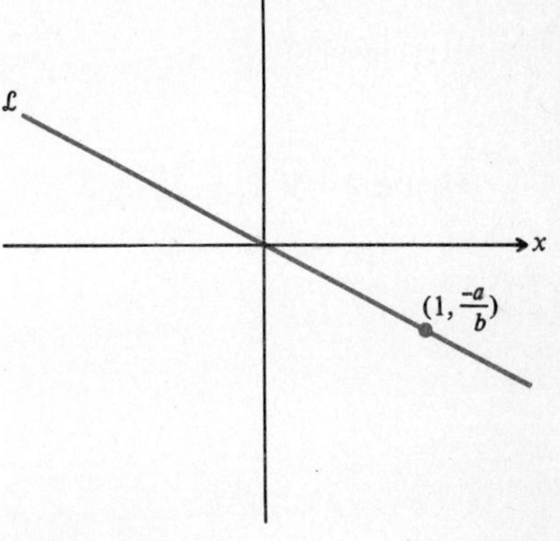

Figure 2.15

Let (x, y) be any point in the plane. If $(x, y) \neq (0, 0)$ and $(x, y) \neq (1, -a/b)$, then (x, y) is on $\mathscr{L}$ if and only if $\triangle OAB \approx \triangle OCD$. See Figure 2.16. Hence (x, y) is on $\mathscr{L}$ if and only if

$$\frac{y}{a/b} = \frac{-x}{1}$$

$$\frac{by}{a} = -x$$

$$ax + by = 0.$$

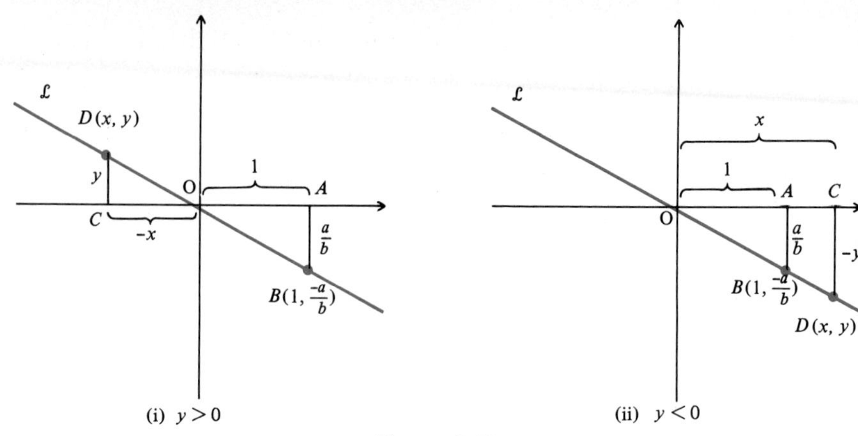

Figure 2.16

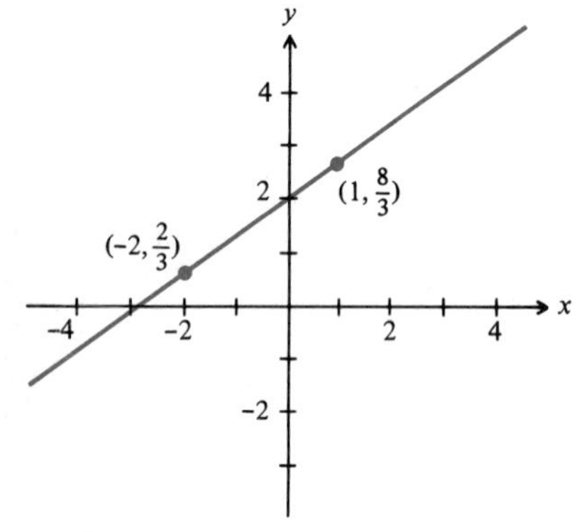

Figure 2.17 $2x - 3y + 6 = 0$

Example 1
Graph $2x - 3y + 6 = 0$.

Solution. Because a line is determined by any two of its points, it suffices to find any two solutions. When $x = 1$, $y = \frac{8}{3}$, and when $x = -2$, $y = \frac{2}{3}$. Thus $(1, \frac{8}{3})$ and $(-2, \frac{2}{3})$ are solutions. The graph is shown in Figure 2.17. ■

The y-coordinate of the point of intersection of a nonvertical line with the y-axis is called the **y-intercept.** Graphically, it is the value on the y-axis where the line crosses. Algebraically, it is the value of y when $x = 0$. Similarly, the x-coordinate of the point of intersection of a nonhorizontal line with the x-axis is called the **x-intercept.** Graphically, it is the value on the x-axis where the line crosses. Algebraically, it is the value of x when $y = 0$. Note that the concept of y-intercept is defined for nonvertical lines only. Thus no vertical line (not even the y-axis) has a y-intercept. Similarly, no horizontal line (not even the x-axis) has an x-intercept. As seen in Example 2, locating the intercepts provides a quick means of determining the graph of a linear equation.

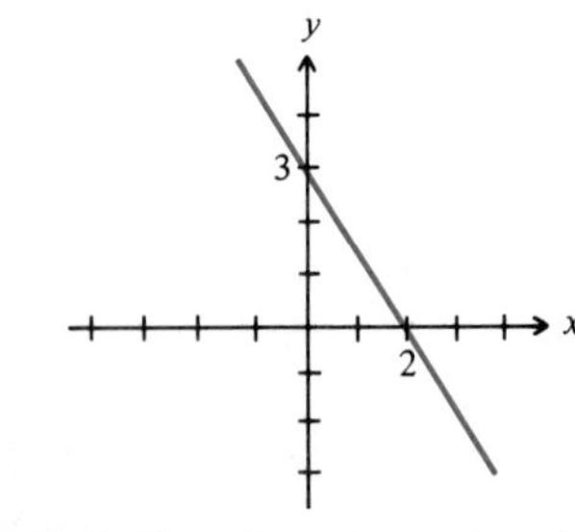

Figure 2.18 $3x + 2y - 6 = 0$

Example 2
Graph $3x + 2y - 6 = 0$.

Solution. We will locate the intercepts. When $x = 0$, $y = 3$, and when $y = 0$, $x = 2$. Noting that the y-intercept is 3 and that the x-intercept is 2, we get the graph in Figure 2.18. ■

Example 3
Graph $y = 2$.

Solution. To determine the graph it suffices to find two solutions, like (0, 2) and (3, 2). The graph is in Figure 2.19. ■

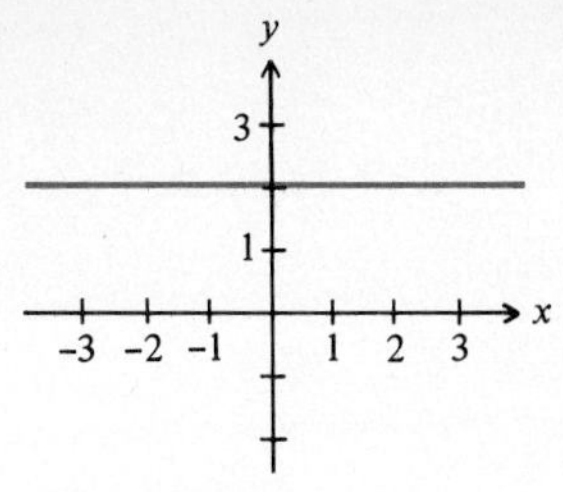

Figure 2.19 $y = 2$

Perhaps some clarification is needed about the equation $y = 2$. Is it really an equation in x and y since it doesn't contain the variable x? The answer is "yes." Because we are working with equations in two variables, we are thinking of this equation as

$$0x + y = 2.$$

Similarly, an equation like $x = -1$ may be thought of as

$$0y + x = -1.$$

Viewing an equation like $x = -1$ as an equation in two variables, instead of one, significantly alters the solutions and the graph. When considered an equation in two variables, the solutions are ordered pairs, and the solution set consists of all ordered pairs whose first coordinate is -1. The graph is a straight line parallel to the y-axis. If the equation were viewed as an equation in a single variable, the solution would be the single real number -1. Its graph would be a single point on a number line. Thus when working with equations in two variables, we are working in the plane and solutions are ordered pairs, even if one of the variables doesn't appear.

Exercise

Graph each of the following.
(a) $2x + y - 4 = 0$ (b) $x + 2 = 0$

Answers

(a)

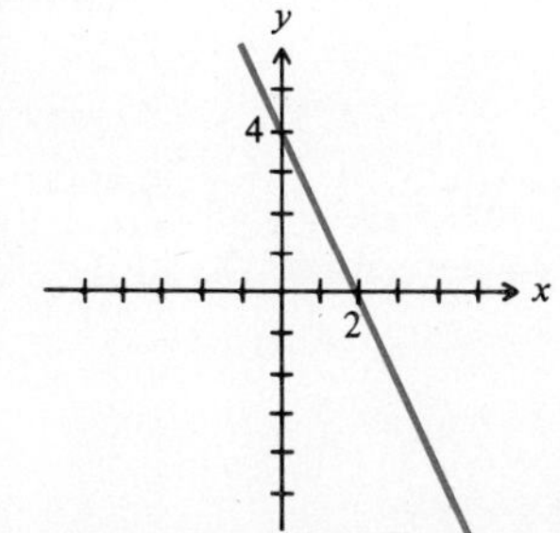

(b)

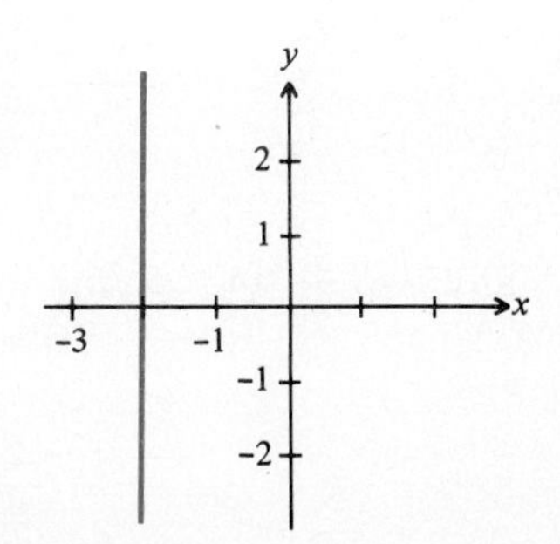

■

Before considering the problem of finding the equation of a given line, we need some definitions. Let $\mathscr{L}$ be a nonvertical line, and let $P_1(x_1, y_1)$ and $P_2(x_2, y_2)$ be two distinct points on $\mathscr{L}$. The **slope** of $\mathscr{L}$, usually denoted by m, is the ratio

$$m = \frac{y_2 - y_1}{x_2 - x_1}. \tag{2-3}$$

The slope of a given line is independent of the points chosen. Suppose that we consider four points on the line: $P_1(x_1, y_1)$, $P_2(x_2, y_2)$, $P_3(x_3, y_3)$, and $P_4(x_4, y_4)$. Now let Q be the point with ordered pair (x_2, y_1) and let R be the point with ordered pair (x_4, y_3). See Figure 2.20.

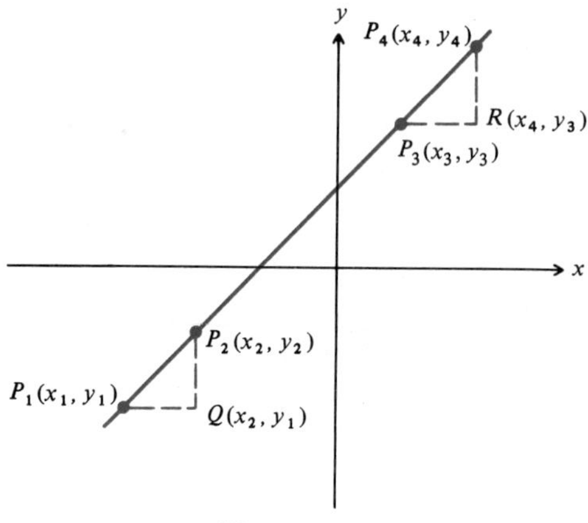

Figure 2.20

Now triangles P_1QP_2 and P_3RP_4 are similar because corresponding angles are equal. Because corresponding sides of similar triangles are proportional,

$$\frac{P_2Q}{P_1Q} = \frac{P_4R}{P_3R}.$$

Upon calculating these distances, we have

$$\frac{y_2 - y_1}{x_2 - x_1} = \frac{y_4 - y_3}{x_4 - x_3}.$$

Thus we get the same ratio whether we use P_1 and P_2 or P_3 and P_4. Consequently, any two distinct points may be used to determine the slope. Furthermore, since

$$\frac{y_2 - y_1}{x_2 - x_1} = \frac{y_1 - y_2}{x_1 - x_2},$$

the order in which the two points are taken is immaterial. (Note, however, that the order in the numerator must agree with the order in the denominator.)

Example 4
Determine the slope of the line passing through (2, 3) and (−1, 5).

Solution.

$$m = \frac{3-5}{2-(-1)} = \frac{-2}{2+1} = -\frac{2}{3}. \quad ■$$

Example 5
Determine the slope of the line whose equation is $2x - 3y + 6 = 0$.

Solution. We first need to find any two points on the graph. Now the x-intercept is -3 and the y-intercept is 2. Thus $(-3, 0)$ and $(0, 2)$ are two points on the line. Therefore,

$$m = \frac{0-2}{-3-0} = \frac{-2}{-3} = \frac{2}{3}. \quad ■$$

Exercise
Determine the slope of the line
(a) passing through $(-1, 2)$ and $(3, 4)$
(b) whose equation is $3x + 2y - 6 = 0$

Answers. (a) $\frac{1}{2}$ (b) $\frac{-3}{2}$ ■

Observe that the definition of slope applies to nonvertical lines only. For a vertical line, equation (2-3) will give us a zero in the denominator. Hence slope is not defined for vertical lines, and we shall say that such lines have slope undefined.

Slope is a measure of the change in y relative to the change in x. To say that the slope of a line is $\frac{3}{2}$ means that moving along the line, y increases 3 units as x increases 2 units. [Or, since $\frac{3}{2} = (-3)/(-2)$, y decreases 3 units as x decreases 2.] A line with slope $-\frac{3}{5}$ has the property that y decreases 3 units as x increases 5 (or y increases 3 as x decreases 5). As the x values increase, a line with positive slope rises, while one with negative slope falls. A line with zero slope is horizontal (perpendicular to the y-axis).

Example 6

Graph the line which passes through (1, 2) and has slope $\frac{-3}{4}$.

Solution. To graph the line we need to locate two of its points. We already have (1, 2), and the slope will give us another. Because y decreases 3 as x increases 4, the point $(5, -1)$, three units below and four units to the right of (1, 2), is on the line. We get the graph shown in Figure 2.21. ■

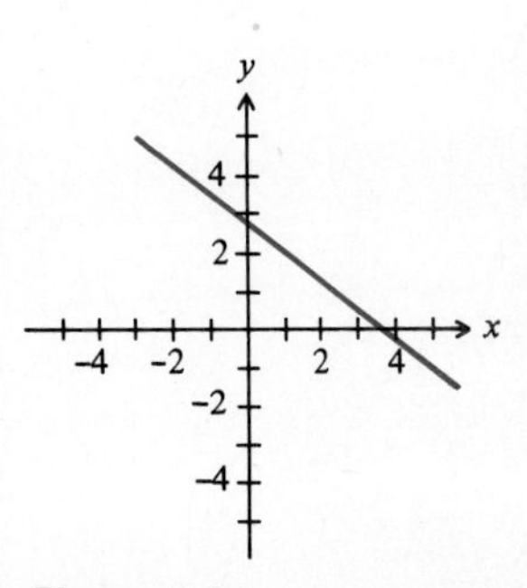

Figure 2.21

Exercise

Graph the line which passes through $(-2, 1)$ and has slope $\frac{-2}{3}$.

Answer

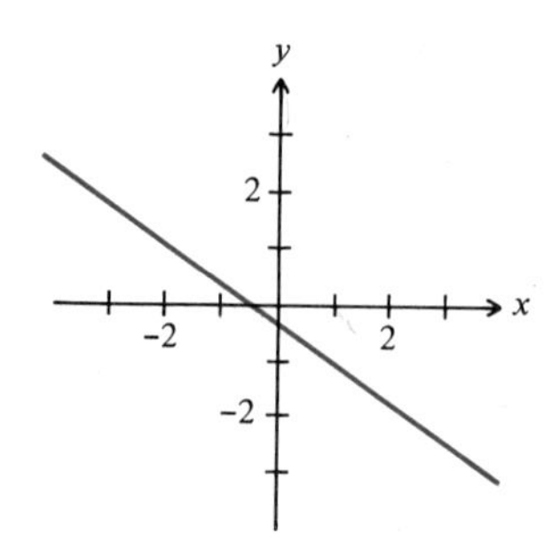

■

As seen in Example 5, given a linear equation we can find the slope of the line by finding two solutions and applying the definition. However, as the next theorem indicates, there is another technique which is sometimes easier.

Theorem 2.3

Let $\mathscr{L}$ be a line whose equation is $y = ax + b$, where a and b are constants. Then the slope of $\mathscr{L}$ is a, and the y-intercept is b.

Proof. Clearly, the y-intercept is b (when $x = 0$, $y = b$). To determine the slope we need one additional point on the line. When $x = 1$, $y = a + b$. Hence two points are $(0, b)$ and $(1, a + b)$. The slope is thus

$$m = \frac{(a + b) - b}{1 - 0} = a.$$

■

Because $m = a$, the given equation may be written

$$y = mx + b.$$

This form is called the **slope-intercept form** of a linear equation.

This theorem gives us an alternative method of finding the slope and y-intercept. If the equation is solved for y in terms of x, then the coefficient of x is the slope, and the constant term is the y-intercept.

Example 7

Determine the slope and y-intercept of $5x - 2y + 6 = 0$.

Solution. First we must put this equation in slope-intercept form. Solving for y we get

$$-2y = -5x - 6$$
$$y = \frac{5}{2}x + 3.$$

Thus the slope is $\frac{5}{2}$, and the y-intercept is 3. ■

Exercise
Determine the slope and y-intercept of each of the following.
(a) $y = -3x + 5$ (b) $-2x + 7y - 3 = 0$

Answers. (a) Slope is -3; y-intercept is 5.

(b) Slope is $\frac{2}{7}$; y-intercept is $\frac{3}{7}$. ■

Theorems 2.4 and 2.5 describe the relationship between slopes of parallel and perpendicular lines.

Theorem 2.4

Two nonvertical lines are parallel if and only if they have the same slope.

The proof of this theorem is based on similar triangles and is almost as easy as the result is obvious. (Try to prove it as an exercise.) Theorem 2.5 and its proof, however, are not so evident.

Theorem 2.5

Two nonvertical lines are perpendicular if and only if the slope of each is the negative reciprocal of the slope of the other.

Proof. From Theorem 2.4 we know that lines whose slopes are negative reciprocals must intersect. For simplicity we may assume that they intersect at the origin. Let m_1 and m_2 be the slopes of $\mathscr{L}_1$ and $\mathscr{L}_2$, respectively. See Figure 2.22.

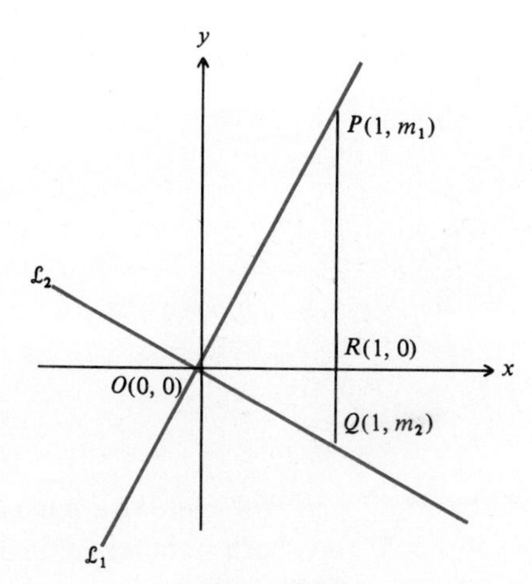

Figure 2.22

(if) Assuming that $m_1m_2 = -1$, we must show that $\angle POQ$ is a right angle. Now $\angle PRO$ is a right angle. If we can show that $\triangle POQ \approx \triangle PRO$, then we can conclude that $\angle POQ$ (which would correspond to $\angle PRO$) is also a right angle. To this end let's note that

$$PR = m_1,$$
$$PO = \sqrt{1 + m_1^2},$$

and

$$PQ = m_1 - m_2.$$

Thus

$$\frac{PR}{PO} = \frac{m_1}{\sqrt{1 + m_1^2}} = \frac{m_1\sqrt{1 + m_1^2}}{1 + m_1^2}$$

and

$$\frac{PO}{PQ} = \frac{\sqrt{1 + m_1^2}}{m_1 - m_2} = \frac{\sqrt{1 + m_1^2}}{m_1 + \dfrac{1}{m_1}} = \frac{m_1\sqrt{1 + m_1^2}}{m_1^2 + 1}.$$

Therefore,

$$\frac{PR}{PO} = \frac{PO}{PQ}.$$

Since $\angle P$ is common to both triangles,

$$\triangle POQ \approx \triangle PRO$$

by side–angle–side. Hence $\angle POQ$ is a right angle (i.e., $\mathscr{L}_1$ is perpendicular to $\mathscr{L}_2$).

(only if) Assuming that $\angle POQ$ is a right angle, we must show that $m_1 m_2 = -1$. Because $\angle POQ$ and $\angle PRO$ are both right angles, and $\angle P$ is common,

$$\triangle POQ \approx \triangle PRO$$

by angle–angle. And because $\angle POQ$ and $\angle QRO$ are both right angles, and

$\angle Q$ is common,

$$\triangle POQ \approx \triangle ORQ$$

by angle–angle. Thus

$$\triangle PRO \approx \triangle ORQ.$$

Therefore,

$$\frac{PR}{RO} = \frac{OR}{RQ}$$

$$\frac{m_1}{1} = \frac{1}{-m_2}$$

$$m_1 m_2 = -1. \qquad \blacksquare$$

Theorems 2.4 and 2.5 are stated for nonvertical lines because slope is not defined for vertical lines. But two lines which have slope undefined are parallel (since each is perpendicular to the x-axis), and a line with slope undefined (a vertical line) is perpendicular to one with zero slope (a horizontal line).

Example 8
Determine the slope of a line which is (a) parallel and (b) perpendicular to $3x + 8y - 4 = 0$.

Solution. We must first find the slope of the given line. Putting the given equation in slope-intercept form, we get

$$y = \frac{-3}{8}x + \frac{1}{2}.$$

Thus the slope of the given line is $-\frac{3}{8}$. Hence a parallel line has a slope $-\frac{3}{8}$, and a perpendicular line has slope $\frac{8}{3}$. ■

Exercise
Determine the slope of a line which is (a) parallel and (b) perpendicular to $5x - 2y + 6 = 0$.

Answers. (a) $\frac{5}{2}$ (b) $\frac{-2}{5}$ ■

Problem Set 2.3

In Problems 1–6, determine the x- and y-intercepts and the slope of the line.

1. $-4x + 5y - 20 = 0$
2. $4x + 3y - 12 = 0$
3. $5x - 2y + 3 = 0$
4. $3x + 5y + 4 = 0$
5. $3x = 0$
6. $2y + 3 = 0$

In Problems 7–12, graph the equation.

7. $2x + 3y - 6 = 0$
8. $3x - y - 3 = 0$
9. $4x - 5y - 10 = 0$
10. $-x - 2y + 2 = 0$
11. $2x - 5 = 0$
12. $y + 3 = 0$

In Problems 13–16, graph the line described.

13. passing through $(-3, 1)$ with slope $\frac{1}{3}$

14. passing through $(-2, 5)$ with slope -2

15. with slope 3 and y-intercept -2

16. with slope $\frac{-1}{2}$ and y-intercept 5

In Problems 17–26, determine the slope of the line passing through the given points.

17. $(2, 3), (1, 5)$
18. $(-3, 5), (-1, 11)$
19. $(-8, -7), (2, 1)$
20. $(-1, -5), (-5, -2)$
21. $\left(\frac{2}{5}, 1\right), \left(2, \frac{3}{4}\right)$
22. $\left(\frac{-2}{3}, \frac{3}{5}\right), (0, 0)$
23. $(\sqrt{2}, -2\sqrt{5}), (3\sqrt{2}, 2\sqrt{5})$
24. $(-\sqrt{6}, 3\sqrt{5}), (-2\sqrt{6}, \sqrt{5})$
25. $(3, 5), (-2, 5)$
26. $(1, 4), (1, 6)$

In Problems 27–34, determine the slope and y-intercept.

27. $y = 2x - 5$
28. $y = \frac{-1}{2}x + 2$
29. $2x + 5y - 1 = 0$
30. $3x - 2y + 3 = 0$
31. $-5x - 3y + 2 = 0$
32. $-x + y - 3 = 0$
33. $x = 2$
34. $y = 0$

In Problems 35–44, determine the slope of the line described.

35. parallel to $-2x + y + 5 = 0$
36. parallel to $x + 3y - 7 = 0$
37. perpendicular to $-x + 5y - 6 = 0$
38. perpendicular to $3x + 2y + 8 = 0$
39. parallel to the x-axis
40. perpendicular to the x-axis
41. passing through $(2, 1)$ and the midpoint of the segment from $(-1, 5)$ to $(1, 6)$
42. passing through $(-3, 5)$ and the midpoint of the segment from $(3, 7)$ to $(-1, -4)$
43. with x-intercept -3 and y-intercept 4
44. with x-intercept $\frac{3}{5}$ and y-intercept $\frac{5}{2}$

45. Use slopes to show that $A(-2, -3)$, $B(1, 1)$, $C(5, 2)$, and $D(2, -2)$ are the vertices of a parallelogram.
46. Use slopes to show that $A(-5, 3)$, $B(-1, 5)$, $C(2, -1)$, and $D(-2, -3)$ are the vertices of a rectangle.
47. Use slopes to show that $A(0, 3)$, $B(2, 2)$, and $C(-4, 5)$ lie on a straight line.

2.4 Lines Continued

We now turn our attention to the problem of finding the equation of a line given specific information. As mentioned earlier, the equation will be linear. Since we can graph a line if we know the slope and y-intercept, it is reasonable to expect that we can find its equation. This is proven in the next theorem, which is the converse of Theorem 2.3.

Theorem 2.6

The equation of a line with slope m and y-intercept b is

$$y = mx + b.$$

Proof. Let $\mathscr{L}$ be the line with slope m and y-intercept b, and let (x, y) represent each point in the plane. Then (x, y) is on $\mathscr{L}$ if and only if $(x, y) = (0, b)$ or $(y - b)/x = m$. Therefore, (x, y) is on $\mathscr{L}$ if and only if $y = mx + b$. ■

Example 1

Determine the equation of a line with slope -2 and y-intercept $-\frac{3}{2}$.

Solution. By Theorem 2.6, we get

$$y = -2x - \frac{3}{2}.$$ ■

Example 2
Determine the equation of the line with y-intercept 3 and x-intercept -4.

Solution. To use Theorem 2.6 we must know the slope. We can calculate it using the definition because we know that $(0, 3)$ and $(-4, 0)$ are on the line. Therefore,

$$m = \frac{3 - 0}{0 - (-4)} = \frac{3}{4}.$$

Since the slope is $\frac{3}{4}$ and the y-intercept is 3, the desired equation is

$$y = \frac{3}{4}x + 3.$$ ■

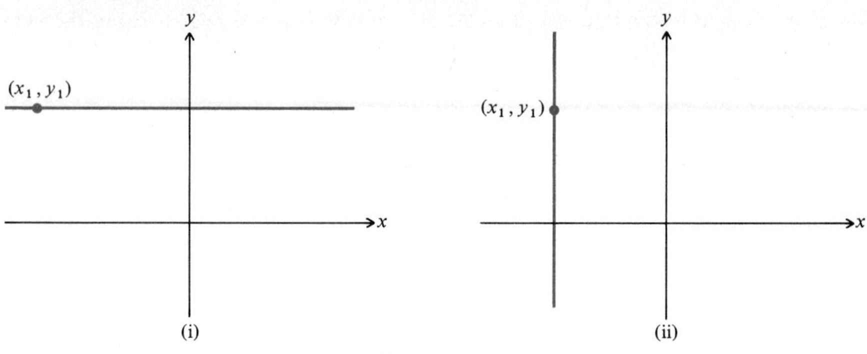

Figure 2.23

What is the equation of the horizontal line passing through (x_1, y_1)? [See Figure 2.23 (i).] Because its slope is 0 and its y-intercept is y_1, its equation is $y = y_1$. (This seems reasonable since the y-coordinates of every point on a horizontal line are the same.) Of course, since a vertical line has slope undefined, its equation does not have a slope-intercept form. But because every point on a vertical line has the same first coordinate [see Figure 2.23 (ii)], the equation of the vertical line passing through (x_1, y_1) is $x = x_1$.

Now the equation of a vertical line, $x = k$ for some constant k, is linear $(x + 0y - k = 0)$, and from Theorem 2.6 we see that the equation of a nonvertical line, $y = mx + b$ for some slope m and y-intercept b, is also linear $(-mx + y - b = 0)$. Hence the equation of every line is a linear equation.

We can graph a line if we know one point on the line and the slope. Accordingly, we can find the equation of such a line.

Theorem 2.7

The equation of the line passing through (x_1, y_1) *with slope* m *is*

$$y - y_1 = m(x - x_1). \tag{2-4}$$

Proof. Let $\mathscr{L}$ be the line passing through (x_1, y_1) with slope m, and let (x, y) represent each point in the plane. Then (x, y) is on $\mathscr{L}$ if and only if $(x, y) = (x_1, y_1)$ or $(y - y_1)/(x - x_1) = m$. Therefore, (x, y) is on $\mathscr{L}$ if and only if $y - y_1 = m(x - x_1)$. ■

Equation (2-4) is called the **point-slope form** of the equation of a line. The slope-intercept form, however, is perhaps a more useful format for a linear equation. Hence, although our initial equations in Examples 3, 4, and 5 will be in point-slope form, we shall put the final result in slope-intercept form.

Example 3

Determine the equation (in slope-intercept form) of the line passing through $(-1, 3)$ with slope $\frac{2}{3}$.

Solution. By Theorem 2.7, the equation of this line (in point-slope form) is

$$y - 3 = \frac{2}{3}[x - (-1)].$$

Putting this equation in slope-intercept form, we get

$$y = \frac{2}{3}x + \frac{11}{3}. \qquad \blacksquare$$

Example 4
Determine the equation (in slope-intercept form) of the line passing through $(2, -1)$ and $(-3, 2)$.

Solution. We can determine the equation of a line if we know the slope and one point on the line. Although we aren't given the slope here, we can use the two points to calculate it. We have

$$m = \frac{-1 - 2}{2 - (-3)} = \frac{-3}{5}.$$

Now we use the slope and one of the points to write the equation in point-slope form. Using $(2, -1)$, we get

$$y + 1 = \frac{-3}{5}(x - 2). \tag{2-5}$$

In slope-intercept form this equation is

$$y = \frac{-3}{5}x + \frac{1}{5}. \tag{2-6}$$ $\blacksquare$

To get equation (2-5) in Example 4, we used the point $(2, -1)$. Had we used $(-3, 2)$ instead, we would have obtained

$$y - 2 = \frac{-3}{5}(x + 3).$$

But this equation is equivalent to equation (2-5). This can be easily seen by writing this equation in slope-intercept form. We get equation (2-6).

Example 5
Determine the equation (in slope-intercept form) of the line passing through $(3, -2)$ which is (a) parallel and (b) perpendicular to $3x - 4y + 6 = 0$.

Solution (a). We first need to find the slope of $3x - 4y + 6 = 0$. After writing this equation in slope-intercept form,

$$y = \frac{3}{4}x + \frac{3}{2},$$

we see that the slope is $\frac{3}{4}$. Thus a line parallel to $3x - 4y + 6 = 0$ has slope $\frac{3}{4}$. Consequently, the equation of the parallel line passing through $(3, -2)$ is

$$y + 2 = \frac{3}{4}(x - 3).$$

In slope-intercept form it is

$$y = \frac{3}{4}x - \frac{17}{4}.$$

(*b*) A line perpendicular to $3x - 4y + 6 = 0$ has slope $-\frac{4}{3}$. Hence the equation of the perpendicular line passing through $(3, -2)$ is

$$y + 2 = \frac{-4}{3}(x - 3).$$

In slope-intercept form it is

$$y = \frac{-4}{3}x + 2.$$ ■

Exercise

Determine the equation (in slope-intercept form) of the line described.

(a) passing through $(-4, -6)$ with slope $\frac{1}{4}$

(b) passing through $(-2, 3)$ and $(4, 1)$
(c) passing through $(-4, 3)$ and parallel to $2x + y - 3 = 0$
(d) passing through $(3, -1)$ and perpendicular to $x - 3y + 5 = 0$

Answers. (a) $y = \frac{1}{4}x - 5$ (b) $y = \frac{-1}{3}x + \frac{7}{3}$ (c) $y = -2x - 5$

(d) $y = -3x + 8$ ■

Problem Set 2.4

In Problems 1–6, determine the equation of the line with the given slope and intercept.

1. slope 3, y-intercept -2

2. slope -1, y-intercept 3

3. slope $\frac{-1}{2}$, y-intercept $\frac{3}{2}$

4. slope $\frac{3}{5}$, y-intercept $\frac{-2}{5}$

5. slope 0, y-intercept -2

6. slope undefined, x-intercept 5

In Problems 7–10, determine the equation (in slope-intercept form) of the line with the given intercepts.

7. x-intercept 5, y-intercept 3

8. x-intercept -1, y-intercept 2

9. x-intercept $\frac{-2}{5}$, y-intercept $\frac{3}{5}$

10. x-intercept $\frac{3}{5}$, y-intercept $\frac{-2}{3}$

In Problems 11–20, determine the equation (in slope-intercept form if possible) of the line passing through the given point with the specified slope m.

11. $(-3, 5)$, $m = -2$

12. $(3, 4)$, $m = -1$

13. $(-3, -6)$, $m = \frac{2}{3}$

14. $(4, -8)$, $m = \frac{-3}{4}$

15. $\left(\frac{2}{5}, \frac{7}{5}\right)$, $m = -2$

16. $\left(\frac{3}{4}, \frac{1}{4}\right)$, $m = 3$

17. $\left(\frac{-3}{4}, \frac{1}{5}\right)$, $m = \frac{1}{2}$

18. $\left(\frac{1}{2}, \frac{-2}{3}\right)$, $m = \frac{-1}{5}$

19. $(3, -5)$, $m = 0$

20. $(-2, 1)$, slope undefined

In Problems 21–30, determine the equation (in slope-intercept form if possible) of the line passing through the given points.

21. $(7, 6)$, $(8, 4)$

22. $(3, 5)$, $(1, 11)$

23. $(-2, -3)$, $(1, -2)$

24. $(-1, 0)$, $(1, -5)$

25. $\left(\frac{7}{2}, \frac{-1}{2}\right)$, $(5, -3)$

26. $\left(\frac{2}{3}, \frac{-5}{3}\right)$, $\left(\frac{5}{3}, \frac{-1}{3}\right)$

27. $\left(\frac{5}{2}, \frac{1}{3}\right)$, $\left(\frac{2}{3}, \frac{-7}{4}\right)$

28. $\left(\frac{3}{4}, \frac{1}{6}\right)$, $\left(\frac{-7}{3}, \frac{11}{2}\right)$

29. $(1, 5)$, $(1, -3)$

30. $(-3, 6)$, $(2, 6)$

In Problems 31–35, determine the equation (in slope-intercept form if possible) of the line passing through the given point which is (a) parallel and (b) perpendicular to the given line.

31. $(-2, 5)$, $2x + y + 5 = 0$

32. $(3, -2)$, $x - 3y + 6 = 0$

33. $\left(\frac{1}{2}, \frac{5}{3}\right)$, $3x - 5y + 7 = 0$

34. $\left(-2, \frac{7}{3}\right)$, $y - 6 = 0$

35. $\left(\frac{5}{2}, -3\right)$, $x = -3$

36. Determine the equation of the line with slope -2, passing through the midpoint of the segment connecting $(-2, 3)$ and $(4, 6)$.
37. Determine the equation of the line passing through $(-3, -5)$ which is parallel to the line passing through $(2, 1)$ and $(6, 5)$.
38. Determine the equation of the line passing through $(-2, 1)$ which is perpendicular to the line passing through $(1, 5)$ and $(3, \frac{5}{2})$.

2.5 Circles

A **circle** is the set of all points in a plane which are equidistant from a fixed point. The fixed distance is called the **radius,** and the fixed point is called the **center.** (Note that the center is *not* on the circle.) Theorem 2.8 tells us how to determine the equation of a circle when we know the center and the radius.

Theorem 2.8

The equation of the circle with radius r (where $r > 0$) and center (h, k) is

$$(x - h)^2 + (y - k)^2 = r^2. \tag{2-7}$$

Proof. Let (x, y) represent each point in the plane. Now (x, y) is on the circle if and only if the distance between (x, y) and (h, k) is r. See Figure 2.24. Thus, (x, y) is on the circle if and only if

$$\sqrt{(x - h)^2 + (y - k)^2} = r.$$

Squaring both sides gives

$$(x - h)^2 + (y - k)^2 = r^2,$$

which is equivalent since $r > 0$. ■

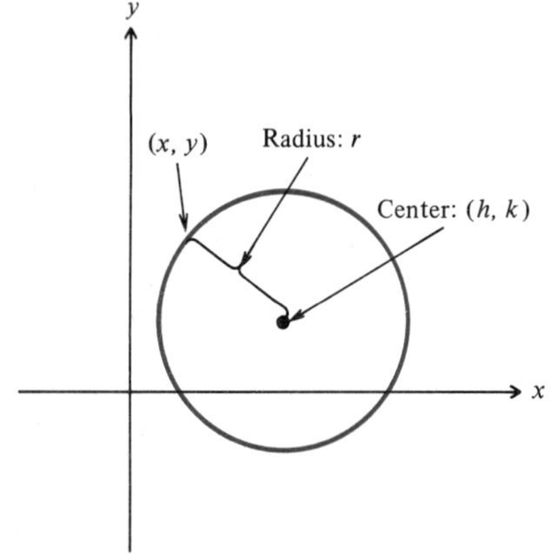

Figure 2.24

Example 1
Determine the equation of the circle described.
(a) radius 2, center at $(-1, 2)$
(b) radius 3, center at origin

Solution. (a) By Theorem 2.8, we get

$$(x - (-1))^2 + (y - 2)^2 = 2^2.$$

Simplified, we have

$$(x + 1)^2 + (y - 2)^2 = 4.$$

(b) Again using Theorem 2.8, we get

$$(x - 0)^2 + (y - 0)^2 = 3^2.$$

Simplified we have

$$x^2 + y^2 = 9. \qquad \blacksquare$$

Equation (2-7) is called the **center–radius form** for the equation of a circle. As seen in Example 1(b), if the center is the origin, the equation becomes

$$x^2 + y^2 = r^2.$$

Exercise
Determine the equation of the circle described.
(a) radius 3, center at $(2, -3)$
(b) radius 4, center at origin

Answers. (a) $(x - 2)^2 + (y + 3)^2 = 9$ (b) $x^2 + y^2 = 16$ ■

From Theorem 2.8 we know that the graph of equation (2-7) is a circle of radius r $(r > 0)$ with center (h, k). For example, the graph of

$$(x - 2)^2 + (y + 1)^2 = 4$$

is a circle of radius 2 centered at $(2, -1)$. However, it is difficult to recognize that the graph of

$$x^2 + y^2 - 4x + 2y + 1 = 0$$

is the same circle. (The two equations are equivalent.) To do this we must employ a method called **completing the square.** This procedure involves converting a binomial of the form $x^2 + kx$, where k is a constant, to a trinomial which is a perfect square. This is done by adding $(k/2)^2$, since

$$x^2 + kx + \left(\frac{k}{2}\right)^2 = \left(x + \frac{k}{2}\right)^2.$$

Example 2
Complete the square on each of the following. That is, add a constant to each of the following algebraic expressions so that the resulting trinomial is a perfect square, and write the result as the square of a binomial.

(a) $x^2 + 2x$ (b) $x^2 - 5x$ (c) $x^2 + \frac{1}{3}x$

Solution

(a) $x^2 + 2x + \left(\frac{2}{2}\right)^2 = x^2 + 2x + 1 = (x + 1)^2$

(b) $x^2 - 5x + \left(\frac{-5}{2}\right)^2 = x^2 - 5x + \frac{25}{4} = \left(x - \frac{5}{2}\right)^2$

(c) $x^2 + \frac{1}{3}x + \left(\frac{1}{6}\right)^2 = x^2 + \frac{1}{3}x + \frac{1}{36} = \left(x + \frac{1}{6}\right)^2$ ■

Exercise
Complete the square.

(a) $x^2 + 4x$ (b) $x^2 - 3x$ (c) $x^2 - \frac{2}{5}x$

Answers. (a) $(x + 2)^2$ (b) $\left(x - \frac{3}{2}\right)^2$ (c) $\left(x - \frac{1}{5}\right)^2$ ■

The method of completing the square is used in Examples 3 through 6.

Example 3
Describe the graph of $x^2 + y^2 + 6x - 4y + 9 = 0$.

Solution. We first rearrange some terms and write

$$x^2 + 6x + y^2 - 4y = -9.$$

Now, to complete the square on $x^2 + 6x$ we must add 3^2. To complete the square on $y^2 - 4y$ we must add $(-2)^2$. In order to maintain an equivalent equation, we must add these two quantities to the right side as well. We obtain the following:

$$(x^2 + 6x + 3^2) + (y^2 - 4y + (-2)^2) = -9 + 3^2 + (-2)^2$$
$$(x + 3)^2 + (y - 2)^2 = 4.$$

Hence the graph is a circle of radius 2 centered at $(-3, 2)$. ■

Example 4
Describe the graph of $4x^2 + 4y^2 - 12x - 27 = 0$.

Solution. As in Example 3 we first write

$$4x^2 - 12x + 4y^2 = 27.$$

Before completing the square we want to divide both sides by 4, so that the coefficients of x^2 and y^2 are 1. We get

$$x^2 - 3x + y^2 = \frac{27}{4}.$$

To complete the square on $x^2 - 3x$, we must add $(-\frac{3}{2})^2$. Since there is no y term, we need not complete the square on y^2. (It's already a perfect square.) We have

$$x^2 - 3x + \left(\frac{-3}{2}\right)^2 + y^2 = \frac{27}{4} + \left(\frac{-3}{2}\right)^2$$

$$\left(x - \frac{3}{2}\right)^2 + y^2 = 9.$$

Now $y^2 = (y - 0)^2$. So the graph is a circle of radius 3 centered at $(\frac{3}{2}, 0)$. ■

Example 5
Describe the graph of $x^2 + y^2 + 2x + 4y + 5 = 0$.

Solution. We first write

$$x^2 + 2x + y^2 + 4y = -5.$$

To complete the square on $x^2 + 2x$ we add 1^2, and to complete the square on $y^2 + 4y$ we add 2^2. We get the following:

$$x^2 + 2x + 1^2 + y^2 + 4y + 2^2 = -5 + 1^2 + 2^2$$

$$(x + 1)^2 + (y + 2)^2 = 0.$$

This equation can be true only when both $(x + 1)^2$ and $(y + 2)^2$ are zero (i.e., only when $x + 1 = 0$ and $y + 2 = 0$). Thus there is but one solution: $(-1, -2)$. The graph consists of a single point. ■

Example 6
Describe the graph of $x^2 + y^2 - 4x - 6y + 17 = 0$.

Solution. We first write

$$x^2 - 4x + y^2 - 6y = -17.$$

Upon completing the square, we have the following:

$$x^2 - 4x + (-2)^2 + y^2 - 6y + (-3)^2 = -17 + (-2)^2 + (-3)^2$$

$$(x - 2)^2 + (y - 3)^2 = -4.$$

Because the left side is always nonnegative, there are no solutions. Thus the solution set is $\varnothing$, and there is no graph. ■

Exercise
Describe the graph of each of the following.
(a) $x^2 + y^2 - 8x + 2y + 1 = 0$
(b) $x^2 + y^2 - 4y + 5 = 0$
(c) $x^2 + y^2 - 4x - 6y + 13 = 0$
(d) $4x^2 + 4y^2 - 4y - 15 = 0$

Answers. (a) circle of radius 4, centered at $(4, -1)$
(b) no graph
(c) the point $(2, 3)$
(d) circle of radius 2, centered at $\left(0, \frac{1}{2}\right)$ ■

We can conclude from the procedure used in Examples 3 through 6 that the graph of

$$Ax^2 + Ay^2 + Dx + Ey + F = 0,$$

if there is one, where A, D, E, and F are constants, with $A \neq 0$, is either a circle or a point. (We know that it's possible that there are no solutions and thus no graph.) By proceeding as we did in Examples 3, 4, 5, and 6, we can observe the constant on the right in the final equation and discover the precise situation. If that constant is positive, then the graph is a circle. If that constant is zero, then the graph is a point. If that constant is negative, then there is no graph.

Problem Set 2.5

In Problems 1–8, determine the equation of the circle described.

1. radius 3, center at $(2, -5)$

2. radius 4, center at $(-1, 3)$

3. radius $\frac{5}{2}$, center at origin

4. radius $\frac{1}{4}$, center at origin

5. radius $\frac{3}{2}$, center at $\left(\frac{5}{2}, \frac{5}{3}\right)$

6. radius $\frac{3}{4}$, center at $(0, -2)$

7. center at $(1, 1)$, passing through $(6, 5)$

8. center at $\left(3, \frac{-5}{2}\right)$, passing through $\left(-2, \frac{-1}{2}\right)$

In Problems 9–12, complete the square.

9. $x^2 + 6x$

10. $x^2 - 8x$

11. $x^2 - \frac{5x}{2}$

12. $x^2 + \frac{2x}{5}$

In Problems 13–22, describe the graph.

13. $x^2 + y^2 - 4x - 6y + 4 = 0$

14. $x^2 + y^2 - 2x - 8y - 8 = 0$

15. $x^2 + y^2 + 4x - 2y + 3 = 0$

16. $x^2 + y^2 + 10x + 2y + 21 = 0$

17. $4x^2 + 4y^2 - 4y - 15 = 0$

18. $9x^2 + 9y^2 + 12x - 5 = 0$

19. $36x^2 + 36y^2 - 36x + 168y + 124 = 0$

20. $16x^2 + 16y^2 - 32x + 24y + 41 = 0$

21. $x^2 + y^2 + 10x - 6y + 34 = 0$

22. $4x^2 + 4y^2 + 20x - 8y + 29 = 0$

23. Find the equation of the line tangent to the circle $x^2 + y^2 = 16$ at the point $(-2\sqrt{3}, 2)$.

24. Find the equation of the line tangent to the circle $4x^2 - 16x + 4y^2 + 20y = 59$ at the point $(5, \frac{3}{2})$.

25. Find the equation of the line containing the centers of $x^2 + y^2 - 4x + 6y = 7$ and $4x^2 + 4y^2 + 4x - 12y = 30$.

26. Let $P = (-5/2, 2)$ and $Q = (-3/2, 5)$. Find the equation of the circle which has $\overline{PQ}$ as a diameter.

2.6 Parabolas

A **parabola** is the set of all points in a plane equidistant from a fixed point and a fixed line (not containing the point). The fixed point is called the **focus** of the parabola, and the fixed line is called the **directrix.**

Theorem 2.9

The equation of the parabola with focus $(0, a)$ *and directrix* $y = -a$ *is*

$$y = \frac{x^2}{4a}.$$

Proof. Let (x, y) represent each point in the plane. Then (x, y) is on the parabola if and only if the distance between (x, y) and $(0, a)$ equals the distance between (x, y) and $y = -a$. See Figure 2.25. Now the distance from (x, y) to $(0, a)$ is

$$\sqrt{x^2 + (y - a)^2},$$

and the distance from (x, y) to $y = -a$ is

$$|y - (-a)|.$$

Thus (x, y) is on the parabola if and only if

$$|y - (-a)| = \sqrt{x^2 + (y - a)^2}.$$

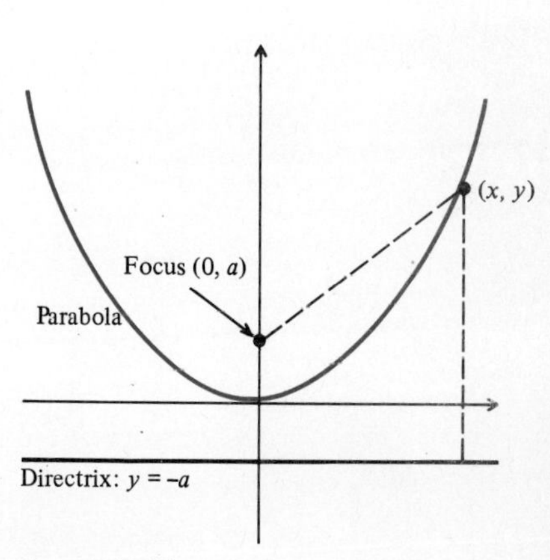

Figure 2.25

Squaring both sides, we get

$$(y + a)^2 = x^2 + (y - a)^2,$$

which is equivalent since $|y + a| \geq 0$. Simplifying, we have the following:

$$\begin{aligned} y^2 + 2ay + a^2 &= x^2 + y^2 - 2ay + a^2 \\ 4ay &= x^2 \\ y &= \frac{x^2}{4a}. \end{aligned}$$

■

The line passing through the focus, perpendicular to the directrix, is called the **axis** of the parabola. (In this case it is the y-axis.) The intersection of the parabola with its axis is called the **vertex.** (In this case it is the origin.) Note that the vertex is the midpoint of the perpendicular line segment from the focus to the directrix. The relationships among the focus, vertex, directrix, axis, and parabola are pictured in Figure 2.26. (Note that the focus is *not* on the parabola and that the parabola does *not* intersect its directrix.)

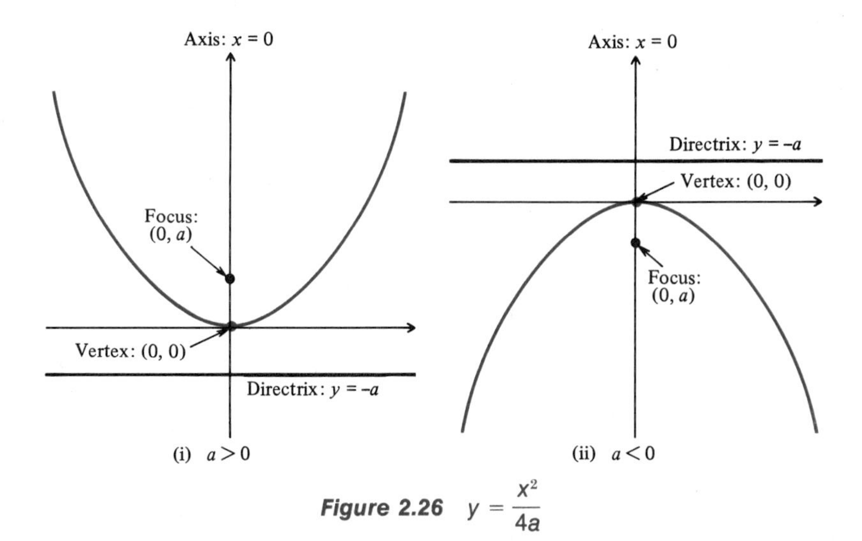

Figure 2.26 $y = \dfrac{x^2}{4a}$

A parabola opens up if and only if $a > 0$ and down if and only if $a < 0$. If $a > 0$, then the vertex will be the lowest point on the graph. If $a < 0$, then the vertex will be the highest point. Note that the distance between the focus and vertex and between the vertex and directrix is $|a|$.

Even though a parabola is defined in terms of its focus and directrix, the most evident characteristics of a parabola are its vertex and axis. Hence we state the following theorem.

Theorem 2.10

The equation of a parabola with vertex at the origin and axis vertical is

$$y = \frac{x^2}{4a},$$

where $|a|$ = *distance from vertex to the focus* = *distance from vertex to the directrix.*

Proof. Because the vertex is (0, 0) and the axis is vertical, the focus is (0, a). Similarly, the directrix is $y = -a$. By Theorem 2.9 the equation is

$$y = \frac{x^2}{4a}. \quad \blacksquare$$

In using Theorem 2.10 to get the equation of a parabola, we must calculate the constant a. This is seen in Examples 1 and 2.

Example 1
Determine the equation of the parabola with focus (0, 3) and vertex at the origin.

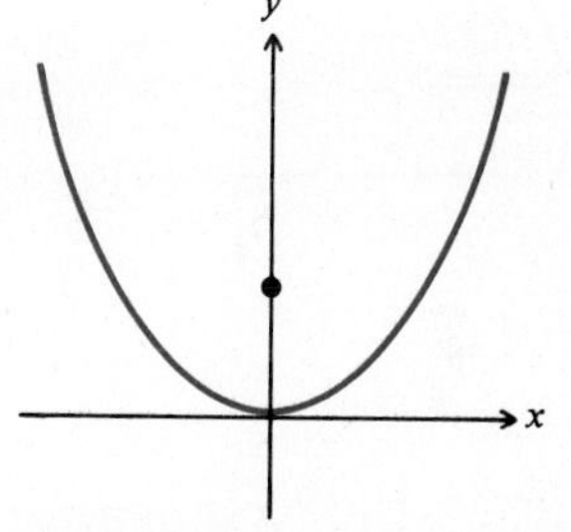

Figure 2.27

Solution. From a quick sketch of the graph (Figure 2.27), we see that such a parabola has a vertical axis and opens upward. Thus the general equation is $y = x^2/(4a)$, where a is positive. Since the distance from the focus (0, 3) to the vertex (0, 0) is 3, we have

$$a = 3.$$

Hence the desired equation is

$$y = \frac{x^2}{12}. \quad \blacksquare$$

Example 2
Determine the equation of the parabola with directrix $y = 4$ and vertex at the origin.

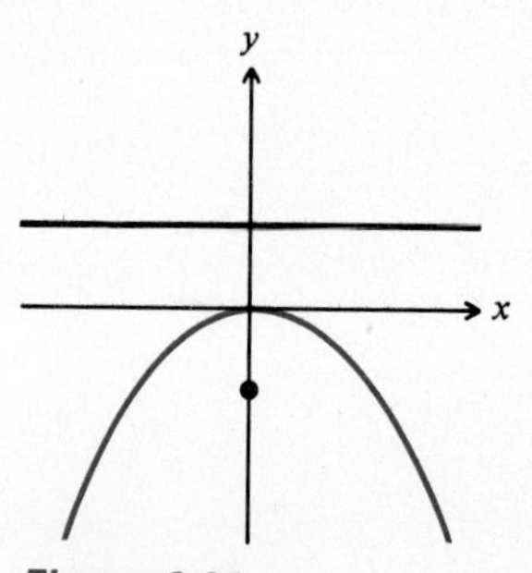

Figure 2.28

Solution. From a quick sketch of the graph (Figure 2.28), we see that such a parabola has a vertical axis and opens downward. Thus the general equation is $y = x^2/(4a)$, where a is negative. Since the distance from $y = 4$ to the origin is 4, we have

$$a = -4.$$

Hence the desired equation is

$$y = \frac{x^2}{-16}. \qquad \blacksquare$$

Now the vertex of a parabola doesn't have to be at the origin. Thus Theorem 2.11 takes into account other such vertices.

Theorem 2.11

The equation of a parabola with vertex (h, k) and axis vertical is

$$y - k = \frac{(x - h)^2}{4a},$$

where $|a|$ = distance from the vertex to the focus = distance from the vertex to the directrix.

Proof. Suppose that the parabola opens upward. Then a is positive, the focus is $(h, k + a)$, and the directrix is $y = k - a$, as seen in Figure 2.29.

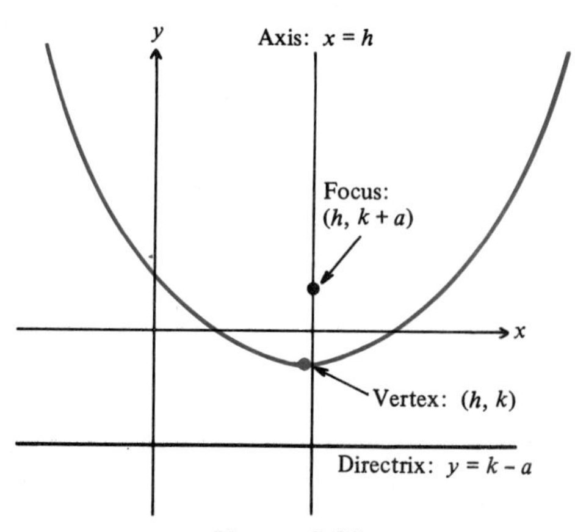

Figure 2.29

Let (x, y) represent each point in the plane. Then (x, y) is on the parabola if and only if the distance between (x, y) and $(h, k + a)$ equals the distance between (x, y) and $y = k - a$. Now the distance from (x, y) to $(h, k + a)$ is

$$\sqrt{(x - h)^2 + [y - (k + a)]^2},$$

and the distance from (x, y) to $y = k - a$ is

$$|y - (k - a)|.$$

Thus (x, y) is on the parabola if and only if

$$|y - (k - a)| = \sqrt{(x - h)^2 + [y - (k + a)]^2}.$$

Squaring both sides, we get

$$y^2 - 2y(k - a) + (k - a)^2 = (x - h)^2 + y^2 - 2y(k + a) + (k + a)^2,$$

which is equivalent since $|y - (k - a)| \geq 0$. Simplifying, we have the following:

$$\begin{aligned} y^2 - 2ky + 2ay + k^2 - 2ka + a^2 & \\ = (x - h)^2 + y^2 - 2ky - 2ay + k^2 + 2ka + a^2 & \\ 4ay - 4ka = (x - h)^2 & \\ 4a(y - k) = (x - h)^2 & \\ y - k = \frac{(x - h)^2}{4a}. & \end{aligned}$$

If the parabola opens downward, then we may choose a to be negative and still represent the focus as $(h, k + a)$ and the directrix as $y = k - a$. Hence we will get the same equations. ■

In using Theorem 2.11 to get the equation of a parabola, we must calculate the vertex and the constant a. This is seen in Example 3.

Example 3
Determine the equation of the parabola with focus (2, 3) and directrix $y = -1$.

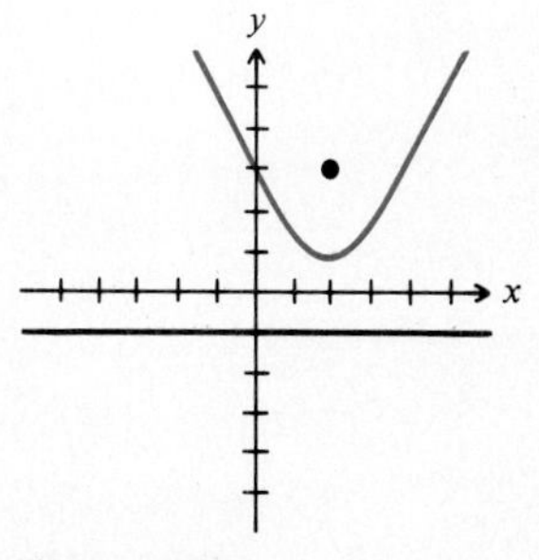

Figure 2.30

Solution. From a quick sketch of the graph (Figure 2.30), we see that such a parabola opens upward. Thus the general equation is $y - k = (x - h)^2/(4a)$, where a is positive. The vertex, being the midpoint of the segment from (2, 3) to (2, −1), is (2, 1). Since the distance between the focus and the vertex is $|a|$, we have $|a| = 2$. Thus $a = 2$ since a is positive. So by Theorem 2.11, the equation is

$$y - 1 = \frac{(x - 2)^2}{8}. \quad ■$$

Exercise
Determine the equation of the parabola described.
(a) vertex at the origin and focus (0, −3)
(b) vertex at the origin and directrix $y = -2$
(c) focus (3, −5) and directrix $y = 1$

Answers. (a) $y = \dfrac{x^2}{-12}$ (b) $y = \dfrac{x^2}{8}$ (c) $y + 2 = \dfrac{(x - 3)^2}{-12}$ ■

Theorem 2.12 tells us what the equation will be when the axis is horizontal instead of vertical. The proof is analogous to that of Theorem 2.11.

Theorem 2.12

The equation of the parabola with vertex (h, k) and axis horizontal is

$$x - h = \frac{(y - k)^2}{4a},$$

where $|a|$ = distance from vertex to focus = distance from vertex to directrix.

Although the vertex is still (h, k), the focus, axis, and directrix are different. See Figure 2.31.

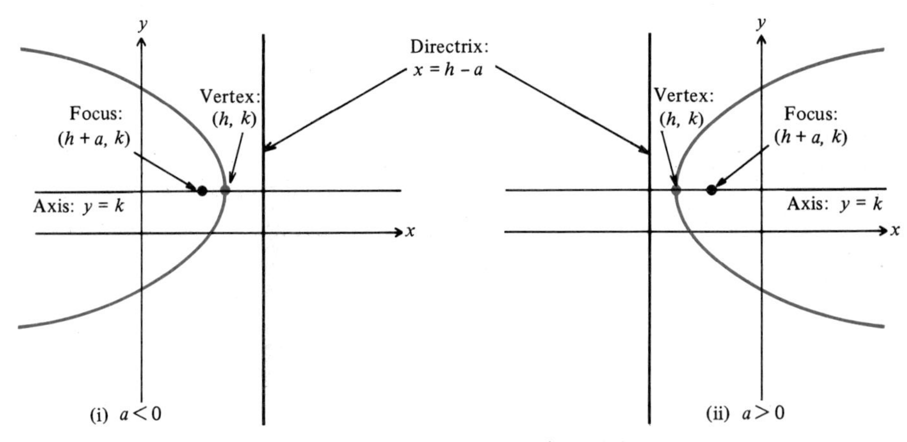

Figure 2.31 $x - h = \frac{(y - k)^2}{4a}$

This parabola opens to the right if and only if $a > 0$ and to the left if and only if $a < 0$. The vertex will be the leftmost (if $a > 0$) or rightmost (if $a < 0$) point on the graph.

Example 4
Determine the equation of the parabola with focus $(-1, 2)$ and vertex $(2, 2)$.

Solution. The first thing to determine is whether the general equation is

$$y - k = \frac{(x - h)^2}{4a} \quad \text{or} \quad x - h = \frac{(y - k)^2}{4a}.$$

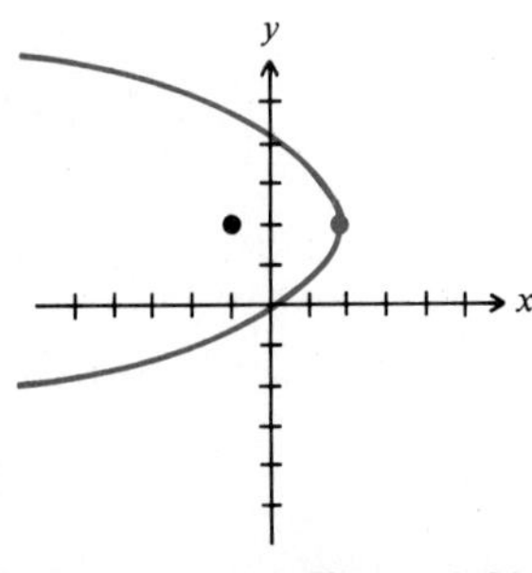

Figure 2.32

To this end, it is best to sketch the graph, as in Figure 2.32. Since the parabola has a horizontal axis and opens to the left, we see that the general equation is $x - h = (y - k)^2/(4a)$, where a is negative. Since the distance between the focus and the vertex is $|a|$, we have $|a| = 3$. Thus $a = -3$. Since the vertex is (2, 2), we have, by Theorem 2.12,

$$x - 2 = \frac{(y - 2)^2}{-12}. \qquad \blacksquare$$

Note that if $h = 0$ and $k = 0$, the equation of Theorem 2.12 becomes

$$x = \frac{y^2}{4a}.$$

This, then, is the equation of a parabola with vertex at the origin and axis horizontal. Similarly, we may note that Theorem 2.10 is just a special case of Theorem 2.11.

Example 5
Determine the equation of the parabola with directrix $x = 2$ and vertex at the origin.

Solution. From a quick sketch of the graph (Figure 2.33), we see that the general equation is $x = y^2/(4a)$, where a is negative. Since the distance from the vertex to the directrix is 2, we have $a = -2$. Hence the equation is

$$x = \frac{y^2}{-8}. \qquad \blacksquare$$

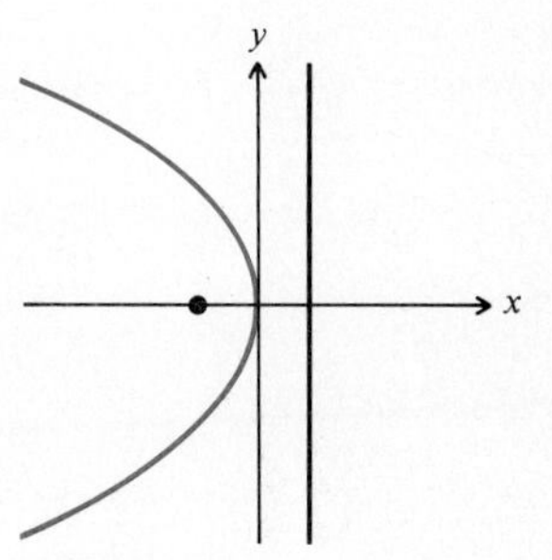

Figure 2.33

Example 6
Determine the equation of the parabola passing through $(1, -3)$ with vertex $(-3, -1)$ and axis vertical.

Solution. From a rough sketch of the graph (Figure 2.34), we see that the general equation is $y - k = (x - h)^2/(4a)$, where a is negative. Because the vertex is $(-3, -1)$, the equation is

$$y + 1 = \frac{(x + 3)^2}{4a}. \tag{2-8}$$

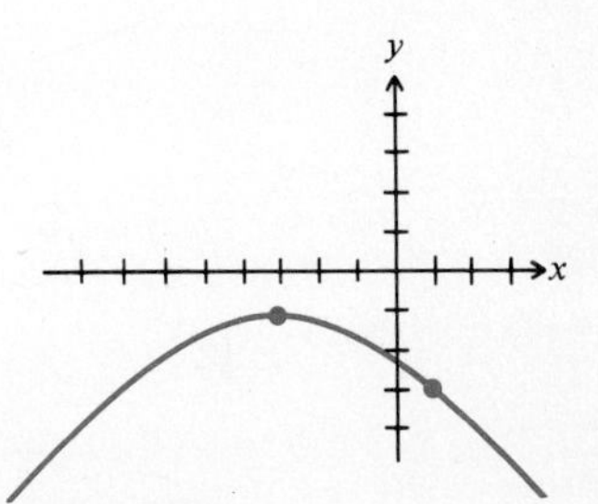

Figure 2.34

All that remains is to determine the constant a. Because $(1, -3)$ is on the parabola, it must be a solution of equation (2-8). Therefore,

$$-3 + 1 = \frac{(1 + 3)^2}{4a}.$$

Solving for a we get $a = -2$. Thus the equation is

$$y + 1 = \frac{(x + 3)^2}{-8}. \qquad \blacksquare$$

Exercise
Determine the equation of the parabola described.
(a) with focus (4, 3) and directrix $x = 1$
(b) with focus $(-2, -3)$ and vertex $(-2, 1)$
(c) passing through $(4, -8)$ with vertex $(1, -2)$ and axis horizontal

Answers. (a) $x - \frac{5}{2} = \frac{(y - 3)^2}{6}$ (b) $y - 1 = \frac{(x + 2)^2}{-16}$

(c) $x - 1 = \frac{(y + 2)^2}{12}$ ■

Now the graph of

$$y = Ax^2 + Bx + C$$

and

$$x = Ay^2 + By + C,$$

where A, B, and C are constants with $A \neq 0$, is a parabola. The vertex and graph are determined by completing the square, as illustrated in Example 7.

Example 7
Determine the vertex of the parabola $y = 2x^2 - 12x + 17$ and graph.

Solution. Subtracting 17 from both sides, we have

$$y - 17 = 2x^2 - 12x. \qquad (2\text{-}9)$$

Now we want to complete the square on the right side. Since this is more easily accomplished when the coefficient of x^2 is 1, we'll factor out a 2. We get

$$y - 17 = 2(x^2 - 6x).$$

To complete the square on the binomial $x^2 - 6x$, we must add 9 $[(-6/2)^2 = 9]$. Because we will be adding 9 within a set of parentheses which is being multiplied by 2, we are adding 18 to the right side of the equation. We must add 18 to the left side in order to obtain the following equivalent equations:

$$y - 17 + 18 = 2(x^2 - 6x + 9)$$
$$y + 1 = 2(x - 3)^2. \qquad (2\text{-}10)$$

Thus the vertex is $(3, -1)$. [Note that the parabola opens up because the coefficient of $(x - 3)^2$ is positive.] To graph we will simply plot some additional points by choosing some values for x and finding the corresponding values for y. To this end, it might be easiest to write equation (2-10) as

$$y = 2(x - 3)^2 - 1.$$

Our table of solutions:

x	0	1	2	3	4	5	6
y	17	7	1	−1	1	7	17

The graph is in Figure 2.35. ■

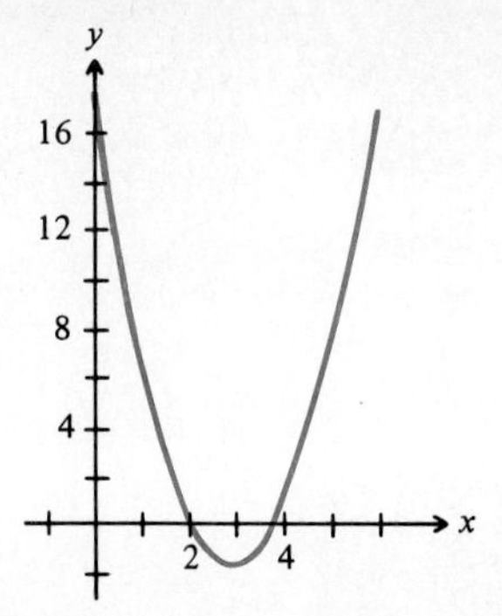

Figure 2.35 $y = 2x^2 - 12x + 17$

In Example 7 note that we located solutions on each side of the vertex and that when $x = 4$, 5, or 6, we got the same y-value as when $x = 2$, 1, or 0, respectively. This is because the axis of a parabola separates it into two halves, each of which is the mirror image of the other. In this sense we say that a parabola is **symmetric** about its axis. Also note that in completing the square on the right side of equation (2-9), we chose not to divide by 2. Although this would have made the coefficient of x equal 1, we would also have had to divide the left side by 2, thus making the coefficient of y equal $\frac{1}{2}$. We want its coefficient to remain 1.

Exercise
Determine the vertex of the parabola $y = 3x^2 - 6x + 1$ and graph.

Answer. Vertex $(1, -2)$.

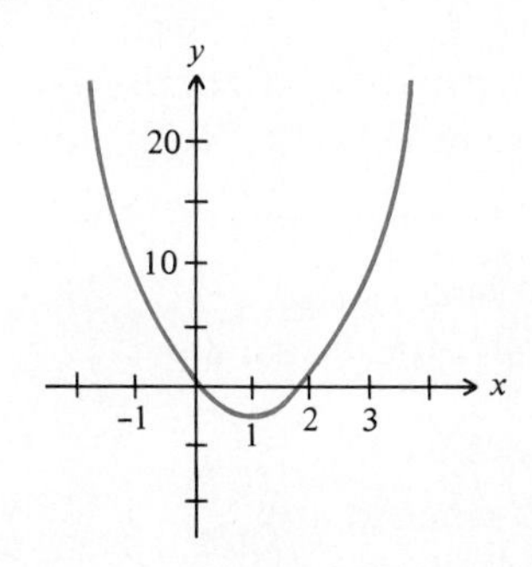

■

Examples 8 and 9 illustrate important applications of parabolas to applied problems.

Example 8
Suppose that the cost C, in dollars, of producing x items is given by $C = 800 + 3x + 0.01x^2$. How many should be manufactured to maximize profit if the selling price of each item is \$10? What is the maximum profit?

Solution. We begin by letting

$$P = \text{profit (in dollars)}.$$

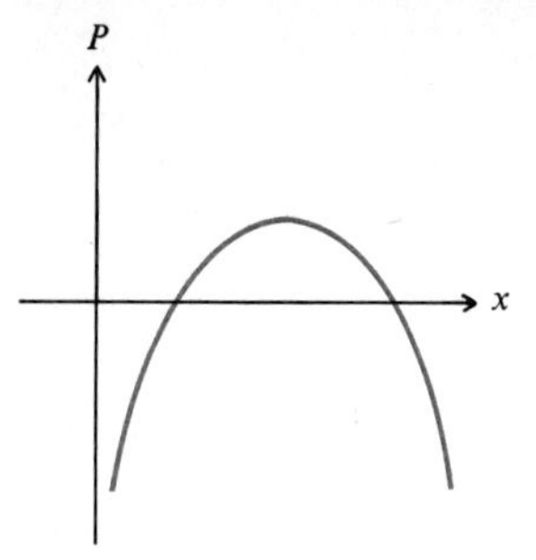

Figure 2.36

Now the profit will be total revenue (selling price times the number of items sold) minus cost. Thus

$$\begin{aligned} P &= 10x - C \\ P &= 10x - (800 + 3x + 0.01x^2) \\ P &= -0.01x^2 + 7x - 800. \end{aligned} \qquad (2\text{-}11)$$

Note that if we relabel the y-axis as the P axis and sketch the graph of equation (2-11), we will have a parabola that opens downward (see Figure 2.36). Hence the maximum value of P occurs at the vertex, which must be found.

$$\begin{aligned} P + 800 &= -0.01(x^2 - 700x) \\ P + 800 - 0.01(350)^2 &= -0.01(x^2 - 700x + 350^2) \\ P - 425 &= -0.01(x - 350)^2. \end{aligned}$$

The vertex is (350, 425). Hence 350 should be produced to maximize profit, and the maximum profit is \$425. ■

Example 9

Suppose that a projectile is fired from the origin so that it reaches a maximum altitude of $32\sqrt{3}$ feet, 16 feet down range (horizontal distance). What is the equation of its parabolic path?

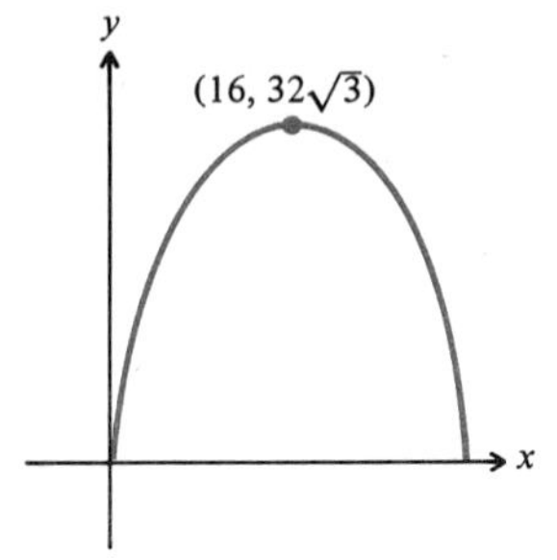

Figure 2.37

Solution. We have the top part of a parabola which passes through the origin and has a vertex of $(16, 32\sqrt{3})$. See Figure 2.37. The general equation of such a parabola is

$$y - 32\sqrt{3} = \frac{(x - 16)^2}{4a},$$

where $a < 0$. Since $y = 0$ when $x = 0$,

$$-32\sqrt{3} = \frac{(-16)^2}{4a}$$

$$4a = \frac{256}{-32\sqrt{3}} = \frac{-8}{\sqrt{3}}.$$

Thus the equation is

$$y - 32\sqrt{3} = \frac{-\sqrt{3}(x - 16)^2}{8}.$$ ■

Problem Set 2.6

In Problems 1–22, determine the equation of the parabola described.

1. focus (0, 3), directrix $y = -3$
2. focus (0, −2), directrix $y = 2$
3. vertex (0, 0), focus (0, −5)
4. vertex (0, 0), directrix $y = -4$
5. vertex (2, 8), directrix $y = 11$
6. vertex (−1, 1), directrix $y = -1$
7. vertex (1, −4), directrix $x = \frac{-1}{3}$
8. vertex (−2, −3), directrix $x = 0$
9. vertex $\left(\frac{-1}{2}, -2\right)$, focus (−3, −2)
10. vertex (0, 3), focus (4, 3)
11. vertex (3, 3), focus (3, 2)
12. vertex (1, −1), focus $\left(1, \frac{2}{3}\right)$
13. focus (−1, 3), directrix $y = \frac{31}{5}$
14. focus (2, 1), directrix $x = -4$
15. focus (−5, −3), directrix $x = -2$
16. focus (2, 5) directrix $y = -5$
17. passing through (6, 3) with vertex at the origin and axis vertical
18. passing through $\left(4, \frac{8}{3}\right)$ with vertex at the origin and axis vertical
19. passing through (−7, 9) with vertex (−1, 3) and axis vertical
20. passing through (−1, 6) with vertex (−3, 2) and axis horizontal
21. passing through $\left(\frac{-7}{4}, 0\right)$ with vertex (−1, −3) and axis horizontal
22. passing through $\left(3, \frac{18}{5}\right)$ with vertex (5, 2) and axis vertical

In Problems 23–30, determine the vertex and graph.

23. $y = x^2 - 4x + 1$
24. $x = -y^2 - 2y + 1$
25. $x = 2y^2 + 16y + 31$
26. $y = -4x^2 + 24x - 34$
27. $y = \frac{1}{4}x^2 - 2x + 6$
28. $x = \frac{2}{3}y^2 + 4y + 7$
29. $x = \frac{-1}{2}y^2 - 2y + 1$
30. $y = \frac{-7}{9}x^2 + \frac{14}{3}x - \frac{17}{2}$

31. Determine the focus and directrix of the parabola whose equation is $y = 4x^2 - 24x + 35$.
32. Determine the focus and directrix of the parabola whose equation is

$$x = \frac{-1}{4}y^2 - \frac{1}{2}y + \frac{7}{4}.$$

33. Let $P = (0, -2)$ and $Q = (4, 1)$. Find the equation of the parabola whose focus is the midpoint of $\overline{PQ}$ and whose vertex is the center of the circle $x^2 + y^2 - 4x + 6y = -9$.
34. Let $P = (4/5, -3)$ and $Q = (-3/5, 5)$. Find the equation of the parabola which has a horizontal axis, passes through the midpoint of $\overline{PQ}$, and whose vertex is the center of the circle $4x^2 + 4y^2 - 4x - 16y = 3$.
35. A radio manufacturer determines that in order to sell x radios, the selling price p, in dollars, must be $p = 60 - 0.01x$. If the cost C, in dollars, of manufacturing x radios is $C = 0.004x^2 + 25x + 4000$, how many should be produced to maximize profit?
36. A manufacturer of a certain item determines that in order to sell x units of his product, the selling price p, in dollars, must be $p = 100 - 0.01x$. If the total cost C, in dollars, of producing x units is $C = 50x + 10{,}000$, how many units must be produced to maximize profit?
37. The records of a retail store indicate that if the retail price of a certain book is p, in dollars, $500 - 100p$ of them are sold. If the cost to the store is \$2 per book, what price maximizes profit?
38. A teapot manufacturer has fixed costs of \$500 per week and variable costs of \$8 per teapot. Records indicate that if the retail price is p, in dollars, then $400 - 20p$ are sold. What price maximizes profit?
39. Suppose that a farmer has 1000 feet of fencing and wishes to fence off a rectangular plot next to a river. What is the maximum area that he can enclose if the side next to the river doesn't need fence?
40. Find the point on the line $x + 3y = 5$ that is closest to the origin.
41. The equation of the path of a projectile fired from the origin with an initial speed of 64 ft/sec at an angle of 45° to the x-axis is

$$y = x - \frac{x^2}{128},$$

where x and y are in feet. What is the maximum height attained by the projectile? How far down range (horizontal distance) is it when it reaches its maximum altitude?
42. If an object is thrown from the origin with an initial speed of 128 ft/sec at angle of 60° to the x-axis, the equation of its path is

$$y = \sqrt{3}x - \frac{x^2}{256},$$

where x and y are in feet. What is the maximum altitude of the object? What is the range (total horizontal distance traveled) of the object?
43. Suppose that an object is thrown from the origin so that it reaches a maximum height of 72 feet, 144 feet down range. What is the parabolic equation of its path?
44. Suppose that a projectile is fired from the origin so that its maximum altitude is

768 feet and its range is $1024\sqrt{3}$ feet. What is the parabolic equation of its path?

45. If a projectile is fired from the origin at an angle of 30° to the x-axis, the equation of its path is

$$y = \frac{x}{\sqrt{3}} - \frac{64x^2}{3v_0^2},$$

where x and y are in feet, and v_0 is the initial velocity in ft/sec. What must the initial velocity be in order for the maximum height to be 36 feet?

46. If a golfer hits a golf ball at an angle of 45° to the ground, and we assume that the ball is resting on level ground at the origin and that it is hit in the positive x direction, the equation of its path is

$$y = x - \frac{32x^2}{v_0^2},$$

where x and y are in feet, and v_0 is the initial velocity in ft/sec. What must the initial velocity be in order to hit the front edge of a 15-foot elevated green, 80 yards away?

47. If a projectile is fired from the origin at an angle of 0° to the x-axis, the equation of its path is

$$y = \frac{-16x^2}{v_0^2},$$

where x and y are in feet, and v_0 is the initial velocity in ft/sec. If the initial velocity of a bullet leaving a rifle is 3000 ft/sec, how far will it drop in 50 yards?

48. Water squirts from the end of a horizontal pipe 18 feet above the ground. At a point 6 feet below the pipe, the stream of water is 8 feet beyond the end of the pipe. What horizontal distance does the stream travel before it strikes the ground? (The path is parabolic.)

If an object is thrown upward, then its distance above ground s, in feet, after t seconds, is given by

$$s = -16t^2 + v_0t + s_0,$$

where v_0 is the initial velocity (velocity when $t = 0$) and s_0 is the initial position (distance above ground when $t = 0$). This information is needed for Problems 49 and 50.

49. If an object is thrown upward from ground level with an initial velocity of 48 ft/sec, what is the maximum altitude attained, and how long does it take to reach this maximum height?

50. If an object is thrown upward from a 48-foot building with an initial velocity of 32 ft/sec, what is the maximum altitude attained, and how long does it take the object to hit the ground?

The parabola has an important optical property. Consider a cup-shaped mirror with parabolic cross sections. If a light source is placed at the focus, the reflected rays are parallel to the axis. Further, if parallel light rays strike a parabolic mirror, they will be directed to the focus of the parabola. This is the basis of the design of reflecting and radio telescopes. This information is needed for Problems 51–54.

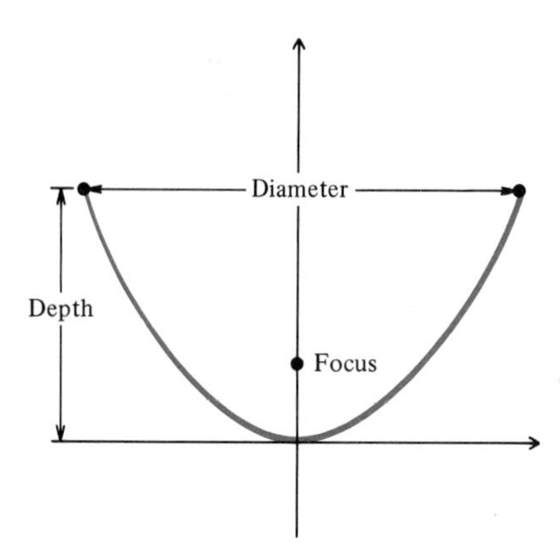

51. If the diameter of a parabolic mirror is 10 inches and if the mirror is 5 inches deep at the center, how far from the center (vertex) should the light source be placed so that reflected rays are parallel?

52. A headlight is a parabolic mirror. If the light source is $\frac{3}{4}$ of an inch from the center, and the mirror is 8 inches in diameter, how deep is it?

53. If the parabolic mirror of a reflecting telescope is 6 feet in diameter and 3 inches deep, how far from the center are the light rays focused? (The rays may be assumed to be parallel.)

54. If the parabolic reflector of a radio telescope is 20 feet deep and 100 feet in diameter, how far from the center is the antenna located? (The antenna is located at the focus.)

A suspended cable tends to hang in the shape of a parabola if the weight of the cable is small compared to the weight it supports, and if the weight is uniformly distributed along the cable. Bridges supported in this fashion are called suspension bridges. All of the larger bridges of the world are suspension bridges. For such bridges, the horizontal distance between the main supports is called the span, and the vertical distance between the points where the cables are attached to the supports and the lowest point of the cable is called the sag. This information is needed in Problems 55–57.

55. The George Washington Bridge across the Hudson River from New York to New Jersey has a span of 3500 feet and a sag of 316 feet. Find the length of a vertical supporting cable which is 875 feet from the center if the roadway is 12 feet below the lowest point of the cable.

56. A certain suspension bridge has a span of 400 meters. The cable is 100 meters above the roadway at the supports and 4 meters above the roadway at the center. Find the length of a supporting cable that is 150 meters from the support.

57. Assuming that the lowest point is at the origin, determine the equation of the cable of a suspension bridge with a span of a and a sag of b.

2.7 Ellipses

An **ellipse** is the set of all points in a plane, the sum of whose distances to two fixed points is constant. The fixed points are called the **foci** of the ellipse.

Theorem 2.13

The equation of the ellipse with foci $(c, 0)$ *and* $(-c, 0)$ *and constant sum* $2a$ *(where* $0 < c < a$*) is*

$$\frac{x^2}{a^2} + \frac{y^2}{b^2} = 1,$$

where $b^2 = a^2 - c^2$.

Proof. Let (x, y) represent each point in the plane. Then (x, y) will be on the ellipse if and only if the distance from (x, y) to $(c, 0)$ plus the distance from (x, y) to $(-c, 0)$ is $2a$. See Figure 2.38. Now the distance between (x, y) and $(c, 0)$ is

$$\sqrt{(x - c)^2 + y^2},$$

and the distance between (x, y) and $(-c, 0)$ is

$$\sqrt{(x + c)^2 + y^2}.$$

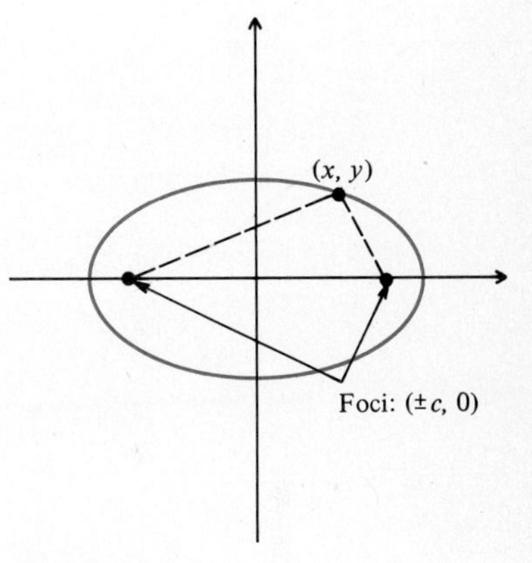

Figure 2.38

Thus (x, y) is on the ellipse if and only if

$$\sqrt{(x - c)^2 + y^2} + \sqrt{(x + c)^2 + y^2} = 2a.$$

After taking the second radical to the right and squaring, we have the following:

$$\begin{aligned} (x - c)^2 + y^2 &= 4a^2 - 4a\sqrt{(x + c)^2 + y^2} + (x + c)^2 + y^2 \\ -4cx - 4a^2 &= -4a\sqrt{(x + c)^2 + y^2} \\ cx + a^2 &= a\sqrt{(x + c)^2 + y^2}. \end{aligned}$$

Upon squaring both sides again, we have

$$\begin{aligned} c^2x^2 + 2a^2cx + a^4 &= a^2((x + c)^2 + y^2) \\ c^2x^2 + 2a^2cx + a^4 &= a^2x^2 + 2a^2cx + a^2c^2 + a^2y^2 \\ c^2x^2 - a^2x^2 - a^2y^2 &= a^2c^2 - a^4. \end{aligned}$$

After multiplying both sides by -1 and factoring, we get

$$x^2(a^2 - c^2) + a^2y^2 = a^2(a^2 - c^2).$$

If we let $b^2 = a^2 - c^2$ and then divide both sides by a^2b^2, we get

$$\frac{x^2}{a^2} + \frac{y^2}{b^2} = 1. \tag{2-12}$$

Because these steps are reversible, (x, y) is on the ellipse if and only if it is a solution of equation (2-12). ■

When $y = 0$, equation (2-12) gives

$$\frac{x^2}{a^2} = 1$$

$$x^2 = a^2$$

$$x = \pm a.$$

Thus such an ellipse has x-intercepts of a and $-a$. Similarly, the y-intercepts are b and $-b$. Since $b^2 = a^2 - c^2$, $b^2 < a^2$, and thus the distance between the x-intercepts is greater than the distance between the y-intercepts. If the foci are on the y-axis as $(0, c)$, and $(0, -c)$, with the constant distance still $2a$, then the equation is

$$\frac{x^2}{b^2} + \frac{y^2}{a^2} = 1, \tag{2-13}$$

and the distance between the y-intercepts is greater than the distance between the x-intercepts. Equations (2-12) and (2-13) are called the **standard form** of the equation of an ellipse.

Hence given an equation of the form

$$\frac{x^2}{k_1} + \frac{y^2}{k_2} = 1,$$

where k_1 and k_2 are positive constants with $k_1 \neq k_2$, we know that the graph is an ellipse with the foci on the x-axis if the first denominator is larger and on the y-axis if the second denominator is larger. (If the denominators are equal, then the graph is a circle.)

Example 1
Graph each of the following.

(a) $\dfrac{x^2}{16} + \dfrac{y^2}{9} = 1$ (b) $\dfrac{x^2}{9} + \dfrac{y^2}{16} = 1$

Solution. (a) The x-intercepts are ± 4 and the y-intercepts are ± 3. These four values give us a reasonable sketch, and we could plot additional points for more accuracy. For instance, when $x = \pm 2$, $y = \pm 3\sqrt{3}/2 \cong \pm 2.6$. The graph is shown in Figure 2.39.

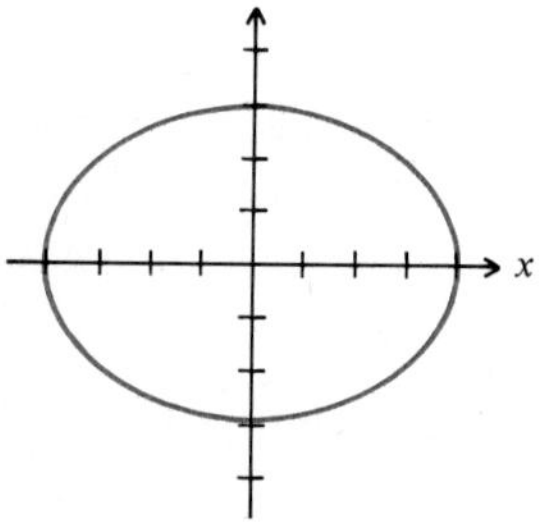

Figure 2.39 $\dfrac{x^2}{16} + \dfrac{y^2}{9} = 1$

(b) The graph, shown in Figure 2.40, is an ellipse with x-intercepts ± 3 and y-intercepts ± 4. ■

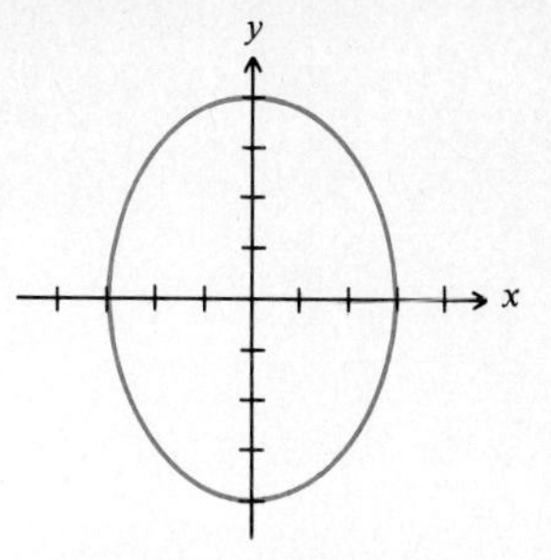

Figure 2.40 $\frac{x^2}{9} + \frac{y^2}{16} = 1$

The graph of

$$Ax^2 + By^2 = C,$$

where A, B, and C are constants with like signs, and $A \neq B$, is an ellipse. This is seen upon dividing by C. We get

$$\frac{Ax^2}{C} + \frac{By^2}{C} = 1,$$

which is equivalent to

$$\frac{x^2}{C/A} + \frac{y^2}{C/B} = 1.$$

Since A, B, and C have like signs, the denominators of this equation are positive. And because $A \neq B$, the denominators are unequal. Thus this equation is the standard form of an ellipse.

Example 2
Put $4x^2 + 9y^2 = 25$ in standard form and then graph.

Solution. Upon dividing by 25, we have

$$\frac{4x^2}{25} + \frac{9y^2}{25} = 1.$$

But this is equivalent to the following:

$$\frac{x^2}{\frac{25}{4}} + \frac{y^2}{\frac{25}{9}} = 1$$

$$\frac{x^2}{(\frac{5}{2})^2} + \frac{y^2}{(\frac{5}{3})^2} = 1.$$

This is the standard form with $a = \frac{5}{2}$ and $b = \frac{5}{3}$. The graph is seen in Figure 2.41. ■

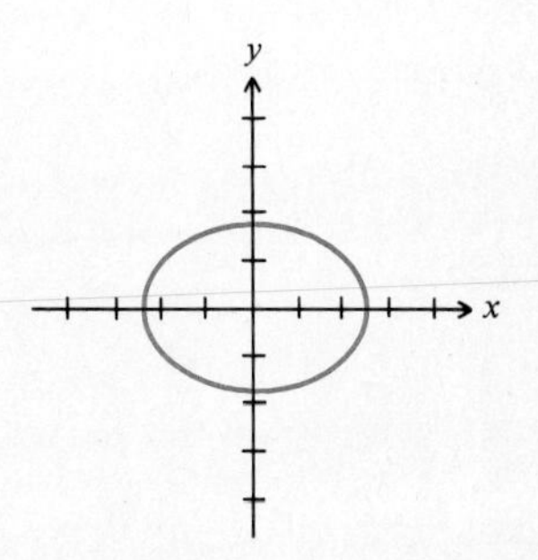

Figure 2.41 $4x^2 + 9y^2 = 25$

Exercise
Put $9x^2 + 16y^2 = 25$ in standard form and then graph.

Answer. Standard form: $\dfrac{x^2}{(\frac{5}{3})^2} + \dfrac{y^2}{(\frac{5}{4})^2} = 1.$

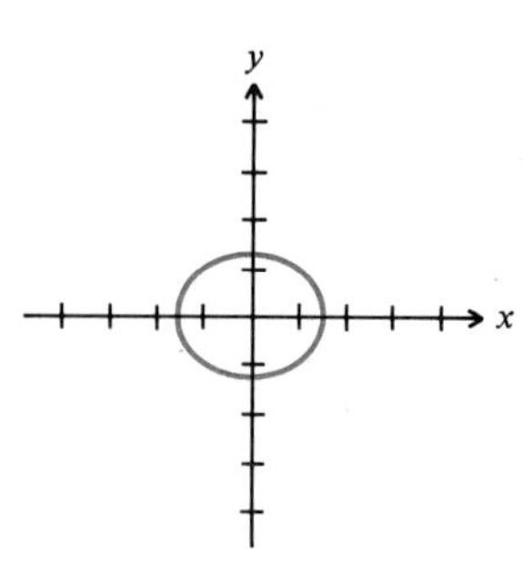

■

Example 3

Determine the equation (in standard form) of the ellipse whose foci are $(\pm 3\sqrt{2}, 0)$ and whose constant sum is 10.

Solution. Since the foci are on the x-axis, the general equation is

$$\frac{x^2}{a^2} + \frac{y^2}{b^2} = 1.$$

Now $2a = 10$, so $a = 5$. And it remains only to find b^2. Since $b^2 = a^2 - c^2$, we have

$$b^2 = 5^2 - (3\sqrt{2})^2 = 25 - 18 = 7.$$

Hence the equation is

$$\frac{x^2}{25} + \frac{y^2}{7} = 1.$$

■

Exercise

Determine the equation (in standard form) of the ellipse whose foci are $(\pm 2\sqrt{2}, 0)$ and whose constant sum is 12.

Answer. $\dfrac{x^2}{36} + \dfrac{y^2}{28} = 1$ ■

The following example illustrates a practical application of ellipses.

Figure 2.42

Example 4

The arch of a bridge is in the shape of a semiellipse with a horizontal span of 50 feet and a height of 10 feet at the center. How high is the arch 5 feet to the right of the center?

Solution. Placing the semiellipse in the plane as shown in Figure 2.42, we see that it is the upper half of an ellipse with x-intercepts of ± 25 and y-intercepts of ± 10. Thus the equation of this semiellipse is

$$\frac{x^2}{25^2} + \frac{y^2}{10^2} = 1,$$

where $y \geq 0$ (upper half). To find the height 5 feet to the right of the center we must find the y-value when $x = 5$. We have

$$\frac{5^2}{25^2} + \frac{y^2}{10^2} = 1$$

$$\frac{1}{25} + \frac{y^2}{100} = 1$$

$$4 + y^2 = 100$$

$$y^2 = 96$$

$$y = 4\sqrt{6} \qquad \text{(since } y \geq 0\text{)}.$$

So the height at the desired point is $4\sqrt{6}$ (≈ 9.8) feet. ■

Problem Set 2.7

In Problems 1–6, graph the equation.

1. $\frac{x^2}{9} + \frac{y^2}{4} = 1$
2. $\frac{x^2}{16} + \frac{y^2}{4} = 1$
3. $\frac{x^2}{4} + \frac{y^2}{25} = 1$
4. $x^2 + \frac{y^2}{9} = 1$
5. $\frac{x^2}{6} + \frac{y^2}{4} = 1$
6. $\frac{x^2}{9} + \frac{y^2}{9} = 1$

In Problems 7–12, put the equation in standard form and then graph.

7. $16x^2 + 49y^2 = 196$
8. $16x^2 + 25y^2 = 100$
9. $49x^2 + 36y^2 = 441$
10. $25x^2 + 9y^2 = 25$
11. $5x^2 + 8y^2 = 30$
12. $25x^2 + 12y^2 = 75$

In Problems 13–22, determine the equation (in standard form) of the ellipse with the given properties.

13. foci: $(\pm 2\sqrt{5}, 0)$, sum of distances from foci is 12
14. foci: $(0, \pm 4\sqrt{3})$, sum of distances from foci is 14
15. foci: $(\pm 2\sqrt{3}, 0)$, x-intercepts: ± 4
16. foci: $(\pm 4, 0)$, y-intercepts: ± 3
17. foci: $(0, \pm 4)$, y-intercepts: ± 5
18. foci: $(0, \pm \sqrt{11})$, x-intercepts: ± 1

19. passing through $\left(\frac{\sqrt{15}}{2}, 2\right)$, y-intercepts: ± 4

20. passing through $\left(-4, \frac{3\sqrt{10}}{5}\right)$, x-intercepts: ± 5

21. passing through (3, 2) and $(\sqrt{15}, 0)$.

22. passing through $\left(\frac{2\sqrt{15}}{3}, -2\right)$ and $(0, -3)$

23. What are the foci of an ellipse whose equation is $5x^2 + 9y^2 = 180$?
24. What are the foci of an ellipse whose equation is $16x^2 + 9y^2 = 36$?

If the ends of a piece of string are fixed and a pencil is used to draw the string tight, then the figure traced while moving the pencil and keeping the string taut is an ellipse. This is because the sum of the distances from the pencil tip to the two fixed points is a constant—the length of the string. This information is needed for Problems 25 and 26.

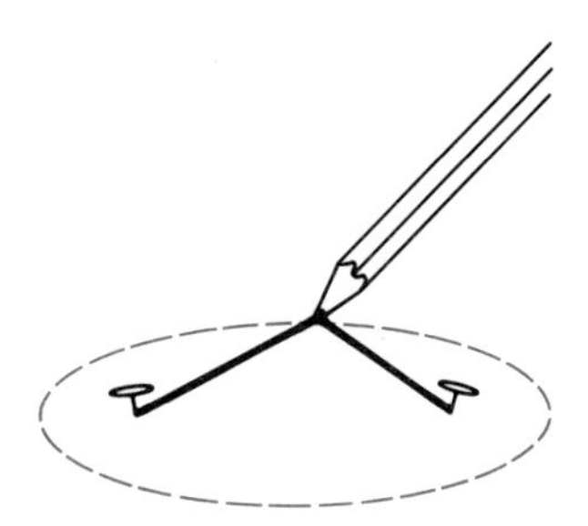

25. A rectangular piece of wood 8 feet by 4 feet is to be cut to form an ellipse. If we want to use the full length and width of the board, where should the foci be located so that with pencil and string we can outline the border? How long should the string be?
26. How long must a piece of string be in order to lay out an elliptical flower bed 20 feet wide and 60 feet long? How far apart should the two stakes (foci) be?
27. The arch of a bridge is in the shape of a semiellipse with a horizontal span of 40 feet and a height of 16 feet at the center. How high is the arch 5 feet to the left of the center?
28. An arch in the shape of a semiellipse is to support a roadway over a river 100 feet wide. The center of the arch will be 25 feet above the base, and at the center, the road will rest on the arch. What will be the vertical distance between the roadway and the arch 10 feet from the end of the arch?
29. The orbit of the planet Mars is an ellipse whose equation is approximately

$$\frac{x^2}{(228)^2} + \frac{y^2}{(227)^2} = 1,$$

where x and y are in millions of kilometers, with center at the origin and the sun at one focus. How close does Mars come to the sun?
30. A whispering gallery is a building with an elliptically shaped dome roof. Such a building has the acoustical property that a whisper at one focus can be heard at

the other. If the gallery is 100 feet wide and 40 feet high at the center, how far from the walls are the foci?

31. A satellite orbiting the earth moves in an ellipse with the center of the earth at one focus. Suppose that a satellite at perigee (nearest point to the center of the earth) is 100 miles from the earth's surface and at apogee (farthest point to the center of the earth) is 700 miles from the earth's surface. Assuming that the earth is a sphere of radius 4000 miles, what is the approximate equation of this path if the earth's center is placed on the x-axis and the center of the ellipse is the origin?

32. The *center* of an ellipse is the midpoint of the segment between the two foci. Show that the equation of an ellipse with center (h, k), foci $(h + c, k)$ and $(h - c, k)$, and constant sum $2a$ is

$$\frac{(x - h)^2}{a^2} + \frac{(y - k)^2}{b^2} = 1,$$

where $b^2 = a^2 - c^2$.

2.8 Hyperbolas

A **hyperbola** is the set of all points in a plane, the difference of whose distances to two fixed points is constant. The two fixed points are called the **foci** of the hyperbola.

Theorem 2.14

The equation of the hyperbola with foci $(c, 0)$ and $(-c, 0)$ and constant difference $2a$ (where $0 < a < c$) is

$$\frac{x^2}{a^2} - \frac{y^2}{b^2} = 1,$$

where $b^2 = c^2 - a^2$.

Proof. Let $P(x, y)$ represent each point in the plane, and let $F_1 = (-c, 0)$ and $F_2 = (c, 0)$. Then P will be on the hyperbola if and only if

$$PF_1 - PF_2 = 2a \quad \text{or} \quad PF_2 - PF_1 = 2a.$$

See Figure 2.43. Thus P will be on the hyperbola if and only if

$$PF_2 - PF_1 = \pm 2a.$$

Now the distance between $P(x, y)$ and $F_2(c, 0)$ is

$$\sqrt{(x - c)^2 + y^2},$$

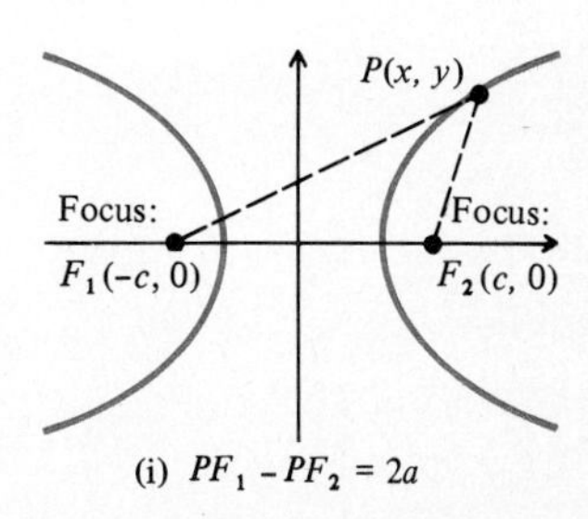

(i) $PF_1 - PF_2 = 2a$

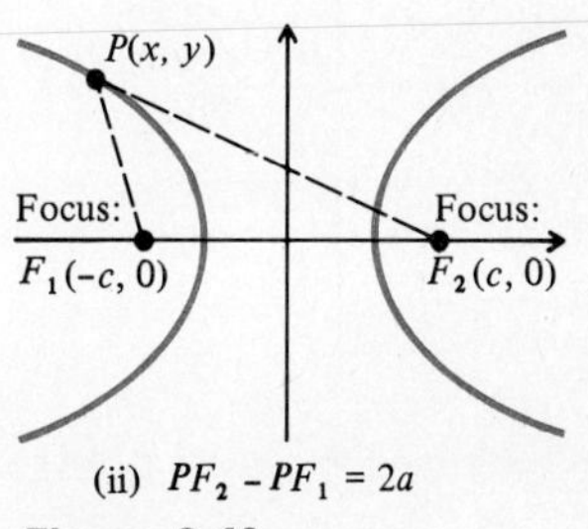

(ii) $PF_2 - PF_1 = 2a$

Figure 2.43

and the distance between $P(x, y)$ and $F_1(-c, 0)$ is

$$\sqrt{(x + c)^2 + y^2}.$$

Thus (x, y) is on the hyperbola if and only if

$$\sqrt{(x - c)^2 + y^2} - \sqrt{(x + c)^2 + y^2} = \pm 2a.$$

After taking the second radical to the right and squaring, we have the following:

$$\begin{aligned}
(x - c)^2 + y^2 &= 4a^2 \pm 4a\sqrt{(x + c)^2 + y^2} + (x + c)^2 + y^2 \\
-4cx - 4a^2 &= \pm 4a\sqrt{(x + c)^2 + y^2} \\
cx + a^2 &= \pm a\sqrt{(x + c)^2 + y^2}.
\end{aligned}$$

Upon squaring both sides again, we have

$$\begin{aligned}
c^2x^2 + 2a^2cx + a^4 &= a^2((x + c)^2 + y^2) \\
c^2x^2 + 2a^2cx + a^4 &= a^2x^2 + 2a^2cx + a^2c^2 + a^2y^2 \\
c^2x^2 - a^2x^2 - a^2y^2 &= a^2c^2 - a^4 \\
x^2(c^2 - a^2) - a^2y^2 &= a^2(c^2 - a^2).
\end{aligned}$$

If we let $b^2 = c^2 - a^2$ and then divide both sides by a^2b^2, we get

$$\frac{x^2}{a^2} - \frac{y^2}{b^2} = 1. \qquad (2\text{-}14)$$

Because these steps are reversible, (x, y) is on the hyperbola if and only if it is a solution of equation (2-14). ■

Such a hyperbola has x-intercepts of a and $-a$ but has no y-intercepts since setting $x = 0$ yields $y^2 = -b^2$, an equation with no solutions.

Solving equation (2-14) for $|y|$, we get

$$y^2 = \frac{b^2}{a^2}(x^2 - a^2)$$

$$|y| = \frac{b}{a}\sqrt{x^2 - a^2}.$$

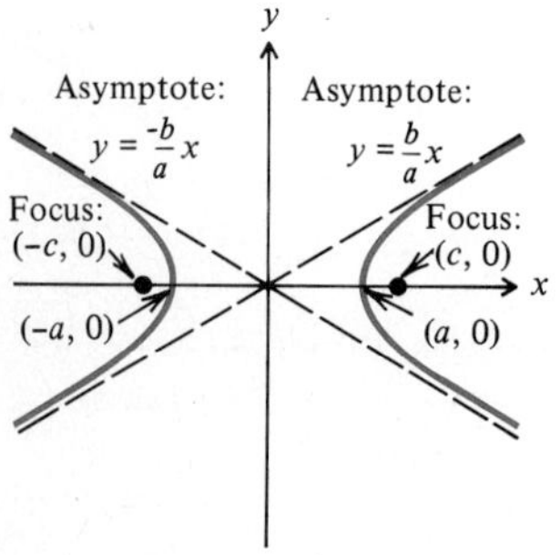

Figure 2.44 $\frac{x^2}{a^2} - \frac{y^2}{b^2} = 1$

But a is a constant. Therefore, as $|x|$ becomes infinitely large, $\sqrt{x^2 - a^2}$ approaches $\sqrt{x^2}$. Thus as $|x|$ gets large without bound, $|y|$ gets closer and closer to $\frac{b}{a}|x|$. With respect to the graph, this means that as $|x|$ increases, the graph approaches the lines $y = (b/a)x$ and $y = (-b/a)x$. See Figure 2.44. A line with the property that the distance from a point P to the line approaches zero as P moves away from the origin along some part of the graph is called an **asymptote.** Thus the lines with equations

$$y = \frac{b}{a}x \text{ and } y = \frac{-b}{a}x$$

are asymptotes of the hyperbola.

If the foci are on the y-axis as $(0, c)$ and $(0, -c)$, with the constant difference still $2a$ (where $0 < a < c$), then the equation is

$$\frac{y^2}{a^2} - \frac{x^2}{b^2} = 1, \qquad (2\text{-}15)$$

where $b^2 = c^2 - a^2$. Here the y-intercepts are a and $-a$, and the asymptotes are

$$y = \frac{a}{b}x \text{ and } y = \frac{-a}{b}x.$$

Equations (2-14) and (2-15) are called the **standard form** of the equation of a hyperbola. Hence, given an equation of the form

$$\frac{x^2}{k_1} - \frac{y^2}{k_2} = 1 \quad \text{or} \quad \frac{y^2}{k_1} - \frac{x^2}{k_2} = 1,$$

where k_1 and k_2 are positive constants, we know that the graph is a hyperbola.

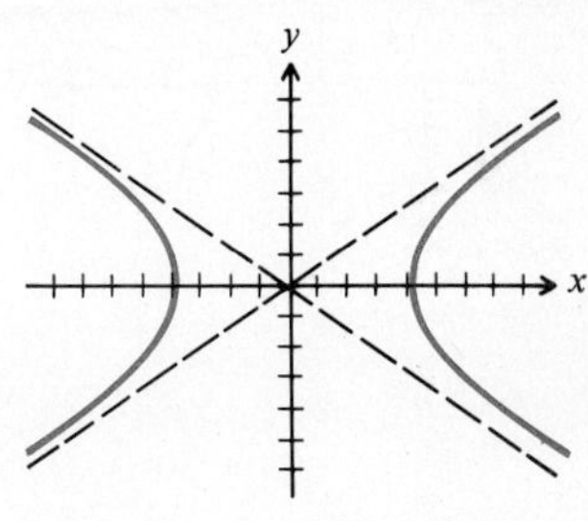

Figure 2.45 $\frac{x^2}{16} - \frac{y^2}{9} = 1$

Example 1

Graph each of the following.

(a) $\frac{x^2}{16} - \frac{y^2}{9} = 1$ (b) $\frac{y^2}{9} - \frac{x^2}{16} = 1$

Solution. (a) The x-intercepts are ± 4, and the asymptotes are $y = \pm\frac{3}{4}x$. This gives us a reasonable sketch, and we could plot additional points for more accuracy. For instance, when $x = \pm 8$, $y = \pm 3\sqrt{3} \approx \pm 5.2$. The graph is shown in Figure 2.45.

(b) The graph, shown in Figure 2.46, is a hyperbola with y-intercepts ± 3 and asymptotes $y = \pm\frac{3}{4}x$. ■

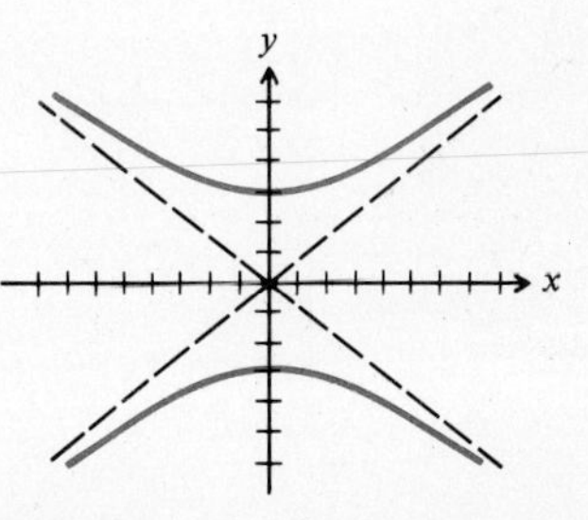

Figure 2.46 $\frac{y^2}{9} - \frac{x^2}{16} = 1$

Note that each of the graphs of Example 1 consists of two pieces, called branches. Also observe that each graph is symmetric about each axis.

The graph of

$$Ax^2 + By^2 = C,$$

where A, B, and C are constants with A and B having opposite signs and $C \neq 0$, is a hyperbola. This is because it may be rewritten as

$$\frac{x^2}{C/A} + \frac{y^2}{C/B} = 1,$$

where C/A and C/B differ in sign.

Example 2
Put $4x^2 - 9y^2 = 25$ in standard form and then graph.

Solution. Upon dividing by 25, we have

$$\frac{4x^2}{25} - \frac{9y^2}{25} = 1,$$

which is equivalent to the following.

$$\frac{x^2}{\frac{25}{4}} - \frac{y^2}{\frac{25}{9}} = 1$$

$$\frac{x^2}{(\frac{5}{2})^2} - \frac{y^2}{(\frac{5}{3})^2} = 1.$$

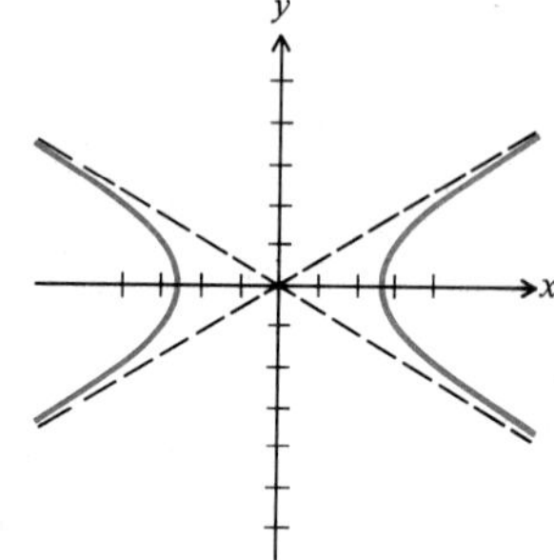

Figure 2.47 $4x^2 - 9y^2 = 25$

This is the standard form with $a = \frac{5}{2}$ and $b = \frac{5}{3}$. Thus the x-intercepts are $\pm\frac{5}{2}$, and the asymptotes are $y = \pm\frac{2}{3}x$. The graph is seen in Figure 2.47. ■

Exercise
Put $4y^2 - 9x^2 = 25$ in standard form and then graph.

Answer. Standard form: $\frac{y^2}{(\frac{5}{2})^2} - \frac{x^2}{(\frac{5}{3})^2} = 1$

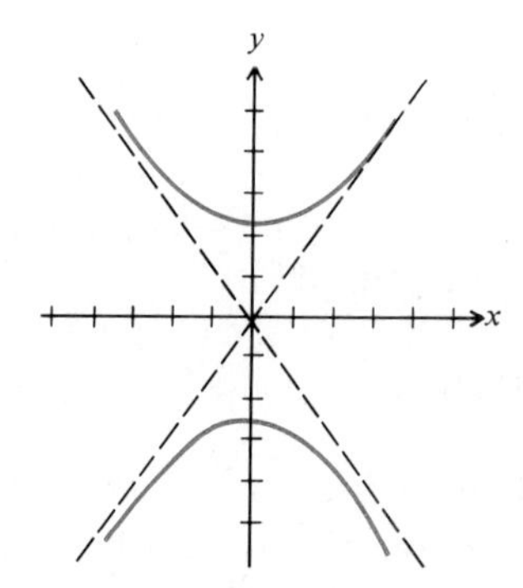

■

Example 3
Determine the equation (in standard form) of the hyperbola whose foci are $(0, \pm 2\sqrt{5})$ and whose constant difference is 6.

Solution. Since the foci are on the y-axis, the general equation is

$$\frac{y^2}{a^2} - \frac{x^2}{b^2} = 1.$$

Now $2a = 6$, so $a = 3$. And it remains only to find b^2. Since $b^2 = c^2 - a^2$, we have

$$b^2 = (2\sqrt{5})^2 - 3^2 = 20 - 9 = 11.$$

Hence the equation is

$$\frac{y^2}{9} - \frac{x^2}{11} = 1. \quad \blacksquare$$

Exercise
Determine the equation (in standard form) of the hyperbola whose foci are $(0, \pm 2\sqrt{6})$ and whose constant difference is 8.

Answer. $\frac{y^2}{16} - \frac{x^2}{8} = 1$ $\blacksquare$

The following example illustrates a practical application of hyperbolas.

Example 4
Points A and B are 1000 feet apart, and it is determined from the sound of an explosion heard at these points that the explosion occurred 600 feet closer to A than to B. Find an equation of the possible locations of the explosion.

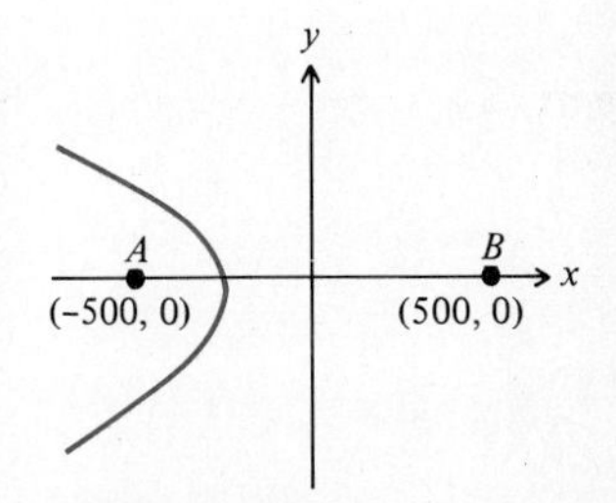

Figure 2.48

Solution. The possible locations will consist of all points P such that $PB - PA = 600$. Such points P form one branch of a hyperbola with foci A and B. Letting $A = (-500, 0)$ and $B = (500, 0)$, the branch shown in Figure 2.48 is the set of possible locations. We have $c = 500$ and $2a = 600$. Thus $a = 300$ and

$$b^2 = c^2 - a^2 = 500^2 - 300^2 = 400^2.$$

The equation of such a hyperbola is

$$\frac{x^2}{300^2} - \frac{y^2}{400^2} = 1.$$

Since we want the left branch only, we'll require x to be less than or equal to the left x-intercept. Hence the equation is

$$\frac{x^2}{300^2} - \frac{y^2}{400^2} = 1, \qquad x \leq -300. \qquad \blacksquare$$

Problem Set 2.8

In Problems 1–6, graph the equation.

1. $\frac{x^2}{9} - \frac{y^2}{4} = 1$
2. $\frac{x^2}{4} - \frac{y^2}{16} = 1$
3. $y^2 - \frac{x^2}{9} = 1$
4. $\frac{y^2}{25} - \frac{x^2}{4} = 1$
5. $\frac{x^2}{8} - \frac{y^2}{4} = 1$
6. $\frac{x^2}{9} - \frac{y^2}{9} = 1$

In Problems 7–12, put the equation in standard form and graph.

7. $16x^2 - 49y^2 = 196$
8. $25x^2 - 16y^2 = 100$
9. $49y^2 - 36x^2 = 441$
10. $9y^2 - 25x^2 = 25$
11. $8y^2 - 5x^2 = 30$
12. $25x^2 - 12y^2 = 75$

In Problems 13–22, determine the equation (in standard form) of the hyperbola with the given properties.

13. foci: $(\pm 6, 0)$, difference of distances from foci is $4\sqrt{5}$
14. foci: $(0, \pm\sqrt{5})$, difference of distances from foci is 2
15. foci: $(0, \pm 5)$, y-intercepts: ± 3
16. foci: $(0, \pm 2\sqrt{3})$, y-intercepts: $\pm 2\sqrt{2}$
17. foci: $(\pm 6, 0)$, x-intercepts: $\pm\sqrt{11}$
18. foci: $(\pm 2\sqrt{7}, 0)$, x-intercepts: $\pm 2\sqrt{5}$
19. passing through $(2\sqrt{3}, 2)$, x-intercepts: $\pm\sqrt{6}$
20. passing through $(-2, -\sqrt{5})$ and $(0, -2)$
21. asymptotes: $y = \pm\frac{\sqrt{5}}{2}x$, x-intercepts: ± 4
22. asymptotes: $y = \pm\frac{\sqrt{15}}{3}x$, y-intercepts: $\pm\sqrt{10}$
23. What are the foci of a hyperbola whose equation is $x^2 - y^2 = 20$?
24. What are the foci of a hyperbola whose equation is $3y^2 - 5x^2 = 30$?
25. An explosion is heard by Mike and Jim, who are talking to each other by telephone, 8800 feet apart. Mike hears the explosion 4 seconds before Jim does. Find an equation of the possible locations of the explosion, assuming that the speed of sound is 1100 ft/sec.
26. Mark fires a rifle 1100 feet from the base of a large rock. John hears the echo of the shot 1.5 seconds after he hears the shot. If John is closer to Mark than to the

rock, find the equation of his possible locations assuming that the speed of sound is 1100 ft/sec. (*Hint:* It takes 1 second for the sound to get to the rock.)

27. The production cost of an item is \$12 less at a point A than it is at a point B, and the distance between A and B is 100 miles. Assuming that the route of the delivery of the item is along a straight line (items are delivered separately) and that the delivery cost is 20 cents per mile, find the equation of the curve, at any point of which the item can be supplied from A or B at the same total cost.

28. Given the hyperbola $\frac{x^2}{a^2} - \frac{y^2}{b^2} = 1$, show that the asymptotes contain the diagonals of the rectangle determined by the four points $(\pm a, \pm b)$.

29. Show that the hyperbola $\frac{x^2}{a^2} - \frac{y^2}{b^2} = 1$ will have asymptotes given by $\frac{x^2}{a^2} - \frac{y^2}{b^2} = 0$.

30. The *center* of a hyperbola is the midpoint of the segment between the two foci. Show that the equation of the hyperbola with center (h, k), foci $(h + c, k)$ and $(h - c, k)$, and constant difference $2a$ is

$$\frac{(x - h)^2}{a^2} - \frac{(y - k)^2}{b^2} = 1,$$

where $b^2 = c^2 - a^2$.

Chapter 2 Review

1. Let $P = \left(-3, \frac{5}{2}\right)$ and $Q = \left(\frac{-4}{3}, 1\right)$. Determine PQ and the midpoint of $\overline{PQ}$.

2. Determine the domain and range given the relation or the graph.

(a) $r = \{(x, 0) | x \in Z^+\}$

(b) $r = \{(x, y) | x + y = 0\}$

(c)

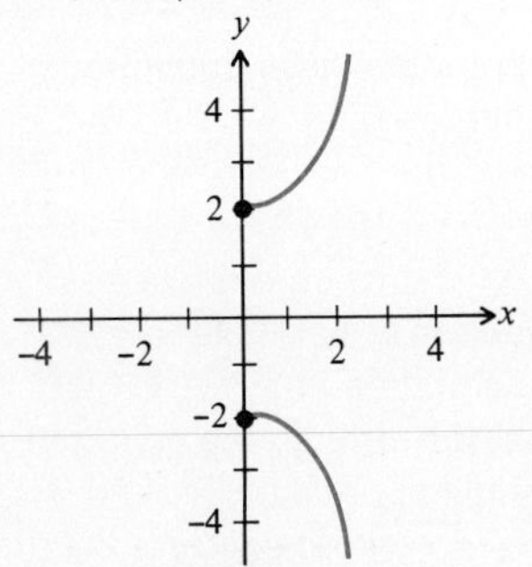

In Problems 3–9, graph the given equation.

3. $3x - 4y + 12 = 0$

4. $y = -3$

5. $y = 2x^2 - 4x - 1$

6. $x = -y^2 + 4y - 1$

7. $x^2 + y^2 - 4x + 2y - 11 = 0$

8. $16x^2 + 25y^2 = 100$

9. $100y^2 - 3x^2 = 36$

In Problems 10–16, determine the equation (in slope-intercept form if possible) of the line described.

10. slope $\frac{1}{2}$ and y-intercept -2
11. passing through $(-1, 2)$ with slope 5
12. passing through $(5, 2)$ and $(-1, 3)$
13. passing through $(2, 1)$ and parallel to $2x + y = 3$
14. passing through $(-1, -3)$ and perpendicular to $2x + 3y = 3$
15. passing through $(-2, -1)$ and parallel to the y-axis
16. with x-intercept 2 and y-intercept -3

In Problems 17–27, determine the equation of the relation described.

17. a circle of radius 2, centered at $(-3, 0)$
18. a circle centered at $(2, -3)$, passing through $(4, 1)$
19. a parabola with focus $(-3, 2)$ and directrix $x = 2$

20. a parabola passing through $(4, 6)$ with vertex $\left(1, \frac{3}{2}\right)$ and axis vertical

21. a parabola with vertex $(-2, -3)$ and focus $(-2, -5)$

22. a parabola with vertex $(4, -3)$ and directrix $x = \frac{8}{3}$

23. an ellipse with foci $\left(\frac{\pm\sqrt{39}}{4}, 0\right)$ and sum of distances from foci is 4

24. an ellipse passing through $\left(1, \frac{-9}{4}\right)$ with x-intercepts $\frac{\pm 5}{4}$

25. a hyperbola with foci $(0, \pm 4)$ and y-intercepts ± 2

26. a hyperbola with asymptotes $y = \pm \frac{\sqrt{6}}{2} x$ and x-intercepts $\pm 2\sqrt{2}$

27. a hyperbola passing through $(-4, -6)$ and $(0, -2)$

28. Determine the focus and directrix of the parabola whose equation is

$$y = \frac{9x^2 + 12x + 40}{72}.$$

29. Determine the foci of the ellipse whose equation is $6x^2 + 22y^2 = 33$.
30. Determine the foci of the hyperbola whose equation is $2y^2 - 4x^2 = 5$.
31. Determine the equation of the line which passes through the center of the circle $x^2 + y^2 + 8x - 2y = -15$ and the vertex of the parabola $y = -5x^2 + 20x - 17$.
32. Determine the equation of the circle of radius $\sqrt{5}$ whose center is the focus of the parabola $6x = y^2 - 6y - 9$.
33. The surface of a roadway over a creek is in the shape of a parabola with the vertex in the middle. The span is 60 feet, and the road surface is 1 foot higher in the middle than at the ends. How much higher than the end of the roadway is a point 9 feet from an end?
34. A parabolic arch has a base of 20 feet and a height of 10 feet. Find the height of the arch at a point on the base 4 feet from the center.

35. Until recently hot dogs at a basketball arena sold for \$1 each. At that price an average of 10,000 were sold on game nights. When the price was increased to \$1.20, the average number sold decreased to 8000. The concessions have a fixed cost of \$1000 per night and a variable cost of 30 cents per hot dog. If the relationship between the number sold and the price is linear, what price maximizes the nightly hot dog profit?
36. The Golden Gate Bridge has a span of 4200 feet and a sag of 490 feet. Find the length of a vertical supporting cable 1500 feet from the center if the roadway is 36 feet below the lowest point of the suspension cable.
37. A tunnel is to be cut such that the opening is a semielliptic arch. If it is to be 40 feet wide and 15 feet high at the center, how high will the tunnel be 4 feet from the center?
38. The orbit of the planet Mercury is an ellipse whose equation is approximately

$$\frac{x^2}{18^2} + \frac{4y^2}{35^2} = 1,$$

where x and y are in millions of miles, with center at the origin and the sun at one focus. What is the greatest distance between Mercury and the sun?
39. Two towers, 20 miles apart, are located on a coast which runs north-south. Each tower sends out signals which travel 2 miles per second. A ship to the east receives a signal from the northern tower 2 seconds sooner than it receives a simultaneous signal from the southern tower. Assign coordinates to each tower and write an equation which represents the ship's possible locations.

Functions

3

In this chapter we study one of the most important concepts in mathematics—the idea of a function. It is a means of expressing the association between quantities.

3.1 A Special Type of Relation

A **function** is a relation in which no two different ordered pairs have the same first coordinate. In other words, a function is a relation in which there do *not* exist two ordered pairs with the same first coordinates and different second coordinates. The relation

$$\{(1, 2), (2, 4), (3, 4)\}$$

is a function with domain $\{1, 2, 3\}$ and a range $\{2, 4\}$. (Recall the definition of domain and range of relations, and hence of functions. The set of all first coordinates of the ordered pairs of a function is the domain, and the range is the set of all second coordinates of the ordered pairs.)

But not every relation is a function. The relation

$$\{(1, 2), (2, 3), (1, 5)\}$$

is *not* a function because there are two different ordered pairs, in particular (1, 2) and (1, 5), with the same first coordinate. (Note that in a function it is permissible to have two different ordered pairs with the same *second* coordinate.)

Exercise
Determine whether the relation is a function.
(a) $\{(1, 2), (2, 2), (3, 2)\}$ (b) $\{(1, 5), (1, 6)\}$ (c) $\{(0, 0)\}$

Answer. (a) and (c) are functions. ■

We can also look at the graph of a relation to determine whether it is the graph of a function. A relation is a function provided that no vertical line intersects the graph of the relation in more than one point.

Exercise
Determine whether the graph is the graph of a function.

(a)

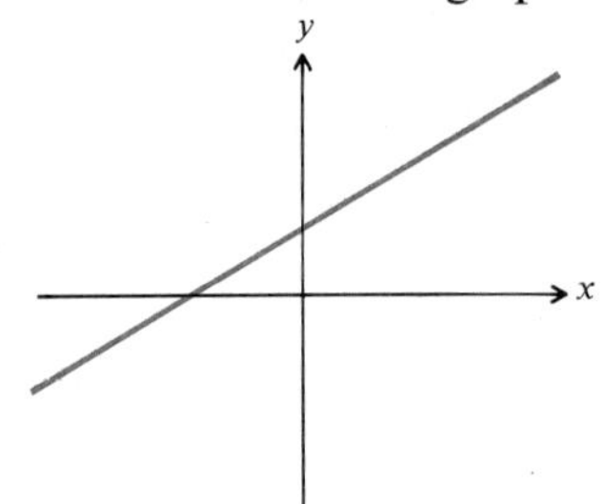

(b)

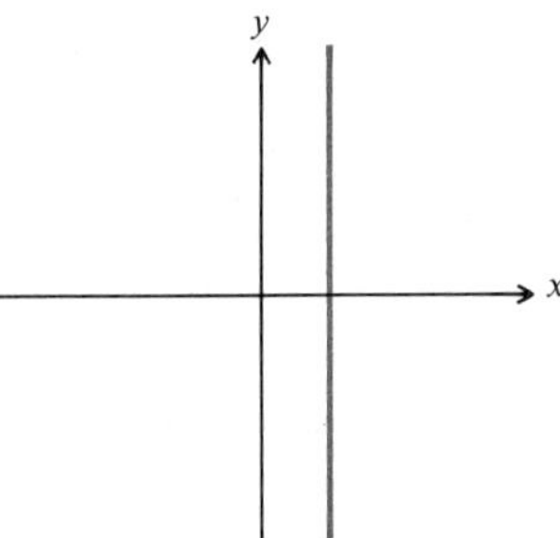

(c)

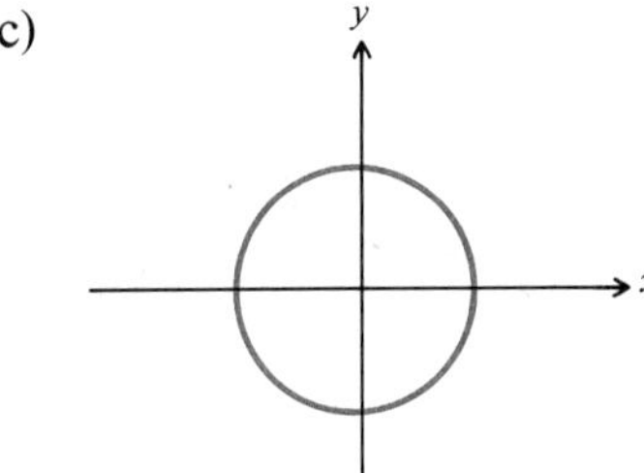

(d)

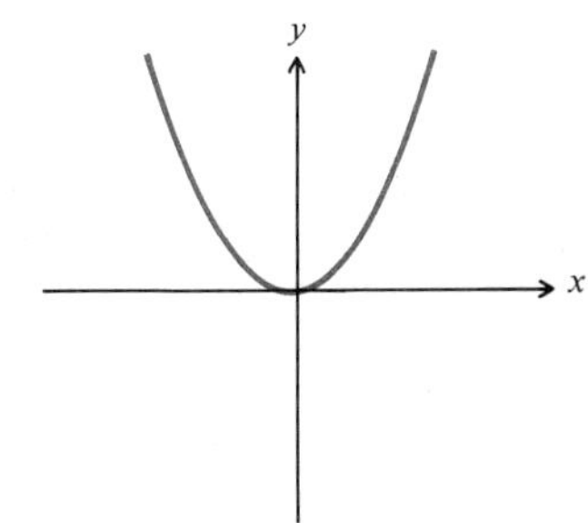

(e)

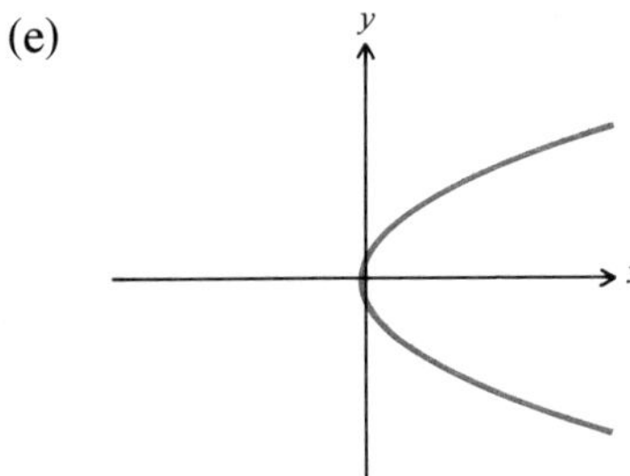

(f)

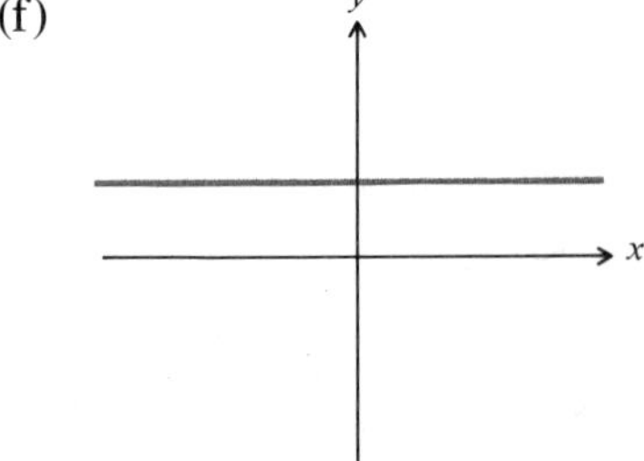

Answer. (a), (d), and (f) are the graphs of functions. ■

We may think of a function as an association or correspondence. For each element in the domain, there is *one and only one* element in the range that is associated with it. That is, each element of the domain corresponds to a *unique* element of the range.

One could visualize the function {(1, 2), (2, 4), (3, 4)} as indicated in Figure 3.1. Viewed in this fashion, a function will *not* have two arrows beginning at the same point in the domain. (But it may have two or more arrows

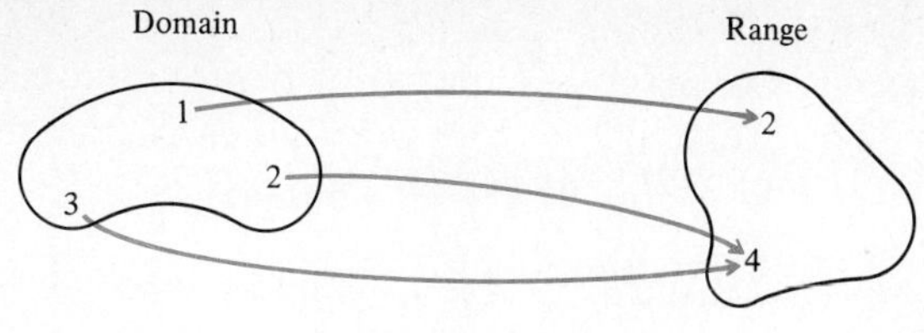

Figure 3.1

ending at the same point in the range.) Thus the relation pictured in Figure 3.2 is not a function because there is an element of the domain, -2, which doesn't correspond to a unique element of the range.

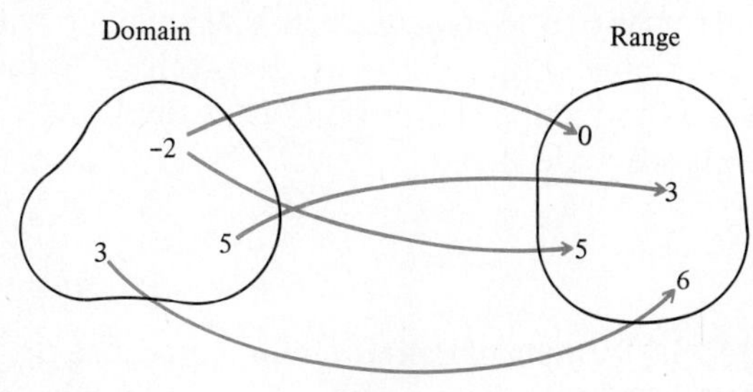

Figure 3.2

Many functions are defined by equations in x and y. (Recall that a relation thus defined is the solution set of the equation.) The relation defined by $y = x^2$ is a function because for each x in the domain there is only one y in the range that is associated with it. But the relation defined by $y^2 = x$ is not a function. Each positive number x in the domain corresponds to two elements y in the range. (For example, when $x = 4$, $y = 2$ or $y = -2$.)

Exercise

Determine whether the relation is a function.

(a) $\{(x, y) \mid y = x^3\}$ (b) $\{(x, y) \mid y^2 = x^2\}$ (c) $\{(x, y) \mid x^2 + y^2 = 4\}$

Answer. (a) is a function. ■

Let f be a function, and let a and b be constants. If $(a, b) \in f$, then we shall say that b is the **value of f at a** (or that b is the **image of a under f**). We shall denote this value by $f(a)$, read "f of a" or "f at a." (In this context, the parentheses do *not* mean multiplication.) Thus $f(a)$ is the second coordinate of the ordered pair whose first coordinate is a. Here, of course, we have $f(a) = b$.

If $f = \{(-1, 2), (-2, 4), (3, 7)\}$ and $g = \{(x, y) \mid y = x^2\}$, then $f(-1) = 2$, $f(3) = 7$, $g(-1) = 1$, and $g(2) = 4$.

Exercise
Let $f = \{(0, 3), (-1, 2), (3, 6)\}$, $g = \{(x, y) \mid y = x^3\}$. Find each of the following.
(a) $f(0)$ (b) $f(-1)$ (c) $g(2)$ (d) $g(-3)$

Answers. (a) $f(0) = 3$ (b) $f(-1) = 2$ (c) $g(2) = 8$
(d) $g(-3) = -27$ ■

We have already encountered two ways of defining a function. An equation in the variables x and y can define a function, and, of course, we can simply list the ordered pairs that belong to the function. There is yet an additional method, very similar to using an equation in x and y, which is called writing a **formula of definition.**

Suppose that we want to associate each real number with its absolute value. Let x be a variable with universe R. Hence $|x|$ is to correspond to x. This association is a function. Let's call it f. Thus the value of f at x is $|x|$. In mathematical terms we state that

$$f(x) = |x|. \qquad (3\text{-}1)$$

Equation (3-1) is the formula of definition for f. Because the variable x represents each real number, statement (3-1) says that the image of each real number under f is its absolute value. Thus the function is defined.

To find $f(-2)$, the value of the function at -2, we use the formula of definition and substitute. We have

$$f(-2) = |-2| = 2.$$

When a function is defined with a formula, the universe of the variable chosen and the domain of the function are the same. And although the variable chosen is frequently x, it does not have to be. For the previous function our formula of definition could have been

$$f(t) = |t| \qquad \text{or} \qquad f(y) = |y|.$$

Each formula determines the same set of ordered pairs.

Example 1
Let f be defined by $f(x) = 2x + 1$. Determine each of the following.
(a) $f(3)$ (b) $f(-2)$ (c) $f(a)$ (d) $f(x + h)$
(e) $f(x) + f(h)$ (f) $f(x) + h$

Solution. (a) To calculate $f(3)$, we simply use the formula of definition and replace x by 3. Thus

$$f(3) = 2 \cdot 3 + 1 = 7.$$

Similarly,
(b) $f(-2) = 2(-2) + 1 = -3$
(c) $f(a) = 2a + 1$
(d) To find $f(x + h)$ we do the same. That is, for each x in the formula we substitute $x + h$. Hence

$$f(x + h) = 2(x + h) + 1 = 2x + 2h + 1.$$

(e) $f(x) + f(h) = 2x + 1 + 2h + 1 = 2x + 2h + 2$
(f) $f(x) + h = 2x + 1 + h$ ■

Note that for the function of Example 1, $f(x + h) \neq f(x) + f(h)$ and $f(x + h) \neq f(x) + h$.

Exercise
Let f be defined by $f(x) = 3x - 2$. Determine each of the following.
(a) $f(-2)$ (b) $f(1)$ (c) $f(a)$ (d) $f(x + h)$
(e) $f(x) + f(h)$ (f) $f(x) + h$

Answers. (a) $f(-2) = -8$ (b) $f(1) = 1$ (c) $f(a) = 3a - 2$
(d) $f(x + h) = 3x + 3h - 2$ (e) $f(x) + f(h) = 3x + 3h - 4$
(f) $f(x) + h = 3x - 2 + h$ ■

When the domain of a function is not explicitly specified, it is assumed to be the set of all real numbers whose images under the function are defined and real. Thus if $h(x) = 1/x$, the domain of h is $\mathsf{R} - \{0\}$ since $1/x$ is not defined if x is zero. And if $g(x) = \sqrt{x}$, then the domain of g is $[0, \infty)$ because $\sqrt{x}$ is not real if x is negative.

Exercise
Determine the domain of each of the following functions.

(a) $f(x) = \dfrac{1}{x + 2}$ (b) $g(x) = \sqrt{x - 1}$

Answers. (a) $\mathsf{R} - \{-2\}$ (b) $[1, \infty)$ ■

In some cases the domain must be specified. For example, suppose that we want to associate each integer with its absolute value. We could let

$$g(x) = |x| \qquad \text{for } x \in \mathsf{Z},$$

where the specification "for $x \in \mathsf{Z}$" tells us that x represents integers only. That is, the domain of g is Z. Alternatively, we could write the following:

Let $g\colon \mathsf{Z} \to \mathsf{R}$ be defined by $g(x) = |x|$.

The notation "$g\colon \mathsf{Z} \to \mathsf{R}$" is read "$g$, a function from Z into R" and means that g is a function with domain Z and range a subset of R. The latter set listed can be any set which *contains* the range. Thus, instead of

$$g: \mathsf{Z} \to \mathsf{R},$$

we could have written

$$g: \mathsf{Z} \to \mathsf{Z} \qquad \text{or} \qquad g: \mathsf{Z} \to [0, \infty).$$

We have already noted that some equations in x and y define functions. For example, let f be the function defined by

$$2x + y = 5.$$

Since f is the solution set of this equation, we have

$$f = \{(x, y) \mid 2x + y = 5\}.$$

If we want to write the formula of definition for f, we solve the equation for y in terms of x. Here we get

$$y = -2x + 5.$$

The right-hand side of the equation will be the right-hand side of our formula if we use the variable x. Thus

$$f(x) = -2x + 5.$$

If an equation in the variables x and y defines a function, then we can solve the equation for y in terms of x. And because $y = f(x)$ for some function f, we shall say that **y is a function of x**. Furthermore, y will be referred to as the dependent variable and x will be called the independent variable.

Exercise
For each of the following, let f be the function defined by the equation and determine $f(x)$.

(a) $3x + y = 2$ (b) $x^2 - 2y + 3x = 5$ (c) $\dfrac{y - 2}{3} = x^2$

Answers. (a) $f(x) = 2 - 3x$ (b) $f(x) = \dfrac{x^2 + 3x - 5}{2}$

(c) $f(x) = 3x^2 + 2$ ■

Problem Set 3.1

In Problems 1–12, determine whether the relation is a function.

1. $\{(-1, 2), (-2, 4), (-3, 6)\}$

2. $\{(5, 6), (5, -1)\}$

3. $\{(1, 2), (1, 2)\}$

4. $\{(1, 7), (-1, 7), (0, 7)\}$

5. $\{(x, y) \mid y = -x^3\}$

6. $\{(x, y) \mid y^2 = -x\}$

7. $\{(x, y) \mid x^2 + y^2 = 4\}$

8. $\{(x, y) \mid y = 2x^2 + 8x - 1\}$

9. $\{(x, y) \mid x^2 + 4y^2 = 1\}$

10. $\{(x, y) \mid x^2 - 4y^2 = 1\}$

11. $\{(x, y) \mid y = |x|\}$

12. $\{(x, y) \mid x = y^2\}$

In Problems 13–26, determine whether the graph is the graph of a function.

13.

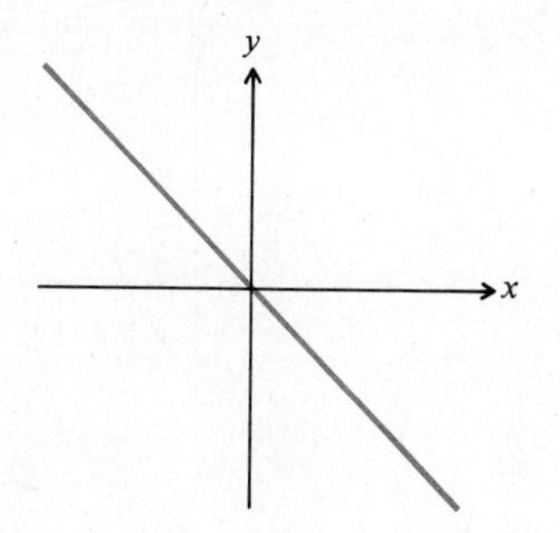

14.

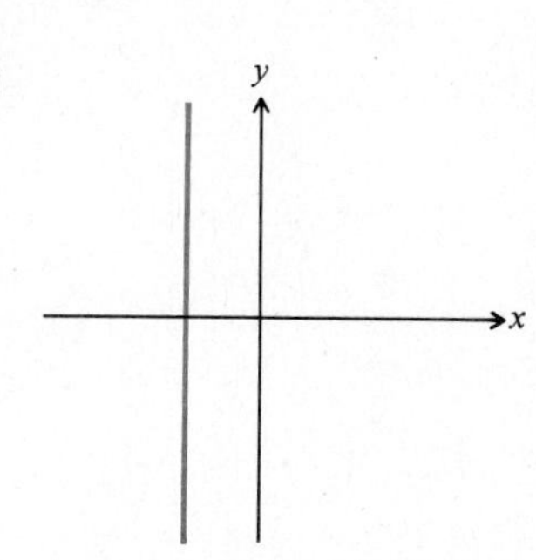

15.

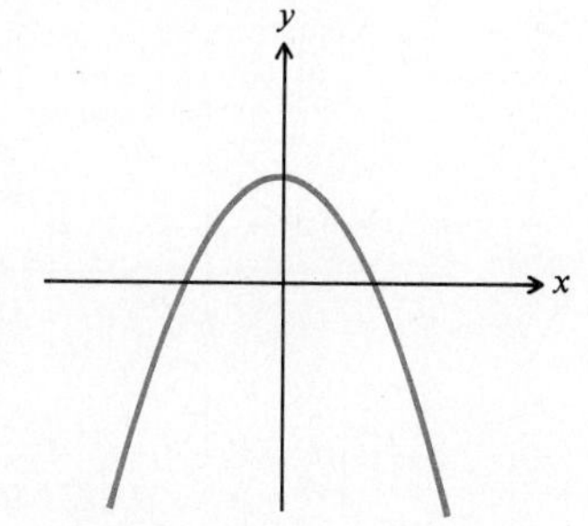

16.

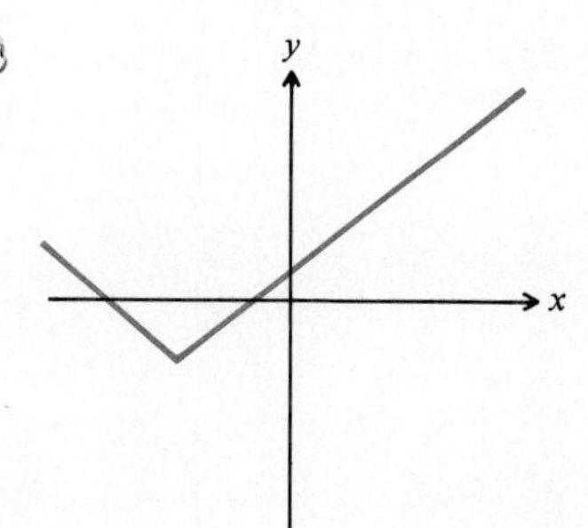

17.

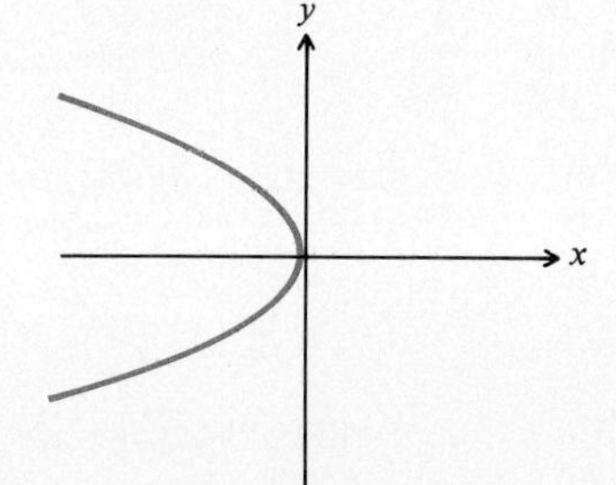

18.

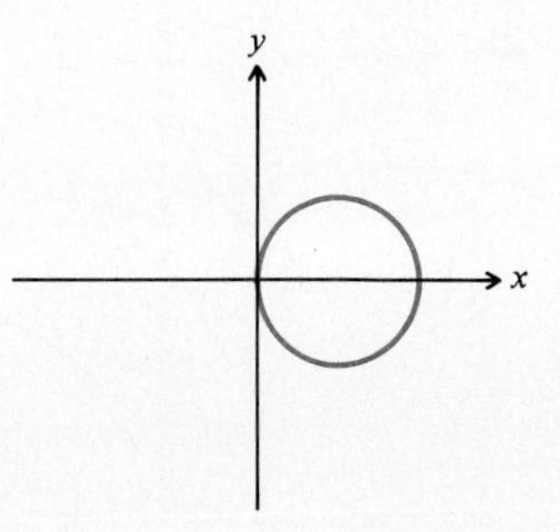

19.

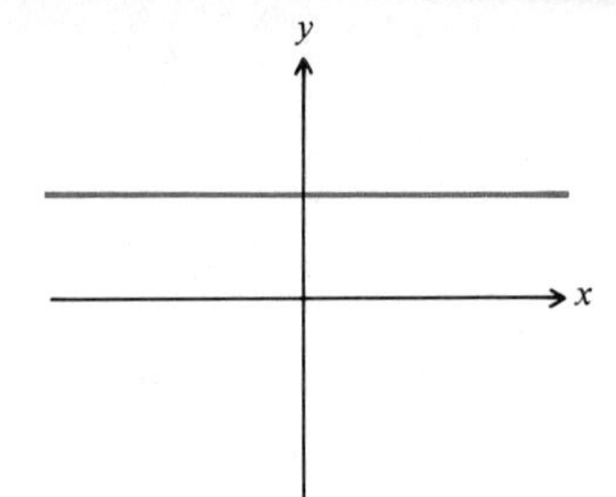

20.

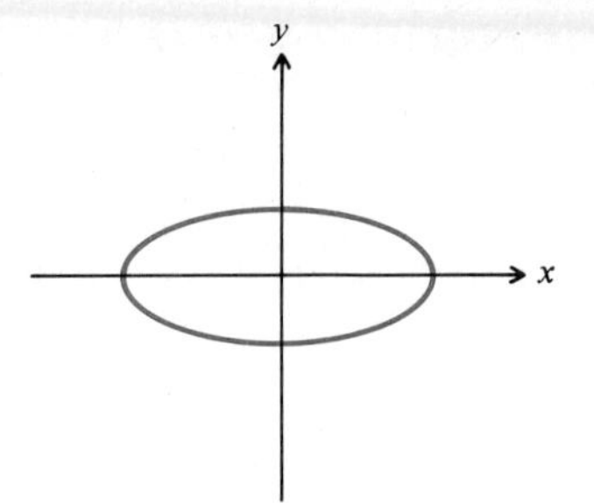

21.

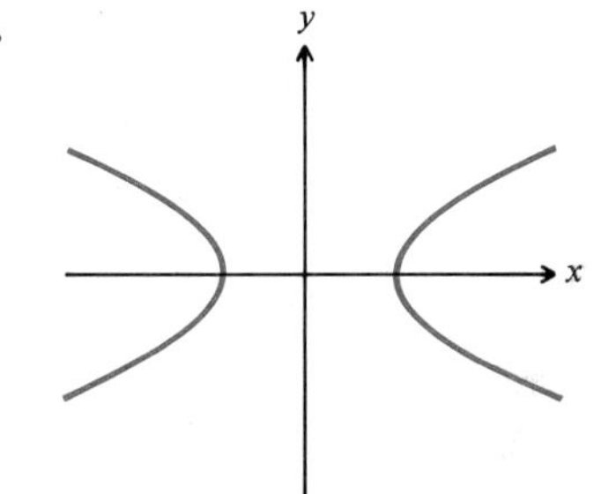

22.

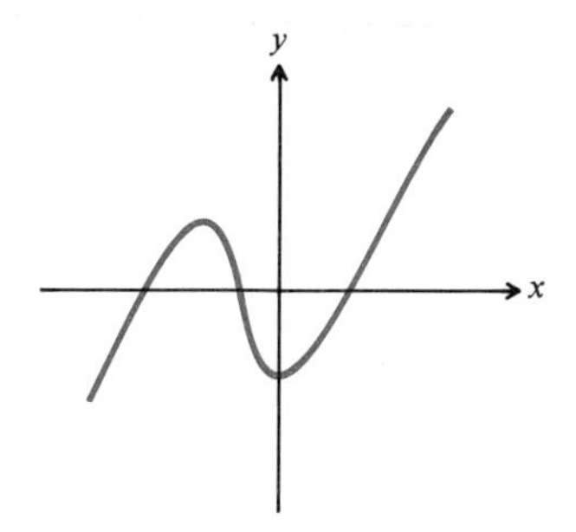

23.

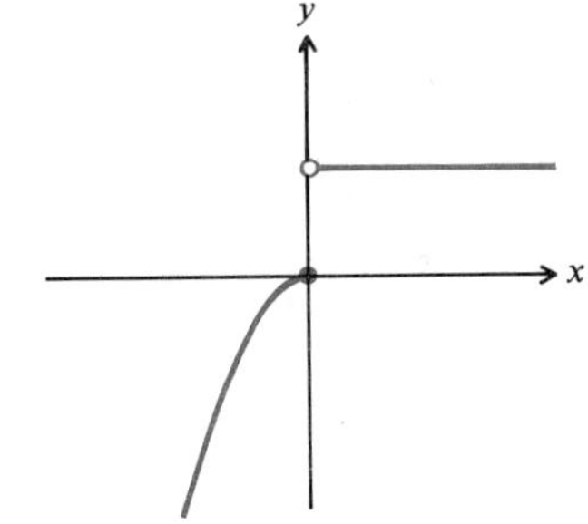

24.

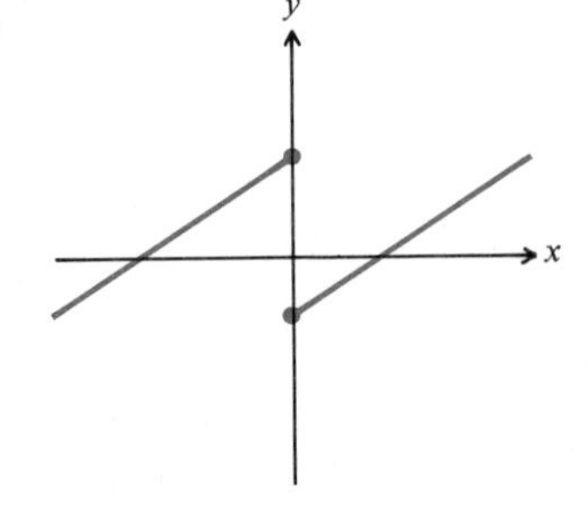

25.

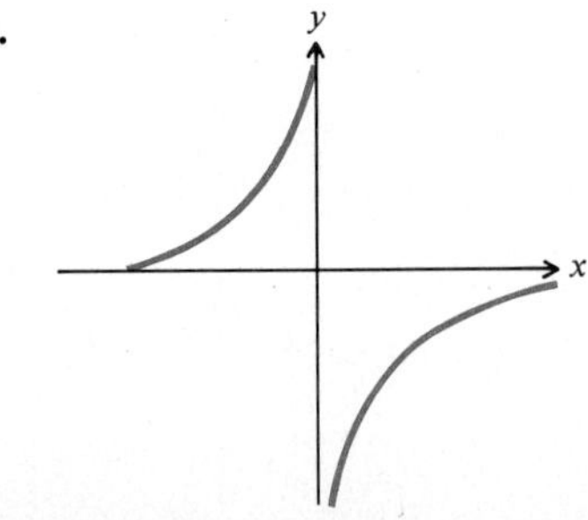

26.

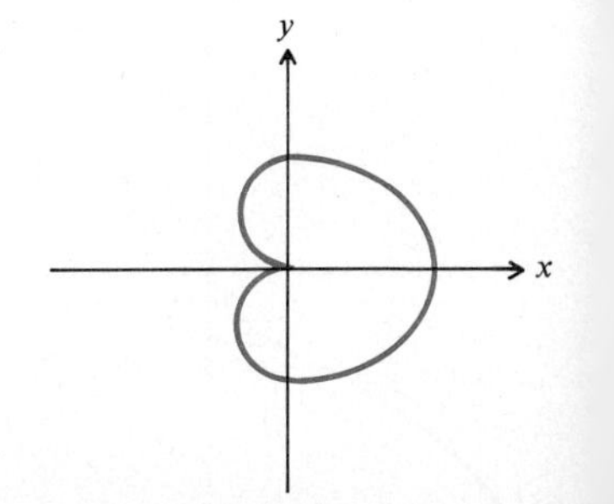

In Problems 27–36, determine the domain of the function f.

27. $f(x) = |x|$

28. $f(x) = -x^3$

29. $f(x) = \dfrac{1}{x^2}$

30. $f(x) = \dfrac{2}{x-3}$

31. $f(x) = -\sqrt{x}$

32. $f(x) = \sqrt{-x}$

33. $f(x) = \sqrt[3]{2x+3}$

34. $f(x) = \sqrt{2x+3}$

35. $f(x) = \sqrt{3 - x^2 - 2x}$

36. $f(x) = \sqrt{\dfrac{2x-1}{x+2}}$

37. Let $f = \{(-1, 3), (-2, 5), (3, 6)\}$ and find (a) $f(-2)$, (b) $f(3)$, and (c) $f(-1)$.

38. Let $f = \{(x, y) \mid y = x^2 - 1\}$ and find (a) $f(-1)$, (b) $f(2)$, (c) $f(a)$, and (d) $\dfrac{f(t+h) - f(t)}{h}$.

39. Let $f(x) = 2x^2 - x + 3$ and find (a) $f(0)$, (b) $f(a)$, (c) $f(\sqrt{2})$, (d) $f(1)$, (e) $f(f(1))$, and (f) $\dfrac{f(t+h) - f(t)}{h}$.

40. Let $g(x) = \dfrac{1}{x^2}$ and find (a) $g(a)$, (b) $\dfrac{1}{g(a)}$, (c) $g\left(\dfrac{1}{a}\right)$, (d) $g(\sqrt{a})$, (e) $\sqrt{g(a)}$, (f) $g(2)$, and (g) $g(g(2))$.

41. Let $f(x) = x^2 + 2x - 1$ and find (a) $\dfrac{f(x+h) - f(x)}{h}$, (b) $f(x^2)$, and (c) $f(f(x))$.

42. Let $f(x) = 5$ and find (a) $f(0)$, (b) $f(-3)$, (c) $f(5)$, (d) $f(x^2)$, and (e) $f(f(x))$.

In Problems 43–46, let f be the function defined by the given equation and find f(x).

43. $2x - y + 5 = 0$

44. $x - 3y + 2 = 0$

45. $2x^2 + x - 2y = 7$

46. $\dfrac{y-1}{2} = x^2$

47. Let f be the function with the following graph:

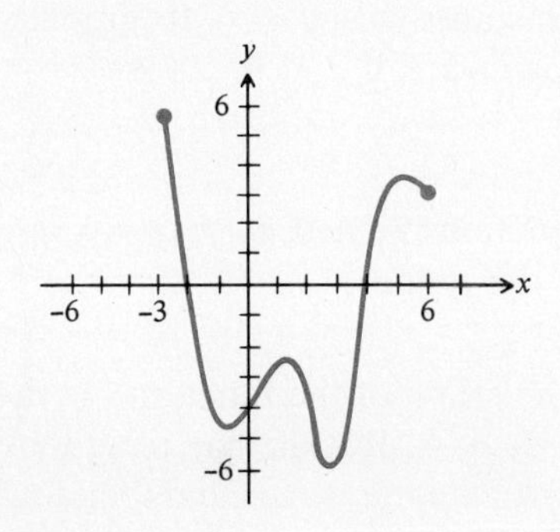

(a) What is the domain of f? (b) What is the range of f?
(c) $f(0) = ?$ (d) $f(5) = ?$
(e) If $f(a) = 0$, then $a = ?$

48. A function f is said to be **odd** provided that $f(-x) = -f(x)$. It is said to be **even** provided that $f(-x) = f(x)$. Show that g is odd and h is even if

$$g(x) = x^3 - 2x$$
$$h(x) = x^4 - 3x^2 + 5.$$

3.2 Graphs of Functions

Of the relations studied in Chapter 2, only nonvertical lines and parabolas with a vertical axis are functions. The domain of each of these is R, and the range is the projection of the graph onto the y-axis.

Horizontal lines are the graphs of the so-called **constant functions.** Since the equation of a horizontal line is $y = c$ for some constant c, the formula of definition for a constant function is

$$f(x) = c$$

for all $x \in$ R. The range of such a function is, of course, $\{c\}$.

The remaining nonvertical lines are the graphs of *linear* functions. A function f: R $\to$ R is called a **linear function** provided that its formula of definition is

$$f(x) = mx + b,$$

where $m \neq 0$. Of course, the graph of this function is the graph of $y = mx + b$, a line with slope m and y-intercept b. Its range is R.

A function f: R $\to$ R is called a **quadratic function** provided that its formula of definition is

$$f(x) = ax^2 + bx + c,$$

where a, b, and c are constants with $a \neq 0$. Its graph is a parabola that opens upward if $a > 0$ and downward if $a < 0$.

Example 1
Graph the quadratic function defined by $f(x) = x^2 - 2x + 4$, and then use the graph to determine the range.

Solution. Of course, the graph of the function f is the same as the graph of the equation $y = x^2 - 2x + 4$. Recall that to graph such an equation we completed the square to locate the vertex and then found some additional solutions. Doing the same here, we get

$$y - 4 = x^2 - 2x$$
$$y - 4 + (-1)^2 = x^2 - 2x + (-1)^2$$
$$y - 3 = (x - 1)^2.$$

Thus the vertex is (1, 3). Rewriting the formula of definition as

$$f(x) = (x - 1)^2 + 3$$

and finding additional solutions, we get the following table:

x	−2	−1	0	1	2	3	4
f(x)	12	7	4	3	4	7	12

Plotting these solutions we get the graph shown in Figure 3.3. ■

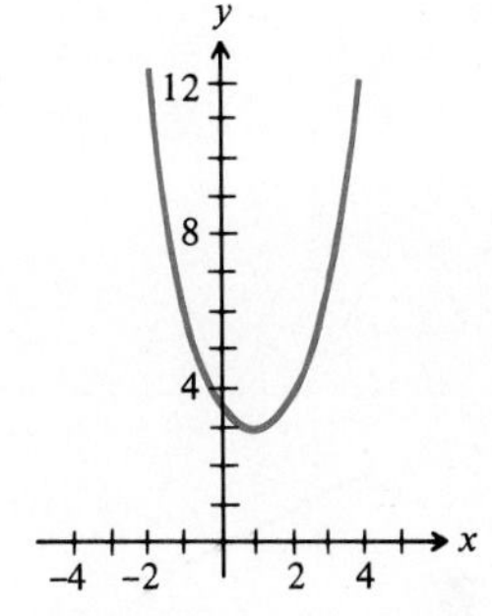

Figure 3.3 $f(x) = x^2 - 2x + 4$

Exercise
Graph the following functions.

(a) $f(x) = \frac{3}{2}$ (b) $f(x) = 2x - 1$ (c) $f(x) = x^2 - 2x - 1$

Answers
(a)

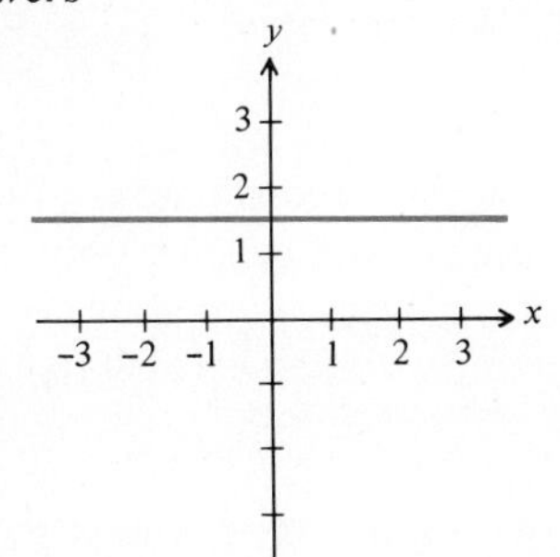

(b)

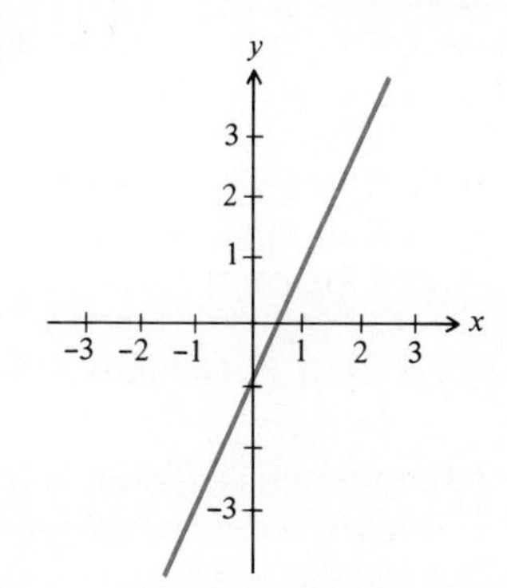

(c)

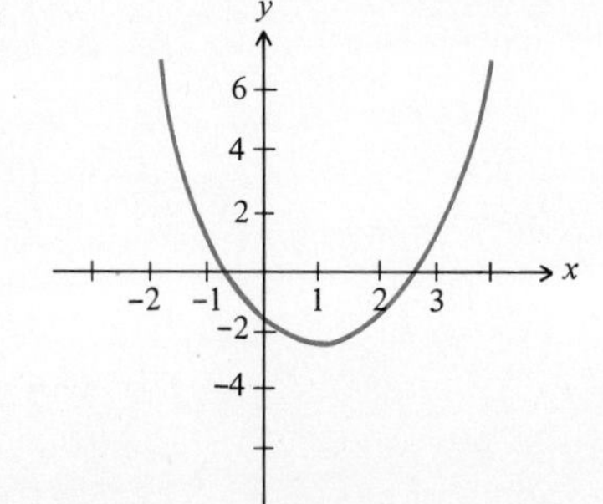

■

Suppose that we want to graph the function defined by

$$y = \sqrt{x}. \tag{3-2}$$

If we square both sides, we have

$$y^2 = x. \tag{3-3}$$

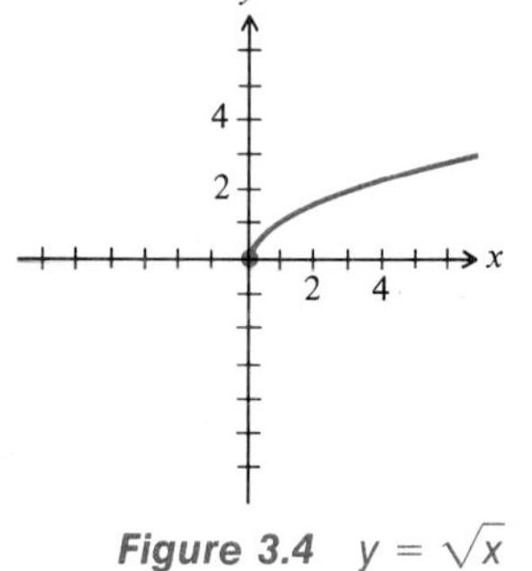

Figure 3.4 $y = \sqrt{x}$

Of course, the graph of equation (3-3) is a parabola, with vertex at the origin, which opens to the right. But recall that in squaring both sides of a given equation, the derived equation is not necessarily equivalent. There may be extraneous solutions. Note that for a given positive value of x, equation (3-2) has a single positive value of y which will be a solution. But for the same positive value of x, equation (3-3) has *two* values of y, one positive as before and the other its negative, which produce solutions. [For example, (4, 2) is a solution of equation (3-2), but (4, 2) and (4, -2) are solutions of equation (3-3).] Thus the extraneous solutions of equation (3-3) are those with negative second coordinates. Hence the graph of $y = \sqrt{x}$, shown in Figure 3.4, is that part of the parabola which lies on or above the x-axis. (The graph of $y = -\sqrt{x}$ is the bottom half of the parabola $y^2 = x$.)

Example 2
Graph $y = -\sqrt{4 - x^2}$.

Solution. Upon squaring both sides, we get

$$y^2 = 4 - x^2. \tag{3-4}$$

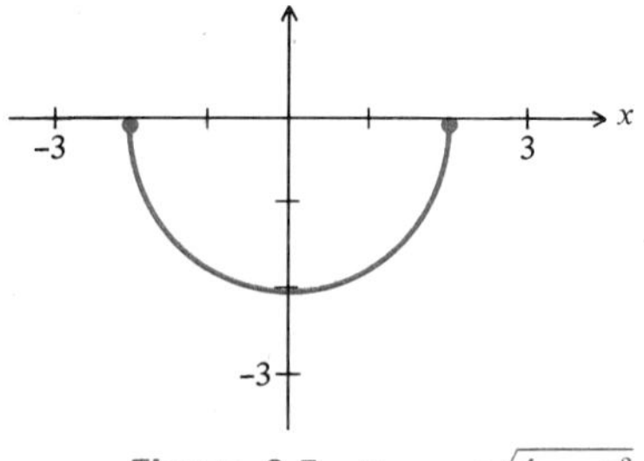

Figure 3.5 $y = -\sqrt{4 - x^2}$

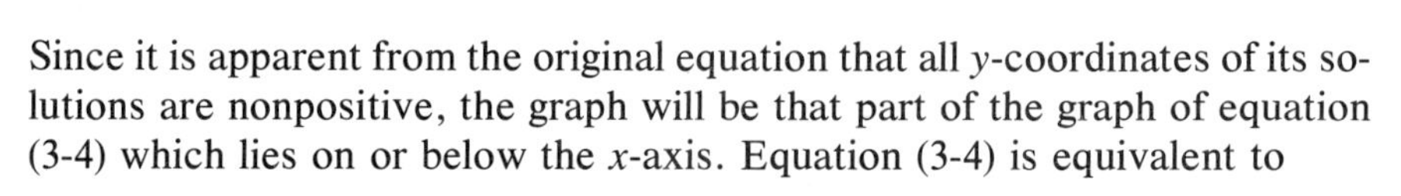

Since it is apparent from the original equation that all y-coordinates of its solutions are nonpositive, the graph will be that part of the graph of equation (3-4) which lies on or below the x-axis. Equation (3-4) is equivalent to

$$x^2 + y^2 = 4,$$

whose graph is a circle of radius 2 centered at the origin. The graph is shown in Figure 3.5. ■

Example 3
Graph $y = -2\sqrt{x + 3} + 2$.

Solution. This equation is equivalent to

$$y - 2 = -2\sqrt{x + 3}. \tag{3-5}$$

After squaring both sides, we have

$$(y - 2)^2 = 4(x + 3),$$

whose graph is a parabola that opens to the right and has $(-3, 2)$ as a vertex.

From equation (3-5) we see that the desired graph is the lower half of this parabola. We need to locate some other points on the graph. When $x = 1$, $y = -2$, and when $x = -2$, $y = 0$. The graph is shown in Figure 3.6. ■

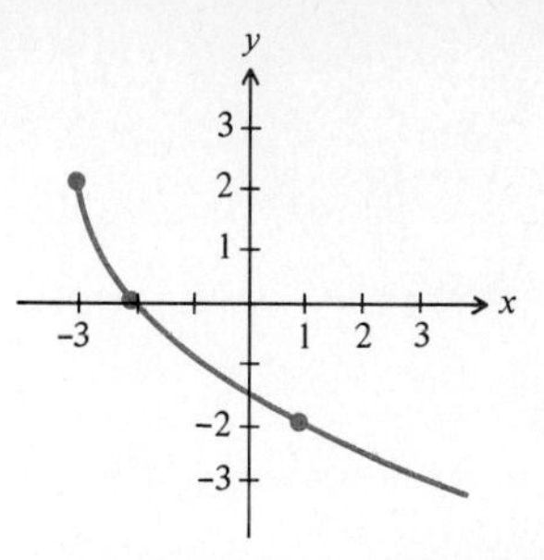

Figure 3.6 $y = -2\sqrt{x + 3} + 2$

Exercise
Graph $y = \sqrt{9 - x^2} + 3$.

Answer

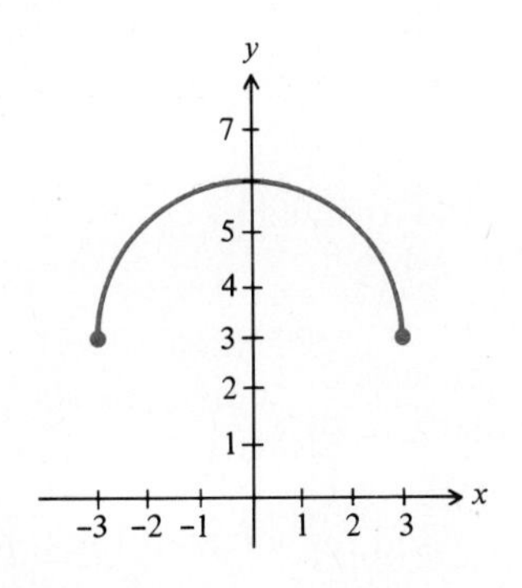

■

Problem Set 3.2

In Problems 1–12, graph the function, and then use the graph to determine the range.

1. $f(x) = -2$
2. $f(x) = 3$
3. $f(x) = 3x + 2$
4. $f(x) = -2x + 1$
5. $f(x) = \dfrac{-1}{2}x + 1$
6. $f(x) = \dfrac{2}{3}x - 2$
7. $f(x) = x^2$
8. $f(x) = -x^2 + 2$
9. $f(x) = x^2 + 4x + 3$
10. $f(x) = -x^2 - 2x + 1$
11. $f(x) = \dfrac{2x - x^2 + 5}{2}$
12. $f(x) = 2x^2 + 4x - 1$

In Problems 13–28, graph the equation.

13. $y = \sqrt{9 - x^2}$
14. $y = -\sqrt{25 - x^2}$
15. $y = -\sqrt{2x + 6}$
16. $y = 2\sqrt{2 - x}$
17. $y = \dfrac{3\sqrt{4 - x^2}}{2}$
18. $y = \dfrac{-3\sqrt{16 - x^2}}{4}$
19. $y = \dfrac{\sqrt{x^2 - 4}}{2}$
20. $y = \dfrac{-2\sqrt{x^2 + 9}}{3}$
21. $y = \sqrt{-2x} + 2$
22. $y = -2\sqrt{x - 1} - 3$
23. $y = \sqrt{5 - x^2 + 4x} - 1$
24. $y = \dfrac{\sqrt{55 - 9x^2 + 18x} - 9}{3}$

25. $y = \dfrac{-7\sqrt{25 - 4x^2}}{10}$

26. $y = \dfrac{3\sqrt{9 - 25x^2}}{5}$

27. $y = \dfrac{-5\sqrt{4 + x^2}}{4}$

28. $y = \dfrac{\sqrt{2x^2 - 4}}{3}$

3.3 Special Functions

For some functions it is necessary to split the formula of definition. For example, we may encounter a function f defined by

$$f(x) = \begin{cases} x^2 & \text{when } x \geq 0 \\ x & \text{when } x < 0. \end{cases}$$

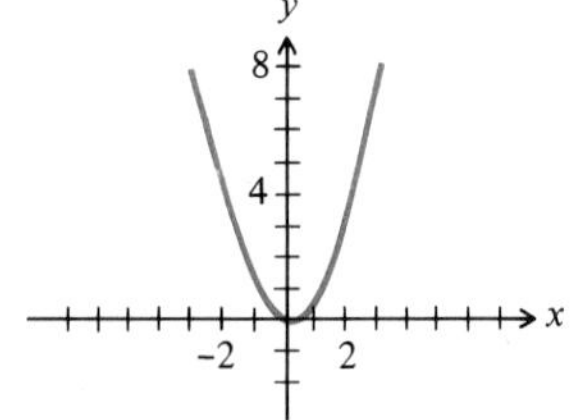

Figure 3.7 $y = x^2$

Now the domain of this function is R. To find the value of the function at any point in $[0, \infty)$ we use the formula $f(x) = x^2$. To find the value of the function at any point in $(-\infty, 0)$ we use the formula $f(x) = x$. Thus we have

$$f(-3) = -3, \qquad f(2) = 4, \qquad f(-1) = -1, \qquad \text{and} \qquad f(3) = 9.$$

We may graph f by first graphing $y = x^2$, as in Figure 3.7. Now the graph of f coincides with this parabola *when* $x \geq 0$. Hence the graph of f on and to the right of the y-axis (when $x \geq 0$) is that of Figure 3.8. Next we graph $y = x$(Figure 3.9). The graph of f coincides with this line *when* $x < 0$. Thus the graph of f to the left of the y-axis (when $x < 0$) is that of Figure 3.10. Putting the pieces together, we get the graph of f, shown in Figure 3.11.

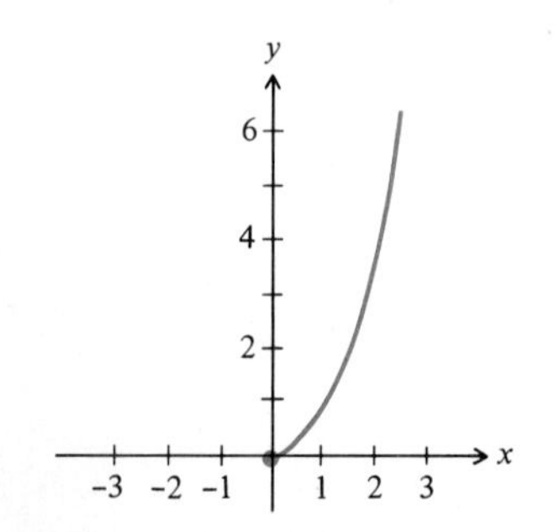

Figure 3.8 $y = x^2$ for $x \geq 0$

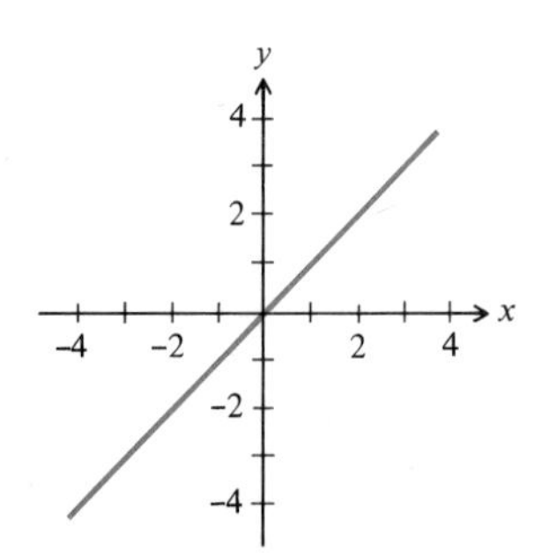

Figure 3.9 $y = x$

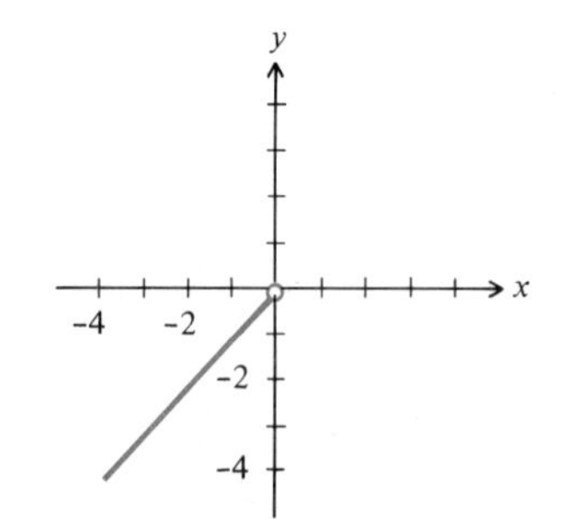

Figure 3.10 $y = x$ for $x < 0$

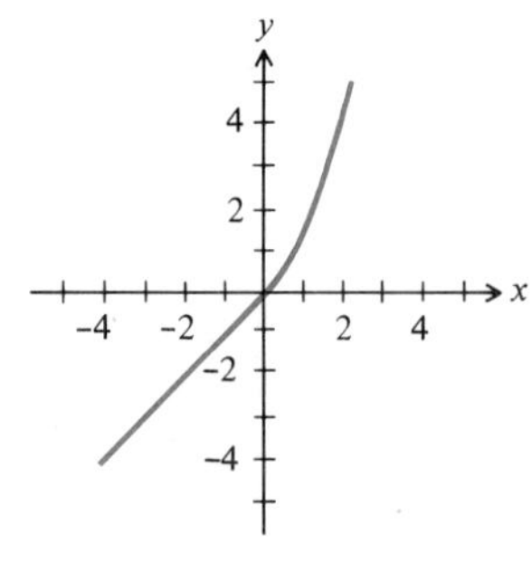

Figure 3.11

Example 1

Graph f, where $f(x) = \begin{cases} -x^2 & \text{when } x \leq 1 \\ x + 2 & \text{when } x > 1. \end{cases}$

Solution. First we graph that part of the parabola $y = -x^2$ for $x \leq 1$. The result is shown in Figure 3.12. To this we add that part of the line $y = x + 2$ for $x > 1$. We get the graph shown in Figure 3.13. ■

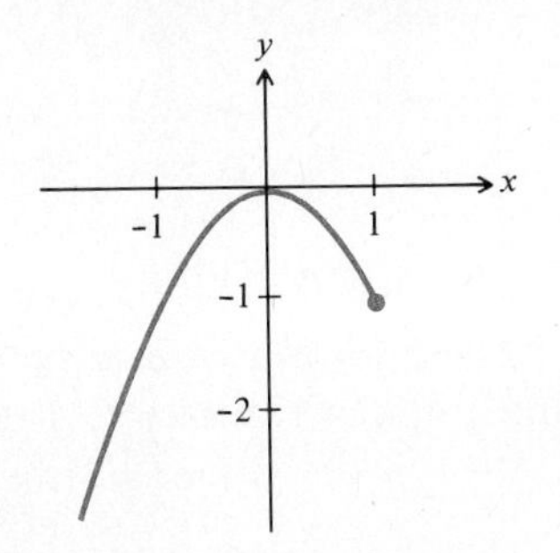

Figure 3.12 $y = -x^2$ for $x \leq 1$

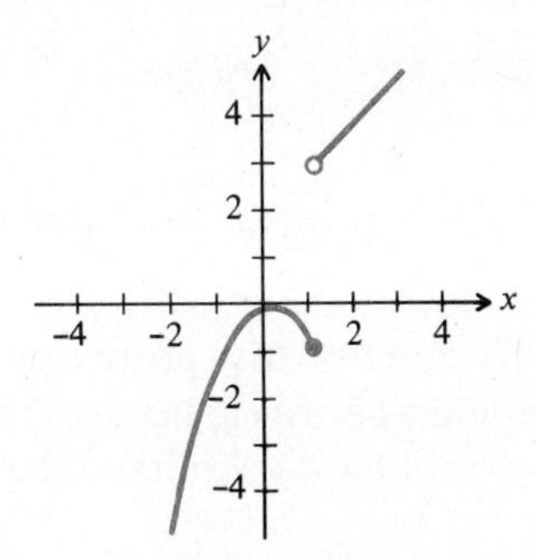

Figure 3.13

Exercise

Graph f, where $f(x) = \begin{cases} x^2 & \text{when } x \leq -1 \\ -x - 1 & \text{when } x > -1. \end{cases}$

Answer.

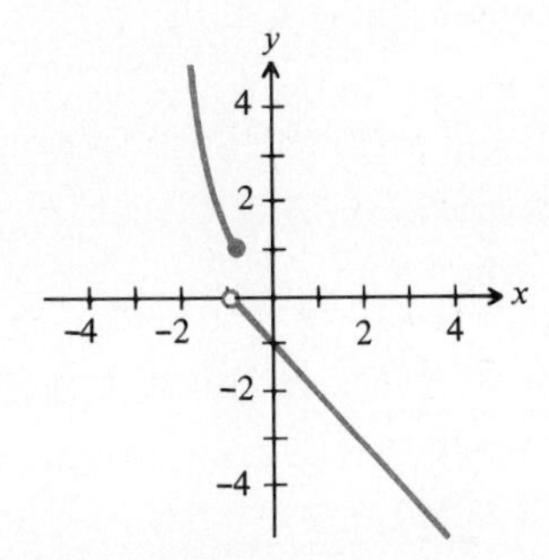

■

From the definition of absolute value, we see that if p_x is an algebraic expression in x, then

$$|p_x| = \begin{cases} p_x & \text{when } p_x \geq 0 \\ -p_x & \text{when } p_x < 0. \end{cases}$$

Thus the function f, defined by $f(x) = |x|$, could be written with a split formula of definition as

$$f(x) = |x| = \begin{cases} x & \text{when } x \geq 0 \\ -x & \text{when } x < 0. \end{cases}$$

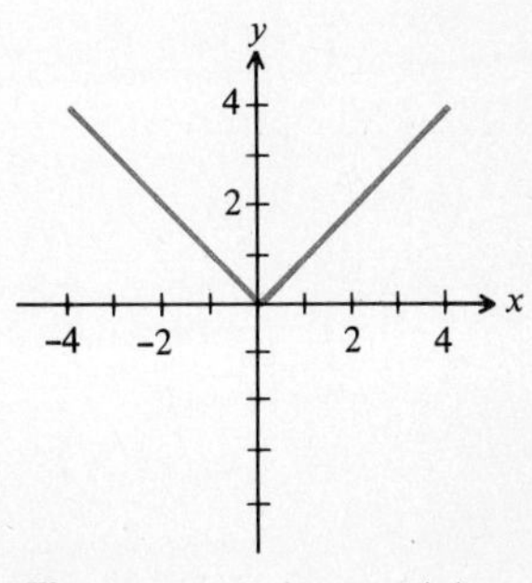

Figure 3.14 $f(x) = |x|$

Graphing the line $y = x$, for $x \geq 0$, and adding the graph of $y = -x$, for $x < 0$, we get the graph in Figure 3.14.

As another example, let's consider the function g defined by $g(x) =$

$|x - 2|$. It may be written

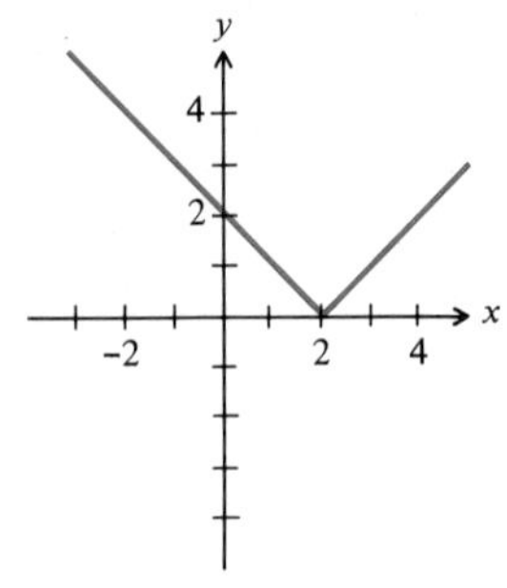

Figure 3.15 $g(x) = |x - 2|$

$$g(x) = \begin{cases} x - 2 & \text{when } x - 2 \geq 0 \\ -(x - 2) & \text{when } x - 2 < 0. \end{cases}$$

Simplifying, we get

$$g(x) = \begin{cases} x - 2 & \text{when } x \geq 2 \\ -x + 2 & \text{when } x < 2. \end{cases}$$

By graphing the proper parts of the lines $y = x - 2$ and $y = -x + 2$, we get the graph in Figure 3.15. Note that it is simply the graph of f (where $f(x) = |x|$) shifted two units to the right.

Example 2
Graph $y = |2x - 1| + 3$.

Solution. By using the definition of absolute value we split the formula of definition.

$$y = \begin{cases} (2x - 1) + 3 & \text{when } 2x - 1 \geq 0 \\ -(2x - 1) + 3 & \text{when } 2x - 1 < 0. \end{cases}$$

After simplifying, we have

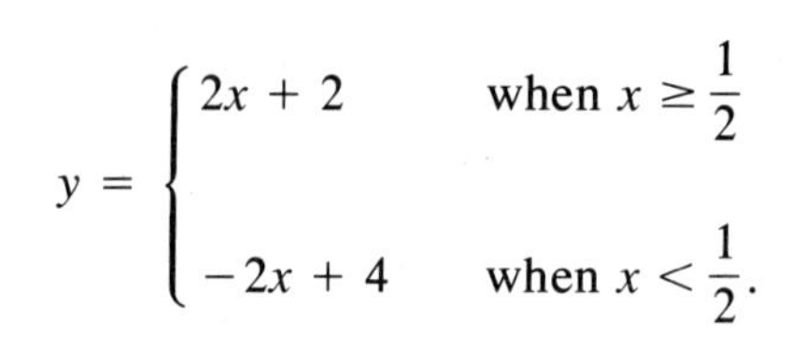

$$y = \begin{cases} 2x + 2 & \text{when } x \geq \dfrac{1}{2} \\ -2x + 4 & \text{when } x < \dfrac{1}{2}. \end{cases}$$

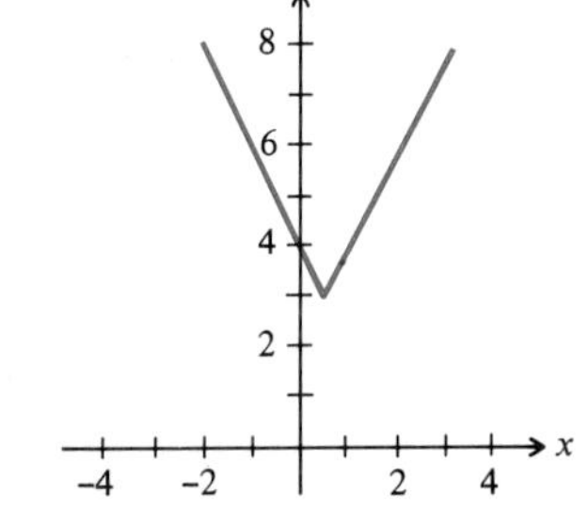

Figure 3.16 $y = |2x - 1| + 3$

By graphing the proper parts of $y = 2x + 2$ and $y = -2x + 4$, we get the graph shown in Figure 3.16. ■

Example 3
Graph $y = |x - 1| + x$.

Solution. Using the definition of absolute value we split the formula of definition.

$$y = \begin{cases} (x - 1) + x & \text{when } x - 1 \geq 0 \\ -(x - 1) + x & \text{when } x - 1 < 0. \end{cases}$$

Simplifying, we get

$$y = \begin{cases} 2x - 1 & \text{when } x \geq 1 \\ 1 & \text{when } x < 1. \end{cases}$$

Graphing the line $y = 2x - 1$, for $x \geq 1$, and adding the graph of $y = 1$, for $x < 1$, we get the graph shown in Figure 3.17. ■

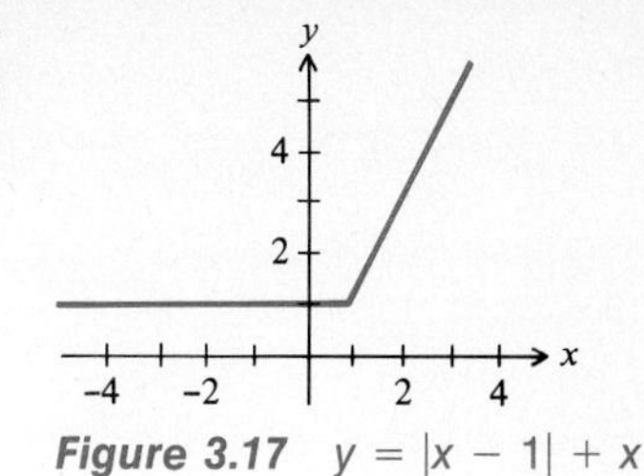

Figure 3.17 $y = |x - 1| + x$

Exercise

Graph each of the following.

(a) $y = |2x + 1| - 3$ (b) $y = |x + 1| - x$

Answers

(a)

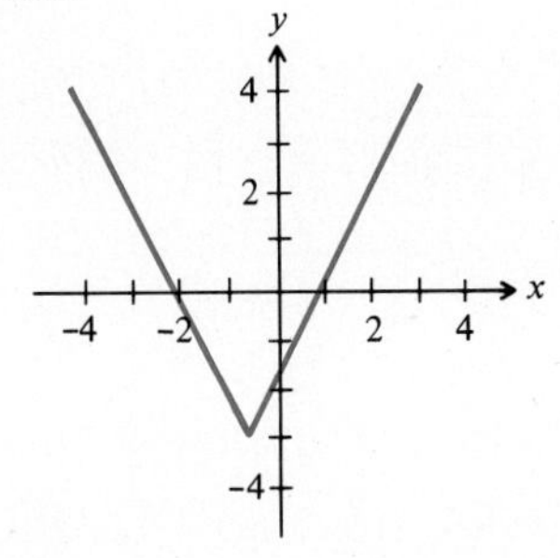

(b)

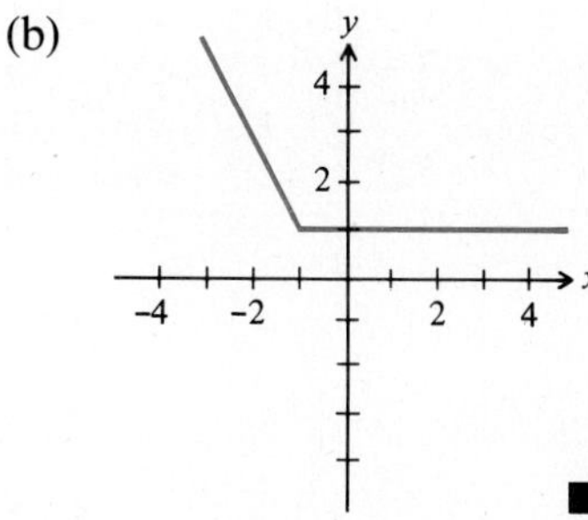

■

Problem Set 3.3

In Problems 1–8, graph f.

1. $f(x) = \begin{cases} 2x + 5 & \text{when } x \leq -1 \\ x - 2 & \text{when } x > -1 \end{cases}$

2. $f(x) = \begin{cases} -2x + 1 & \text{when } x < 2 \\ 3x - 9 & \text{when } x \geq 2 \end{cases}$

3. $f(x) = \begin{cases} -x^2 & \text{when } x \leq 0 \\ x + 2 & \text{when } x > 0 \end{cases}$

4. $f(x) = \begin{cases} (x + 1)^2 & \text{when } x < -1 \\ 2x + 3 & \text{when } x \geq -1 \end{cases}$

5. $f(x) = \begin{cases} x & \text{when } x < 0 \\ \sqrt{x} & \text{when } x \geq 0 \end{cases}$

6. $f(x) = \begin{cases} -x & \text{when } x < 2 \\ \dfrac{\sqrt{x^2 - 4}}{2} & \text{when } x \geq 2 \end{cases}$

7. $f(x) = \begin{cases} -2x - 4 & \text{when } x < -2 \\ \sqrt{4 - x^2} & \text{when } -2 \leq x \leq 2 \\ x + 1 & \text{when } x > 2 \end{cases}$

8. $f(x) = \begin{cases} 2 & \text{when } x \leq -3 \\ \sqrt{4 - x^2} & \text{when } -3 < x \leq 0 \\ x - 2 & \text{when } x > 0 \end{cases}$

In Problems 9–20, graph the function defined.

9. $y = |x - 3| + 1$

10. $y = -3|x + 2| + 5$

11. $y = -|3x + 5| - 3$

12. $y = |2x - 3| + 2$

13. $y = |x + 2| + x$

14. $y = |2 - x| - x$

15. $y = -3|2x + 3| - 2x$

16. $y = 2|2x - 5| + 3x$

17. $y = |x - 2| + |x|$

18. $y = |3 - x| - |2x|$

19. $y = |x^2 - 4x - 5| + x^2$

20. $y = 2|3 - x| + x^2$

The symbol $[x]$ denotes the largest integer less than or equal to x. For example, $[2.5] = 2$, $[-\frac{3}{2}] = -2$, $[5] = 5$, $[\sqrt{2}] = 1$, and so on. The function f: R → R defined by $f(x) = [x]$ is called the **greatest integer function.** Using a split formula of definition, it may be written

$$f(x) = [x] = \begin{cases} \ \vdots & \\ -2 & \text{when } -2 \le x < -1 \\ -1 & \text{when } -1 \le x < 0 \\ 0 & \text{when } 0 \le x < 1 \\ 1 & \text{when } 1 \le x < 2 \\ 2 & \text{when } 2 \le x < 3 \\ \ \vdots & \end{cases}$$

In Problems 21–26, graph the function.

21. $f(x) = [x]$

22. $f(x) = 2[x]$

23. $f(x) = [-x]$

24. $f(x) = -[x]$

25. $f(x) = [x - 2]$

26. $f(x) = \left[\frac{x}{2}\right]$

27. Let f be the function which rounds off a number to the nearest integer, and let g be the function which rounds off a number to the nearest tenth. Use the greatest integer symbol to write the formulas of definition for f and g.

28. A utility company charges its residential customers a minimum monthly amount of \$2.25 to include 20 kilowatt-hours (KWH) used per month. For the next 30 KWH used the charge is 8.5 cents per KWH. For the next 50 KWH used, the charge is 6.5 cents per KWH. And for all KWH used in excess of 100, the charge is 5 cents per KWH. Write the formula for the monthly charge, in dollars, for x kilowatts used.

3.4 Combining Functions

We will now consider the operations of addition, subtraction, multiplication, and division of functions. The definitions are straightforward and are precisely what we might expect.

Let f and g be functions with domains D_f and D_g, respectively.

1. The sum of f and g, denoted $f + g$, is defined by

$$(f + g)(x) = f(x) + g(x) \qquad \text{for } x \in D_f \cap D_g.$$

2. The difference of f and g, denoted $f - g$, is defined by

$$(f - g)(x) = f(x) - g(x) \qquad \text{for } x \in D_f \cap D_g.$$

3. The product of f and g, denoted $f \cdot g$ (usually written fg), is defined by

$$(f \cdot g)(x) = f(x) \cdot g(x) \qquad \text{for } x \in D_f \cap D_g.$$

4. The quotient of f by g, denoted f/g, is defined by

$$\left(\frac{f}{g}\right)(x) = \frac{f(x)}{g(x)} \qquad \text{for all } x \in D_f \cap D_g \quad \text{such that } g(x) \neq 0.$$

Note that these combinations are not defined at points outside the intersection of the domains. These combinations exist only at values of x which are in the domains of both functions. And further, in the case of the quotient, we must exclude all values in the intersection which make the dividing function zero.

Example 1
Let $f(x) = x^2 - x + 1$ for $x \in (-2, 2)$, $g(x) = x^2 + x$ for $x \in (-\infty, 0]$. Define $f + g, f - g, fg$, and f/g.

Solution. Let's first concern ourselves with the domains of the combinations. For the sum, difference, and product, we intersect the domains of f and g.

$$(-2, 2) \cap (-\infty, 0] = (-2, 0].$$

This interval, excluding those values which make $g(x)$ equal zero, will be the domain of f divided by g. Since the solution set of $x^2 + x = 0$ is $\{0, -1\}$, the domain of f/g is

$$(-2, 0] - \{0, -1\} = (-2, -1) \cup (-1, 0).$$

Next we simply apply the appropriate definition and note the domain.

$$(f + g)(x) = f(x) + g(x) = x^2 - x + 1 + x^2 + x = 2x^2 + 1 \qquad \text{for } x \in (-2, 0]$$

$$(f - g)(x) = f(x) - g(x) = x^2 - x + 1 - (x^2 + x) = -2x + 1 \qquad \text{for } x \in (-2, 0]$$

$$(fg)(x) = f(x) \cdot g(x) = (x^2 - x + 1)(x^2 + x) = x^4 + x \qquad \text{for } x \in (-2, 0]$$

$$\left(\frac{f}{g}\right)(x) = \frac{f(x)}{g(x)} = \frac{x^2 - x + 1}{x^2 + x} \qquad \text{for } x \in (-2, -1) \cup (-1, 0) \qquad ■$$

Exercise
Let $f(x) = 2x^2 - x + 3$ for $x \in (-3, 4)$, $g(x) = x^2 - 2x$ for $x \in [0, \infty)$. Define $f + g, f - g, fg$, and f/g.

Answers. $(f+g)(x) = 3x^2 - 3x + 3$ for $x \in [0, 4)$

$(f-g)(x) = x^2 + x + 3$ for $x \in [0, 4)$

$(fg)(x) = 2x^4 - 5x^3 + 5x^2 - 6x$ for $x \in [0, 4)$

$$\left(\frac{f}{g}\right)(x) = \frac{2x^2 - x + 3}{x^2 - 2x} \quad \text{for } x \in (0, 2) \cup (2, 4)$$

■

There is another very important combination of functions. Let f and g be functions, and suppose that there exists an element c in the domain of f such that $f(c)$ is in the domain of g. Then $g(f(c))$ is an element of the range of g.

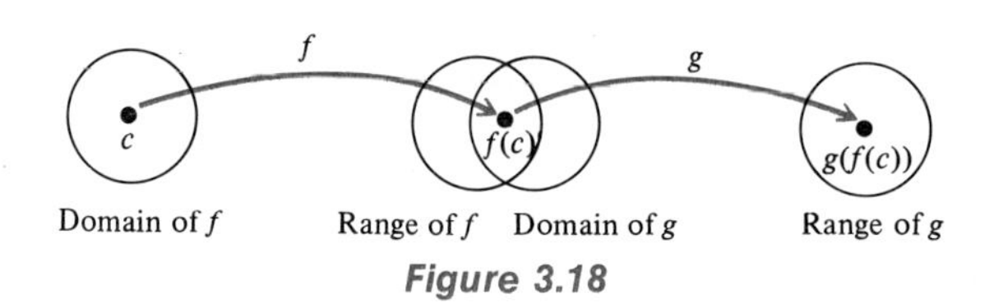

Figure 3.18

See Figure 3.18. By linking f and g together in this fashion, we are constructing a new function. This function is called g **composition** f and is denoted $g \circ f$. Its formula of definition is

$$(g \circ f)(x) = g(f(x)).$$

Its domain is the set of all elements in the domain of f whose images under f are in the domain of g. Note that this is not necessarily the same as the domain of f. If a is in the domain of f but its image under f, $f(a)$, is not in the domain of g, then $g(f(a))$ doesn't make sense.

Example 2

Let $f(x) = 2x - 5$ and $g(x) = \sqrt{x}$. Find each of the following.
(a) $(f \circ g)(4)$ (b) $(f \circ g)(1)$ (c) $(g \circ f)(4)$ (d) $(g \circ f)(1)$

Solution. (a) By the definition,

$$(f \circ g)(4) = f(g(4)).$$

Using the formula of definition for g, we get

$$g(4) = \sqrt{4} = 2.$$

Thus

$$\begin{aligned}(f \circ g)(4) &= f(g(4)) \\ &= f(2) \\ &= -1.\end{aligned}$$

(b) We have

$$
\begin{aligned}
(f \circ g)(1) &= f(g(1)) \\
&= f(1) \\
&= -3.
\end{aligned}
$$

(c) We get

$$
\begin{aligned}
(g \circ f)(4) &= g(f(4)) \\
&= g(3) \\
&= \sqrt{3}.
\end{aligned}
$$

(d) We have

$$
\begin{aligned}
(g \circ f)(1) &= g(f(1)) \\
&= g(-3).
\end{aligned}
$$

But $g(-3)$ is not defined. Thus $(g \circ f)(1)$ is not defined. ■

Exercise
Let $f(x) = 3x - 4$ and $g(x) = \sqrt{x}$. Find each of the following.
(a) $(f \circ g)(1)$ (b) $(f \circ g)(9)$ (c) $(g \circ f)(2)$ (d) $(g \circ f)(1)$

Answers. (a) -1 (b) 5 (c) $\sqrt{2}$ (d) not defined ■

In part (d) of Example 2 we found that $(g \circ f)(1)$ is not defined because $f(1)$ is not in the domain of g. What is the domain of $g \circ f$ of Example 2? It consists of all elements in the domain of f whose images under f are in the domain of g. Since the domain of f is R, and the domain of g is $[0, \infty)$, the domain of $g \circ f$ consists of all x such that

$$x \in \mathsf{R} \quad \text{and} \quad f(x) \in [0, \infty).$$

Since $f(x) = 2x - 5$, this is the solution set of

$$2x - 5 \geq 0,$$

which is $[\frac{5}{2}, \infty)$.

What is the domain of $f \circ g$? It is the set of all elements in the domain of g whose images under g are in the domain of f. That is, all x such that

$$x \in [0, \infty) \quad \text{and} \quad g(x) \in \mathsf{R}.$$

This is just $[0, \infty)$.

Example 3
Let $f(x) = x^2 + x$ for $x \geq 0$, and $g(x) = 2x - 1$ for $x \leq 6$. Define $g \circ f$.

Solution. Now the domain of $g \circ f$ consists of all elements in the domain of f whose images under f are in the domain of g. Thus the domain of $g \circ f$ consists of all x such that

$$x \geq 0 \qquad \text{and} \qquad f(x) \leq 6.$$

Since $f(x) = x^2 + x$, we must find the solution set of

$$x^2 + x \leq 6.$$

Using the critical number method (or considering cases), we get a solution set of $[-3, 2]$. Hence the domain of $g \circ f$ is the set of all x such that

$$x \geq 0 \qquad \text{and} \qquad x \in [-3, 2].$$

Consequently, the domain is $[0, 2]$. To find the formula of definition, we replace x by $f(x)$ in the formula for g. We get

$$\begin{aligned}(g \circ f)(x) &= g(f(x)) \\ &= g(x^2 + x) \\ &= 2(x^2 + x) - 1 \\ &= 2x^2 + 2x - 1.\end{aligned}$$

Therefore,

$$(g \circ f)(x) = 2x^2 + 2x - 1 \qquad \text{for } x \in [0, 2].$$ ■

Exercise
Let $f(x) = x^2 + x$ for $x \geq 0$, $g(x) = 2x - 1$ for $x \leq 6$. Define $f \circ g$.

Answer. $(f \circ g)(x) = 4x^2 - 2x$ for $x \in [\frac{1}{2}, 6]$ ■

From the preceding exercise and example, we see that $f \circ g$ and $g \circ f$ are, in general, different functions. Hence the order of the composition is extremely important.

Problem Set 3.4

In Problems 1–8, define $f + g$, $f - g$, fg, and f/g.

1. $f(x) = x^2 - x$
   $g(x) = x - 2$
2. $f(x) = x^2 + 3x$
   $g(x) = x + 1$
3. $f(x) = x^2 + 4$ for $x \in [-4, \infty)$
   $g(x) = 2x^2 - x - 1$ for $x \in (0, \infty)$
4. $f(x) = x^2 - x + 1$ for $x \in (-3, 3)$
   $g(x) = x^2 - 1$ for $x \in (-\infty, 0]$

5. $f(x) = 2x^2 - 3x - 2$

$g(x) = \dfrac{x}{x - 2}$

6. $f(x) = \dfrac{x}{2x + 1}$

$g(x) = 2x^2 - 9x - 5$

7. $f = \{(1, 6), (2, -5), (3, -2)\}$
$g = \{(0, 5), (1, -4), (2, 0)\}$

8. $f = \{(-1, 3), (-2, 4), (-3, -7)\}$
$g = \{(-1, 0), (-2, 5), (3, -7)\}$

In Problems 9–16, define $f \circ g$ and $g \circ f$.

9. $f(x) = x^2$, $g(x) = -x + 1$
10. $f(x) = x^2 + 2x$, $g(x) = 2x - 1$
11. $f(x) = 2x^2 + x$, $g(x) = x - 2$ for $x > 3$
12. $f(x) = 2x + 1$ for $x \leq 3$, $g(x) = 6x^2 + 7x$
13. $f(x) = 3x^2 + 1$, $g(x) = \sqrt{x + 2}$
14. $f(x) = \dfrac{1}{x^2 - 1}$, $g(x) = \sqrt{x + 1}$
15. $f = \{(-2, 5), (5, 7), (0, 6)\}$, $g = \{(5, 7), (6, 5), (7, -2)\}$
16. $f = \{(0, 5), (-3, 8), (2, 6)\}$, $g = \{(-1, -3), (0, 2), (6, 0)\}$

17. Define $f \circ f$ if $f(x) = -2x + 5$ for $x \geq 1$.
18. Define $f \circ f$ if $f(x) = x^2 - 1$.

3.5 Inverses

Let A and B be nonempty sets, and let f be a function from A to B. The **inverse** of f, denoted f^{-1}, is the relation from B to A defined by

$$f^{-1} = \{(y, x) \mid (x, y) \in f\}.$$

Thus, given a function, we get its inverse by simply interchanging the first and second coordinates of each ordered pair. For example, if

$$f = \{(1, 4), (2, 4), (3, 2)\},$$

then

$$f^{-1} = \{(4, 1), (4, 2), (2, 3)\}.$$

Note that although f is a function, f^{-1} is not. Those functions whose inverses are also functions are called one-to-one. A function f is said to be a **one-to-one function** provided that no two different ordered pairs of f have the same *second* coordinate.

A function associates each element of the domain with a unique element of the range. But an element in the range may be paired with more than one element of the domain. However, *for one-to-one functions*, each element of the range corresponds to exactly one element in the domain. Thus, if f is a

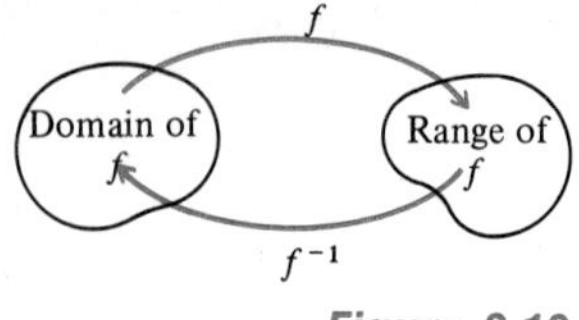

Figure 3.19

one-to-one function, then f^{-1} is a function from the range of f into the domain of f. See Figure 3.19.

Examination of the graph of a function will indicate whether it is one-to-one. A function is one-to-one provided that no horizontal line intersects the graph in more than one point.

Example 1
Determine which of the following functions are one-to-one.
(a) $\{(-1, 3), (-2, 4), (-3, 5)\}$ (b) $\{(x, y) \mid y = x^2\}$
(c) $\{(x, y) \mid y = 2x - 1\}$.

Solution. (a) This function is clearly one-to-one by the definition, as no two different ordered pairs have the same second coordinate.
(b) This function is not one-to-one because each nonzero value of y corresponds to two distinct values of x. This may be seen by noting that every horizontal line above the x-axis intersects the graph of $y = x^2$ in exactly two points. (However, to show that a function is not one-to-one, it suffices to find only two ordered pairs of the function with the same second coordinate. This corresponds to one horizontal line which intersects the graph in more than one point.)
(c) This function is one-to-one as each value of y arises from only one value of x. (That is, two different values of x always generate two different values of y.) Again we could examine the graph, and note that no horizontal line intersects the graph in more than one point. ■

Exercise
Determine which of the following functions are one-to-one.
(a) $\{(-1, 3), (-2, 7), (-3, 3)\}$ (b) $\{(x, y) \mid y = 1/x^2\}$
(c) $\{(x, y) \mid y = x^3\}$

Answer. (c) ■

Let g be the function defined by $y = 3x + 2$. What equation defines g^{-1}? Since g^{-1} is derived by switching the first and second coordinates of the ordered pairs in g, the equation which defines g^{-1} is obtained by interchanging the variables x and y in the equation which defines g. We get $x = 3y + 2$.

Example 2
Determine the inverse of each of the following one-to-one functions:
(a) $f(x) = 2x - 1$ (b) $h(x) = -x^2, x \leq 0$

Solution. (a) The function f is defined by the equation

$$y = 2x - 1.$$

As previously indicated, the inverse function can be found by switching the variables x and y. Thus f^{-1} is defined by

$$x = 2y - 1.$$

Solving this equation for y, we have

$$y = \frac{x + 1}{2}.$$

Therefore,

$$f^{-1}(x) = \frac{x + 1}{2}.$$

(b) The function h is defined by

$$y = -x^2, \qquad x \leq 0.$$

Thus its inverse is defined by

$$x = -y^2, \qquad y \leq 0,$$

which is equivalent to

$$y^2 = -x, \qquad y \leq 0.$$

Taking the square root of both sides, we get

$$|y| = \sqrt{-x}, \qquad y \leq 0.$$

To simplify $|y|$, we must note that $y \leq 0$ (since $x \leq 0$ before we switched x and y). So we have $|y| = -y$, and

$$\begin{aligned} -y &= \sqrt{-x} \\ y &= -\sqrt{-x}. \end{aligned}$$

Thus

$$h^{-1}(x) = -\sqrt{-x}. \qquad \blacksquare$$

Exercise

Determine the inverse of each of the following one-to-one functions.
(a) $f(x) = 3x + 2$ (b) $h(x) = -(x - 1)^2$, $x \leq 1$

Answers. (a) $f^{-1}(x) = \dfrac{x - 2}{3}$ (b) $h^{-1}(x) = 1 - \sqrt{-x}$

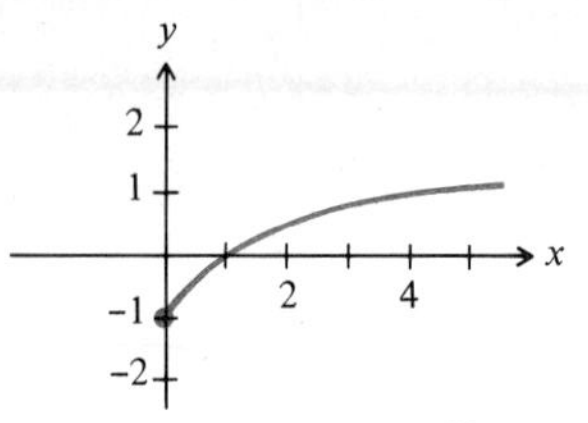

Figure 3.20 $g(x) = \sqrt{x} - 1$

Example 3
Sketch the graph of $g(x) = \sqrt{x} - 1$, and if one-to-one, determine the inverse.

Solution. From the graph (Figure 3.20) we see that g is one-to-one. Since the function g is defined by

$$y = \sqrt{x} - 1, \tag{3-6}$$

the inverse is defined by

$$x = \sqrt{y} - 1. \tag{3-7}$$

To solve this equation for y we square both sides after adding 1 to each side and get

$$y = (x + 1)^2. \tag{3-8}$$

But squaring both sides does not, in general, produce an equivalent equation. [Note, for example, that $(-2, 1)$ is a solution of (3-8) but is not a solution of (3-7).] We must specify the restrictions on x. We see that the range of g, and thus the domain of g^{-1}, is $[-1, \infty)$. Hence equation (3-8) is equivalent to (3-7) provided that $x \geq -1$. Therefore,

$$g^{-1}(x) = (x + 1)^2, \qquad x \geq -1. \qquad \blacksquare$$

Exercise
Sketch the graph of $g(x) = \sqrt{x - 1}$, and if one-to-one, determine the inverse.

Answer.

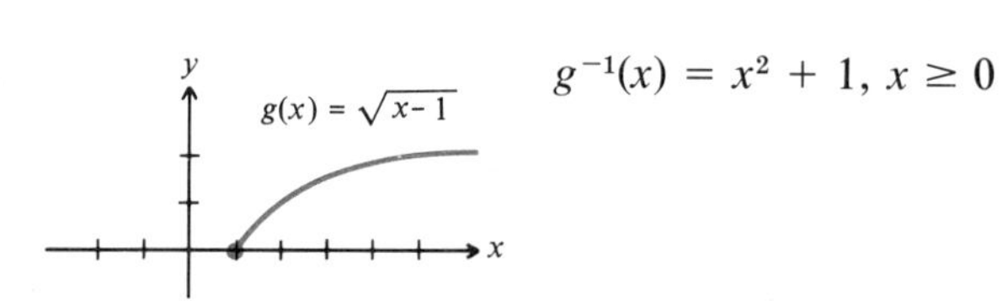

$g^{-1}(x) = x^2 + 1, x \geq 0$ $\blacksquare$

Now (a, b) is an element of a one-to-one function f if and only if (b, a) is an element of the function f^{-1}. In other words, $f^{-1}(b) = a$ if and only if $f(a) = b$. This fact forms the basis of Theorem 3.1.

Theorem 3.1

Let f be a one-to-one function. Then
(i) $f(f^{-1}(x)) = x$ *for all x in the domain of f^{-1}.*
(ii) $f^{-1}(f(x)) = x$ *for all x in the domain of f.*

Theorem 3.1 means that, in a sense, the inverse function "undoes" what the function does. That is, if we start with x in the domain of f, apply f and then f^{-1}, we end up with x. If we start with x in the domain of f^{-1}, apply f^{-1} and then f, we end up with x. See Figure 3.21.

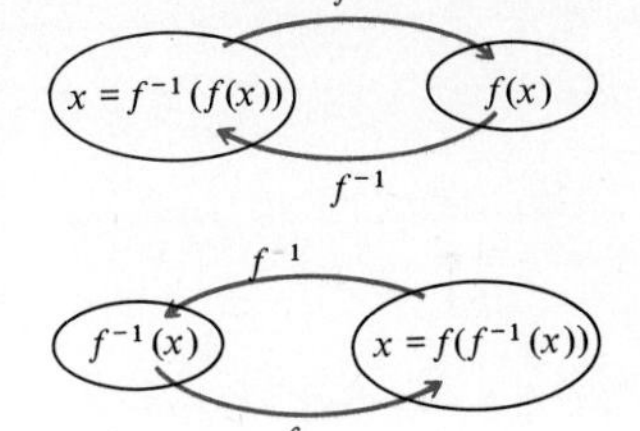

Theorem 3.1 serves as a convenient means of checking the inverse. Let's look at the function g from Example 3:

$$g(x) = \sqrt{x} - 1.$$

We determined the inverse to be

$$g^{-1}(x) = (x + 1)^2, \qquad x \geq -1.$$

We need to show that $g(g^{-1}(x)) = x$ and $g^{-1}(g(x)) = x$. We have

$$\begin{aligned} g(g^{-1}(x)) &= g((x + 1)^2) \\ &= \sqrt{(x + 1)^2} - 1 \\ &= |x + 1| - 1 \\ &= x + 1 - 1 \qquad \text{(since } x + 1 \geq 0) \\ &= x \end{aligned}$$

and

$$\begin{aligned} g^{-1}(g(x)) &= g^{-1}(\sqrt{x} - 1) \\ &= (\sqrt{x} - 1 + 1)^2 \\ &= x. \end{aligned}$$

As seen in Example 4, Theorem 3.1 may also be used to determine the formula of the inverse.

Example 4

Determine the inverse of $f(x) = -2x + 1$ by using Theorem 3.1.

Solution. First we evaluate the function f at $f^{-1}(x)$. We get

$$f(f^{-1}(x)) = -2(f^{-1}(x)) + 1.$$

But according to Theorem 3.1, $f(f^{-1}(x)) = x$. Therefore,

$$-2(f^{-1}(x)) + 1 = x.$$

Upon solving this equation for $f^{-1}(x)$, we have

$$f^{-1}(x) = \frac{1 - x}{2}.$$

■

Exercise
Determine the inverse of $f(x) = -3x + 2$ by using Theorem 3.1.

Answer. $f^{-1}(x) = \dfrac{2 - x}{3}$ ■

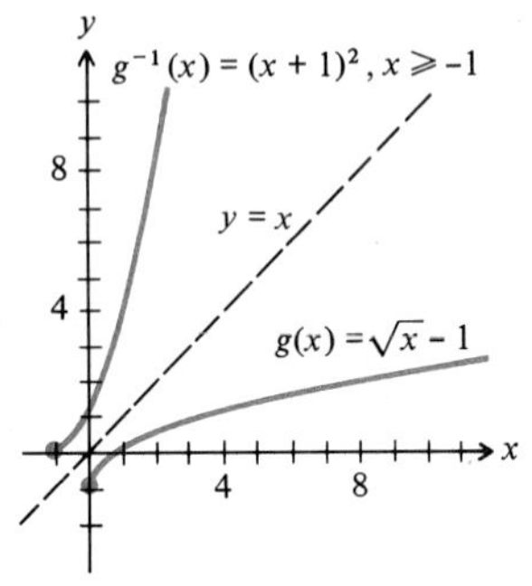

Figure 3.22

There is an interesting relationship between the graphs of a function and its inverse. Let's graph the functions g and g^{-1} of Example 3 on the same coordinate system, with the same scale on each axis. See Figure 3.22.

Now (a, b) is on the graph of g if and only if (b, a) is on the graph of g^{-1}. We see that the points (1, 0), (4, 1), and (9, 2) are on the graph of g, while (0, 1), (1, 4), and (2, 9) are on the graph of g^{-1}. Consequently, if this page were folded along the dashed line $y = x$, the graphs of g and g^{-1} would coincide. In this sense, we say that each graph is the reflection of the other through the line $y = x$.

It follows that, given any one-to-one function f, for each point $P(a, b)$ on the graph of f, there is a point $Q(b, a)$ on the graph of f^{-1}, such that the line $y = x$ is the perpendicular bisector of $\overline{PQ}$.

Problem Set 3.5

In Problems 1–10, determine whether the function is one-to-one.

1. $\{(0, -3), (-2, -5), (-3, -6)\}$
2. $\{(1, -3), (2, -5), (3, -3)\}$
3. $\{(x, y) \mid 2x + y = 5\}$
4. $\{(x, y) \mid y = -3\}$
5. $\{(x, y) \mid y = x^4\}$
6. $\left\{(x, y) \mid y = \dfrac{1}{x}\right\}$
7. $\{(x, y) \mid y = \sqrt{x}\}$
8. $\{(x, y) \mid y = \sqrt{4 - x^2}\}$
9. $\left\{(x, y) \mid y = \dfrac{-3\sqrt{4 - x^2}}{2}\right\}$
10. $\{(x, y) \mid y = |x - 1| + x\}$

In Problems 11–14, determine whether the graph is the graph of a one-to-one function.

11.

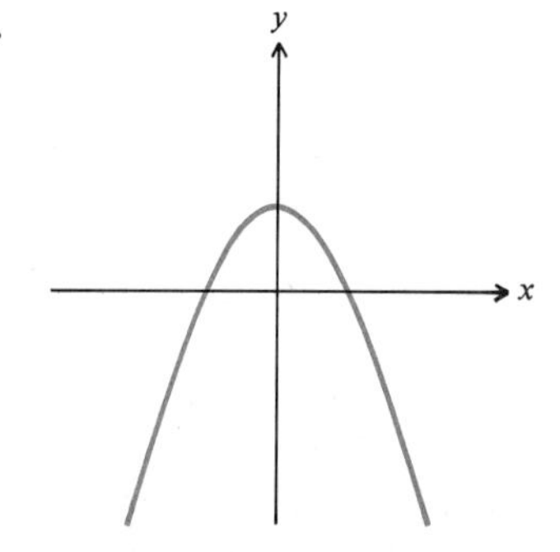

12.

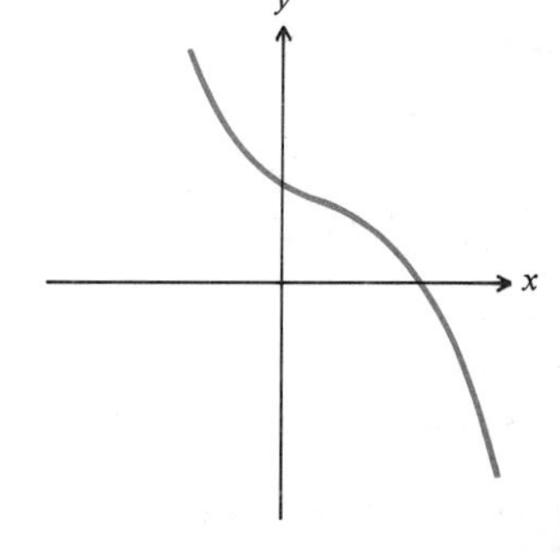

13.

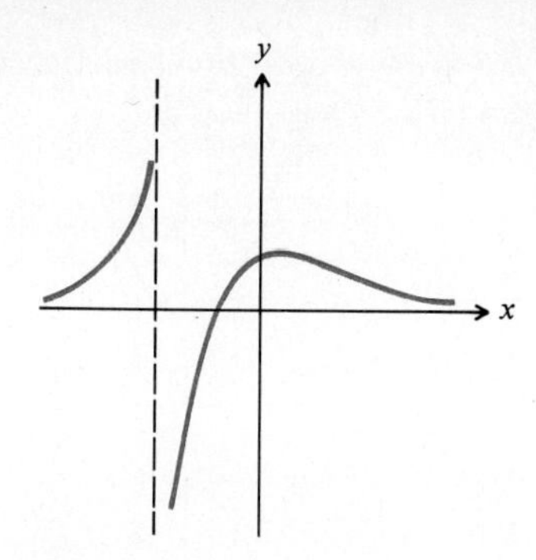

14.

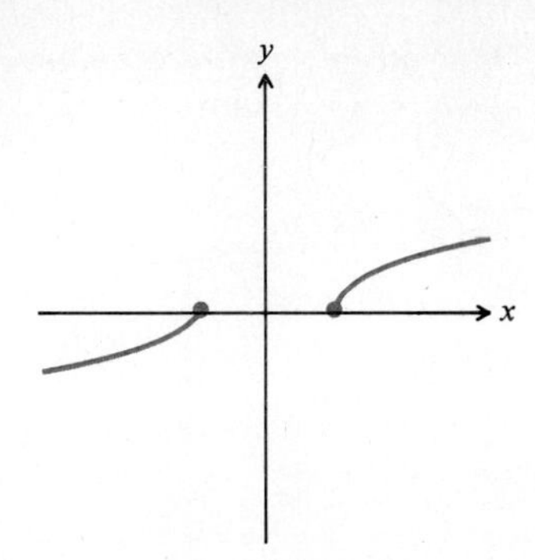

In Problems 15–26, determine the inverse of the one-to-one function.

15. $f(x) = 3x - 2$

16. $f(x) = 2 - 5x$

17. $f(x) = -x^2 + 2,\ x \le 0$

18. $f(x) = (x + 2)^2,\ x \ge -2$

19. $f(x) = x^2 + 4x + 3,\ x \le -2$

20. $f(x) = x^2 - 2x + 5,\ x \ge 1$

21. $f(x) = \sqrt{2x + 1}$

22. $f(x) = -\sqrt{2x - 3} + 1$

23. $f(x) = \dfrac{2x - 1}{x - 1}$

24. $f(x) = \dfrac{-2x - 3}{2x + 4}$

25. $f(x) = \begin{cases} 3x + 1 & \text{when } x < 0 \\ 2x^2 + 1 & \text{when } x \ge 0 \end{cases}$

26. $f(x) = \begin{cases} -(x + 1)^3 & \text{when } x < 0 \\ -2x - 1 & \text{when } x \ge 0 \end{cases}$

In Problems 27–30, sketch the graph, and if one-to-one, determine the inverse.

27. $f(x) = \sqrt{4 - x^2},\ 0 \le x \le 2$

28. $f(x) = \dfrac{-\sqrt{9 - x^2}}{3},\ -3 \le x \le 0$

29. $f(x) = x^2 + 4x + 4,\ x \le 0$

30. $f(x) = \begin{cases} x & \text{when } x \le 1 \\ (x - 1)^2 & \text{when } x > 1 \end{cases}$

31. Let $f(x) = \sqrt{3 - 2x} - 2$, and find $f^{-1}(-2)$.

32. Let $f(x) = 2x^2 - x - 5,\ x \ge \frac{1}{2}$, and find $f^{-1}(5)$.

33. Sketch the graph of $f(x) = \sqrt{3 - x} - 1$, and then, on the same set of axes, sketch the graph of f^{-1} without finding its formula of definition.

Chapter 3 Review

In Problems 1–4, determine whether the relation is a function.

1. $\{(x, y) \mid y = |x|\}$

2. $\{(x, y) \mid |y| = |x|\}$

3. $\{(x, y) \mid x = \sqrt{y}\}$

4. $\{(x, y) \mid \sqrt{x} = \sqrt{y}\}$

In Problems 5–8, determine whether the graph is the graph of a function.

5.

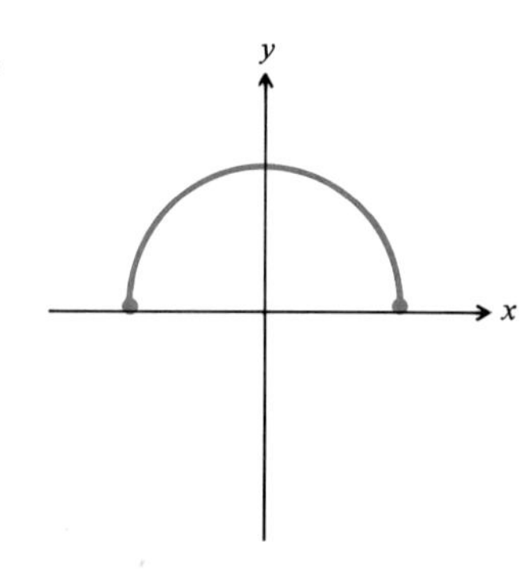

6.

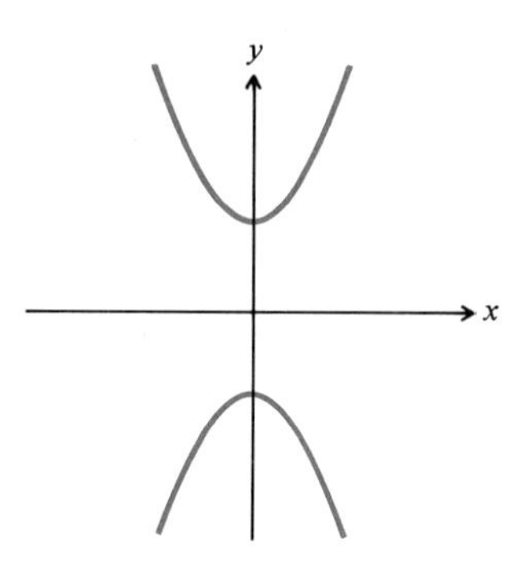

7.

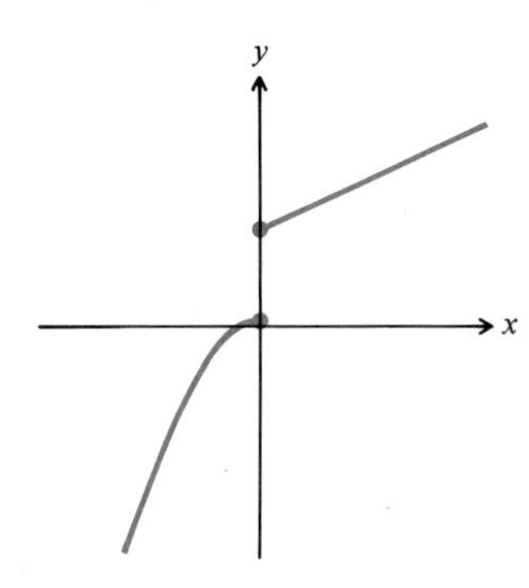

8.

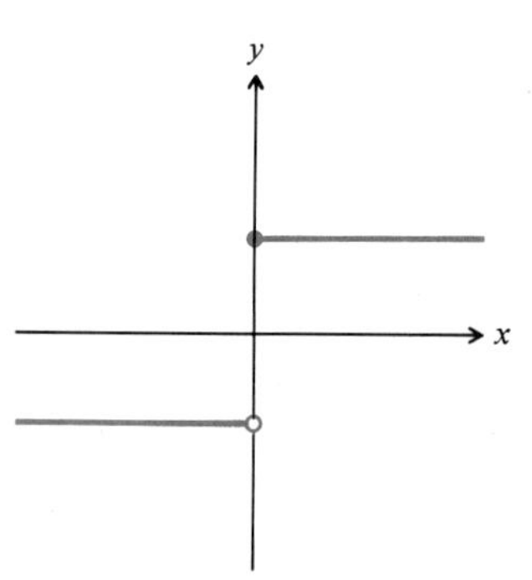

In Problems 9–12, determine the domain of the function f.

9. $f(x) = \dfrac{1}{x - 1}$

10. $f(x) = \sqrt{|x^2 - 2x|}$

11. $f(x) = \sqrt{\dfrac{x + 3}{x}}$

12. $f(x) = \dfrac{\sqrt{2 + x - x^2}}{x - 1}$

13. Let $f = \{(-1, 0), (-2, 5), (3, 3)\}$ and find (a) $f(-1)$, (b) $f(3)$, and (c) $f(f(3))$.

14. Let $f(x) = \dfrac{1}{\sqrt{x}}$ and find (a) $f(1)$, (b) $f(4)$, (c) $f(a^2)$, and (d) $f(f(16))$.

15. Let $f(x) = 2x^2 - x + 3$ and find a if $f(a) = 18$.

16. Let f be the function with the following graph:

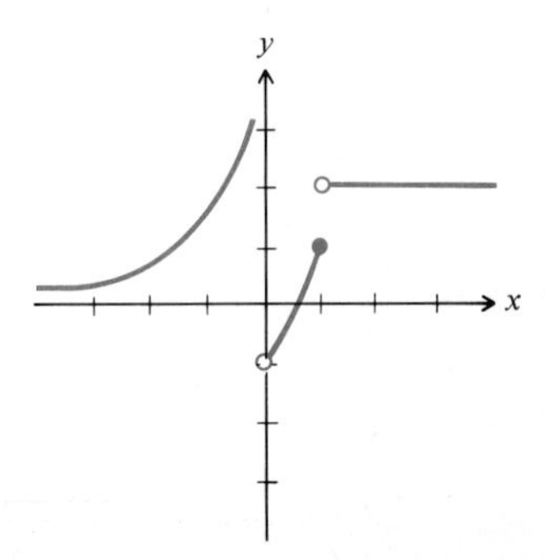

(a) What is the domain of f?
(b) What is the range of f?
(c) $f(1) = ?$
(d) $f(0) = ?$

In Problems 17–20, graph the function, and then use the graph to determine the range.

17. $y = 2x - 3$

18. $y = 2$

19. $y = x^2 - 6x + 3$

20. $y = 8x - 4x^2$

In Problems 21–33, graph the function defined.

21. $y = 3 - \sqrt{12 - x^2 + 4x}$

22. $y = \sqrt{2x - 6}$

23. $y = 2\sqrt{1 - x^2}$

24. $y = \dfrac{\sqrt{71 + 12x - 4x^2} - 5}{2}$

25. $y = \sqrt{6x^2 - 36}$

26. $y = \dfrac{-\sqrt{9 - x^2}}{3}$

27. $y = 1 - \sqrt{2 - x}$

28. $y = \dfrac{-2\sqrt{12x^2 + 27}}{3}$

29. $y = -|2x - 3| + 1$

30. $y = |x^2 - 4x + 5| - |4x|$

31. $f(x) = \begin{cases} \sqrt{-x} & \text{when } x \le 0 \\ \dfrac{x}{2} & \text{when } x > 0 \end{cases}$

32. $f(x) = \begin{cases} -x - 3 & \text{when } x < -3 \\ -\sqrt{9 - x^2} & \text{when } -3 \le x \le \sqrt{5} \\ \sqrt{x} & \text{when } x > \sqrt{5} \end{cases}$

33. $f(x) = \begin{cases} \dfrac{\sqrt{9 - x^2}}{6} & \text{when } -3 \le x \le 3 \\ \dfrac{-\sqrt{x^2 - 9}}{6} & \text{when } x > 3 \end{cases}$

34. Let $f(x) = \dfrac{1}{x}$ and $g(x) = \dfrac{x(x - 1)}{2x + 1}$. Define $f + g, f - g, fg$, and f/g.

35. Let $f(x) = \dfrac{x^2}{x^2 + 2}$ and $g(x) = \sqrt{1 - x}$. Define $f \circ g$ and $g \circ f$.

In Problems 36–38, determine whether the function is one-to-one.

36. $\left\{(x, y) \mid y = \dfrac{1}{x^2}\right\}$

37. $\{(x, y) \mid y = -(x - 3)^3\}$

38. $\{(x, y) | y = |2x + 1| - x\}$

In Problems 39–44, determine the inverse of the given one-to-one function.

39. $f(x) = 2x + 5$

40. $f(x) = \sqrt{3x - 2}$

41. $f(x) = \dfrac{2x - 1}{x + 5}$

42. $f(x) = -\sqrt{4x - x^2},\ 0 \le x \le 1$

43. $f(x) = \sqrt{3x - 1} + 2$

44. $f(x) = -x^2 + 6x - 8,\ x \ge 3$

45. Sketch the graph of $f(x) = -x^2 - 2x + 1,\ x \ge -1$, and then, on the same set of axes, sketch the graph of f^{-1} without finding its formula of definition.

Polynomial and Rational Functions

4

In this chapter we study two of the most important classes of functions, emphasizing techniques for sketching their graphs.

4.1 Polynomial Functions

A function f defined by

$$f(x) = a_n x^n + a_{n-1}x^{n-1} + \cdot \cdot \cdot + a_1 x + a_0,$$

where n is a nonnegative integer and $a_n, a_{n-1}, \ldots, a_1, a_0$ are constants with $a_n \neq 0$, is called a **polynomial function of degree n**. The term $a_n x^n$ is called the **leading term,** and the constant a_n is referred to as the **leading coefficient.** For example,

$$f(x) = -2x^4 + 4x^3 - 5 \qquad \text{and} \qquad g(x) = 7x^3 + x^2 - 2x + 3$$

are polynomials of degrees 4 and 3, respectively. The leading terms are $-2x^4$ and $7x^3$, and the leading coefficients are -2 and 7, respectively. On the other hand,

$$h(x) = \sqrt{x} \qquad \text{and} \qquad r(x) = \frac{2}{x - 3}$$

are *not* polynomial functions.

We have already encountered some polynomial functions in Chapter 3. A polynomial function of degree 0,

$$f(x) = a_0,$$

where $a_0 \neq 0$, is a constant function. [The constant function $f(x) = 0$ is also a polynomial, but by convention it is not assigned a degree.] A polynomial function of degree 1,

$$f(x) = a_1x + a_0,$$

where $a_1 \neq 0$, is a linear function. And a polynomial function of degree 2,

$$f(x) = a_2x^2 + a_1x + a_0,$$

where $a_2 \neq 0$, is a quadratic function.

Polynomials of degrees ≥ 3 can be quite complex. Graphing such polynomials is, in general, more difficult than graphing constant, linear, or quadratic functions. A complete analysis requires methods studied in calculus. We begin by simply plotting points.

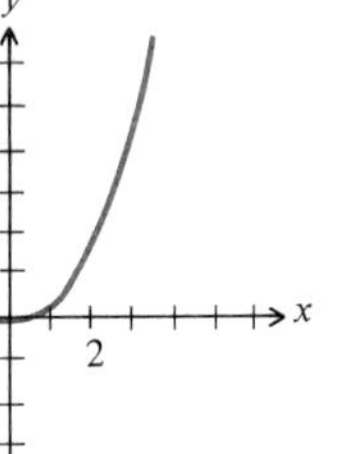

Figure 4.1 $f(x) = x^3 + 3x^2 - 2x + 1$

Example 1
Graph f, where $f(x) = x^3 + 3x^2 - 2x + 1$.

Solution. Using the formula of definition and substituting some values of x, we get the following table:

x	−5	−4	−3	−2	−1	0	1	2	3
$f(x)$	−39	−7	7	9	5	1	3	17	49

After plotting these points, we get the graph shown in Figure 4.1. ■

Note that this is just a *sketch* of the graph. Without additional information, any of the graphs in Figure 4.2 are possible. Some additional information, however, will tell us that none of these alternative sketches is reasonable.

First, the graphs of polynomials are smooth. That is, there are no sharp turns like the ones in Figure 4.2(i).

Second, all polynomials are continuous. Although a formal definition will not be given, a function is **continuous** (on its domain) provided that the only "gaps" in its graph are at "gaps" in its domain. Hence the graph of a continuous function with domain R, like polynomials, consists of one piece. Thus the graph of Figure 4.2(ii) may be excluded.

Third, the number of turning points (locally high or locally low points) in the graph of a polynomial of degree n is at most $n - 1$, and if not $n - 1$, is less than $n - 1$ by an even number. For instance, a polynomial of degree

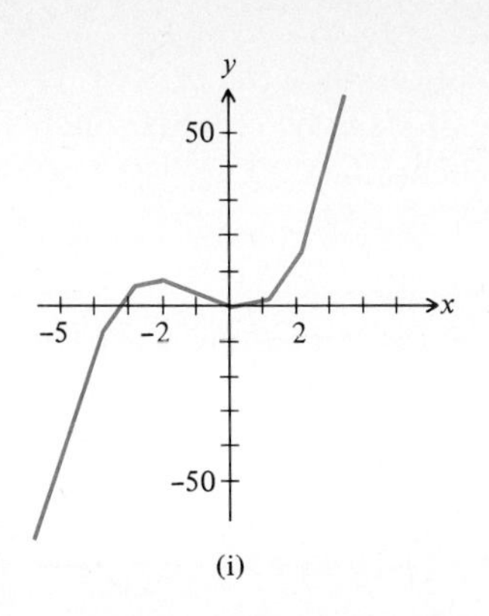

(i)

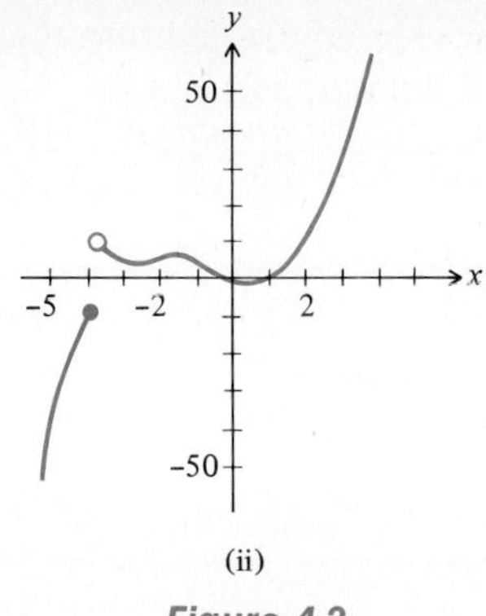

(ii)

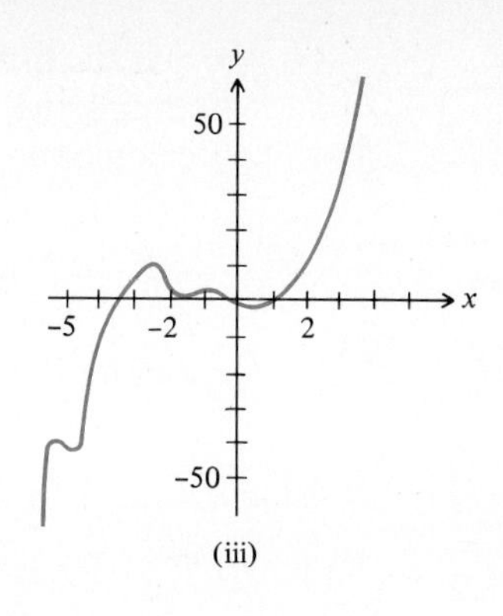

(iii)

Figure 4.2

four can have three or one turning points. A polynomial of degree three (like that of Example 1) has two or none. Hence the sketch in Figure 4.2(iii) may be excluded because it has six turning points.

Thus, in sketching polynomials after plotting points, we will connect them into a smooth, one-piece curve. And we will include only those turning points that we know must exist. This still leaves alternative possible sketches, however. See Figure 4.3, where the turning points are different. But the differences are minor. To obtain a more accurate sketch we would have to plot additional points or use some calculus concepts.

When graphing polynomials by plotting points, how can we be certain that our particular selection of x-coordinates will give us an accurate sketch? The fact is that we cannot unless we have a general idea of what the graph looks like before we begin. In the next section we learn the general shapes of some common polynomials. Until then, suggested values for plotting points will be included with the problems.

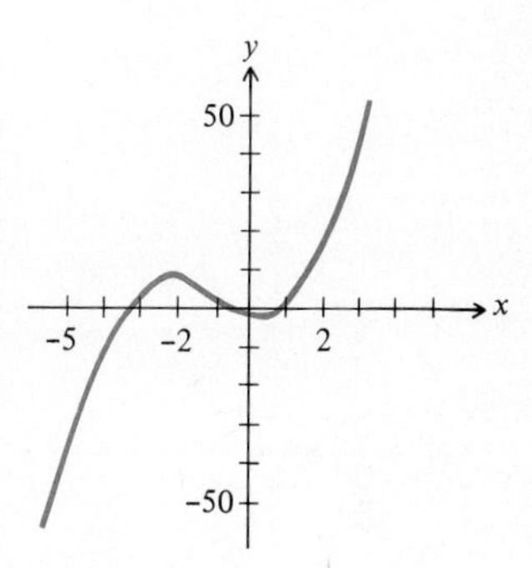

Figure 4.3

Exercise

Graph f, where $f(x) = -x^3 + 2x^2 + 4x + 5$, after plotting points whose x-coordinates are $-3, -2, -1, 0, 1, 2, 3, 4$, and 5.

Answer

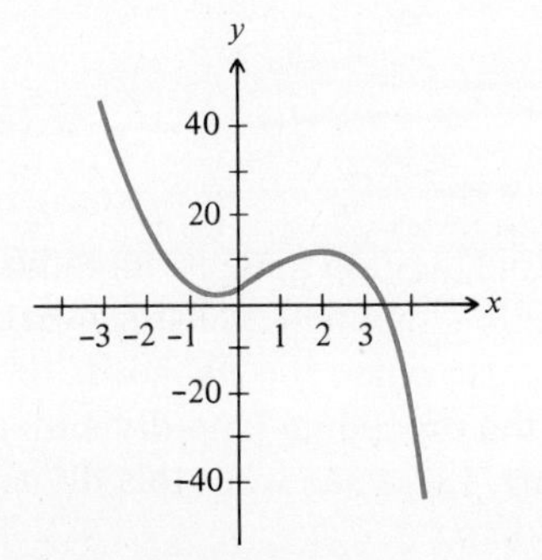

■

If the function has a complicated formula of definition, then the process of finding the value of the function at a selected value of x can be tedious. Such calculations will be simplified with the following theorem.

Theorem 4.1 (Remainder Theorem)

Let P be a polynomial function. For every real number c, there exists a unique polynoial Q such that

$$P(x) = (x - c)Q(x) + P(c).$$

This theorem tells us that if $P(x)$ is divided by $x - c$, where c is a real number, then the remainder is $P(c)$, the value of P at c.

Consider the polynomial P defined by

$$P(x) = 2x^3 + 13x^2 + 6x.$$

The value of P at -5, of course, can be determined by substitution. We have

$$P(-5) = 2(-5)^3 + 13(-5)^2 + 6(-5) = 45.$$

According to the Remainder Theorem, $P(-5)$ will also be the remainder when $P(x)$ is divided by $x + 5$. Consequently, we may calculate $P(-5)$ by long division.

$$
\begin{array}{rllll}
 & & 2x^2 & +\ 3x & -\ 9 \\ \cline{2-5}
x+5\,\big) & 2x^3 & +\ 13x^2 & +\ 6x & \\
 & 2x^3 & +\ 10x^2 & & \\ \cline{2-3}
 & & 3x^2 & +\ 6x & \\
 & & 3x^2 & +\ 15x & \\ \cline{3-5}
 & & & -\ 9x & \\
 & & & -\ 9x & -\ 45 \\ \cline{4-5}
 & & & & 45
\end{array}
$$

The remainder, 45, is $P(-5)$. Although this method appears as tedious as the substitution, this calculation is significantly simplified by using **synthetic division,** a simplified process which we may use when the divisor is of the form $x - c$ for some constant c. Note that in the preceding long-division problem, there is a considerable repetition of terms. Let's see what this division looks like when we omit such repetition.

$$
\begin{array}{r|l}
 & 2x^2 + \ 3x - 9 \\
\hline
x + 5 & 2x^3 + 13x^2 + \ 6x \\
 & \underline{+ 10x^2} \\
 & \quad 3x^2 \\
 & \quad\quad \underline{+ 15x} \\
 & \quad\quad - \ 9x \\
 & \quad\quad\quad \underline{- 45} \\
 & \quad\quad\quad 45
\end{array}
$$

If we keep the same powers of x beneath one another, we may use the coefficients only.

$$
\begin{array}{r|llll}
 & 2 & +3 & -9 & \\
\hline
1 + 5 & 2 & 13 & 6 & 0 \\
 & & \underline{10} & & \\
 & & 3 & & \\
 & & & \underline{15} & \\
 & & & -9 & \\
 & & & & \underline{-45} \\
 & & & & 45
\end{array}
$$

Since this procedure is being applied only when the divisor is of the form $x - c$, the coefficients are always $1 - c$, and the 1 may be assumed. Furthermore, we can make our notation more compact.

$$
\begin{array}{r|rrrr}
 & & 2 & +3 & -9 \\
\hline
5 & 2 & 13 & 6 & 0 \\
 & & 10 & 15 & -45 \\
\hline
 & & 3 & -9 & 45
\end{array}
$$

If we insert the leading coefficient of the dividend in the first position of the bottom row, the coefficients of the bottom row become the coefficients of the quotient with the last entry being the remainder. We can then discard the top row and get

$$
\begin{array}{r|rrrr}
5 & 2 & 13 & 6 & 0 \\
 & & 10 & 15 & -45 \\
\hline
 & 2 & 3 & -9 & 45
\end{array}
$$

coefficients of quotient (2, 3, −9) — remainder (45)

We may find the entries in the bottom two rows without even thinking of a division problem. Note that each entry in the second row is the product of 5 (from the divisor) and the entry in the *preceding* column of the bottom row. The first entry of the third row is the same as the leading coefficient of the dividend. For every other entry, we simply subtract the entry of the second row from that of the first. We may avoid the subtractions by replacing $-c$ by $+c$ and *adding* the entries in the first and second rows. We would have the following:

$$\begin{array}{r|rrrr} -5 & 2 & 13 & 6 & 0 \\ & & -10 & -15 & 45 \\ \hline & 2 & 3 & -9 & 45 \end{array}$$

This last process is called synthetic division. The rightmost entry of the bottom row, the remainder, is the value of P at -5.
Similarly, we may find $P(-4)$.

$$\begin{array}{r|rrrr} -4 & 2 & 13 & 6 & 0 \\ & & -8 & -20 & 56 \\ \hline & 2 & 5 & -14 & 56 \end{array}$$

We followed these steps:

1. 2 was "brought down" from the top row to become the first entry of the bottom row.
2. -8, the product of 2 and -4, was entered into the second column of the second row.
3. 5, the sum of 13 and -8, was written below.
4. -20, the product of 5 and -4, became the next entry of the second row.
5. -14, the sum of 6 and -20, was written below.
6. 56, the product of -14 and -4, became the next entry of the second row.
7. 56, the sum of 0 and 56, was written below and is the remainder.

Hence $P(-4) = 56$. (Although we are using synthetic division to find the remainder, note that it is giving us the quotient as well.) Thus

$$\frac{2x^3 + 13x^2 + 6x}{x + 4} = 2x^2 + 5x - 14 + \frac{56}{x + 4}.$$

Exercise
Let $P(x) = 2x^3 + 13x^2 + 6x$. Use synthetic division to calculate each of the following.
(a) $P(2)$ (b) $P(-3)$ (c) $P(-6)$

Answers. (a) 80 (b) 45 (c) 0 ■

Problem Set 4.1

Graph the polynomial function after plotting points with the given x-coordinates.

1. $f(x) = x^3 - 2x^2 - 11x + 12$,
   x-values: $-4, -3, -2, -1, 0, 1, 2, 3, 4$, and 5
2. $f(x) = x^3 - 2x^2 - 10x + 9$,
   x-values: $-4, -3, -2, -1, 0, 1, 2, 3, 4$, and 5
3. $f(x) = 2x^3 + 3x^2 + x - 6$,
   x-values: $-3, -2, -1, 0, 1, 2$, and 3
4. $f(x) = -2x^3 - 3x^2 + x + 5$,
   x values: $-3, -2, -1, 0, 1$, and 2
5. $f(x) = -x^4 + 4x^3 + 9x^2 - 5$,
   x values: $-3, -2, -1, 0, 1, 2, 3, 4, 5$, and 6
6. $f(x) = x^4 - 3x^3 - 12x^2 + x - 5$,
   x-values: $-3, -2, -1, 0, 1, 2, 3, 4, 5$, and 6
7. $f(x) = x^4 + x^2$,
   x-values: $-3, -2, -1, 0, 1, 2$, and 3
8. $f(x) = x^4 - x^3 + 2x^2 - 12$,
   x-values: $-2, -1, 0, 1, 2$, and 3
9. $f(x) = x^5 - 2x^2 + x + 6$,
   x-values: $-2, -1, 0, 1$, and 2
10. $f(x) = -2x^5 + x^3 + x - 10$,
    x-values: $-2, -1, \frac{-1}{2}, 0, \frac{1}{2}, 1$, and 2

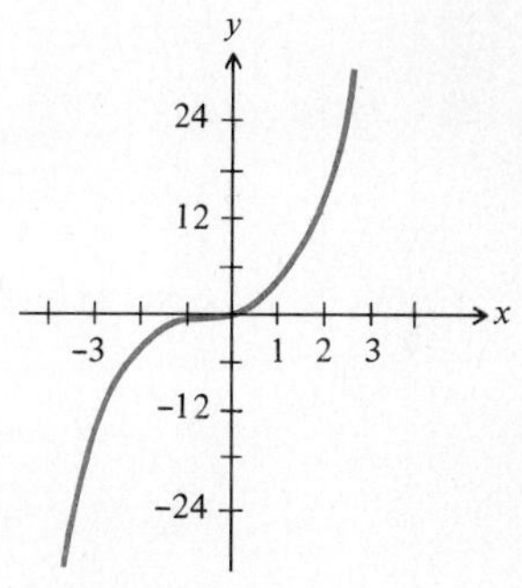

Figure 4.4 $y = x^3$

4.2 Elementary Polynomials

In this section we learn the general shapes of polynomials of the form $y = x^n$ as well as transformations of such polynomials. We begin with $y = x^3$, whose graph is shown in Figure 4.4.

There are several important facts to be noted about this graph. Near the origin the graph is very flat, but the farther away from the origin the graph gets, the steeper it becomes. And note that if we took that part of the graph in the first quadrant and reflected it about the x-axis and then about the y-axis, it would correspond exactly to that part of the graph in the third quadrant. This is because the graph has some symmetry. A graph is said to be **symmetric about the origin** provided that for each point (a, b) on the graph, the point $(-a, -b)$ is on the graph. This means that for each point P on the graph (excluding the origin), the point Q which has the property that $\overline{PQ}$ is bisected by the origin is also on the graph. See Figure 4.5.

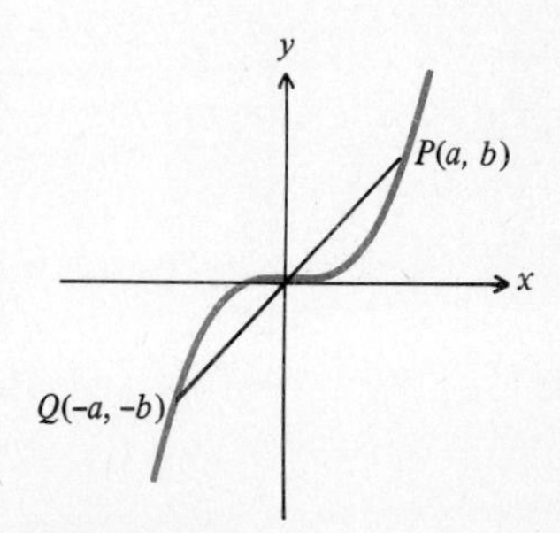

Figure 4.5 Symmetry about the origin

The graph of $y = x^5$ is shown in Figure 4.6. Note that the graph of $y = x^5$ looks a lot like the graph of $y = x^3$ (in Figure 4.4). Each contains the points $(1, 1)$ and $(-1, -1)$, and each is symmetric about the origin. The two principal differences are that the graph of $y = x^5$ is flatter about the origin [on $(-1, 1)$] and is steeper elsewhere [on $(-\infty, -1)$ and $(1, \infty)$].

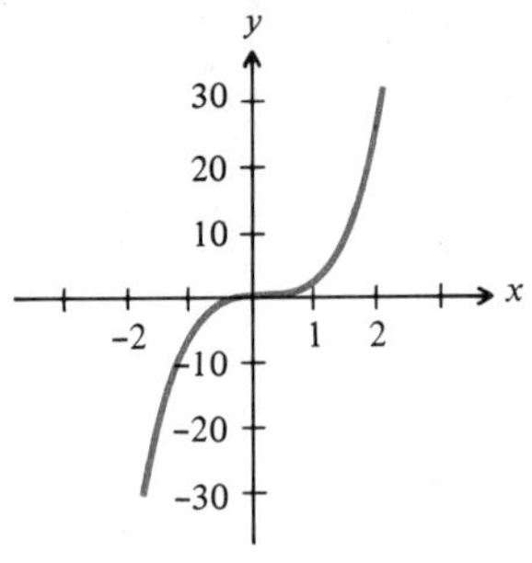

Figure 4.6 $y = x^5$

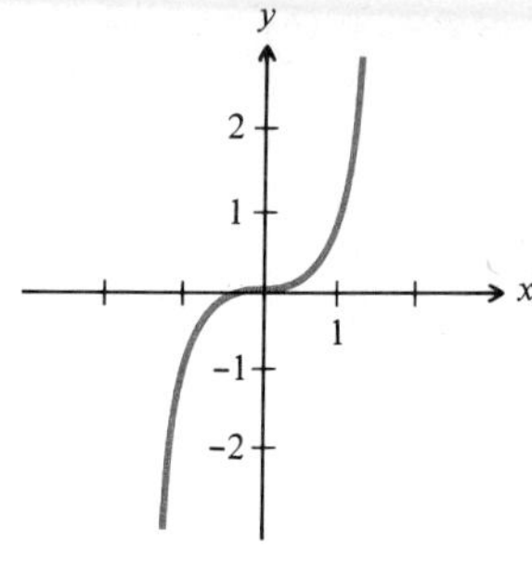

Figure 4.7 $y = x^7$

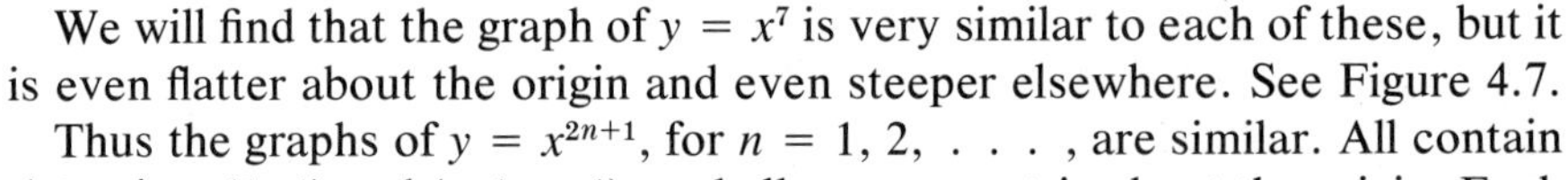

We will find that the graph of $y = x^7$ is very similar to each of these, but it is even flatter about the origin and even steeper elsewhere. See Figure 4.7.

Thus the graphs of $y = x^{2n+1}$, for $n = 1, 2, \ldots$, are similar. All contain the points $(1, 1)$ and $(-1, -1)$, and all are symmetric about the origin. Each is flat about the origin and steep elsewhere. And the greater the exponent, the greater this flatness and steepness.

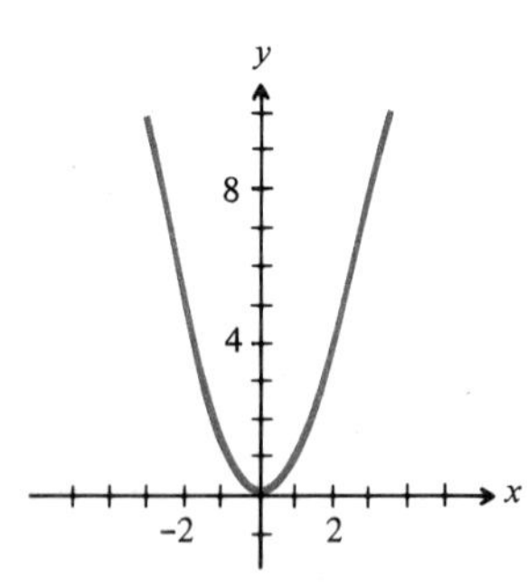

Figure 4.8 $y = x^2$

Of course, we are already familiar with the graph of $y = x^2$, shown in Figure 4.8. We know that it is a parabola with vertex at the origin.

There are some important facts to be noted about this graph. The graph is relatively flat near the origin, but the farther away from the origin the graph gets, the steeper it becomes. And note that if we took that part of the graph in the first quadrant and reflected it about the y-axis, it would correspond exactly to that part of the graph in the second quadrant. This is because the graph has some symmetry. A graph is said to be **symmetric about the y-axis** provided that for each point (a, b) on the graph, the point $(-a, b)$ is on the graph. This means that for each point P on the graph (excluding the origin), the point Q which has the property that $\overline{PQ}$ is perpendicular to and bisected by the y-axis lies on the graph. See Figure 4.9.

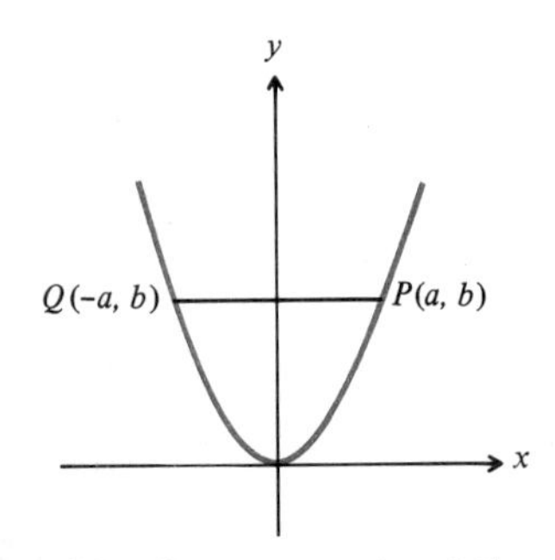

Figure 4.9 Symmetry about the y-axis

The graphs of $y = x^4$ and $y = x^6$ are shown in Figure 4.10. Even though the graphs of $y = x^4$ and of $y = x^6$ are not parabolas, they look a lot like the graph of $y = x^2$. All contain the points $(1, 1)$ and $(-1, 1)$, and all are symmetric about the y-axis. But the latter two graphs are flatter about the origin and

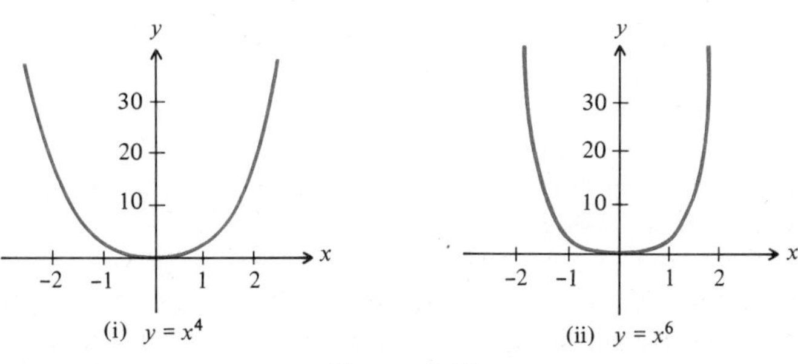

Figure 4.10

steeper elsewhere. Thus the graphs of $y = x^{2n}$, for $n = 1, 2, \ldots$, are similar. And the greater the exponent, the greater the flatness and steepness.

Let c be a positive constant. Then the graph of $y = f(x) + c$ is the graph of $y = f(x)$ shifted up c units. We know that the graph of $y = x^2 + 2$ is a parabola which opens upward and has vertex (0, 2). In fact, as seen in Figure 4.11, it is simply the parabola $y = x^2$ shifted up two units.

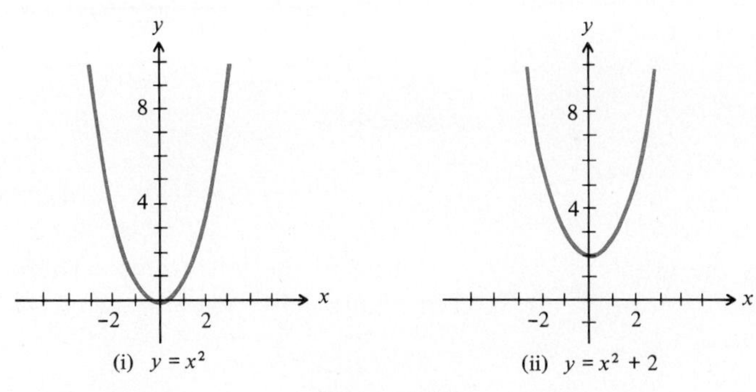

Figure 4.11

Similarly, if c is a positive constant, then the graph of $y = f(x) - c$ is the graph of $y = f(x)$ shifted down c units. The graphs of $y = x^3$ and $y = x^3 - 1$ are shown in Figure 4.12.

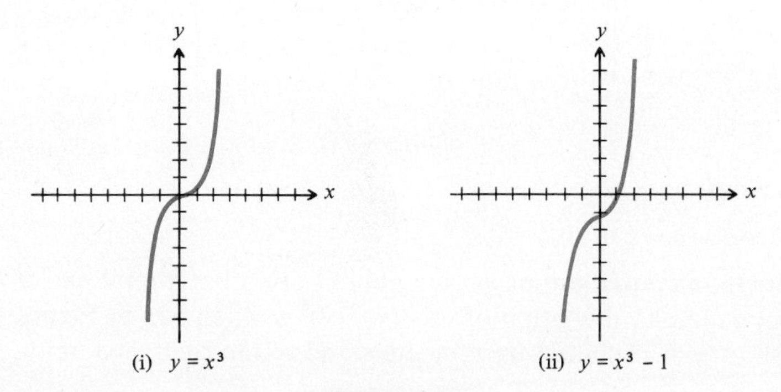

Figure 4.12

Let c be a positive constant. Then the graph of $y = f(x - c)$ is the graph of $y = f(x)$ shifted to the right c units. We know that the graph of $y = (x - 1)^2$ is a parabola that opens upward and has vertex (1, 0). In fact, as noted in Figure 4.13, it is the parabola $y = x^2$ shifted to the right one unit.

In a similar fashion, if c is a positive constant, then the graph of $y =$

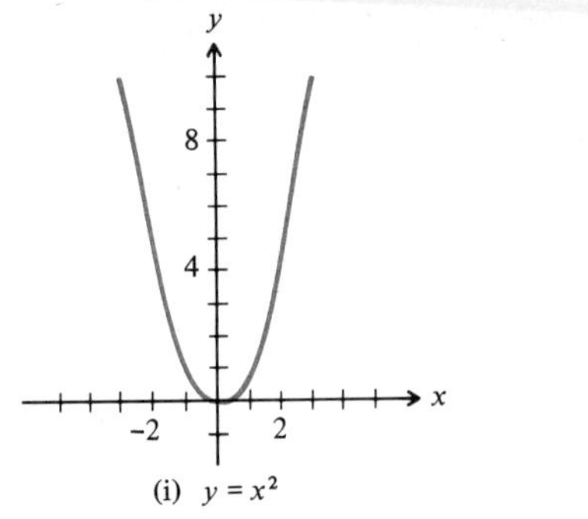

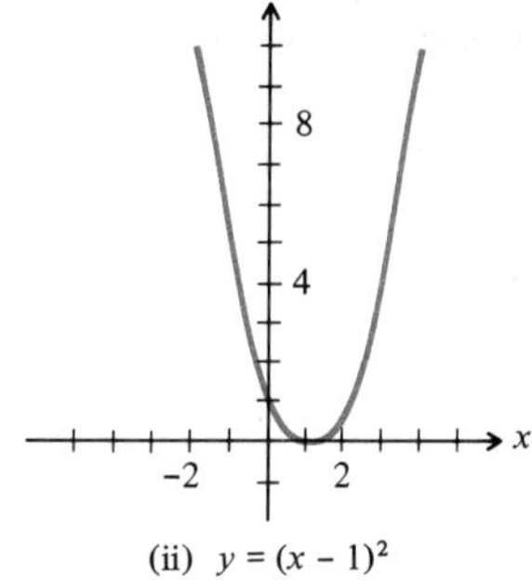

Figure 4.13

$f(x + c)$ is the graph of $y = f(x)$ shifted to the left c units. The graphs of $y = x^3$ and $y = (x + 1)^3$ are seen in Figure 4.14.

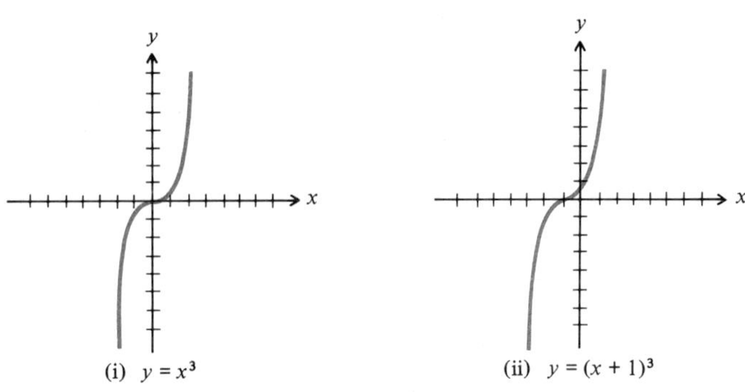

Figure 4.14

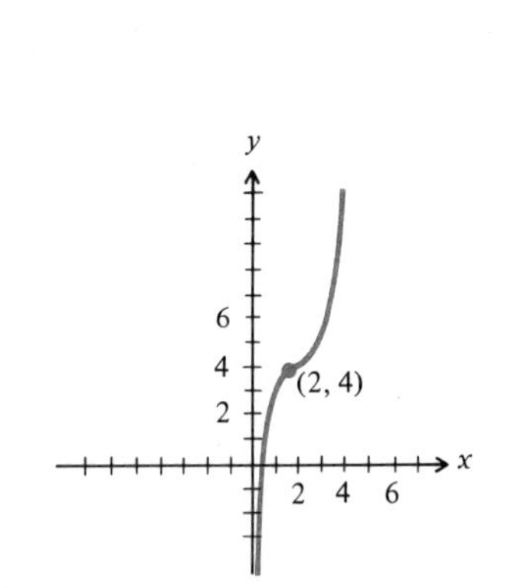

Figure 4.15 $y = (x - 2)^3 + 4$

Of course, an equation may bring about both a horizontal and a vertical shift. For example, the graph of $y = (x - 2)^3 + 4$, shown in Figure 4.15, is the graph of $y = x^3$ shifted up four units and to the right two units.

Example 1
Sketch the graph of $y = (x - 1)^3 + 3$.

Solution. The graph is a transformation of $y = x^3$. It is this basic graph shifted to the right one unit and up three, as shown in Figure 4.16. ■

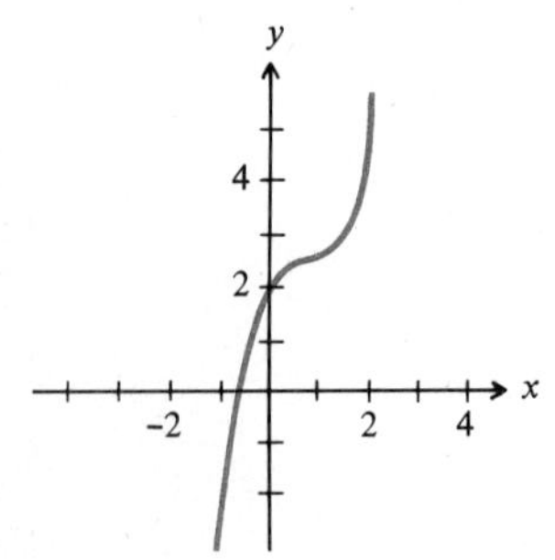

Figure 4.16 $y = (x - 1)^3 + 3$

Exercise
Sketch the graph of $y = (x + 1)^4 - 4$.

Answer.

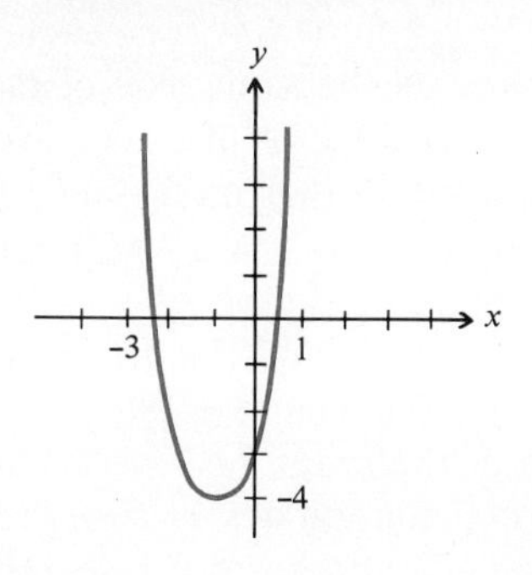

■

Example 2
Choose the equation which *best* describes the graph in Figure 4.17.
(a) $y = (x + 2)^3 - 3$
(b) $y = (x + 2)^2 + 3$
(c) $y = (x + 2)^5 - 3$
(d) $y = (x + 2)^4 - 3$
(e) $y = (x - 2)^2 + 3$

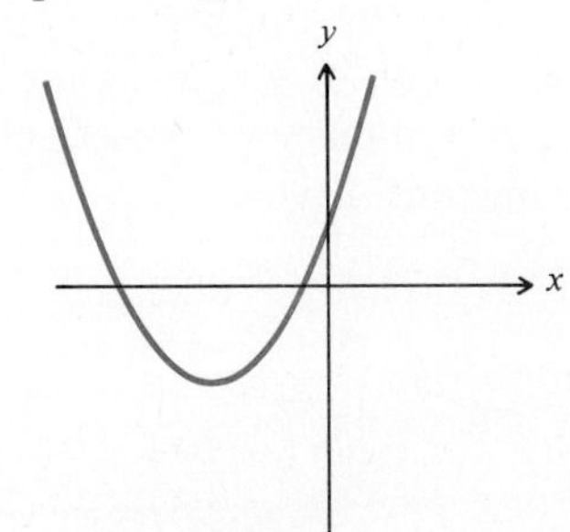

Figure 4.17

Solution. The general shape tells us immediately that this is a shift of a graph whose equation is $y = x^{2n}$ for some positive integer n. Thus answers (a) and (c) are eliminated immediately. Further, since the shift is to the third quadrant, the equation must be of the form $y = (x + c_1)^n - c_2$, where c_1 and c_2 are positive constants. Hence the answer is (d). ■

Exercise
Choose the equation which *best* describes the graph.
(a) $y = (x + 2)^2 + 3$
(b) $y = (x + 2)^5 + 3$
(c) $y = (x - 2)^4 + 3$
(d) $y = (x + 2)^3 - 3$
(e) $y = (x - 2)^3 + 3$

Answer. (b)

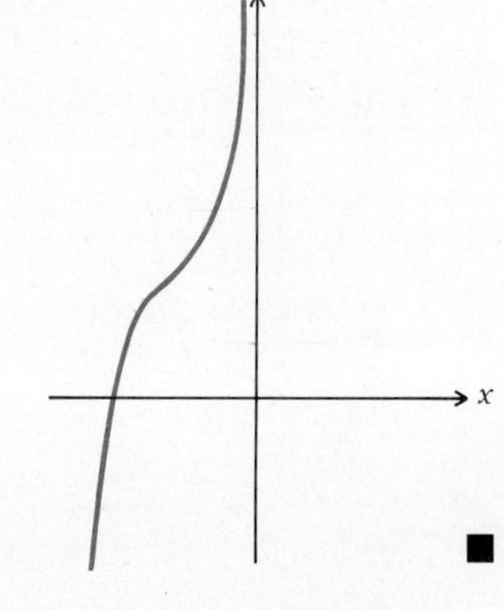

■

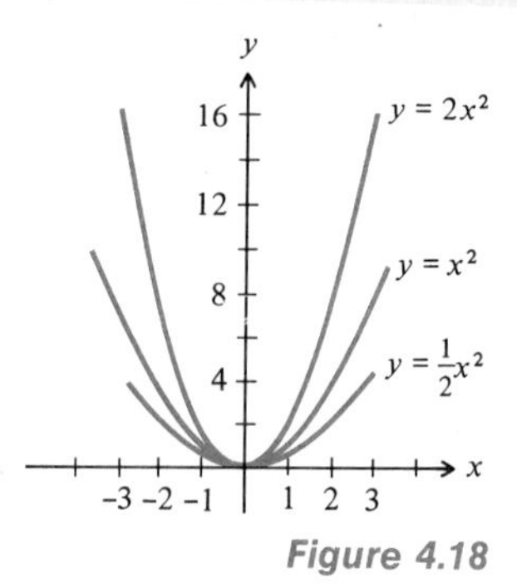

Figure 4.18

If c is a positive constant, then the graph of $y = c \cdot f(x)$ is a vertical stretching (if $c > 1$) or compressing (if $c < 1$) of the graph of $y = f(x)$. That is, to get the graph of $y = c \cdot f(x)$ from the graph of $y = f(x)$, each point on the latter graph is moved so that it remains on the same side of the x-axis, but its distance from the x-axis is adjusted by a factor of c. For example, if each point on the graph of $y = f(x)$ is moved so that its distance from the x-axis is tripled, then we would have the graph of $y = 3 \cdot f(x)$. Of course, points on the x-axis remain stationary. Thus the graph of $y = 2x^2$ can be thought of as a stretching of the graph of $y = x^2$, and the graph of $y = \frac{1}{2}x^2$ may be viewed as a compressing of $y = x^2$. See Figure 4.18.

The graph of $y = -f(x)$ is a reflection of the graph of $y = f(x)$ about the x-axis. This means that for each point P on the graph of $y = f(x)$, the point Q which has the property that $\overline{PQ}$ is perpendicular to and bisected by the x-axis is on the graph of $y = -f(x)$. This relationship is pictured in Figure 4.19.

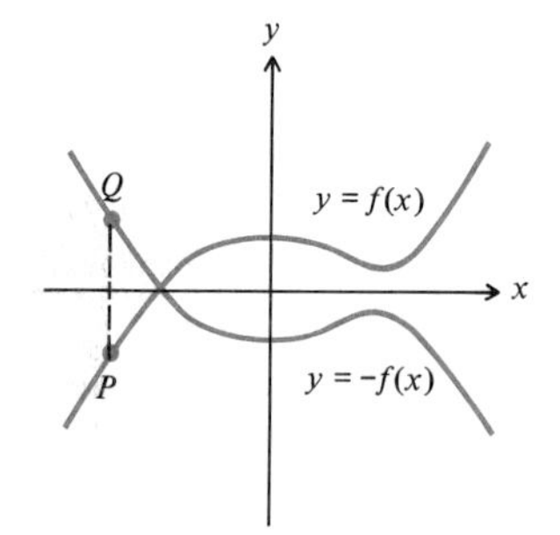

Figure 4.19 Reflection about the x-axis

Thus the graph of $y = -x^3$, shown in Figure 4.20, is simply a reflection of the graph of $y = x^3$ about the x-axis.

A summary of the effects on the graph of $y = f(x)$, where c is a positive constant, follows:

$y =$	*Effect*
$f(x) + c$	Shift up c units
$f(x) - c$	Shift down c units
$f(x - c)$	Shift right c units
$f(x + c)$	Shift left c units
$c \cdot f(x), c > 1$	Vertical stretch
$c \cdot f(x), c < 1$	Vertical compression
$-f(x)$	Reflection about the x-axis

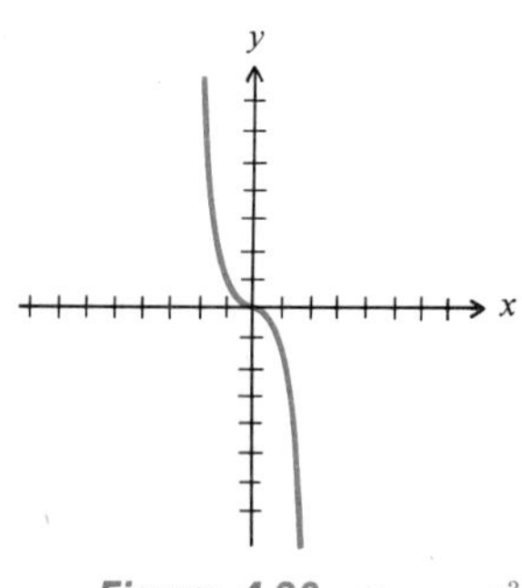

Figure 4.20 $y = -x^3$

Of course, it is possible that a given graph may be a shift, a stretching or compressing, and a reflection of the graph of an elementary polynomial. Let us examine the graph of $y = -2(x - 3)^2 + 5$, noting how each of the constants affects the graph of the elementary polynomial $y = x^2$, whose graph appears in Figure 4.21. Now the graph of $y = (x - 3)^2$ (Figure 4.22) is the

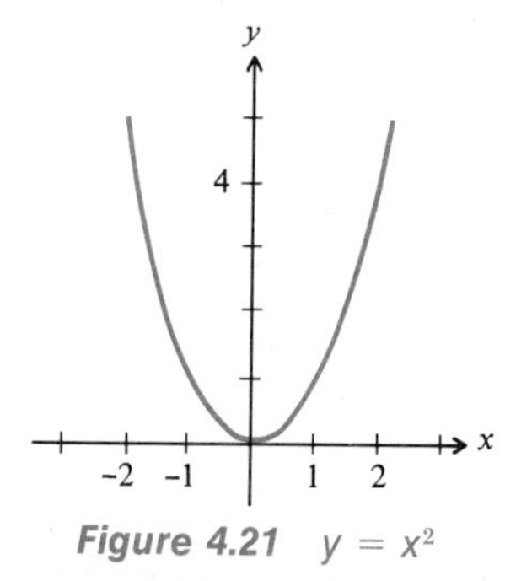

Figure 4.21 $y = x^2$

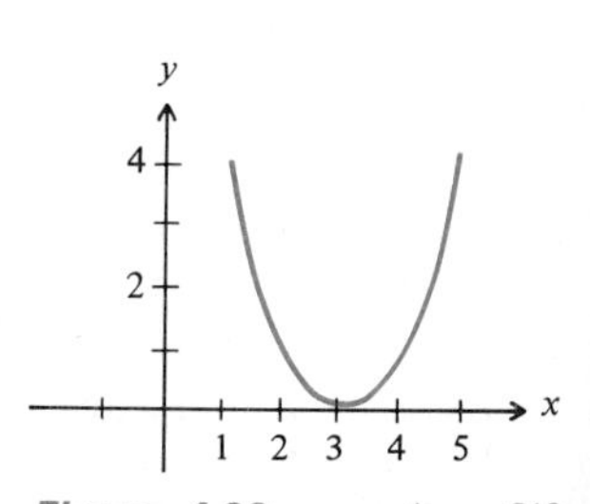

Figure 4.22 $y = (x - 3)^2$

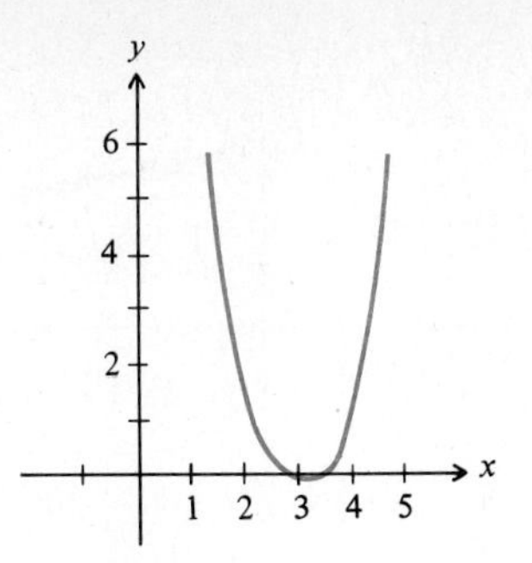

Figure 4.23 $y = 2(x - 3)^2$

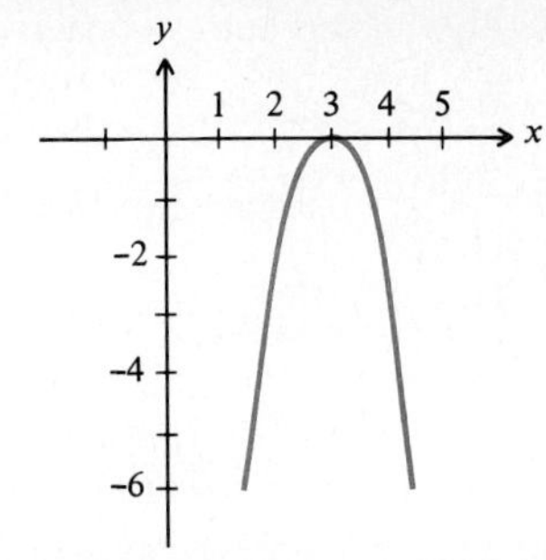

Figure 4.24 $y = -2(x - 3)^2$

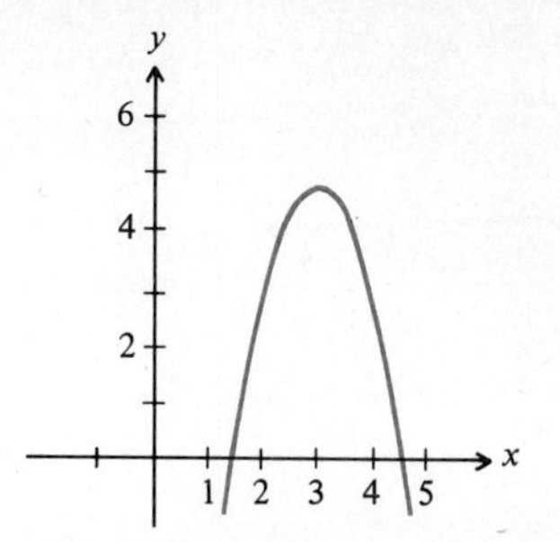

Figure 4.25 $y = -2(x - 3)^2 + 5$

preceding graph shifted to the right three units. The graph of $y = 2(x - 3)^2$ (Figure 4.23) is a stretching of the previous graph. The graph of $y = -2(x - 3)^2$ (Figure 4.24) is a reflection about the x-axis of the preceding graph. And finally, the graph of $y = -2(x - 3)^2 + 5$ (Figure 4.25) is the preceding graph shifted up five units.

Exercise

Choose the equation which *best* describes the graph.

(a) $y = -2(x - 3)^3 - 4$
(b) $y = -2(x - 3)^4 - 4$
(c) $y = 2(x - 3)^2 + 4$
(d) $y = -2(x - 3)^5 + 4$
(e) $y = 2(x - 3)^5 - 4$

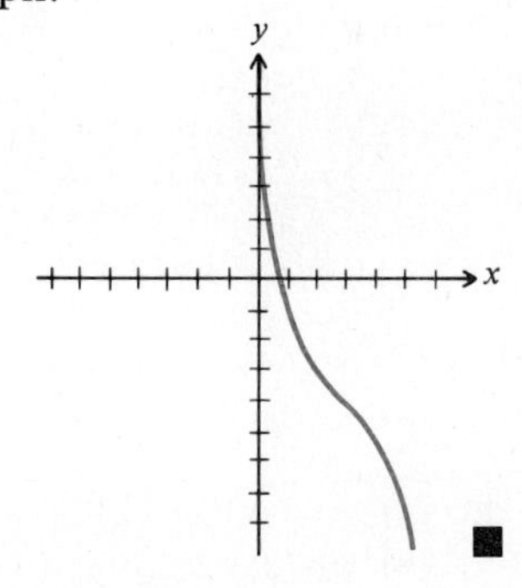

Answer. (a) ■

Problem Set 4.2

In Problems 1–16, sketch the graph of the polynomial.

1. $y = x^2 + 2$ **2.** $y = x^3 - 1$ **3.** $y = (x - 2)^3$

4. $y = (x + 3)^2$ **5.** $y = 3x^2$ **6.** $y = \frac{x^5}{8}$

7. $y = -x^4$ **8.** $y = -2x^2$ **9.** $y = (x - 1)^2 - 3$

10. $y = (x - 2)^5 + 2$ **11.** $y = -x^4 + 2$ **12.** $y = -(x + 2)^5$

13. $y = -(x + 2)^3 + 3$ **14.** $y = \frac{(x - 2)^4}{4} - 1$ **15.** $y = -2(x + 3)^2 - 1$

16. $y = \frac{-(x + 3)^3}{2} - 2$

In Problems 17–20, choose the equation which best *describes the given graph.*

17. **(a)** $y = 2(x - 3)^4 + 5$
(b) $y = 2(x - 3)^5 + 5$
(c) $y = -2(x - 3)^2 + 5$
(d) $y = -2(x - 3)^4 - 5$
(e) $y = -2(x - 3)^3 + 5$

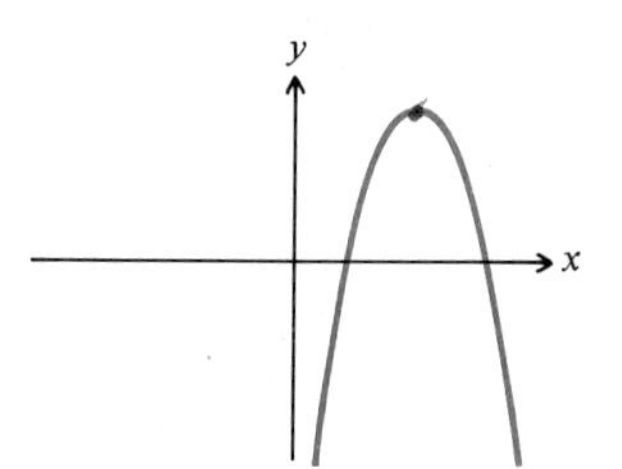

18. **(a)** $y = 2(x - 2)^3 - 3$
(b) $y = -2(x + 2)^3 - 3$
(c) $y = 2(x + 2)^5 + 3$
(d) $y = 2(x + 2)^5 - 3$
(e) $y = 2(x + 2)^4 - 3$

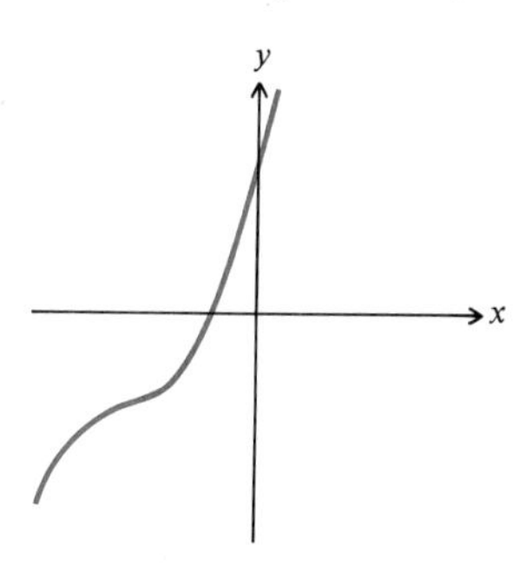

19. **(a)** $y = -(x + 5)^2 + 4$
(b) $y = -(x + 5)^3 + 4$
(c) $y = -(x - 5)^5 + 4$
(d) $y = -(x + 5)^3 - 4$
(e) $y = (x + 5)^5 + 4$

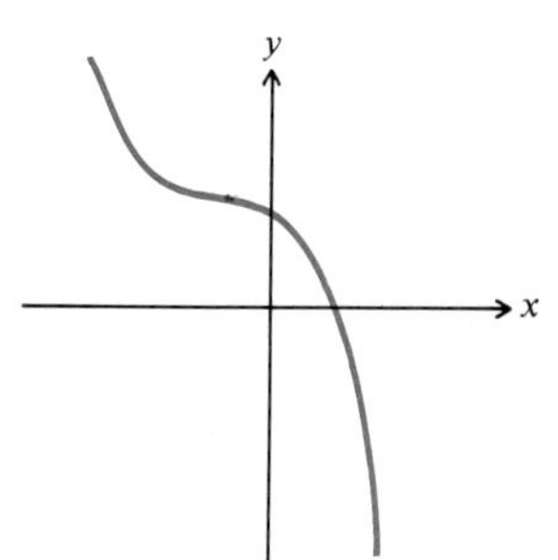

20. **(a)** $y = -3(x - 2)^4 - 4$
(b) $y = 3(x - 2)^3 - 4$
(c) $y = 3(x + 2)^2 - 4$
(d) $y = 3(x - 2)^2 + 4$
(e) $y = 3(x - 2)^4 - 4$

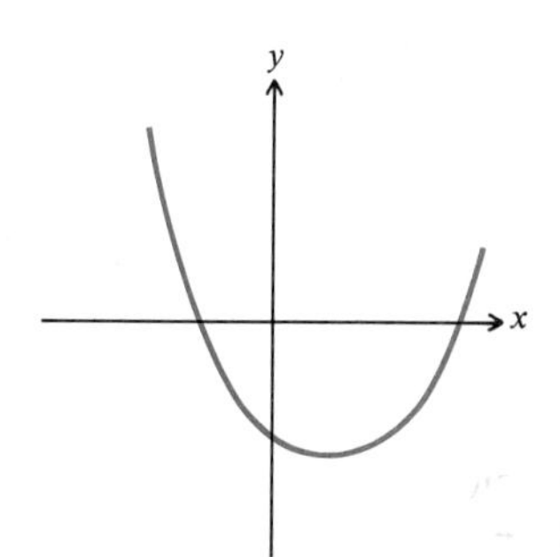

4.3 Polynomials Continued

In this section we want to develop some techniques for sketching the graphs of some other polynomials.

We first consider the graph of $y = x^3$ (see Figure 4.4). Note that as the x

values get larger positively, the graph rises up without bound. This is because the corresponding y values get larger and larger (since we are cubing the values of x). This relationship can also be seen in the following table:

x	1	2	3	4	5	10	100
y	1	8	27	64	125	1,000	1,000,000

How large can the y values get? We can make them as large as we like by choosing x values large enough. This relationship is expressed as follows:

$$\text{as} \quad x \to \infty, \quad y \to \infty.$$

We read "as x approaches infinity, y approaches infinity." Thus this phrase means that as x gets larger and larger (without bound) positively, y gets larger and larger (without bound) positively.

What happens to the values of y as x gets larger negatively? Since these values are being cubed, it is obvious that they are also getting large negatively.

This is also evident from the graph of $y = x^3$ because as the x values get larger negatively, the graph falls without bound. This relationship is expressed as follows:

$$\text{as} \quad x \to -\infty, \quad y \to -\infty.$$

We read "as x approaches minus infinity, y approaches minus infinity." Hence this phrase means that as x gets large negatively (without bound), y gets large negatively (without bound).

Two other relationships which exist for other functions are as follows:

$$\text{as} \quad x \to \infty, \quad y \to -\infty$$
$$\text{as} \quad x \to -\infty, \quad y \to \infty.$$

The former means that as x gets larger and larger positively, y gets larger and larger negatively. Such a relationship exists for $y = -x^2$. The latter expression means that as x gets larger and larger negatively, y gets larger and larger positively. This occurs for $y = x^2$. (The symbols ∞ and $-\infty$ are *not* numbers. They have meaning only when used in conjunction with the other symbols.)

Now let's consider the function defined by

$$y = 2x^3 - 5x^2 - 6x + 1,$$

What happens to y as x approaches infinity? The first term gets large positively, the second and third terms get large negatively, and the fourth term remains constant. Because two terms get large negatively and only one gets large positively, does this mean that their sum, and thus y, gets large negatively? Not necessarily. This can be seen as follows. Suppose that we factor out $2x^3$ and write

$$y = 2x^3\left(1 - \frac{5}{2x} - \frac{3}{x^2} + \frac{1}{2x^3}\right).$$

As x gets large positively, the terms $-5/(2x)$, $-3/x^2$, and $1/(2x^3)$ approach zero. Thus the second factor approaches 1. Since the first factor gets large positively, $y \to \infty$ as $x \to \infty$.

This technique of factoring out the leading term may be used with any polynomial, and the second factor will always approach 1 as $x \to \infty$ and as $x \to -\infty$. Therefore, the behavior of y can be observed by examining the behavior of the leading term.

Example 1
Examine the behavior of y as $x \to \infty$ and as $x \to -\infty$, where $y = 1 + 2x - 3x^4$.

Solution. We merely examine the leading term $-3x^4$. As $x \to \infty$, $-3x^4 \to -\infty$, and as $x \to -\infty$, $-3x^4 \to -\infty$. Therefore, $y \to -\infty$ as $x \to \infty$ and $y \to -\infty$ as $x \to -\infty$. ■

Exercise
Examine the behavior of y as $x \to \infty$ and as $x \to -\infty$, where $y = x - x^2 + 3x^5$.

Answer. $y \to \infty$ as $x \to \infty$, and $y \to -\infty$ as $x \to -\infty$. ■

Before sketching some graphs, we need to recall one special property of polynomials: continuity. Because *polynomials are continuous,* the only values of x at which the values of y can change sign are the x-intercepts. That is, between two consecutive x-intercepts, either all y-values are positive or all are negative. Thus, between two consecutive x-intercepts, the graph of a polynomial lies either entirely above or entirely below the x-axis.

So in sketching the graphs of polynomials of degrees greater than or equal to three which are *not* written as transformations of elementary polynomials, we will do three things before drawing a smooth curve.

1. Locate all x- and y-intercepts.
2. Observe the behavior of y as $x \to \infty$ and as $x \to -\infty$.
3. Observe continuity—note that y can change signs at x-intercepts only.

Example 2
Sketch the graph of $y = (x - 1)^2(x + 2)$.

Solution. The x-intercepts are found by setting y equal to zero and solving for x. Clearly, we get $x = 1$ and $x = -2$. The y-intercept is found by setting

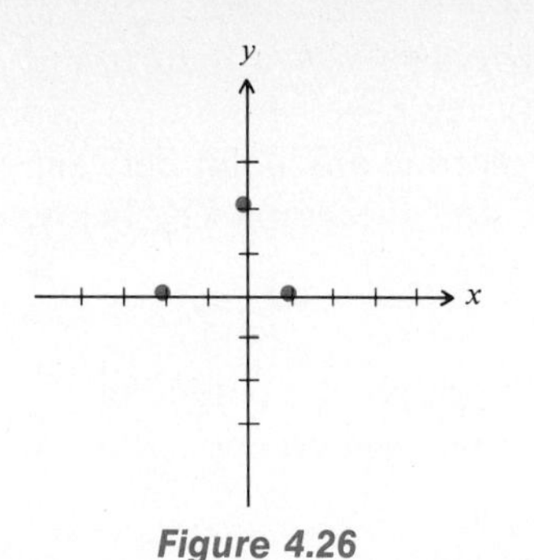

Figure 4.26

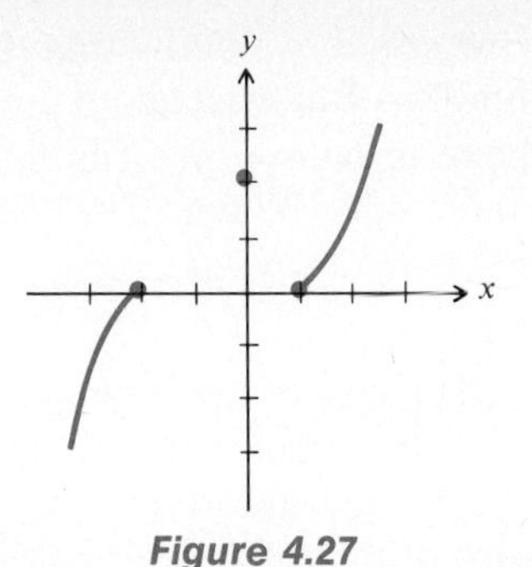

Figure 4.27

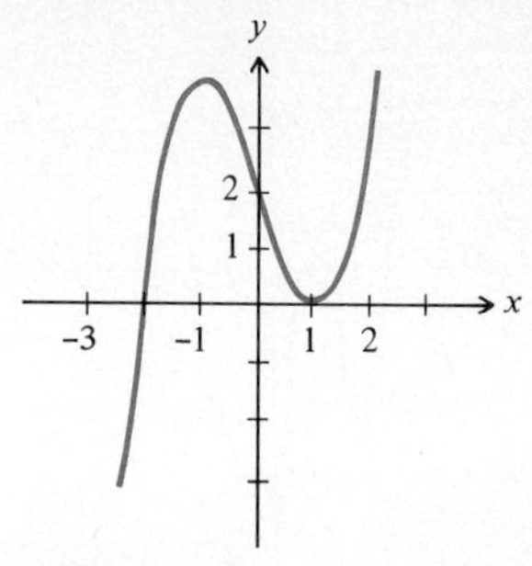

Figure 4.28

x equal to zero. We get $y = 2$. So far, we have three points of our graph (Figure 4.26). If the right side of the given equation were expanded we would see a leading term of x^3. Therefore, as $x \to \infty$, $y \to \infty$, and as $x \to -\infty$, $y \to -\infty$. This tells us that the graph falls without bound to the left of $x = -2$ and rises without bound to the right of $x = 1$. These parts of the graph must look something like the graph in Figure 4.27. Between $x = -2$ and $x = 1$, the graph must lie entirely above or entirely below the x-axis. Because $(0, 2)$ is on the graph it must look something like the graph in Figure 4.28. Note that this is just a *sketch* of the graph. Without additional information, either of the graphs in Figure 4.29 (as well as many others) is just as good a sketch. To obtain a more accurate sketch we would have to plot additional points or use some calculus concepts. ■

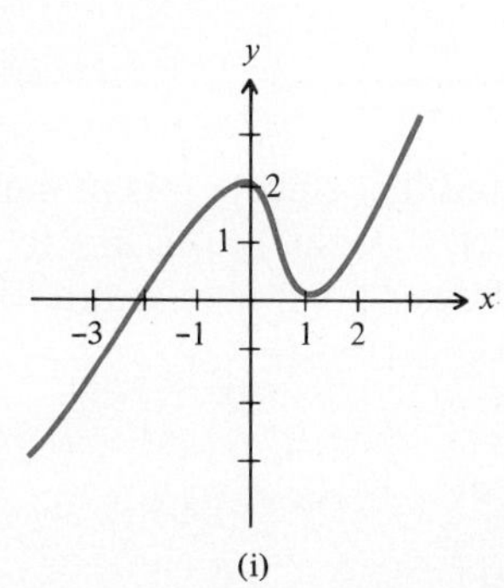

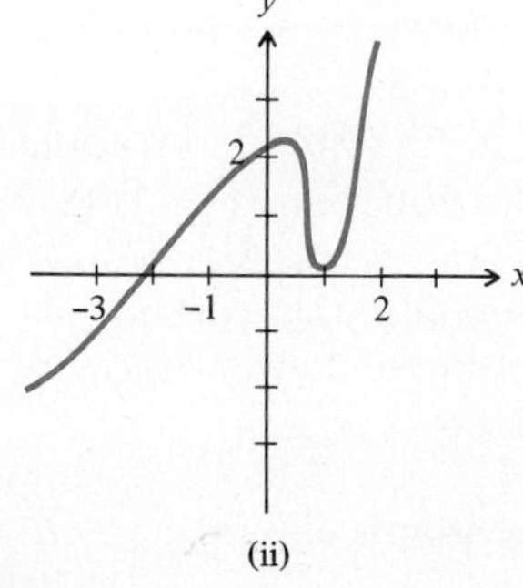

Figure 4.29

Example 3
Sketch the graph of $y = (x + 1)^2(x + 3)(x - 2)$.

Solution. The x-intercepts are -1, -3, and 2. The y-intercept is -6. Since the leading term is x^4, as $x \to \infty$, $y \to \infty$, and as $x \to -\infty$, $y \to \infty$. This information gives us part of the graph. See Figure 4.30. Because the

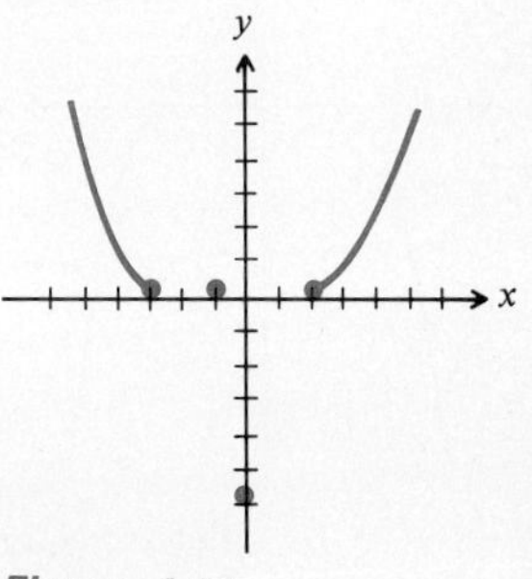

Figure 4.30

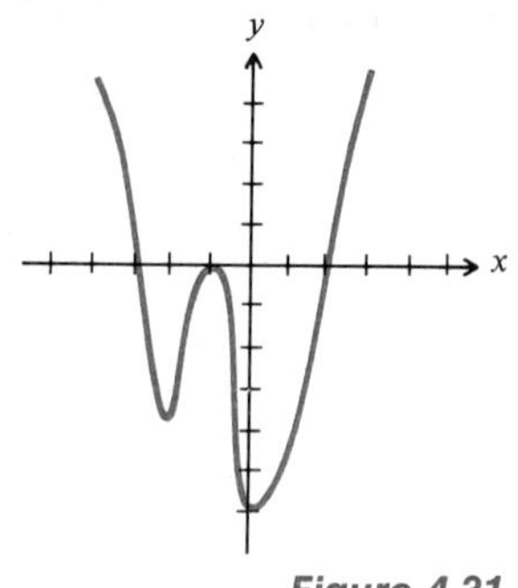

Figure 4.31
$y = (x + 1)^2(x + 3)(x - 2)$

y-intercept is negative, we know that the graph between $x = -1$ and $x = 2$ lies below the x-axis. But what about the graph between $x = -3$ and $x = -1$? Does it lie above or below? We can find out by plotting any point between these two x-intercepts. When $x = -2$, $y = -4$, and the graph lies below (Figure 4.31). ■

There is another way of determining whether the graph of a polynomial is above or below the x-axis between two consecutive x-intercepts. But we must first consider an important consequence of the Remainder Theorem and the concept of multiplicity.

Theorem 4.2 (The Factor Theorem)

Let P be a polynomial, and let a be a constant. Then a is a root of the equation $P(x) = 0$ if and only if $x - a$ is a factor of $P(x)$.

This theorem is a direct result of the Remainder Theorem.

If $x - a$ is a factor of $P(x)$ n times, then we say that a is a root of **multiplicity** n. Thus, for the equation $(x + 2)(x - 1)^2 = 0$, 1 is a root of multiplicity two, and -2 is a root of multiplicity one. (Note that the multiplicities are simply the powers of the factors from which the roots are obtained.)

This leads to the following important theorem.

Theorem 4.3

Let P be a polynomial, and let a be a root of $P(x) = 0$ of multiplicity n. If n is odd, then the graph of P crosses the x-axis at $x = a$. If n is even, then the graph does not cross at $x = a$.

Hence, in Example 3, we could have noted that since -3 is of multiplicity one for the equation $(x + 1)^2(x + 3)(x - 2) = 0$, the graph has to cross at $x = -3$. Or we could have observed that because -1 is of multiplicity two, the graph would not cross at $x = -1$.

Example 4
Sketch the graph of $y = -x^3(2x - 5)(x + 2)^2$.

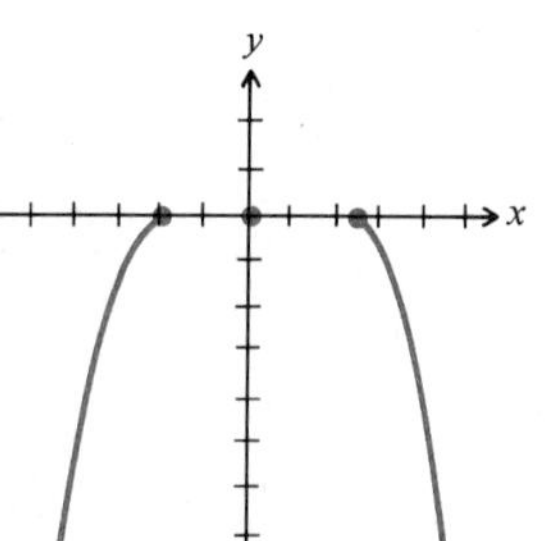

Figure 4.32

Solution. The x-intercepts are 0, $\frac{5}{2}$, and -2. The y-intercept is 0. Since the leading term is $-2x^6$, as $x \to \infty$, $y \to \infty$, and as $x \to -\infty$, $y \to \infty$. This information gives us part of the graph. See Figure 4.32. Now since -2 is of even multiplicity, the graph does not cross at $x = -2$. And since 0 is of odd multiplicity, the graph crosses at $x = 0$. These two facts, together with continuity, force us to conclude that the graph must cross at $x = \frac{5}{2}$, which is consistent with its odd multiplicity. Hence the graph must be something like Figure 4.33. ■

Exercise

Sketch the graph of $y = (x - 3)^2(x - 1)(x + 1)$.

Answer

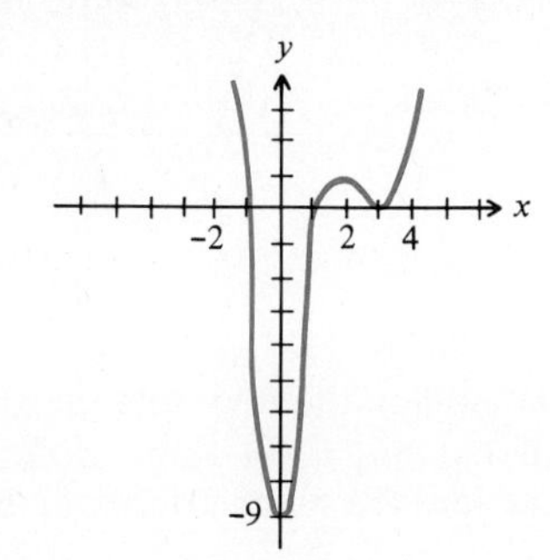

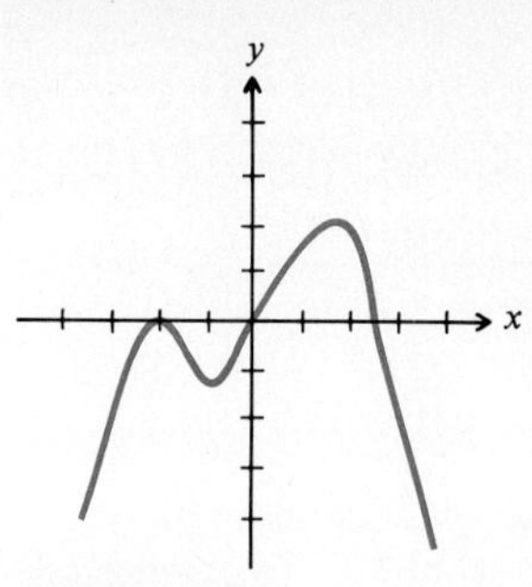

Figure 4.33

$y = -x^3(2x - 5)(x + 2)^2$

■

Problem Set 4.3

In Problem 1–4, examine the behavior of y as $x \to \infty$ and as $x \to -\infty$.

1. $y = 2x^3 - 3x^2 + 5x - 1$

2. $y = 5x^6 + x^3 - 2x + 5$

3. $y = 2x^2 - 5x - 3x^4 + 1$

4. $y = 4x^4 + x^2 - 5x - 2x^5$

In Problems 5–22, sketch the graph of the polynomial.

5. $y = (x - 2)(x + 5)$

6. $y = (x - 3)(x + 2)$

7. $y = (x + 1)^2(x + 3)$

8. $y = (x - 2)^3(x + 3)$

9. $y = (x - 6)(x - 1)(x + 4)$

10. $y = x^2(x - 2)(x + 3)$

11. $y = -2(x - 2)(x + 4)$

12. $y = (2x + 3)(5 - x)$

13. $y = -3(x + 1)^2(x + 5)$

14. $y = -2x(x - 3)(2x + 5)^2$

15. $y = x^2 + 5x + 4$

16. $y = x^2 - x - 6$

17. $y = 2x^2 + 7x - 15$

18. $y = 4x^3 + 4x^2 + x$

19. $y = x^3 - 2 + 2x^2 - x$

20. $y = 4 - x^3 + 4x - x^2$

21. $y = x^3 + 5x^2 - 2x$

22. $y = -x^4 - 2x^3 + 5x^2$

In Problems 23 and 24, choose the equation which best *describes the given graph.*

23. **(a)** $y = (x + 2)^2(x - 1)$
(b) $y = (x - 2)^2(x + 1)$
(c) $y = (x - 2)(x + 1)^2$
(d) $y = (x + 2)^2(x + 1)$
(e) $y = (x - 2)(x + 1)$

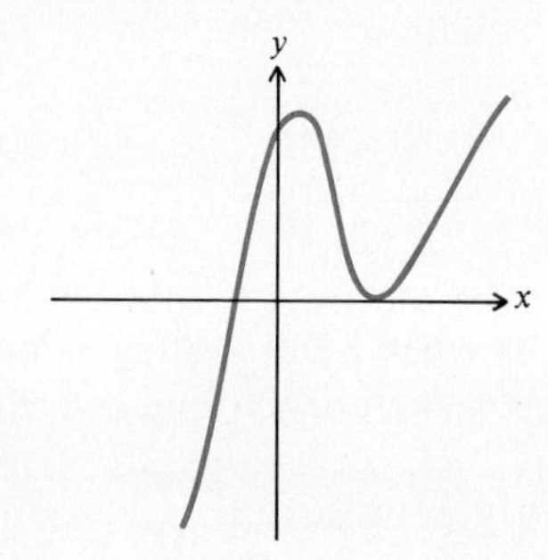

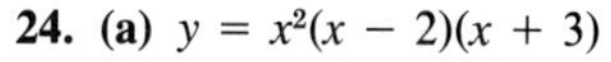

24. (a) $y = x^2(x - 2)(x + 3)$
(b) $y = -x(x + 2)(x - 3)$
(c) $y = x(x - 2)(x + 3)$
(d) $y = x(x + 2)(x - 3)$
(e) $y = -x(x - 2)(x + 3)$

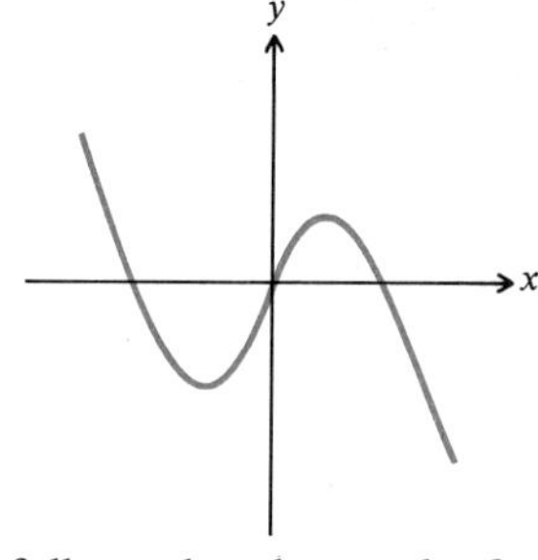

25. From the definition of symmetry about the origin, it follows that the graph of a function f is symmetric about the origin if and only if $f(-x) = -f(x)$. Use this fact to determine whether the following functions are symmetric about the origin.
(a) $f(x) = x^3 - x$ (b) $y = x^3 + x^2$

26. From the definition of symmetry about the y-axis, it follows that the graph of a function f is symmetric about the y-axis if and only if $f(-x) = f(x)$. Use this fact to determine whether the following functions are symmetric about the y-axis.
(a) $f(x) = x^4 - 3x^2 + 1$ (b) $y = x^3 - 3x^2 + 3x$

In Problems 27 and 28, sketch the graph.

27. $y = \begin{cases} x + 5 & \text{when } x < -2 \\ x^2(x^2 - 4) & \text{when } x \geq -2 \end{cases}$

28. $y = |x^2 - 3x|$

4.4 Zeros of Polynomials

A **zero** of a function f is a real number a such that $f(a) = 0$. Such a number a is also called a root of the equation $f(x) = 0$. Hence the study of zeros of functions is also a study of roots of equations. Consider the polynomial P defined by

$$P(x) = x^3 - x^2 - x - 2.$$

To say that 2 is a zero of P is exactly the same as saying that 2 is a root of the equation

$$x^3 - x^2 - x - 2 = 0.$$

Of course, finding the x-intercepts of a polynomial is the same as finding its zeros, and finding x-intercepts is a principle motivation for this section. We begin by noting two facts about the number of zeros of a polynomial.

First, *a polynomial of degree n has at most n zeros.* Thus a polynomial of degree three has no more than three zeros, and a polynomial of degree four has no more than four, and so on.

Second, for every polynomial of odd degree, one of the following two conditions exists:

$$\text{as} \quad x \to \infty,\ y \to \infty, \qquad \text{and as} \qquad x \to -\infty,\ y \to -\infty,$$

or

$$\text{as} \quad x \to \infty,\ y \to -\infty, \qquad \text{and as} \qquad x \to -\infty,\ y \to \infty.$$

Regardless of which occurs, since polynomials are continuous, the graph must cross the x-axis at least once. Hence *a polynomial of odd degree has at least one zero.*

A **variation in sign** in a polynomial is a change in sign between successive coefficients when the terms are written so that the exponents are descending from left to right. For example, the polynomial defined by

$$P(x) = 2x^5 - 3x^3 - x^2 + 2x - 1$$

has three variations in sign. Consider the polynomial defined by

$$Q(x) = x^4 + 2x^2 + 3 - x^3 - 2x.$$

Does it have only one variation in sign? No. Before counting the variations in sign, the terms must be arranged so that the powers decrease from left to right. So we must write

$$Q(x) = x^4 - x^3 + 2x^2 - 2x + 3.$$

Thus this polynomial has four variations in sign.

Theorem 4.4 will give us additional information about the number of zeros.

Theorem 4.4 (Descartes Rule of Signs)

Let P be a polynomial of positive degree and let $Q(x) = P(-x)$. If P has k variations in sign, then the number of positive zeros of P is k or is less than k by an even positive integer. If Q has j variations in sign, then the number of negative zeros of P is j or is less than j by an even positive integer.

This theorem will become clearer when we see it applied, as in Examples 1 and 2.

Example 1

Use Descartes Rule of Signs to analyze the zeros of P, where

$$P(x) = 2x^4 - 3x^3 + x^2 - x - 5.$$

Solution. Because there are three variations in sign, Descartes Rule of Signs says that there must be either three or one positive zeros. To see what Descartes Rule says about the number of negative zeros, we must count the variations in sign of $P(-x)$. Since

$$P(-x) = 2x^4 + 3x^3 + x^2 + x - 5,$$

and there is but one change in sign, there is exactly one negative zero of P. ■

Example 2
Analyze the zeros of P, where $P(x) = 6x^3 - 7x^2 + 1$.

Solution. Because P is of degree 3, it has at most three zeros, and since it is of odd degree it has at least one. By Descartes Rule of Signs, it has either two or no positive zeros. Since

$$P(-x) = -6x^3 - 7x^2 + 1,$$

and there is one variation in sign, there is exactly one negative zero. ■

Exercise
Analyze the zeros of each of the following.
(a) $P(x) = x^4 - 2x^3 - x^2 + x + 3$ (b) $Q(x) = -x^5 + 2x^4 - x^2 - 3x + 2$

Answers. (a) at most four; two or no positive zeros; two or no negative zeros.
(b) at most five; at least one; three or one positive zeros; two or no negative zeros. ■

We shall now return to the problem of finding the zeros of a polynomial. We dealt with this very problem when trying to find x-intercepts. After setting the polynomial equal to zero, we usually tried to factor the nonzero side if the degree of the polynomial was two or more. If that couldn't be done and the polynomial was quadratic, we used the quadratic formula. What do we do when the expression is not readily factorable? The Factor Theorem (Theorem 4.2) will help. It says that, for a polynomial P and a constant a, a is a zero of P if and only if $x - a$ is a factor of $P(x)$. This enables us to simplify the given polynomial if we can find a zero. For example, consider the polynomial P defined by

$$P(x) = 6x^3 - 7x^2 + 1.$$

By inspection we may note that 1 is a zero. Therefore, by the Factor Theorem, $x - 1$ is a factor. Dividing $P(x)$ by $x - 1$, using synthetic division, we get

$$\begin{array}{r|rrrr} 1 & 6 & -7 & 0 & 1 \\ & & 6 & -1 & -1 \\ \hline & 6 & -1 & -1 & 0 \end{array}$$

$\underbrace{6 \quad -1 \quad -1}_{\text{coefficients of quotient}} \quad \underbrace{0}_{\text{remainder}}$

Thus $P(x) = (x - 1)(6x^2 - x - 1)$. To find the remaining zeros, we solve a quadratic equation. We have the following:

$$6x^2 - x - 1 = 0$$
$$(3x + 1)(2x - 1) = 0$$
$$x = \frac{-1}{3} \quad \text{or} \quad x = \frac{1}{2}.$$

Hence the zeros of P are 1, $-\frac{1}{3}$, and $\frac{1}{2}$.

Sometimes it is too difficult to obtain a zero by inspection. In such cases the following theorem is beneficial.

Theorem 4.5 (Rational Roots Theorem)

Let P be defined by

$$P(x) = a_n x^n + \cdots + a_2 x^2 + a_1 x + a_0,$$

where each a_i is an integer with $a_n \neq 0$. If p and q are integers such that p/q, in reduced form, is a zero of P, then p divides a_0 and q divides a_n.

This theorem states that the denominator of a rational zero must be a factor of the leading coefficient, and the numerator must be a factor of the constant term. Hence by listing all the factors of the leading coefficient and of the constant term, we can determine all *possible* rational zeros. This procedure is illustrated in Example 3.

Example 3

Find all rational zeros of P where $P(x) = 3x^4 + 5x^3 + x^2 + 5x - 2$.

Solution. We will begin by analyzing the number of zeros, as such a study may reduce the number of possible rational zeros. We know that P has at most four zeros and that it has exactly one positive zero. And because $P(-x) = 3x^4 - 5x^3 + x^2 - 5x - 2$, we know that it has three or one negative zeros. If p/q is a rational zero, then the possibilities are

$$p: \quad \pm 1, \pm 2$$
$$q: \quad \pm 1, \pm 3$$
$$\frac{p}{q}: \quad \pm 1, \pm 2, \pm\frac{1}{3}, \pm\frac{2}{3}.$$

Since we know that P has both a positive and a negative zero, we cannot rule out any of the values of p/q listed. We begin evaluating P at the possible rational zeros, using synthetic division. Note that we have

$$\begin{array}{r|rrrrr} -2 & 3 & 5 & 1 & 5 & -2 \\ & & -6 & 2 & -6 & 2 \\ \hline & 3 & -1 & 3 & -1 & \end{array}$$

Thus -2 is a zero. Furthermore, we see that

$$P(x) = (x + 2)(3x^3 - x^2 + 3x - 1).$$

To find the rest of the zeros, it suffices to find the zeros of Q, where

$$Q(x) = 3x^3 - x^2 + 3x - 1.$$

Since the expression on the right can be factored, we get the following:

$$\begin{aligned} 3x^3 - x^2 + 3x - 1 &= 0 \\ x^2(3x - 1) + 1(3x - 1) &= 0 \\ (3x - 1)(x^2 + 1) &= 0 \\ 3x = 1 \qquad \text{or} \qquad x^2 &= -1 \\ x = \frac{1}{3}. \end{aligned}$$

Hence the zeros of P are -2 and $\frac{1}{3}$. ■

In evaluating P at the possible rational zeros, one might have tried $\frac{1}{3}$ before trying -2. If that were the case, we would have gotten

$$\begin{array}{r|rrrrr} \frac{1}{3} & 3 & 5 & 1 & 5 & -2 \\ & & 1 & 2 & 1 & 2 \\ \hline & 3 & 6 & 3 & 6 & \end{array}$$

and $P(x)$ would have been written

$$P(x) = \left(x - \frac{1}{3}\right)(3x^3 + 6x^2 + 3x + 6).$$

But the left side of $3x^3 + 6x^2 + 3x + 6 = 0$ is factorable, and we would have obtained the other zero -2.

Example 4
Find all rational zeros of P, where $P(x) = 6x^4 + 17x^3 + 17x^2 + 7x + 1$.

Solution. Now P has at most four zeros, and by Descartes Rule of Signs, there are no positive zeros. Since $P(-x) = 6x^4 - 17x^3 + 17x^2 - 7x + 1$, there are four or two or no negative zeros. By the Rational Roots Theorem the possible rational zeros are

$$\begin{aligned} p&: \pm 1 \\ q&: \pm 1, \pm 2, \pm 3, \pm 6 \\ \frac{p}{q}&: \pm 1, \pm\frac{1}{2}, \pm\frac{1}{3}, \pm\frac{1}{6}. \end{aligned}$$

But in light of what Descartes Rule said about the number of positive zeros, we will look for negative zeros only. Note that -1 is a zero. We have

$$\begin{array}{r|rrrrr} -1 & 6 & 17 & 17 & 7 & 1 \\ & & -6 & -11 & -6 & -1 \\ \hline & 6 & 11 & 6 & 1 & \end{array}$$

Thus we have

$$P(x) = (x + 1)(6x^3 + 11x^2 + 6x + 1).$$

Now we use the Rational Roots Theorem again, on the polynomial Q, where $Q(x) = 6x^3 + 11x^2 + 6x + 1$. Possible rational zeros are

$$\begin{aligned} p&: \pm 1 \\ q&: \pm 1, \pm 2, \pm 3, \pm 6 \\ \frac{p}{q}&: \pm 1, \pm\frac{1}{2}, \pm\frac{1}{3}, \pm\frac{1}{6}. \end{aligned}$$

Again restricting our choices to the negative values, we find that -1 is a zero of Q. We get

$$\begin{array}{r|rrrr} -1 & 6 & 11 & 6 & 1 \\ & & -6 & -5 & -1 \\ \hline & 6 & 5 & 1 & \end{array}$$

Therefore,

$$P(x) = (x + 1)(x + 1)(6x^2 + 5x + 1).$$

Factoring the quadratic gives

$$P(x) = (x + 1)(x + 1)(3x + 1)(2x + 1).$$

Hence the zeros of P are -1, $-\frac{1}{3}$, and $-\frac{1}{2}$. ■

Note that in Example 4, we found three zeros (all negative), but Descartes Rule of Signs said that there were four or two or no negative zeros. This is because the phrase "the number of zeros" as it appears in Descartes Rule really means "the sum of the multiplicities of the zeros." So when counting for Descartes Rule, we count the -1 twice because it has multiplicity two.

Also note that, while looking for a rational root of P, if we had tried $-\frac{1}{6}$ and correctly observed that it was not a zero, then there would have been no need to try it while looking for a rational root of Q, even though it was listed among the possibilities. Clearly, no real number can be a zero of a factor of a polynomial without being a zero of the polynomial.

Exercise

Find all rational zeros of P, where $P(x) = 6x^3 - 7x^2 - 9x - 2$.

Answer. $2, -\dfrac{1}{3}$, and $-\dfrac{1}{2}$ ■

This section has not dealt with the problem of finding the irrational zeros of a polynomial. In general, the best that can be done in this regard is to find rational approximations of these irrational zeros. There are techniques which enable one to approximate these irrational zeros to any degree of accuracy. Programmed with these methods, computers have taken up such tasks, and we shall not study such techniques in this text.

Problem Set 4.4

In Problems 1–6, analyze the zeros.

1. $P(x) = x^4 - 3x^3 + 5x + 1$

2. $P(x) = x^4 + x^3 + 2x^2 - x - 1$

3. $P(x) = -2x^3 + x^2 + 5x - 1$

4. $P(x) = -x^5 + 3x^4 - x^2 - 2x + 5$

5. $P(x) = 2x^5 + x^3 + 3 + 2x$

6. $P(x) = 1 + 5x - 4x^6 + 2x^3$

In Problems 7–12, find all rational zeros.

7. $P(x) = 2x^3 + x^2 - 5x + 2$

8. $P(x) = 3x^3 - 2x^2 - 19x - 6$

9. $P(x) = 3x^4 - 2x^3 - 9x^2 + 12x - 4$

10. $P(x) = 2x^4 - 3x^3 - 27x^2 - 37x - 15$

11. $P(x) = x^3 + 6x^4 + 22x^2 - 8 + 4x$

12. $P(x) = 3x^4 + 12x^2 + 12 - 11x^3 - 22x$

In Problems 13 and 14, find all x-intercepts.

13. $P(x) = 15x^3 + 4x^2 - 14x + 4$

14. $P(x) = 4x^3 + 8x^2 - 5x - 12$

In Problems 15–18, sketch the graph.

15. $y = 3x^4 + 2x^3 - 13x^2 - 8x + 4$

16. $y = 2x^3 - 5x^2 - 14x + 8$

17. $y = 3x^3 - 8x^2 + 7x - 2$

18. $y = x^4 - 8x^3 + 16x^2 - 8x + 15$

19. Show that the polynomial $P(x) = x^{20} + 5x^{10} + 3x^8 + x^2 + 7$ has no real zeros.

20. Show that for any positive integer n and any real number a, $x - a$ is a factor of $x^n - a^n$.
21. Why must $f(x) = x^3 - 5x + 1$ have a zero between 0 and 1?

4.5 Rational Functions

A function f defined by

$$f(x) = \frac{P(x)}{Q(x)},$$

where P and Q are polynomial functions, is called a **rational function.** Thus a rational function is the quotient of two polynomials. For example, f and g, defined by

$$f(x) = \frac{2x + 1}{x - 5} \quad \text{and} \quad g(x) = \frac{1}{x^2 - 4},$$

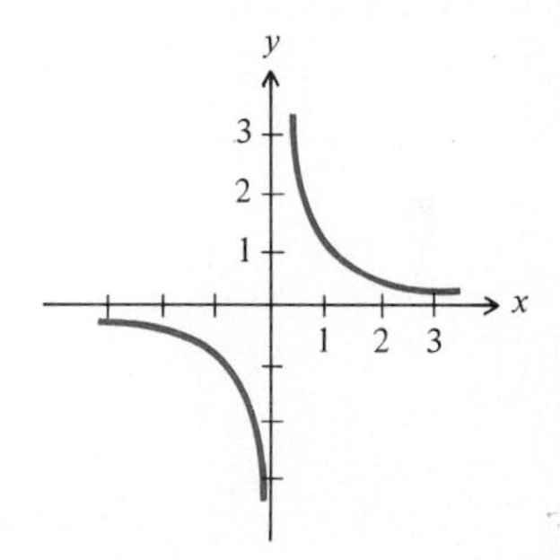

Figure 4.34 $f(x) = \frac{1}{x}$

are rational functions. [All polynomial functions are rational functions, too, with $Q(x) = 1$.] Because a rational function is not defined at values which make the denominator zero, the domain of f is R − {5}, and the domain of g is R − {2, −2}.

We consider the rational function $y = 1/x$. Its graph is shown in Figure 4.34. There are several important facts to be noted about this graph. First, it is symmetric about the origin. Second, although the function is not defined at 0, we can choose values of x as close to 0 as we please. As x gets closer to 0 from the positive side, $f(x)$ gets larger positively. And as x gets closer to 0 from the negative side, $f(x)$ gets larger negatively. Hence, as x approaches 0, the graph of f approaches, but does not intersect, the line $x = 0$ (the y-axis). Recall from the section on hyperbolas that a line with the property that the distance from a point P to the line approaches zero as P moves away from the origin along some part of the graph is called an **asymptote.** In Figure 4.34 the y-axis is a *vertical* asymptote.

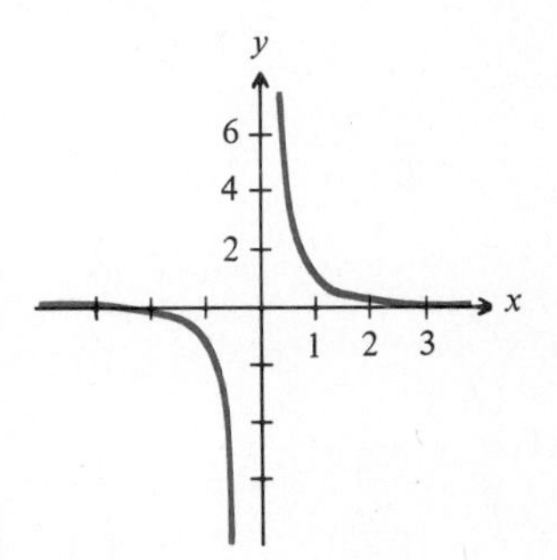

Figure 4.35 $y = \frac{1}{x^3}$

Also note that as x gets larger positively, $f(x)$ gets closer to 0. And $f(x)$ also gets closer to 0 as x gets larger negatively. Hence, as $|x|$ increases, the graph of f approaches the line $y = 0$ (the x-axis). Thus the x-axis is a *horizontal* asymptote in Figure 4.34.

The graph of $y = 1/x^3$ is shown in Figure 4.35. Note that the graph of $y = 1/x^3$ (in Figure 4.35) is similar to the graph of $y = 1/x$ (in Figure 4.34). Each is symmetric about the origin, each contains the points $(-1, -1)$ and $(1, 1)$, each has a vertical asymptote of $x = 0$, and each has a horizontal asymptote of $y = 0$. The principal difference is the rate at which the asymptotes are approached. See Figure 4.36.

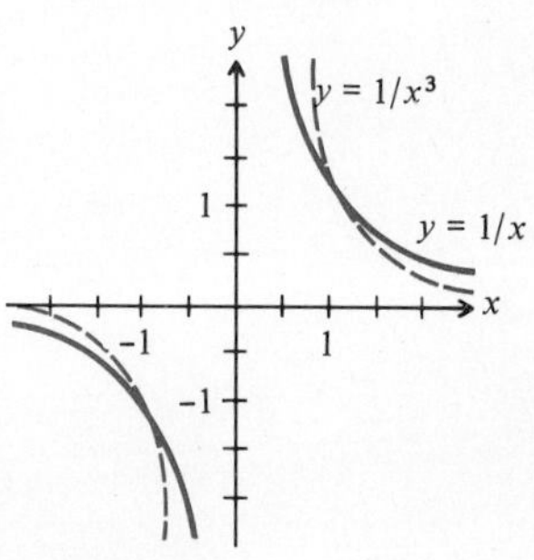

Figure 4.36

Thus the graphs of $y = 1/x^{2n+1}$, for $n = 0, 1, 2, \ldots$, are similar. All are symmetric about the origin. All pass through the points $(-1, -1)$ and $(1, 1)$.

All have a vertical asymptote of $x = 0$, and all have a horizontal asymptote of $y = 0$. But the greater the exponent, the faster the graph approaches the x-axis and the slower it approaches the y-axis.

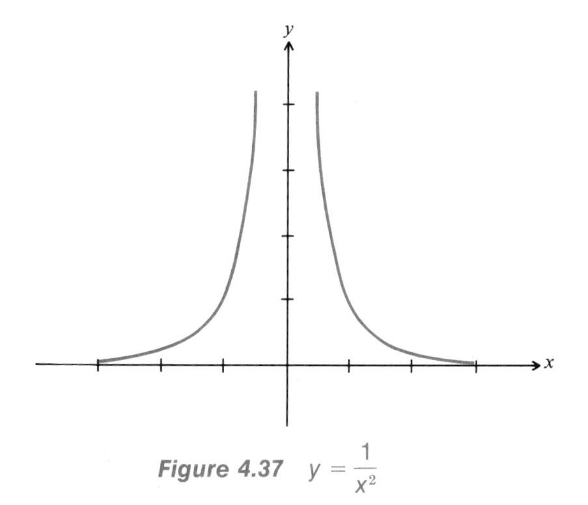

Figure 4.37 $y = \dfrac{1}{x^2}$

The graph shown in Figure 4.37 is the graph of $y = 1/x^2$. There are several important facts to be noted about this graph. First, it is symmetric about the y-axis. Second, the y-axis is a vertical asymptote, and the x-axis is a horizontal asymptote.

In Figure 4.38 we see the graph of $y = 1/x^4$. Note that it looks a lot like the graph of $y = 1/x^2$.

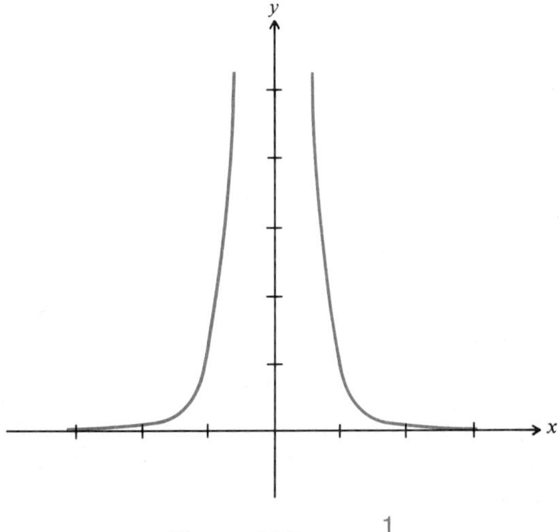

Figure 4.38 $y = \dfrac{1}{x^4}$

We will find that the graphs of $y = 1/x^{2n}$, $n = 1, 2, \ldots$, are similar. All are symmetric about the y-axis. All contain the points $(-1, 1)$ and $(1, 1)$.

And all have a vertical asymptote of $x = 0$ and a horizontal asymptote of $y = 0$. But the greater the exponent, the faster the graph approaches the x-axis and the slower it approaches the y-axis.

Of course, these elementary rational functions can undergo the same types of transformations as the elementary polynomial functions. For example, the graph of $y = (-1/x) + 2$ (shown in Figure 4.39) is simply the graph of $y = 1/x$ reflected about the x-axis and then moved up two units, with the horizontal asymptote being $y = 2$. The graph of $y = 1/(x - 1)^2$ (also in Figure 4.39) is the graph of $y = 1/x^2$ shifted to the right one unit, with the vertical asymptote being $x = 1$. Note that when an asymptote is a line other than the x- or y-axis, it is indicated by a dashed line.

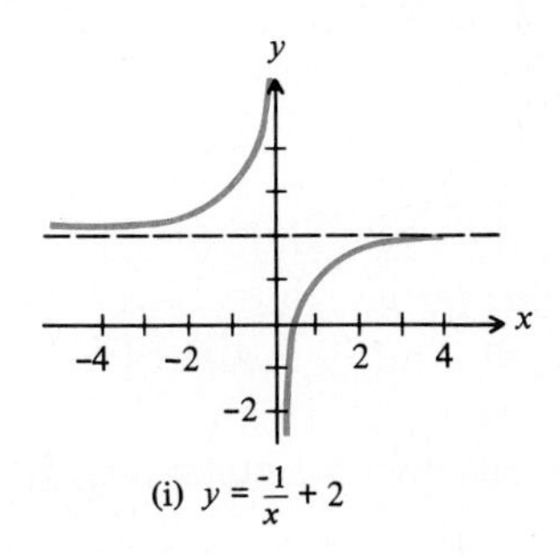

(i) $y = \frac{-1}{x} + 2$

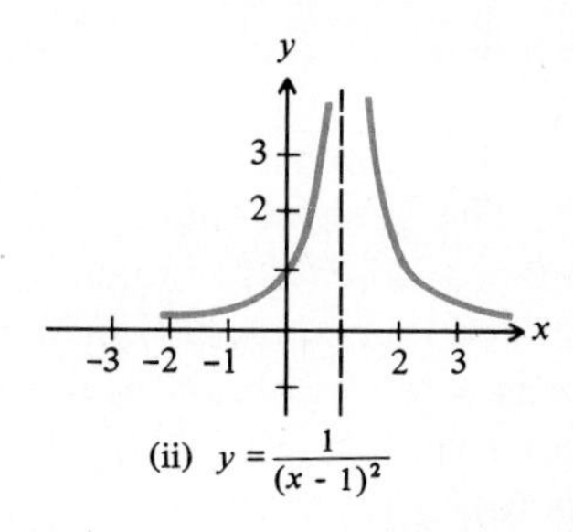

(ii) $y = \frac{1}{(x-1)^2}$

Figure 4.39

Example 1

Sketch the graph of $y = \dfrac{-1}{(x + 2)^3} - 3$.

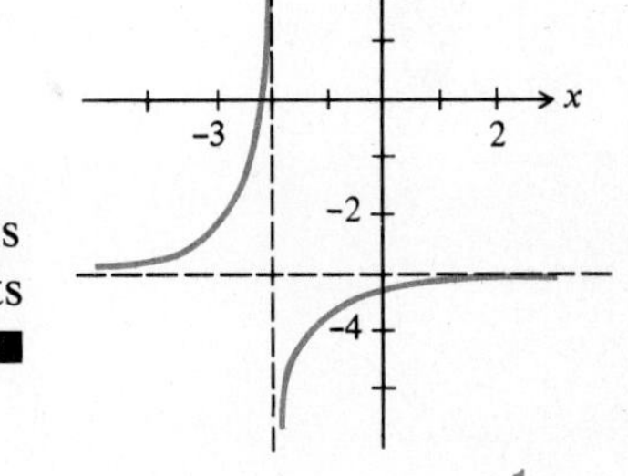

Figure 4.40 $y = \dfrac{-1}{(x + 2)^3} - 3$

Solution. This graph (Figure 4.40) is a transformation of $y = 1/x^3$. It is this basic graph reflected about the x-axis and then shifted to the left two units and down three. ■

Exercise

Sketch the graph of $y = \dfrac{-1}{(x - 2)^2} + 1$.

Answer

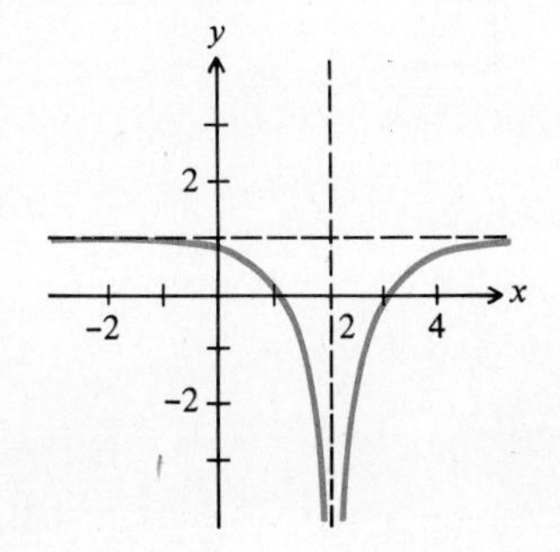

■

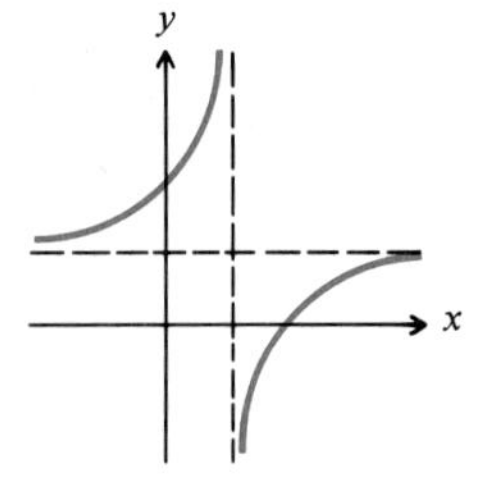

Figure 4.41

Example 2

Choose the equation which *best* describes the graph shown in Figure 4.41.

(a) $y = \dfrac{-1}{x^4} + 1$

(b) $y = \dfrac{1}{x - 1} - 1$

(c) $y = \dfrac{-1}{(x - 1)^3} + 1$

(d) $y = \dfrac{-1}{(x - 1)^2} + 1$

(e) $y = \dfrac{-1}{x + 1} + 1$

Solution. The general shape tells us that this is a shift of a graph whose equation is $y = 1/x^{2n+1}$ for some positive integer n. Thus answers (a) and (d) are eliminated immediately. Further, since the shift is into the first quadrant, and there is a reflection about the x-axis, the equation must be of the form $y = -1/(x - c_1)^{2n+1} + c_2$, where c_1 and c_2 are positive constants. Hence the answer is (c). ■

Exercise

Choose the equation which *best* describes the graph.

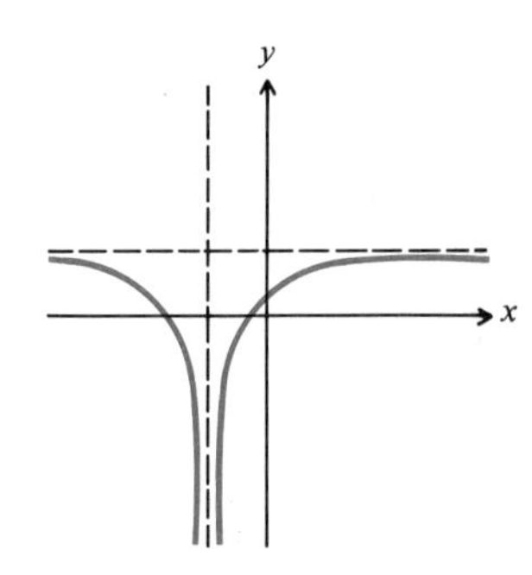

(a) $y = \dfrac{-1}{(x + 1)^2} - 2$

(b) $y = \dfrac{-1}{(x - 1)^4} + 2$

(c) $y = \dfrac{-1}{x + 1} + 2$

(d) $y = \dfrac{-1}{(x - 1)^3} + 2$

(e) $y = \dfrac{-1}{(x + 1)^2} + 2$

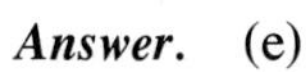

Answer. (e) ■

Problem Set 4.5

In Problems 1–16, sketch the graph of the rational function.

1. $y = \dfrac{1}{x} - 1$ **2.** $y = \dfrac{1}{x^2} + 2$ **3.** $y = \dfrac{1}{(x + 3)^2}$

4. $y = \dfrac{1}{x - 2}$

5. $y = \dfrac{3}{x^2}$

6. $y = \dfrac{1}{8x}$

7. $y = \dfrac{-2}{x^2}$

8. $y = \dfrac{-1}{x^4}$

9. $y = \dfrac{1}{(x - 2)^3} + 2$

10. $y = \dfrac{1}{(x - 1)^2} - 3$

11. $y = \dfrac{-1}{x^2} + 2$

12. $y = \dfrac{-1}{(x + 2)^3}$

13. $y = \dfrac{-1}{x + 2} + 3$

14. $y = \dfrac{1}{4(x - 2)^2} - 1$

15. $y = \dfrac{-1}{2(x + 3)} - 2$

16. $y = \dfrac{-2}{(x + 3)^2} - 1$

In Problems 17–20, choose the equation which best *describes the given graph.*

17. (a) $y = \dfrac{2}{(x + 2)^2} - 3$

(b) $y = \dfrac{2}{x + 2} + 3$

(c) $y = \dfrac{-2}{(x + 2)^3} - 3$

(d) $y = \dfrac{2}{(x + 2)^3} - 3$

(e) $y = \dfrac{2}{(x - 2)^2} - 3$

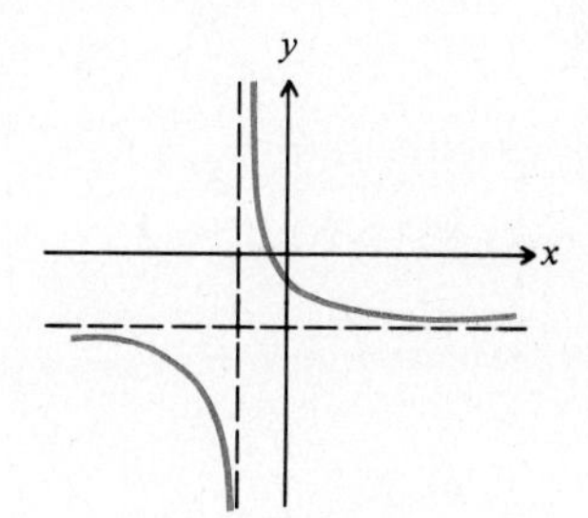

18. (a) $y = \dfrac{2}{(x - 5)^4} + 3$

(b) $y = \dfrac{-2}{x - 5} + 3$

(c) $y = \dfrac{-2}{(x - 5)^2} + 3$

(d) $y = \dfrac{-2}{(x - 5)^3} - 3$

(e) $y = \dfrac{-2}{(x - 5)^2} - 3$

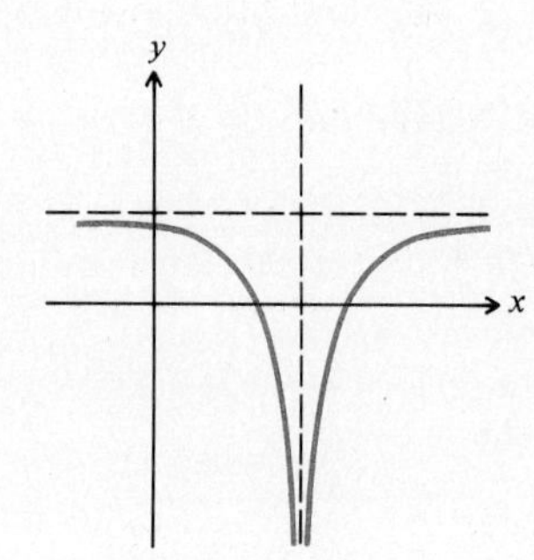

19. **(a)** $y = \dfrac{-1}{(x-2)^4} - 3$

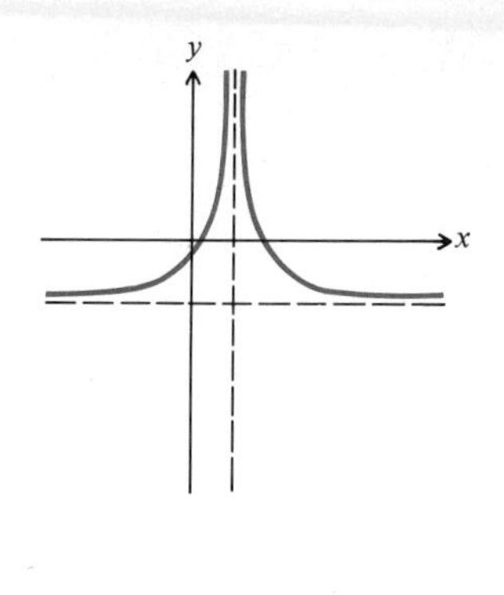

(b) $y = \dfrac{1}{(x-2)^2} + 3$

(c) $y = \dfrac{1}{x-2} - 3$

(d) $y = \dfrac{1}{(x+2)^4} - 3$

(e) $y = \dfrac{1}{(x-2)^4} - 3$

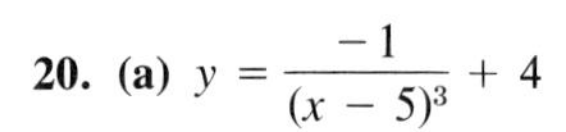

20. **(a)** $y = \dfrac{-1}{(x-5)^3} + 4$

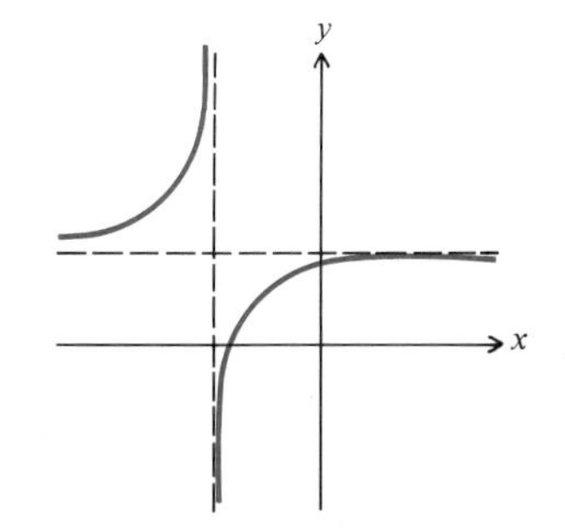

(b) $y = \dfrac{-1}{x+5} + 4$

(c) $y = \dfrac{1}{x+5} + 4$

(d) $y = \dfrac{-1}{x+5} - 4$

(e) $y = \dfrac{-1}{(x+5)^2} + 4$

21. Algebraically show that the graph of $y = \dfrac{3x-1}{x-1}$ is a transformation of the graph of $y = \dfrac{1}{x}$.

22. Algebraically show that the graph of $y = \dfrac{2x^2+4x+3}{(x+1)^2}$ is a transformation of the graph of $y = \dfrac{1}{x^2}$.

In Problems 23–26, sketch the graph.

23. $y = \begin{cases} x^3 & \text{when } x \le 0 \\ \dfrac{1}{x^3} & \text{when } x > 0 \end{cases}$

24. $y = \begin{cases} \dfrac{1}{(x+1)^2} & \text{when } x < -1 \\ -x^2 & \text{when } x \ge -1 \end{cases}$

25. $y = \begin{cases} x^2(x+3) & \text{when } x \le 1 \\ \dfrac{1}{x-1} & \text{when } x > 1 \end{cases}$

26. $y = \left|\dfrac{1}{x-1}\right| + 2$

4.6 Rational Functions Continued

In this section we want to develop some techniques for sketching the graphs of some other rational functions.

We first consider the problem of finding vertical asymptotes. Theorem 4.6 tells us that we can determine the vertical asymptotes of a rational function by finding the values which make the denominator, but not the numerator, equal zero.

Theorem 4.6

Let f be a rational function defined by

$$f(x) = \frac{P(x)}{Q(x)},$$

and let a be a constant. If $Q(a) = 0$ and $P(a) \neq 0$, then the line $x = a$ is a vertical asymptote of the graph of f.

Example 1

Determine the vertical asymptotes of $f(x) = \dfrac{(x+1)(x-3)}{(2x-1)(x+2)}$.

Solution. Since $\frac{1}{2}$ and -2 make the denominator, but not the numerator, equal zero, the vertical asymptotes are

$$x = \frac{1}{2} \quad \text{and} \quad x = -2. \qquad \blacksquare$$

Exercise

Determine the vertical asymptotes of $f(x) = \dfrac{(x+2)(2x-1)}{(x-3)(3x+2)}$.

Answers. $x = 3$ and $x = \dfrac{-2}{3}$ $\blacksquare$

We now direct our attention to the problem of finding the horizontal asymptotes. We noted that the line $y = 0$ is a horizontal asymptote of $y = 1/x$. This is because as x gets larger positively (without bound), y gets closer and closer to zero. Also, as x gets larger negatively (without bound), y approaches zero. These relationships are expressed as follows:

$$\text{as} \quad x \to \infty, \quad y \to 0, \qquad \text{and as} \quad x \to -\infty, \quad y \to 0.$$

Hence, in general, to determine the horizontal asymptotes, we need to examine the behavior of y as x gets large positively and large negatively. To illustrate, let's find the horizontal asymptotes of the graph whose equation is

$$y = \frac{(x + 1)(x - 3)}{(2x - 1)(x + 2)}.$$

Now as x gets larger positively, each factor in both the numerator and in the denominator gets larger positively. Thus we're uncertain about the behavior of y. However, let's rewrite y as follows:

$$y = \frac{x^2 - 2x - 3}{2x^2 + 3x - 2}.$$

Upon dividing the numerator and denominator by x^2, we have

$$y = \frac{1 - \frac{2}{x} - \frac{3}{x^2}}{2 + \frac{3}{x} - \frac{2}{x^2}}.$$

Now as x gets larger positively, the numerator approaches 1, and the denominator approaches 2. Hence y approaches $\frac{1}{2}$. We write

$$\text{as} \quad x \to \infty, \quad y \to \frac{1}{2}.$$

With a similar analysis, we will find that

$$\text{as} \quad x \to -\infty, \quad y \to \frac{1}{2}.$$

Thus the line $y = \frac{1}{2}$ is a horizontal asymptote.

For a rational function, if $y \to a$ as $x \to \infty$, where a is a constant, then as $x \to -\infty$, $y \to a$. The converse is also true. Consequently, to find the horizontal asymptote, we must examine the behavior of the function as $x \to \infty$ or as $x \to -\infty$. And because the technique used in the previous illustration may be applied to any rational function, we merely need to examine the ratio of the leading terms when looking for horizontal asymptotes. Hence if

$$y = \frac{(x + 2)(2x - 1)}{(x - 3)(3x + 2)},$$

the horizontal asymptote is $y = \frac{2}{3}$, because $\frac{2}{3}$ is the ratio of the leading terms.

A rational function does not necessarily have a horizontal asymptote. Consider

$$y = \frac{x(x-1)(x+1)}{(x+2)(x-3)}.$$

The ratio of the leading terms is x $(x^3/x^2 = x)$. Therefore,

$$\text{as} \quad x \to \infty,\ y \to \infty, \qquad \text{and as} \qquad x \to -\infty,\ y \to -\infty.$$

There is no horizontal asymptote. All of this information is summarized in the following theorem.

Theorem 4.7

For a rational function,

(i) *If the degree of the numerator is less than the degree of the denominator, then the line $y = 0$ is a horizontal asymptote.*

(ii) *If the degree of the numerator is greater than the degree of the denominator, then there is no horizontal asymptote.*

(iii) *If the degree of the numerator is equal to the degree of the denominator, then the horizontal asymptote is the horizontal line whose y-intercept is the ratio of the leading coefficients.*

Exercise

Determine the horizontal asymptote (if any) of each of the following.

(a) $y = \dfrac{x}{(x-1)(2x+1)}$ (b) $y = \dfrac{(x-1)(x+2)}{(2x+1)(x+1)}$

(c) $y = \dfrac{x^2(x-1)}{(2x-3)(x-2)}$

Answers. (a) $y = 0$ (b) $y = \dfrac{1}{2}$ (c) none ■

From our study of hyperbolas we know that it's possible for an asymptote to be nonhorizontal and nonvertical. For a rational function this occurs when the degree of the numerator is exactly one more than the degree of the denominator. Consider

$$y = \frac{x(x-1)(x+1)}{(x+2)(x-3)}.$$

If we expand the numerator and denominator and perform long division, we get

$$
\begin{array}{r}
x + 1 \\
x^2 - x - 6 \overline{) x^3 - x} \\
\underline{x^3 - x^2 - 6x} \\
x^2 + 5x \\
\underline{x^2 - x - 6} \\
6x + 6
\end{array}
$$

Therefore,

$$y = x + 1 + \frac{6x + 6}{x^2 - x - 6}. \tag{4-1}$$

As x gets large positive or large negative (without bound), the rightmost term of equation (4-1) approaches zero. Thus y is better and better approximated by $x + 1$, and the line $y = x + 1$ is an asymptote to the graph of the function. Such a nonvertical and nonhorizontal asymptote is called a *slant* asymptote.

Note that regardless of the remainder obtained in the long division, the quotient of the remainder and divisor will approach zero as x gets larger positively or larger negatively. For this reason, calculation of the remainder is unnecessary in determining the slant asymptote.

Example 2

Determine the slant asymptote of $y = \dfrac{2x(x + 5)(x - 2)}{(x - 1)^2}$.

Solution. We expand the numerator and denominator and perform long division, omitting calculation of the remainder.

$$
\begin{array}{r}
2x + 10 \\
x^2 - 2x + 1 \overline{) 2x^3 + 6x^2 - 20x } \\
\underline{2x^3 - 4x^2 + 2x} \\
10x^2 - 22x \\
\underline{10x^2 - 20x + 10}
\end{array}
$$

Thus the slant asymptote is $y = 2x + 10$. ■

Exercise

Determine the slant asymptote of $y = \dfrac{x(x + 2)(x + 3)}{(x + 1)^2}$.

Answer. $y = x + 3$. ■

Before sketching some graphs, we need to observe two special properties of rational functions: smoothness and continuity. Like polynomials, rational functions have no sharp turns in their graphs (i.e., they are smooth). And all rational functions are continuous (on their domains). Thus the only "breaks" in the graph of a rational function occur at "breaks" in the domain. Therefore, the only "gaps" in the graph occur at values of x at which the denominator equals zero.

If the numerator and denominator of a rational function have no common factors, these "breaks" occur only at the vertical asymptotes. So the only values of x at which the values of y can change sign are the x-intercepts and the vertical asymptotes. Thus on an interval bounded on each end by an x-intercept or vertical asymptote, with no other x-intercepts or vertical asymptotes between, either all y values are positive or all are negative. Consequently, the graph lies either entirely above or entirely below the x-axis on such an interval. So in sketching the graphs of rational functions (in reduced form) which are *not* written as transformations of elementary rationals, we will do three things before drawing smooth curves.

1. Locate all x- and y-intercepts.
2. Determine all vertical, horizontal, and slant asymptotes (if there is neither a horizontal nor slant asymptote, we'll examine the behavior of y as $x \to \infty$ and as $x \to -\infty$).
3. Observe continuity—note that y can change signs at x-intercepts and vertical asymptotes only.

Example 3

Sketch the graph of $y = \dfrac{x - 1}{x + 2}$.

Solution. First we find the intercepts. The x-intercept is 1, and the y-intercept is $-\frac{1}{2}$. Now the asymptotes:

vertical: $x = -2$
horizontal: $y = 1$
slant: none

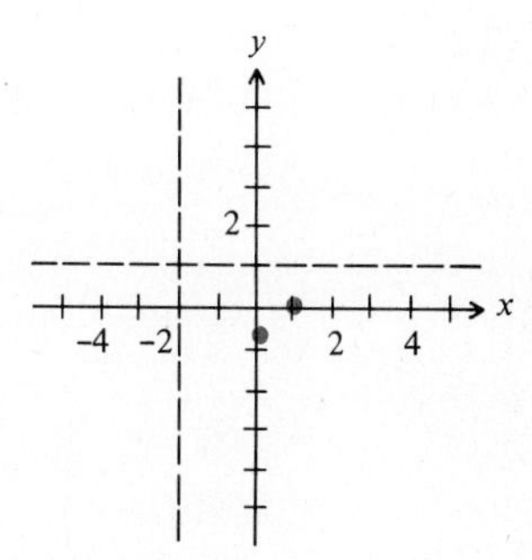

Figure 4.42

(It's not possible for a rational function to have both a horizontal and slant asymptote.) Since there is a horizontal asymptote, we need not check the behavior of y as $x \to \infty$ and as $x \to -\infty$. (We already know that $y \to 1$.) So far, as seen in Figure 4.42, we have two points and two asymptotes. Now that part of the graph to the left of $x = -2$ must lie entirely above or entirely below the x-axis. Since this part of the graph must approach the lines $y = 1$ and $x = -2$, it must lie above. (If it were below, there would be an x-intercept less than -2.) Similarly, since the graph must approach $y = 1$, that part of the graph to the right of the x-intercept 1 must also lie above. These

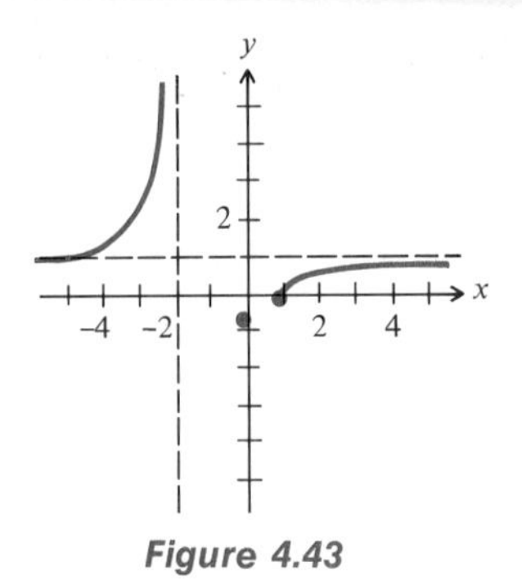

Figure 4.43

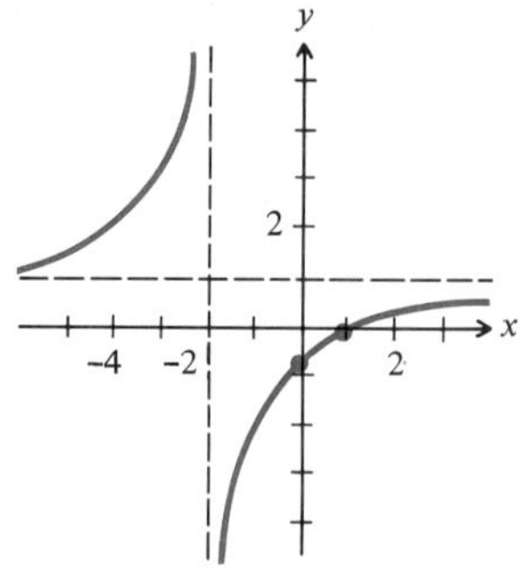

Figure 4.44 $y = \dfrac{x - 1}{x + 2}$

parts of the graph must look something like that shown in Figure 4.43. Between the vertical asymptote and the x-intercept, the graph lies entirely above or below. Since $(0, -\frac{1}{2})$ is on the graph, it must be below. And since $x = -2$ is an asymptote, the graph must be something like that in Figure 4.44. ■

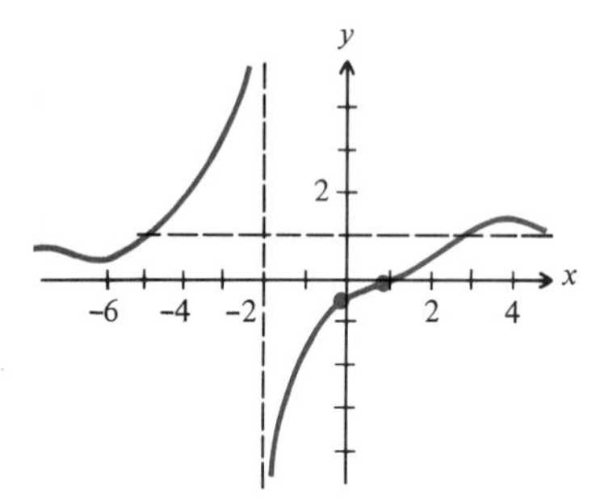

Figure 4.45

Note that this is just a *sketch* of the graph. Without additional information, the graph in Figure 4.45 (as well as many others) is another possible sketch. Note that the graph of Figure 4.45 crosses its horizontal asymptote. We can determine precisely where this occurs by setting the rational function equal to the y-intercept of the horizontal asymptote and finding the solution set. For Example 3 we get

$$\frac{x - 1}{x + 2} = 1.$$

But this equation has no solutions. Hence the graph doesn't cross its horizontal asymptote, and the sketch of Figure 4.44 is better than that of Figure 4.45.

We might note, also, that the given function is a transformation of elementary rationals. We have

$$y = \frac{x - 1}{x + 2} = \frac{x + 2 - 3}{x + 2} = \frac{x + 2}{x + 2} - \frac{3}{x + 2}$$
$$= 1 - \frac{3}{x + 2}.$$

(This result may also be obtained by long division.)

Example 4

Sketch the graph of $y = \dfrac{(x + 5)(x - 1)}{x + 3}$.

Solution. The x-intercepts are -5 and 1, and the y-intercept is $-\frac{5}{3}$. Now the asymptotes:

$$\begin{aligned}&\text{vertical:} \quad x = -3\\&\text{horizontal:} \quad \text{none}\\&\text{slant:} \quad y = x + 1.\end{aligned}$$

Plotting the intercepts and sketching the asymptotes, we get Figure 4.46. To the right of the x-intercept 1, the graph must lie above the x-axis. (It approaches $y = x + 1$.) And between the vertical asymptote $x = -3$ and the x-intercept 1, the graph must lie below the x-axis. (The y-intercept is negative.) To the left of the x-intercept -5 the graph must lie below the x-axis. (It approaches $y = x + 1$.) But what about the graph between $x = -5$ and $x = -3$? We can determine whether it lies above or below by plotting any point between. When $x = -4$, $y = 5$. Hence this part lies above, and the graph must look something like that in Figure 4.47.

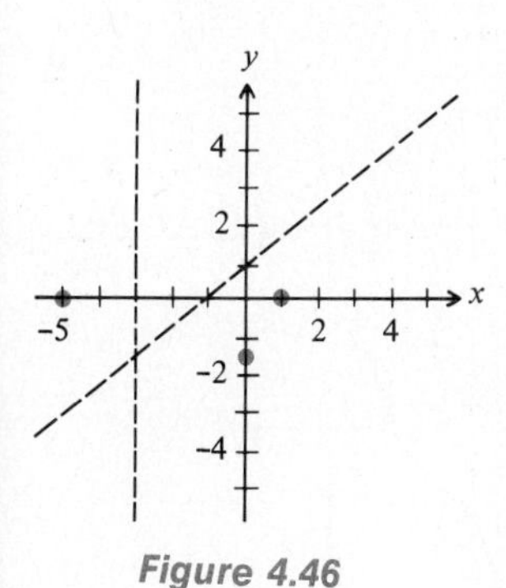

Figure 4.46

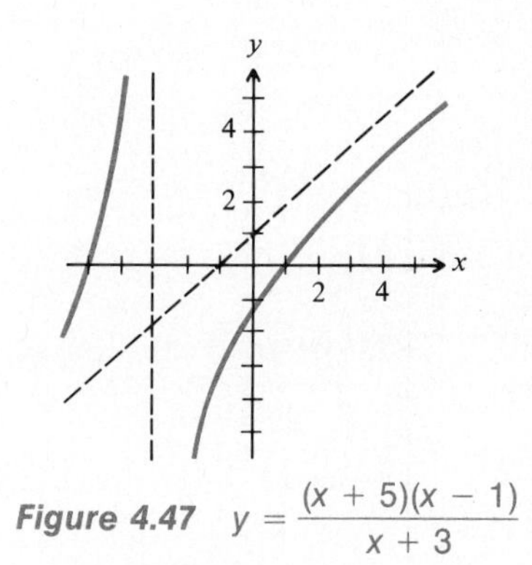

Figure 4.47 $y = \dfrac{(x + 5)(x - 1)}{x + 3}$ ■

Another way of determining whether the graph of a rational function is above or below the x-axis between two consecutive x-intercepts/vertical asymptotes is to employ the following theorem.

Theorem 4.8

Let $y = P(x)/Q(x)$, in reduced form, where P and Q are polynomials, and let a be a root of $P(x) = 0$ or $Q(x) = 0$ of multiplicity n. If n is odd, then y changes sign at $x = a$. If n is even, then y does not change signs at $x = a$.

Hence, in Example 4, we could have noted that since -5 is of multiplicity one, the graph had to cross at $x = -5$. Or since -3 is of multiplicity one, y had to change signs at $x = -3$.

Example 5

Sketch the graph of $y = \dfrac{(2x - 5)(x + 3)}{(5 - x)(x + 1)}$.

Solution. The x-intercepts are $\frac{5}{2}$ and -3, and the y-intercept is -3. The asymptotes:

vertical: $x = 5, x = -1$
horizontal: $y = -2$
slant: none

Plotting the intercepts and sketching the asymptotes, we get Figure 4.48. To the right of the vertical asymptote $x = 5$, the graph must lie below the x-axis. (It approaches $y = -2$.) Because the multiplicity of 5 is odd, y must change signs at $x = 5$. Hence, between the vertical asymptote $x = 5$ and the x-intercept $\frac{5}{2}$, the graph lies above the x-axis. Because the multiplicity of $\frac{5}{2}$ is odd, the graph must cross here. So between the x-intercept $\frac{5}{2}$ and the vertical asymptote $x = -1$, the graph lies below the x-axis. (This is consistent with the location of the y-intercept.) Similarly, y changes signs at $x = -1$ and at the x-intercept -3. See Figure 4.49.

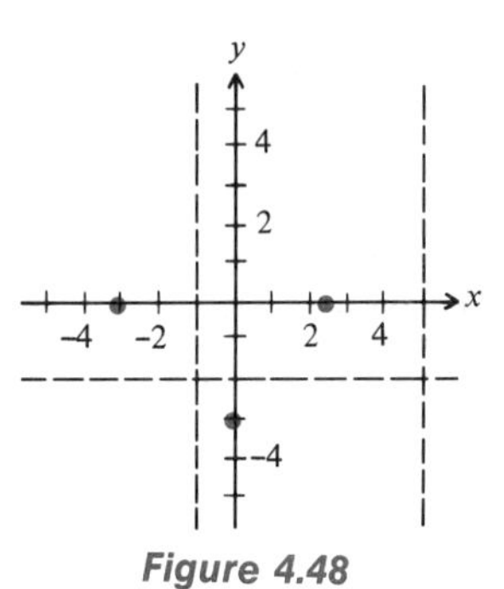

Figure 4.48

Figure 4.49 $y = \dfrac{(2x - 5)(x + 3)}{(5 - x)(x + 1)}$ ■

Exercise

Sketch the graph of $y = \dfrac{(2x - 3)(x + 5)}{(x - 5)(x + 2)}$.

Answer

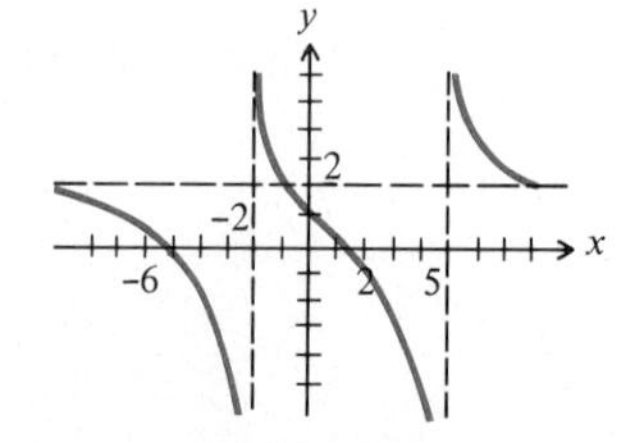

■

Sketch the graph of $y = \frac{x(x - 1)}{x(x + 2)}$.

Solution. The first thing to note is that the fraction on the right is not in reduced form, as x is a factor of both the numerator and denominator. Simplified, we get

$$y = \frac{x - 1}{x + 2}.$$

Now this equation is precisely that of Example 3. Proceeding as we did there, we will get the graph in Figure 4.44. Now we must note one very important difference. The function we have been asked to graph is equivalent to the one in Example 3, with the exception that

$$y = \frac{x(x - 1)}{x(x + 2)}$$

is *not* defined at $x = 0$. (Remember, it's the initial equation which determines the domain.) Hence the graphs are identical with the exception that the graph of this function contains no point with x-coordinate zero. Deleting $(0, -\frac{1}{2})$ from the graph in Figure 4.44, we get the graph in Figure 4.50. ∎

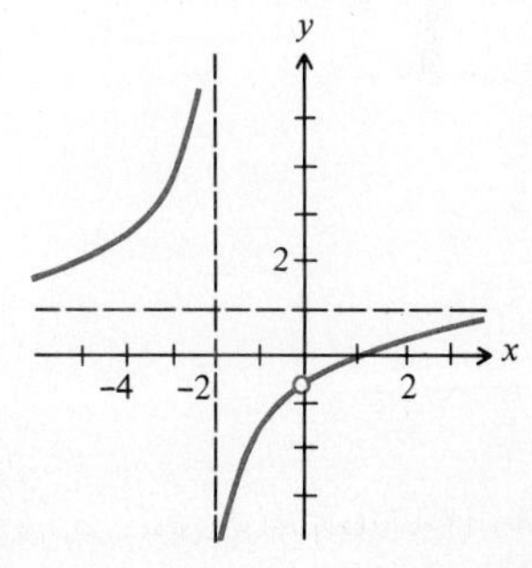

Figure 4.50 $y = \frac{x(x - 1)}{x(x + 2)}$

Problem Set 4.6

In Problems 1–12, determine the vertical, horizontal, and slant asymptotes.

1. $y = \frac{3x - 2}{x + 1}$

2. $y = \frac{5 - 2x}{2x - 3}$

3. $y = \frac{(x + 5)(2 - x)}{(2x - 1)(x + 1)}$

4. $y = \frac{(2x + 1)(3x - 2)}{(x - 5)(2x - 4)}$

5. $y = \frac{(x + 5)^3}{(x - 2)^3(x + 1)}$

6. $y = \frac{1 - 2x}{(x + 2)(x - 2)}$

7. $y = \frac{(x - 2)(x + 3)}{x - 5}$

8. $y = \frac{x^2(x + 1)}{(3 - 2x)^2}$

9. $y = \frac{x^2(x - 3)}{x + 2}$

10. $y = \frac{6x^2 + x - 2}{x^2 - 4}$

11. $y = \frac{-x}{x^2 - 6x + 1}$

12. $y = \frac{-2x^3 + 2x^2 - 3x + 5}{x^2 + 1}$

In Problems 13–28, sketch the graph of the rational function.

13. $y = \dfrac{x+2}{x-3}$

14. $y = \dfrac{x-1}{2x+5}$

15. $y = \dfrac{(2x-1)(x+3)}{(x-1)(x-4)}$

16. $y = \dfrac{(x-1)(x-4)}{(x+3)(1-2x)}$

17. $y = \dfrac{x}{(x+1)(x-3)}$

18. $y = \dfrac{x-2}{(x+1)^2}$

19. $y = \dfrac{(x+3)(x-2)}{x+1}$

20. $y = \dfrac{(1-x)(x-5)}{2x-1}$

21. $y = \dfrac{x(x-3)(x+2)}{2x+5}$

22. $y = \dfrac{15-7x-2x^2}{x^2+2x-3}$

23. $y = \dfrac{x+2}{x+2}$

24. $y = \dfrac{1-x}{x^2-8}$

25. $y = \dfrac{12-2x^2-5x}{2x+5}$

26. $y = \dfrac{-2x^3-x^2+10x}{3x^2+3x-18}$

27. $y = \dfrac{2x^3-5x^2-13x+30}{x^3+6x^2}$

28. $y = \dfrac{2x^2+5x-12}{x^3-7x^2+8x+16}$

In Problems 29 and 30, choose the equation which best *describes the given graph.*

29. (a) $y = \dfrac{(x+6)^3}{(x-3)(x+2)}$

(b) $y = \dfrac{x-6}{(x-3)(x+2)}$

(c) $y = \dfrac{x+6}{(x-3)(x+2)}$

(d) $y = \dfrac{x+6}{(x-3)^2(x+2)}$

(e) $y = \dfrac{x+6}{(x+3)(x-2)}$

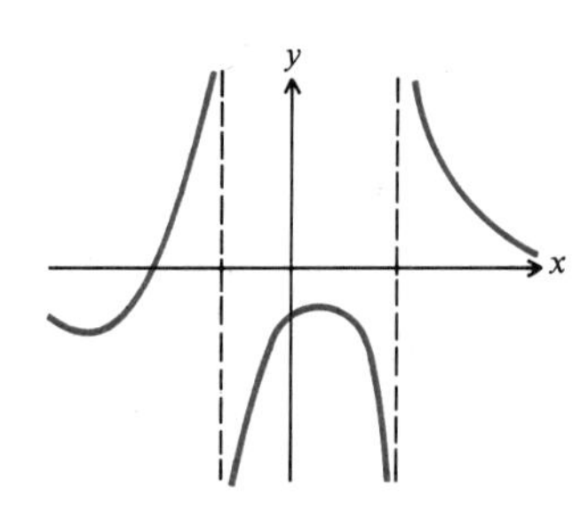

30. (a) $y = \dfrac{2x^2(x-3)}{(x+2)(x-4)}$

(b) $y = \dfrac{2x(x-3)}{(x-2)(x+4)}$

(c) $y = \dfrac{2x(x-3)^2}{(x+2)(x-4)}$

(d) $y = \dfrac{2x(x-3)}{(x+2)(x-4)}$

(e) $y = \dfrac{2x(x-3)}{(x+2)(x-4)^2}$

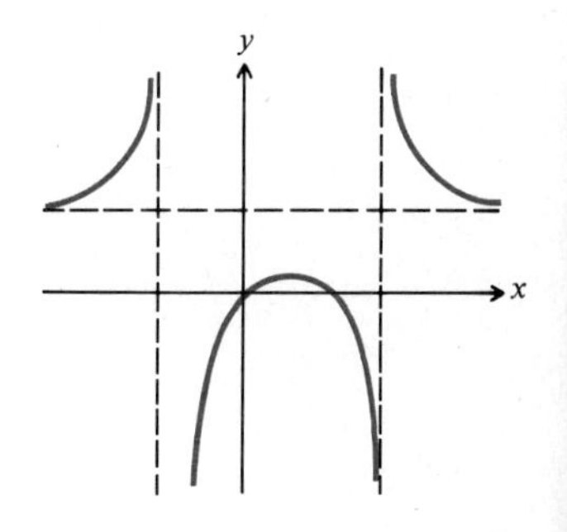

31. At what point does the graph of $y = \dfrac{(2x - 1)(x + 3)}{(x - 1)(x - 4)}$ cross its horizontal asymptote?

32. Show algebraically that the graph of $y = \dfrac{x(x + 2)(x - 1)}{(x - 2)(x^2 + 3)}$ doesn't cross its horizontal asymptote.

33. Construct a rational function which has vertical asymptotes of $x = -3$ and $x = 1$, and a horizontal asymptote of $y = 2$.

34. Show that the graph of $y = \dfrac{2x^4 + x^3 - 5x + 2}{x^4 + x^2 + 3x - 5}$ crosses its horizontal asymptote three times.

35. Sketch the graph of $y = \left|\dfrac{1 - x}{x}\right| - \dfrac{1}{x}$.

Chapter 4 Review

In Problems 1–12, sketch the graph of the polynomial.

1. $f(x) = 3$
2. $f(x) = -2x + 1$
3. $f(x) = -x^2 - 6x - 6$
4. $y = -2(x - 3)^2 + 4$
5. $y = (x + 2)^3 - 3$
6. $y = (x - 2)^4 - 3$
7. $y = (x + 3)(x - 4)$
8. $y = (x - 2)^2(x + 3)$
9. $y = x^2(x + 2)(x - 3)$
10. $y = (x + 2)^3(x - 3)$
11. $y = 2x^3 - 9x^2 + 9x$
12. $y = (x^2 - 4)(x + 2)$

13. Graph $f(x) = -x^4 + 2x^3 + 7x^2 - 10x + 12$ after plotting points with the following x-values: $-3, -2, -1, 0, 1, 2, 3$, and 4.
14. Find all real zeros of $P(x) = 6x^4 + x^3 - 8x^2 - x + 2$.
15. Find all real zeros of $P(x) = 5x^3 + 17x^2 + 11x + 2$.
16. Sketch the graph of $y = 2x^4 - x^3 - 19x^2 + 9x + 9$.
17. Sketch the graph of $y = x^5 + 3x^4 + 2x^3 - x^2 - 3x - 2$.

In Problems 18 and 19, choose the equation which best *describes the given graph.*

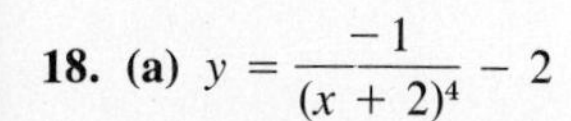

18. (a) $y = \dfrac{-1}{(x + 2)^4} - 2$

(b) $y = \dfrac{1}{(x - 2)^2} - 2$

(c) $y = \dfrac{1}{(x + 2)^2} - 2$

(d) $y = \dfrac{1}{(x - 2)^4} + 2$

(e) $y = \dfrac{-1}{(x - 2)^2} - 2$

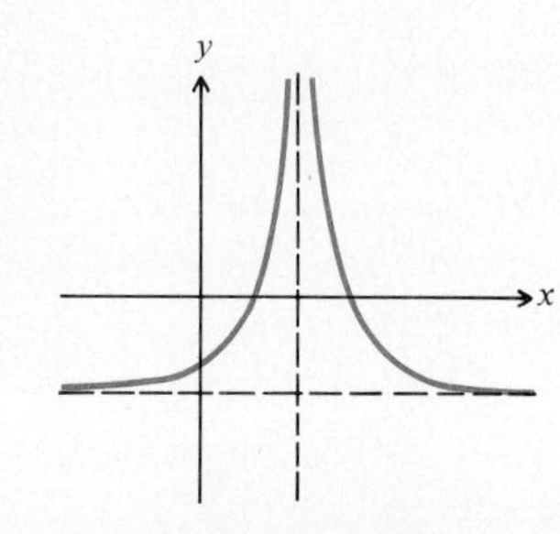

19. (a) $y = \dfrac{-1}{x-2} + 2$

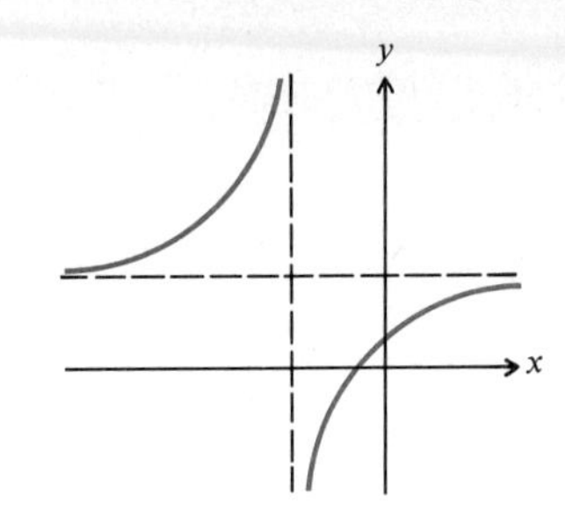

(b) $y = \dfrac{1}{(x+2)^3} + 2$

(c) $y = \dfrac{-1}{(x+2)^3} - 2$

(d) $y = \dfrac{-1}{(x+2)^3} + 2$

(e) $y = \dfrac{1}{x-2} + 2$

In Problems 20–29, sketch the graph of the rational function.

20. $y = \dfrac{-1}{x-2} + 1$

21. $y = \dfrac{1}{(x+1)^2} - 2$

22. $y = \dfrac{-2}{x^3}$

23. $y = \dfrac{1}{2x^2}$

24. $y = \dfrac{2x-3}{x+2}$

25. $y = \dfrac{x-2}{(x+1)(x+3)}$

26. $y = \dfrac{(x-3)(x+2)}{x+4}$

27. $y = \dfrac{x^2-2x}{x^2+6x+8}$

28. $y = \dfrac{x+2}{2x^2+3x-2}$

29. $y = \dfrac{2x+4}{x+2}$

Exponential and Logarithmic Functions

5

The functions of the preceding two chapters are algebraic. That is, their values can be calculated using only finitely many additions, subtractions, multiplications, divisions, and extractions of roots. The subject of this chapter is two important classes of functions which are not algebraic. They have important applications in many scientific fields.

5.1 Exponential Functions

We know how to raise a positive number to any rational power. Let b be a positive number and let p and q be integers with $q > 0$. Recall that

$$b^{p/q} = (\sqrt[q]{b})^p.$$

For example,

$$27^{2/3} = (\sqrt[3]{27})^2 = 3^2 = 9,$$

$$16^{-3/4} = (\sqrt[4]{16})^{-3} = 2^{-3} = \frac{1}{2^3} = \frac{1}{8}.$$

Also remember that if $\sqrt[q]{b}$ is real, then $b^{p/q} = \sqrt[q]{b^p}$. Thus

$$8^{2/3} = \sqrt[3]{8^2} = \sqrt[3]{64} = 4.$$

It is also possible to raise a positive number to an irrational power. Although a formal definition will not be given, we will try to get an intuitive understanding of this concept by considering the meaning of $3^{\sqrt{2}}$. It is possible to obtain a sequence of successively better rational approximations of $\sqrt{2}$:

$$1,\ 1.4,\ 1.41,\ 1.414,\ 1.4142,\ \ldots.$$

Because each approximation is rational, each number in the following sequence is defined.

$$3^1,\ 3^{1.4},\ 3^{1.41},\ 3^{1.414},\ 3^{1.4142},\ \ldots.$$

(Note that $3^{1.4} = 3^{14/10} = (\sqrt[10]{3})^{14}$, etc.) The numbers in this latter sequence get closer and closer to the number which is called $3^{\sqrt{2}}$. (In this text we will not deal with the problem of approximating numbers like $3^{\sqrt{2}}$ with rational numbers. Methods of making such approximations have been programmed into many calculators. If you are interested in a decimal approximation for a number of this type, find a calculator that has this capability.) With this intuitive understanding, we consider another function.

Let $b > 0$, $b \neq 1$, and let f: R → R be defined by

$$f(x) = b^x.$$

This function f is called the **exponential function with base b.** [Because $1^x = 1$ for all real numbers x, a function defined by $f(x) = 1^x$ is simply a constant function. Since such a function differs significantly from exponential functions with bases other than 1, we require that $b \neq 1$.] For example, if f is the exponential function with base 5, then

$$f(x) = 5^x, \qquad f(2) = 25, \qquad f(-1) = \frac{1}{5}, \qquad \text{and} \qquad f\left(\frac{1}{2}\right) = \sqrt{5}.$$

In sketching the graphs of exponential functions, we plot points and use two important properties. Exponential functions are smooth and continuous. See Examples 1 and 2.

Example 1

Sketch the graph of $y = 2^x$ (the exponential function with base 2), after plotting points whose x-coordinates are -3, -2, -1, 0, 1, 2, and 3.

Solution. Substituting the given values, we get the following table.

x	-3	-2	-1	0	1	2	3
y	$\frac{1}{8}$	$\frac{1}{4}$	$\frac{1}{2}$	1	2	4	8

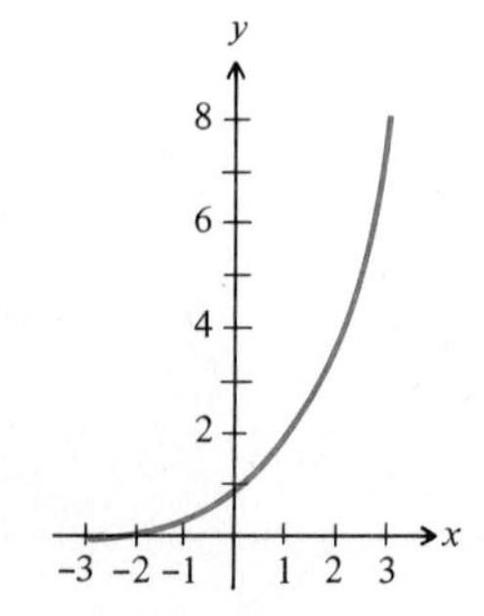

Figure 5.1 $y = 2^x$

After plotting these points, we get the graph in Figure 5.1. ■

Example 2

Sketch the graph of $y = (\frac{1}{2})^x$ (the exponential function with base $\frac{1}{2}$), after plotting points whose x-coordinates are -3, -2, -1, 0, 1, 2, and 3.

Solution. We get the following table:

x	-3	-2	-1	0	1	2	3
y	8	4	2	1	$\frac{1}{2}$	$\frac{1}{4}$	$\frac{1}{8}$

Plotting these points leads to the graph of Figure 5.2. ■

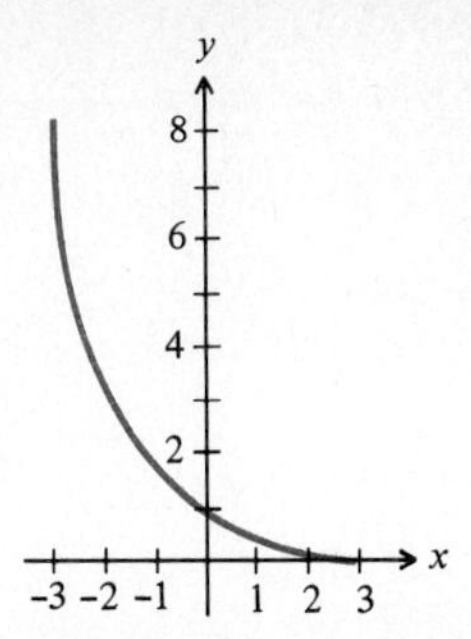

Figure 5.2 $y = (\frac{1}{2})^x$

There are several important facts to be noted about the two exponential functions just graphed. First, the range of each is $(0, \infty)$. Second, each is a one-to-one function (since there is no horizontal line that intersects the graph at more than one point). And third, the x-axis is a horizontal asymptote of each graph. These three observations can be made about the graphs of all exponential functions. That is, regardless of the base, the range of an exponential function is $(0, \infty)$, it is one-to-one, and the x-axis is a horizontal asymptote of its graph.

Note that the horizontal asymptote for an exponential function is somewhat different from the ones for rational functions. For the graph of an exponential function, only one side approaches the asymptote. That is, either the graph approaches the asymptote as $x \to \infty$ (this occurs when $b < 1$) or the graph approaches the asymptote as $x \to -\infty$ (this occurs when $b > 1$). For the graph of a rational function, both sides approach the asymptote. That is, the graph of a rational function approaches the asymptote as $x \to \infty$ and as $x \to -\infty$.

Also note that the graph of $y = (\frac{1}{2})^x$ is the reflection about the y-axis of the graph of $y = 2^x$. In fact, the graph of the exponential function with base $1/b$ $[y = (1/b)^x]$ is the reflection about the y-axis of the graph of the exponential function with base b $(y = b^x)$ for any base b.

Example 3
Sketch the graph of $y = 2^{1-x}$ after plotting points whose x-coordinates are $-2, -1, 0, 1, 2, 3$, and 4.

Solution. We get the following table and the graph of Figure 5.3.

x	-2	-1	0	1	2	3	4
y	8	4	2	1	$\frac{1}{2}$	$\frac{1}{4}$	$\frac{1}{8}$

■

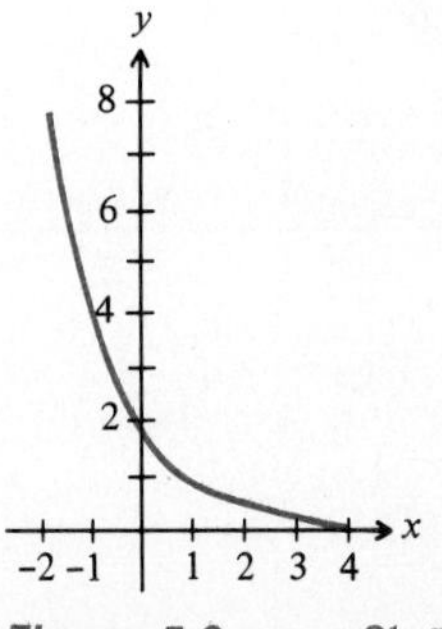

Figure 5.3 $y = 2^{1-x}$

Although the function graphed in Example 3 is not exponential, its graph looks like the graph of an exponential function. This is because it is a transformation of an exponential function. We have

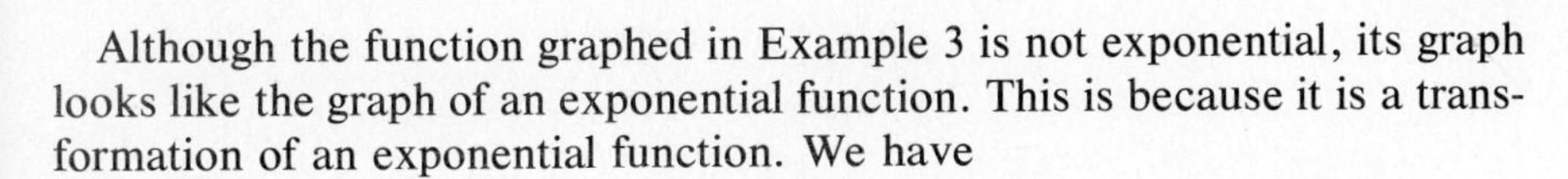

$$y = 2^{1-x} = 2^1 \cdot 2^{-x} = 2 \cdot \left(\frac{1}{2}\right)^x.$$

Hence the graph in Figure 5.3 is merely a stretching of the graph of the exponential function defined by $y = (\frac{1}{2})^x$, whose graph is in Figure 5.2.

Exercise
Sketch the graph of $y = 2^{1+x}$ after plotting points whose x-coordinates are $-4, -3, -2, -1, 0, 1$, and 2.

Answer

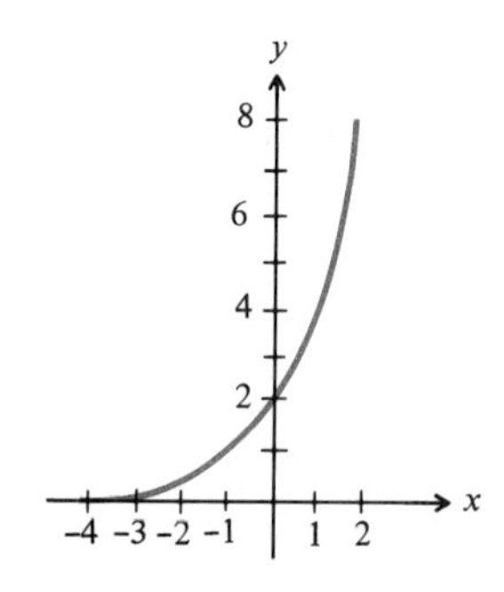

■

Another frequently occurring base is e, an irrational number approximately equal to 2.71828. To approximate values like e^2, e^{-1}, $e^{1/2}$, and so on, we must use a calculator that has the capability or use Table I in the Appendix.

Example 4
Sketch the graph of $y = e^x$ after plotting points whose x-coordinates are -2, -1, 0, 1, and 2.

Solution. To find e^{-1}, for example, using Table I in the Appendix, we first find the entry 1.0 beneath the column labeled x. Moving across until we get beneath the column labeled e^{-x}, we find the entry 0.3679. This is an approximation of e^{-1}. To find e^2, we locate the entry 2.0 beneath the column labeled x. Moving across we find the entry 7.3891 beneath the column labeled e^x. Thus e^2 is approximately 7.3891. Calculating the remaining values and rounding to the nearest tenth, we get the following table:

x	−2	−1	0	1	2
y	0.1	0.4	1.0	2.7	7.4

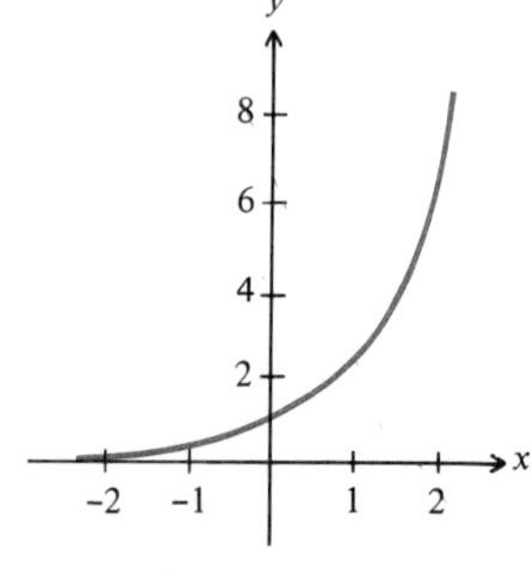

Figure 5.4 $y = e^x$

After plotting these points, we get the graph shown in Figure 5.4. ■

Exercise
Sketch the graph of $y = e^{-x}$ after plotting points whose x-coordinates are -2, -1, 0, 1, and 2.

Answer

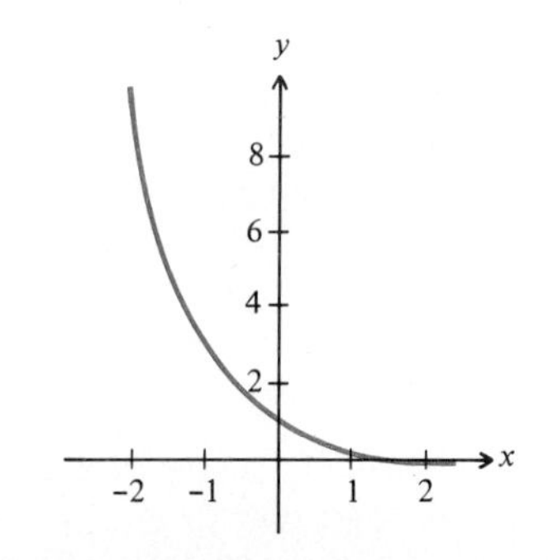

■

Problem Set 5.1

Sketch the graph of the given equation after plotting points with the specified x-coordinates.

1. $y = 3^x$, x-values: $-2, -1, 0, 1$, and 2.
2. $y = 4^x$, x-values: $-2, -1, 0, 1$, and 2.
3. $y = \left(\frac{1}{4}\right)^x$, x-values: $-2, -1, 0, 1$, and 2.
4. $y = \left(\frac{1}{3}\right)^x$, x-values: $-2, -1, 0, 1$, and 2.
5. $y = \left(\frac{3}{2}\right)^x$, x-values: $-2, -1, 0, 1, 2, 3, 4$, and 5.
6. $y = \left(\frac{3}{4}\right)^x$, x-values: $-5, -4, -3, -2, -1, 0, 1$, and 2.
7. $y = 2^{x+2}$, x-values: $-5, -4, -3, -2, -1, 0$, and 1.
8. $y = 3^{x+1}$, x-values: $-3, -2, -1, 0$, and 1.
9. $y = 3^{-x}$, x-values: $-2, -1, 0, 1$, and 2.
10. $y = -3^x$, x-values: $-2, -1, 0, 1$, and 2.
11. $y = e^{2x}$, x-values: $-1, -.5, 0, .5$, and 1.
12. $y = e^{-x/2}$, x-values: $-4, -3, -2, -1, 0, 1$, and 2.
13. $y = 1 + 2e^{-x}$, x-values: $-2, -1, 0, 1$, and 2.
14. $y = 1 - e^x$, x-values: $-2, -1, 0, 1$, and 2.
15. $y = 3^{1-x}$, x-values: $-1, 0, 1, 2$, and 3.
16. $y = 4^{2-x}$, x-values: $-1, 0, 1, 2, 3$, and 4.
17. $y = -2^{x+2}$, x-values: $-4, -3, -2, -1, 0$, and 1.
18. $y = 4^{-(1+x)}$, x-values: $-3, -2, -1, 0$, and 1.
19. $y = 2^{x^2}$, x-values: $-2, -1, 0, 1$, and 2.
20. $y = 3^{1-x^2}$, x-values: $-2, -1, 0, 1$, and 2.

5.2 Logarithmic Functions

Let $b > 0$, $b \neq 1$, and let k be any positive number. Because the exponential function with base b has range $(0, \infty)$ and is one-to-one, there exists a unique number j such that $b^j = k$. This number j is called the **logarithm to the base b of k.** Denoted by $\log_b k$, it is that number which when used as an exponent to b gives k. For example,

$$\begin{array}{lll} \log_{10} 100 = 2 & \text{because} & 10^2 = 100, \\ \log_2 8 = 3 & \text{because} & 2^3 = 8, \\ \log_{10} \frac{1}{10} = -1 & \text{because} & 10^{-1} = \frac{1}{10}, \\ \log_{1/2} 4 = -2 & \text{because} & (\frac{1}{2})^{-2} = 4. \end{array}$$

Also note that for any base b,

$$\log_b 1 = 0 \quad \text{because} \quad b^0 = 1.$$

By definition, then, the equation

$$\log_b k = j$$

means that

$$b^j = k.$$

It is important to realize that these two equations mean the same thing, as we will frequently convert one to the other.

Exercise
Convert logarithmic equations into equivalent exponential equations, and vice versa.
(a) $\log_5 125 = 3$ (b) $\log_3 \frac{1}{9} = -2$ (c) $3^4 = 81$ (d) $4^2 = 16$

Answers. (a) $5^3 = 125$ (b) $3^{-2} = \frac{1}{9}$ (c) $\log_3 81 = 4$
(d) $\log_4 16 = 2$ ■

Exercise
Find the number.
(a) $\log_3 9$ (b) $\log_{10} \frac{1}{100}$ (c) $\log_{1/4} 4$ (d) $\log_\pi 1$

Answers. (a) 2 (b) -2 (c) -1 (d) 0 ■

Logarithms to the base 10 are called **common logarithms.** And as a matter of convenience, $\log_{10} x$ will be written simply as $\log x$. Thus $\log \frac{1}{100} = -2$. Another frequently used logarithmic base is e (an irrational number approximately equal to 2.71828). Logarithms to the base e are called **natural logarithms,** and $\log_e x$ is usually written $\ln x$. It is easy to find the common logarithm of a number which can be readily expressed as a power of 10. But how do we calculate log 3? And how do we find ln 5? To approximate such values, we must use a calculator which has the capacity or use Tables II and III in the Appendix.

If $b > 1$, as it is for common and natural logarithms, then $\log_b x > 0$ when $x > 1$ and $\log_b x < 0$ when $x < 1$. Of course, $\log_b 1 = 0$ regardless of the base.

The next theorem follows directly from definitions.

Theorem 5.1

Let $b > 0$, $b \neq 1$. For all positive numbers x,

$$b^{\log_b x} = x,$$

and for all real numbers t,

$$\log_b b^t = t.$$

From Theorem 5.1 we see that

$$10^{\log x} = x \quad \text{and} \quad \log 10^x = x,$$
$$e^{\ln x} = x \quad \text{and} \quad \ln e^x = x.$$

For example, $10^{\log 2} = 2$ and $\ln e^{-3} = -3$.

Example 1

Simplify.

(a) $5^{-2\log_5 k}$

(b) $\log_3 3^y$

Solution. (a) Using Theorem 5.1 and rules for exponents, we get

$$5^{-2\log_5 k} = (5^{\log_5 k})^{-2} = k^{-2} = \frac{1}{k^2}.$$

(b) By Theorem 5.1 we have

$$\log_3 3^y = y. \qquad \blacksquare$$

Exercise

Simplify.

(a) $2^{3\log_2 x}$

(b) $\log_5 5^k$

Answers. (a) x^3 (b) k $\blacksquare$

Theorem 5.2 lists some very important properties of logarithms, sometimes referred to as laws of logarithms.

Theorem 5.2

Let $b > 0$, $b \neq 1$. Then for all $u > 0$ and all $v > 0$:

(i) $\log_b (uv) = \log_b u + \log_b v$

(ii) $\log_b \left(\frac{u}{v}\right) = \log_b u - \log_b v$

(iii) $\log_b u^k = k \log_b u$ *for all real numbers k*

Proof. (i) Let $m = \log_b u$ and $n = \log_b v$. From the definition of logarithm we get

$$u = b^m,$$
$$v = b^n.$$

Using the laws of exponents, it follows that

$$uv = b^m \cdot b^n = b^{m+n}.$$

Using the definition of logarithm and then substituting, we get

$$\log_b (uv) = m + n = \log_b u + \log_b v.$$

The proofs of (ii) and (iii) are similar. ■

This theorem says that the logarithm of a product is the sum of the logarithms, the logarithm of a quotient is the difference of the logarithms, and the logarithm of a positive number raised to a power is the power times the logarithm of the number. Applications are found in Examples 2 and 3.

Example 2

Express $\log \frac{x\sqrt{y}}{z^2}$ in terms of $\log x$, $\log y$, and $\log z$.

Solution. Since the logarithm of a quotient is the difference of the logarithms, the given expression may be written as

$$\log (x\sqrt{y}) - \log z^2.$$

Because the logarithm of a product is the sum of the logarithms, the first term can be rewritten. The previous expression becomes

$$\log x + \log \sqrt{y} - \log z^2.$$

Now the last two terms may be rewritten because the logarithm of a number raised to a power is the power times the logarithm of the number. (Note that $\sqrt{y} = y^{1/2}$.) Thus we finally get

$$\log x + \tfrac{1}{2} \log y - 2 \log z.$$ ■

Example 3

Express $\log x - \log y + 2 \log z$ as a single logarithm.

Solution. By using the second property of Theorem 5.2, we may combine the first two terms and write the given expression as

$$\log \frac{x}{y} + 2 \log z.$$

Using the third property, we may rewrite the last term. The previous expression becomes

$$\log \frac{x}{y} + \log z^2.$$

And finally, using the first property we get

$$\log \frac{xz^2}{y}. \qquad \blacksquare$$

Exercise

(a) Express $\log \dfrac{x^2 y}{\sqrt[3]{z}}$ in terms of $\log x$, $\log y$, and $\log z$.

(b) Express $\frac{1}{2} \log x - 2 \log y + \log z$ as a single logarithm.

Answers. (a) $2 \log x + \log y - \frac{1}{3} \log z$ (b) $\log \dfrac{z\sqrt{x}}{y^2}$ ■

Let $b > 0$, $b \neq 1$. Define g: $(0, \infty) \to \mathbb{R}$ by

$$g(x) = \log_b x.$$

This function is called the **logarithmic function with base *b*.** Note that the logarithmic functions are defined for *positive* numbers only. We *cannot* take the logarithm of zero or of a negative number.

In sketching the graphs of logarithmic functions, we plot points and use the fact that, like exponential functions, logarithmic functions are smooth and continuous. See Examples 4 and 5.

Example 4

Sketch the graph of $y = \log_2 x$ (the logarithmic function with base 2) after plotting points whose x-coordinates are $\frac{1}{8}$, $\frac{1}{4}$, $\frac{1}{2}$, 1, 2, 4, and 8.

Solution. Substituting the given values, we get the following table:

x	$\frac{1}{8}$	$\frac{1}{4}$	$\frac{1}{2}$	1	2	4	8
y	−3	−2	−1	0	1	2	3

After plotting these points, we get the graph in Figure 5.5. ■

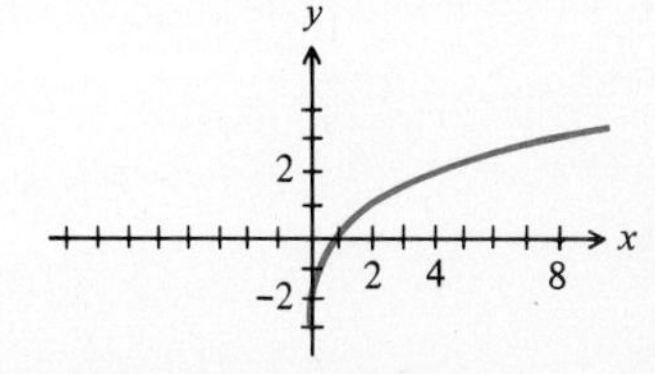

Figure 5.5 $y = \log_2 x$

Example 5

Sketch the graph of $y = \log_{1/2} x$ (the logarithmic function with base $\frac{1}{2}$), after plotting points whose x-coordinates are $\frac{1}{8}$, $\frac{1}{4}$, $\frac{1}{2}$, 1, 2, 4, and 8.

Solution. We get the following table:

x	$\frac{1}{8}$	$\frac{1}{4}$	$\frac{1}{2}$	1	2	4	8
y	3	2	1	0	−1	−2	−3

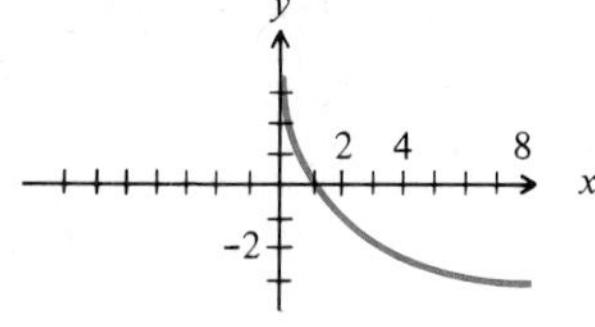

Figure 5.6 $y = \log_{1/2} x$

Plotting these points leads to the graph of Figure 5.6. ■

There are several important facts to be noted about the two logarithmic functions just graphed. First, the range of each is R. Second, each is a one-to-one function (since no horizontal line intersects the graph at more than one point). And third, the y-axis is a vertical asymptote of each graph. These three observations can be made about the graphs of all logarithmic functions. That is, regardless of the base, the range of a logarithmic function is R, it is one-to-one, and the y-axis is a vertical asymptote of its graph.

Also note that the graph of $y = \log_{1/2} x$ is the reflection about the x-axis of the graph of $y = \log_2 x$. In fact, the graph of the logarithmic function with base $1/b$ ($y = \log_{1/b} x$) is the reflection about the x-axis of the graph of the logarithmic function with base b ($y = \log_b x$) for any base b.

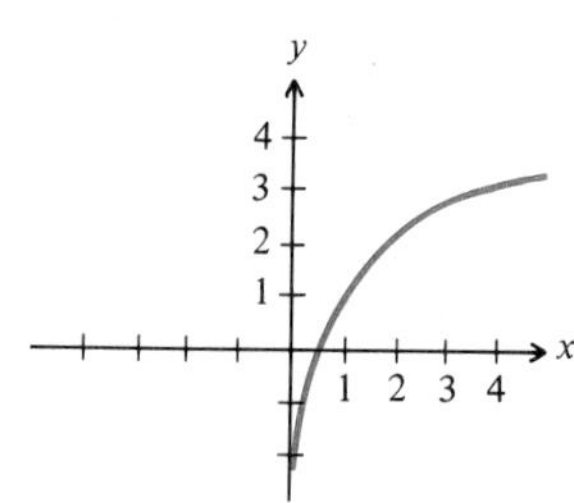

Figure 5.7 $y = \log_2 2x$

Example 6

Sketch the graph of $y = \log_2 2x$ after plotting points whose x-coordinates are $\frac{1}{8}$, $\frac{1}{4}$, $\frac{1}{2}$, 1, 2, and 4.

Solution. We get the following table and the graph of Figure 5.7.

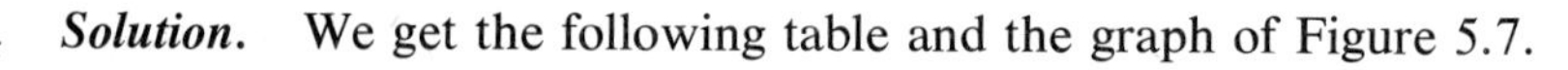

x	$\frac{1}{8}$	$\frac{1}{4}$	$\frac{1}{2}$	1	2	4
y	−2	−1	0	1	2	3

■

The graph in Figure 5.7 looks very similar to the graph of a logarithmic function because it is a transformation of such a function. Using the laws of logarithms, we see that

$$y = \log_2 2x = \log_2 2 + \log_2 x = 1 + \log_2 x.$$

Hence the graph in Figure 5.7 is merely a vertical shifting of the logarithmic function defined by $y = \log_2 x$, whose graph is in Figure 5.5.

Exercise

Sketch the graph of $y = \log_2 \dfrac{x}{2}$ by plotting points whose x-coordinates are $\frac{1}{2}$, 1, 2, 4, and 8.

Answer

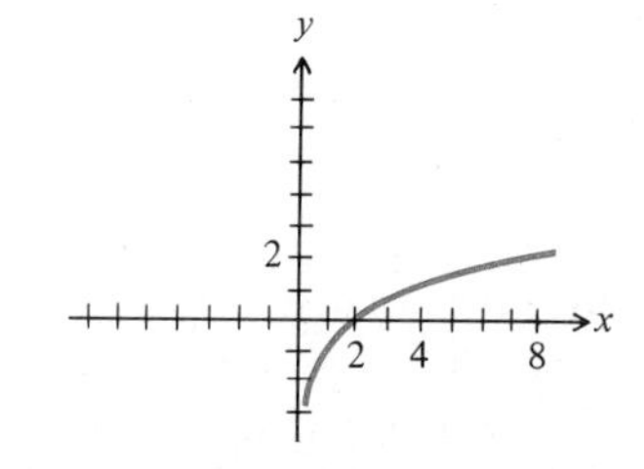

■

Example 7

Sketch the graph of $y = \ln x$ after plotting points whose x-coordinates are $\frac{1}{2}$, 1, 2, 4, 6, and 10.

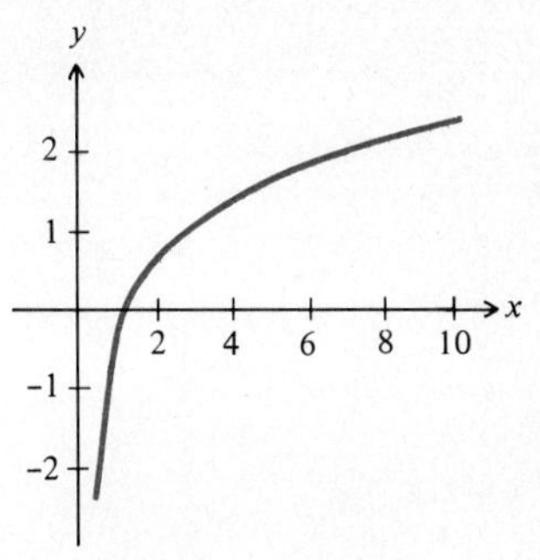

Figure 5.8 $y = \ln x$

Solution. Using Table II in the Appendix and rounding to the nearest tenth, we get the following table:

x	$\frac{1}{2}$	1	2	4	6	10
y	−0.7	0	0.7	1.4	1.8	2.3

After plotting these points, we get the graph shown in Figure 5.8. ■

Exercise

Sketch the graph of $y = \ln \frac{x}{2}$ after plotting points whose x-coordinates are 1, 2, 3, 4, 5, and 10.

Answer

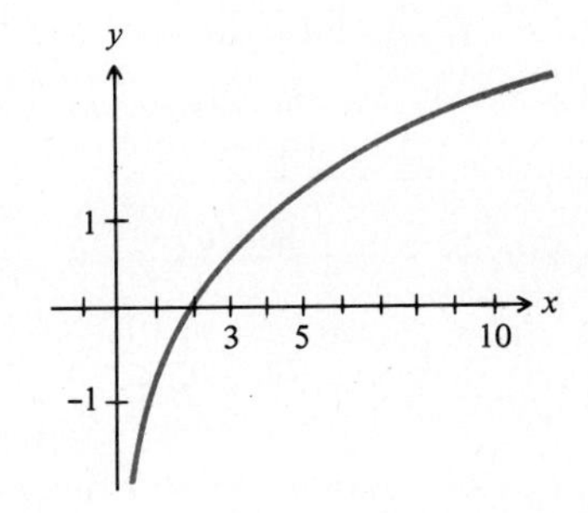

■

Because $\log_b k = j$ means that $b^j = k$, the equation

$$y = \log_b x$$

is equivalent to

$$b^y = x. \tag{5-1}$$

But equation (5-1) defines the inverse of the function defined by

$$b^x = y.$$

Hence we see that the logarithmic function with base b is the *inverse* of the exponential function with base b. In particular, $y = \ln x$ is the inverse of $y = e^x$.

Problem Set 5.2

In Problems 1–6, convert the equation into an equivalent exponential equation.

1. $\log_4 64 = 3$
2. $\log_2 \frac{1}{16} = -4$
3. $\log_{1/3} 9 = -2$
4. $\log_{1/2} \frac{1}{4} = 2$
5. $\log_4 2 = \frac{1}{2}$
6. $\log_8 4 = \frac{2}{3}$

In Problems 7–12, convert the equation into an equivalent logarithmic equation.

7. $10^{-1} = \frac{1}{10}$
8. $3^2 = 9$
9. $(\frac{1}{3})^3 = \frac{1}{27}$
10. $(\frac{1}{2})^{-2} = 4$
11. $16^{-1/2} = \frac{1}{4}$
12. $8^{-2/3} = \frac{1}{4}$

In Problems 13–18, find the number.

13. $\log 100$
14. $\log_2 32$
15. $\log_3 \frac{1}{3}$
16. $\log_{1/4} 64$
17. $\log_9 3$
18. $\log_{16} \frac{1}{8}$

In Problems 19–24, simplify.

19. $6^{2\log_6 k}$
20. $5^{-\log_5 k}$
21. $\log 10^x$
22. $\log_9 \sqrt[k]{9}$
23. $3^{\log_9 x}$
24. $\log_8 4^x$

In Problems 25–28, write the expression in terms of log x, log y, and log z.

25. $\log \dfrac{xy}{\sqrt{z}}$
26. $\log \dfrac{x^2}{yz}$
27. $\log \dfrac{\sqrt{xy}}{z^2}$
28. $\log \sqrt{\dfrac{xy}{z}}$

In Problems 29–32, write the expression as a single logarithm.

29. $\log x - \log y + \frac{1}{2} \log z$
30. $2 \log x + \log y - \log z$
31. $-2 \log x + \frac{1}{2} \log y - \log z$
32. $\log x + 3 \log y - \frac{1}{3} \log z$

In Problems 33–40, sketch the graph of the equation by plotting points with the x-coordinates specified.

33. $y = \log_4 x$, x-values: $\frac{1}{16}$, $\frac{1}{4}$, 1, 4, and 16.
34. $y = \log_{1/3} x$, x-values: $\frac{1}{9}$, $\frac{1}{3}$, 1, 3, and 9.
35. $y = 2 + \log_{1/2} x$, x-values: $\frac{1}{2}$, 1, 2, 4, 8, and 16.
36. $y = -1 + \log_3 x$, x-values: $\frac{1}{9}$, $\frac{1}{3}$, 1, 3, and 9.
37. $y = -\log_3 x$, x-values: $\frac{1}{3}$, 1, 3, and 9.
38. $y = 1 - \log_4 x$, x-values: $\frac{1}{4}$, 1, 4, and 16.
39. $y = \ln 2x$, x-values: .5, 1, 2, 3, 4, and 5.
40. $y = 1 - \ln \dfrac{x}{2}$, x-values: 1, 2, 3, 4, 5, and 10.

In Problems 41–44, sketch the graph of the equation.

41. $y = \log_{3/4} x$
42. $y = \log_{3/2} x$
43. $y = \log x^2$, $x > 0$
44. $y = \log(-x)$

45. Estimate $(526)^{2/3}$ by using common logarithms.
46. Estimate $(780)^{3/4}$ by using common logarithms.

5.3 Exponential and Logarithmic Equations

An **exponential equation** is one in which the variable appears in one or more exponents. For example,

$$3^x = 2 \quad \text{and} \quad 5^{2x+1} = 2^{x-5}$$

are exponential equations. As illustrated in Examples 1, 2, and 3, this type of equation can be solved by taking the logarithm of each side.

Example 1
Find the solution set of $2^x = 5$.

Solution. By taking the common logarithm of each side, we get the equivalent equation

$$\log 2^x = \log 5.$$

After using the laws of logarithms to rewrite the left side, we have

$$x \log 2 = \log 5. \tag{5-2}$$

Now equation (5-2) is a linear equation in x and may be solved accordingly. Dividing both sides by the coefficient of x, we get

$$x = \frac{\log 5}{\log 2}.$$

Since the universe of x is R, the solution set is $\{\log 5/\log 2\}$. ■

Note that in Example 1 we could have used the logarithm to a base other than 10. For instance, we could have used the natural logarithm and written the solution as $\ln 5/\ln 2$. However, since we are more familiar with base 10, we shall generally use the common logarithm. An exception to this is example 2, where the base of the exponential expression is e.

Example 2
Find the solution set of $e^{2x} = 15$.

Solution. By taking the natural logarithm of each side, we get

$$\ln e^{2x} = \ln 15,$$

which is equivalent to each of the following:

$$2x = \ln 15$$

$$x = \frac{\ln 15}{2}.$$

Since the universe is R, the solution set is {(ln 15/2)}.

Example 3
Find the solution set of $2^{x-3} = 7^{x+1}$.

Solution. The given equation is equivalent to each of the following:

$$\log 2^{x-3} = \log 7^{x+1}$$

$$(x - 3) \log 2 = (x + 1) \log 7. \qquad (5\text{-}3)$$

Equation (5-3) is linear and can be solved like any other linear equation. We get

$$x \log 2 - 3 \log 2 = x \log 7 + \log 7$$

$$x \log 2 - x \log 7 = 3 \log 2 + \log 7$$

$$x(\log 2 - \log 7) = 3 \log 2 + \log 7$$

$$x = \frac{3 \log 2 + \log 7}{\log 2 - \log 7}.$$

This result may be simplified to the quotient of two logarithms:

$$x = \frac{\log 2^3 + \log 7}{\log \frac{2}{7}} = \frac{\log 56}{\log \frac{2}{7}}.$$

Since the universe of x is R, the solution set is $\{\log 56/\log \frac{2}{7}\}$. ■

Now the solution of Example 1 is *exactly* log 5/log 2. Using Table III in the Appendix or a calculator with the capability, you can approximate and check this value in the given equation ($\log 5/\log 2 \approx 2.32$). The exact solution may also be checked, but doing so is somewhat more difficult than usual. We must employ the following theorem, which tells us how to convert from one logarithmic base to another.

Theorem 5.3

Let $a, b > 0$ with $a, b \neq 1$. Then for each positive number x,

$$\log_a x = \frac{\log_b x}{\log_b a}.$$

Proof. Let $j = \log_a x$. By definition of logarithm,

$$x = a^j.$$

Taking the logarithm to the base b of both sides, we get

$$\log_b x = \log_b a^j,$$

which is equivalent to

$$\log_b x = j \log_b a.$$

Solving this equation for j, we have

$$j = \frac{\log_b x}{\log_b a}.$$

Substituting for j, we have the desired result:

$$\log_a x = \frac{\log_b x}{\log_b a}. \qquad \blacksquare$$

Using Theorem 5.3, we can check the exact solution $\log 5/\log 2$ of Example 1 by changing these logarithms to base 2. We get

$$\log 5 = \frac{\log_2 5}{\log_2 10} \quad \text{and} \quad \log 2 = \frac{\log_2 2}{\log_2 10} = \frac{1}{\log_2 10}.$$

Therefore,

$$\frac{\log 5}{\log 2} = \frac{\log_2 5}{\log_2 10} \cdot \frac{\log_2 10}{1} = \log_2 5.$$

Our check now proceeds as follows:

$$\begin{aligned} 2^{\log 5/\log 2} &= 5 \\ 2^{\log_2 5} &= 5 \\ 5 &= 5. \end{aligned}$$

It appears that exercising care in deriving equivalent equations and verifying that roots belong to the universe is somewhat easier than checking.

Exercise

Find the solution set of each of the following.

(a) $5^x = 12$ (b) $e^{-3x} = 20$ (c) $5^{x+1} = 3^{2x-1}$

Answers. (a) $\left\{\frac{\log 12}{\log 5}\right\}$ (b) $\left\{\frac{-\ln 20}{3}\right\}$ (c) $\left\{\frac{\log 15}{\log \frac{9}{5}}\right\}$ ■

An equation in which the variable appears in one or more logarithms is called a **logarithmic equation.** For example,

$$\log x = 7 \qquad \text{and} \qquad \ln 5x = 1 + \ln 3x$$

are logarithmic equations. The principal idea in solving such equations is to use the properties of logarithms to get the given equation equivalent to one of the form

$$\log (p_x) = c \qquad \text{or} \qquad \ln (p_x) = c, \tag{5-4}$$

where p_x is an algebraic expression in x and c is a constant. Then, from the definition of logarithm, we have

$$p_x = 10^c \qquad \text{or} \qquad p_x = e^c,$$

which will be an equation of a type already studied. This method is illustrated in Examples 4, 5, and 6.

Example 4

Find the solution set of $4 + \log x = 5 + 2 \log x$.

Solution. Our first goal is to write this equation like the first part of statement (5-4). To this end we have the following:

$$\begin{aligned} \log x - 2 \log x &= 5 - 4 \\ -\log x &= 1 \\ \log x &= -1. \end{aligned} \tag{5-5}$$

Using the definition of logarithm, we see that equation (5-5) is equivalent to

$$x = 10^{-1} = \frac{1}{10}.$$

Since the universe of x is R^+, the solution set is $\{\frac{1}{10}\}$. ■

Example 5
Find the solution set of $\log(x - 15) = 2 - \log x$.

Solution. Again we wish to first write the given equation like the first part of statement (5-4). We have

$$\log(x - 15) + \log x = 2. \tag{5-6}$$

Using the laws of logarithms to write the left side of equation (5-6) as a single logarithm, we have

$$\log((x - 15)x) = 2,$$

which is equivalent to

$$\begin{aligned}(x - 15)x &= 10^2 \\ (x - 15)x &= 100. \end{aligned} \tag{5-7}$$

Equation (5-7) is quadratic and may be solved by the method of factoring. We get

$$\begin{gathered} x^2 - 15x - 100 = 0 \\ (x - 20)(x + 5) = 0 \\ x = 20 \quad \text{or} \quad x = -5. \end{gathered}$$

Since -5 is not in the universe, the solution set is $\{20\}$. ■

You may find that the solution of Example 4 is very easy to check, but the solution of Example 5 is not. We must note that

$$\log 20 = \log \frac{100}{5} = \log 100 - \log 5 = 2 - \log 5.$$

Our check proceeds as follows:

$$\begin{aligned} \log(20 - 15) &= 2 - \log 20 \\ \log 5 &= 2 - (2 - \log 5) \\ \log 5 &= \log 5. \end{aligned}$$

As with exponential equations, it appears that exercising caution in deriving equivalent equations and noting which roots belong to the universe is sometimes easier than checking.

Example 6
Find the solution set of $5 + \ln x = 3 + 2 \ln x$.

Solution. This example uses the techniques outlined in the preceding two examples, except that we are using natural logarithms. The given equation is equivalent to each of the following:

$$\begin{aligned} -2 \ln x + \ln x &= 3 - 5 \\ -\ln x &= -2 \\ \ln x &= 2. \end{aligned}$$

From the definition of logarithm we get

$$x = e^2.$$

Since the universe is R^+, the solution set is $\{e^2\}$. ■

Exercise

Find the solution set of each of the following.

(a) $5 + \log x = 4 + 3 \log x$ (b) $\log (x^2 + 14) = 1 + \log (x - 1)$

(c) $3 \ln x + 5 = \ln x + 2$

Answers. (a) $\{\sqrt{10}\}$ (b) $\{4, 6\}$ (c) $\{e^{-3/2}\}$ ■

Example 7

Find the solution set (with approximate solutions to the nearest hundredth) of $5^{2x} = 20$.

Solution. Because we must make approximations and are already familiar with the natural logarithm table, we will take the natural logarithm of each side instead of the common logarithm. We get

$$\begin{aligned} \ln 5^{2x} &= \ln 20 \\ 2x \ln 5 &= \ln 20 \\ x &= \frac{\ln 20}{2 \ln 5} \end{aligned}$$

Using Table II to approximate ln 20 and ln 5, we have

$$x \approx \frac{2.9957}{2(1.6094)} \approx .93$$

Because the universe of x is R, the solution set is $\{.93\}$. ■

Example 8

Find the solution set (with approximate solutions to the nearest hundredth) of $\ln x + \ln (x + 1) = 3$.

Solution. The given equation is equivalent to each of the following:

$$\ln(x(x+1)) = 3$$
$$\ln(x^2 + x) = 3$$
$$x^2 + x = e^3$$

Solving the preceding quadratic equation gives

$$x^2 + x - e^3 = 0$$
$$x = \frac{-1 \pm \sqrt{1 + 4e^3}}{2}$$

Using Table I, we approximate e^3 and get

$$x \approx \frac{-1 \pm \sqrt{1 + 4(20.086)}}{2}$$
$$\approx \frac{-1 \pm \sqrt{81.344}}{2}$$
$$\approx \frac{-1 \pm 9.02}{2}$$
$$x \approx 4.01 \quad \text{or} \quad x \approx -5.01$$

Since -5.01 is not in the universe, the solution set is $\{4.01\}$. ■

Exercise

Find the solution set (with approximate solutions to the nearest hundredth) of each of the following.

(a) $2^{3x} = 30$ (b) $\ln(x - 2) + \ln x = 5$

Answers. (a) $\{1.64\}$ (b) $\{13.22\}$ ■

Problem Set 5.3

In Problems 1–42, find the solution set (with exact solutions).

1. $6^x = 14$

2. $5^x = 18$

3. $10^x = 2$

4. $10^{2x} = 20$

5. $8^{2x} = 73$

6. $3^{4x} = 12$

7. $5^{-x} = 2$

8. $8^{-2x} = 21$

9. $9^x = \frac{1}{3}$

10. $8^{2x} = 16$

11. $3^{2x} = 4^{2x+1}$

12. $3^{1-2x} = 2^{x+5}$

13. $18^{1+x} = 10^{5-x}$

14. $\left(\frac{1}{2}\right)^x = 100$

15. $3^{-x^2} = 4$

16. $5^{2x^2} = 41^{x^2+1}$

17. $e^{2x+3} = 15$

18. $e^{2-x} = \frac{1}{2}$

19. $2e^{2x} - 5e^x + 2 = 0$

20. $e^{2x} + 5e^x = 6$

21. $\log 2x = -1$

22. $\log 3x = 2$

23. $\log x + \log x^2 = 3$

24. $3 \log x - \log 2x = 2$

25. $\log 3x^2 - \log\left(\frac{-x}{4}\right) = \frac{1}{2}$

26. $2 \log (-x) + \log \left(\frac{-x}{3}\right) = 2$

27. $\log (2x - 17) = 2 - \log x$

28. $\log (3 - 9x) = 1 + \log (x^2 - x)$

29. $1 + \log \frac{5x}{2} = \log 3$

30. $\log \sqrt{x^2 - 1} = 1$

31. $\log (x^2) = (\log x)^2$

32. $\log \sqrt{x} = \sqrt{\log x}$

33. $\log (-2 \log x) = 0$

34. $\log (2 \ln x) = 0$

35. $\sqrt{1 - 2 \log x} = \sqrt{3 \log x - 9}$

36. $1 - 2 \log x = \sqrt{2 \log^2 x - \log x + 3}$

37. $\ln x + \ln \frac{x}{4} = 3$

38. $\ln (x + 1) = 2 + \ln (2x - 3)$

39. $8^{2x+4} - 8^{x+2} - 6 = 0$

40. $2 \cdot 5^x - 7 \cdot 5^{x/2} - 4 = 0$

41. $2x - 5 = x \ln 2$

42. $\ln x = 1 + \ln 2x$

In Problems 43–54, find the solution set (with approximate solutions to the nearest hundredth).

43. $8^x = 15$

44. $6^{-x} = 20$

45. $15^{-2x} = 50$

46. $20^{3x} = 4$

47. $5^{2x-3} = 14$

48. $25^{2-x} = 65$

49. $\ln 3x = 2$

50. $\ln 2x = -1$

51. $2 \ln x - \ln 3x = -.5$

52. $\ln (x + 5) - \ln (x - 2) = 1.5$

53. $\ln (2x + 5) + \ln x = 2.3$

54. $\ln (2 - x) + \ln (1 - x) = 3$

5.4 Applications

When money earns compound interest, the principal is increased at the end of each interest period, the increase being the interest earned during that period. The sum of the principal and the interest becomes the new principal for the next interest period. Suppose that \$100 is invested at 12% interest, compounded quarterly. Then the interest rate for each quarter is 12%/4 = 3%, and the amount of money A_1 at the end of the first interest period is given by

$$A_1 = 100 + 100(0.03) = 100(1.03).$$

The amount A_1 becomes the new principal for the second interest period. So the amount of money at the end of the second period is

$$A_2 = A_1 + A_1(0.03) = A_1(1.03) = 100(1.03)^2.$$

Similarly, we get

$$A_3 = 100(1.03)^3.$$

Hence if money is compounded n times per year at an annual interest rate of r (in decimal form), then the amount of money A after t years is given by

$$A = P\left(1 + \frac{r}{n}\right)^{nt},$$

where P is the original principal invested. Thus, if \$1000 is invested at 8% interest, compounded semiannually for 5 years, then the total amount of money at the end of 5 years is

$$A = 1000\left(1 + \frac{0.08}{2}\right)^{10} = 1000(1.04)^{10}.$$

Using Table IV in the Appendix, we find that $(1.04)^{10} \approx 1.4802$. Thus

$$A \approx \$1480.$$

If the money were compounded quarterly, instead of semiannually, then the total after 5 years would be

$$A = 1000\left(1 + \frac{0.08}{4}\right)^{20} = 1000(1.02)^{20} \approx \$1486.$$

For a constant interest rate and a fixed principal and length of time, the amount accumulated increases as the number of interest periods increases. What would happen if the number of interest periods were increased indefinitely?

It is shown in calculus that if

$$y = \left(1 + \frac{1}{x}\right)^x,$$

$y \to e$ as $x \to \infty$. (Recall that e is an irrational number approximately equal to 2.71828 and is the base of the natural logarithm.) By rewriting the formula,

$$\begin{aligned} A &= P\left(1 + \frac{r}{n}\right)^{nt} \\ &= P\left(1 + \frac{1}{n/r}\right)^{nt} \\ &= P\left[\left(\left(1 + \frac{1}{n/r}\right)^{n/r}\right)^{r}\right]^{t}, \end{aligned}$$

we see that $A \to Pe^{rt}$ as $n \to \infty$. Thus when the formula

$$A = Pe^{rt}$$

is applied, we say that the money is being compounded continuously. So if \$1000 is compounded continuously at 8%, the amount of money at the end of 5 years is

$$A = 1000e^{(0.08)5} = 1000e^{0.4} \approx \$1492.$$

Without increasing the interest rate or lengthening the term, this is the most that can be accumulated from a \$1000 investment.

Example 1
How long will it take a \$1000 investment to amount to \$1500 if compounded continuously at 6%?

Solution. In this particular case we have

$$A = 1000e^{0.06t}.$$

We want to find the value of t which makes A equal 1500. Thus we must solve the following exponential equation for t:

$$1500 = 1000e^{0.06t}.$$

We get

$$\begin{aligned} 1.5 &= e^{0.06t} \\ \ln 1.5 &= 0.06t \\ t &= \frac{\ln 1.5}{0.06}. \end{aligned}$$

By using Table II (or a calculator), we find that this is approximately 6.8 years. ■

Now when interest is compounded continuously, the rate of growth, although increasing, is proportional to the ever-changing amount. If a quantity Q increases with time in such a way that its growth rate is proportional to the value of Q at that time, then

$$Q = Q_0 e^{rt}, \tag{5-8}$$

where Q_0 is the initial value of Q (the value when $t = 0$), r is the constant of proportionality in decimal form, and t represents time. Equation (5-8) is sometimes called the **exponential law of growth.**

In a biological experiment it is frequently found that the rate of growth of the population being studied is proportional to the population size. Under such conditions, the relationship between population size and time satisfies the law of growth. This is the nature of Example 2.

Example 2
If a bacteria population is 10,000 and the proportionality constant is 0.7 per hour, how long will it take for the population to reach 50,000?

Solution. Let Q represent the number of bacteria, and let t represent time in hours. Since the initial population is 10,000, we have $Q_0 = 10{,}000$. Since $r = 0.7$, the law of growth for this problem is

$$Q = 10{,}000e^{0.7t}.$$

We want to determine the value of t which makes $Q = 50{,}000$. So we must solve the following exponential equation for t:

$$50{,}000 = 10{,}000e^{0.7t}.$$

We get

$$\ln 5 = 0.7t$$

$$t = \frac{\ln 5}{0.7}.$$

This is approximately 2.3 hours. ■

In physical chemistry it is found that the rate at which atoms disintegrate is proportional to the number of atoms. Thus the quantity Q of a substance at time t is

$$Q = Q_0 e^{-kt}, \tag{5-9}$$

where Q_0 is the initial amount (the amount present when $t = 0$), and k is a positive constant which depends only on the substance. Equation (5-9) is referred to as the **exponential law of decay.**

In discussing radioactive decay, we customarily work with the half-life, which is the length of time it takes for half of a given amount to decay. For a specified substance, the half-life is constant *regardless of the amount*. Thus the length of time required for 10 grams to disintegrate to 5 is the same as the time it takes for 4 grams to disintegrate to 2. This concept occurs in Example 3.

Example 3
Find the half-life of a radioactive substance which satisfies $Q = Q_0e^{-0.02t}$, where t is in years.

Solution. We want to find the value of t which makes Q equal $Q_0/2$. Thus we have to solve for t in the exponential equation

$$\frac{Q_0}{2} = Q_0e^{-0.02t}, \tag{5-10}$$

which is equivalent to each of the following:

$$\frac{1}{2} = e^{-0.02t}$$

$$\ln\left(\frac{1}{2}\right) = -0.02t$$

$$-\ln 2 = -0.02t$$

$$t = \frac{\ln 2}{.02}$$

This is approximately 34.7 years. ■

Note that in the equation we had to solve in Example 3, equation (5-10), there is an unknown constant Q_0, the initial amount. The fact that we were able to solve the equation without knowing the value of Q_0 is consistent with the comments made about half-life. The half-life is independent of the initial amount.

Problem Set 5.4

Growth: $Q = Q_0 e^{rt}$

Decay: $Q = Q_0 e^{-rt}$

1. How long will it take a \$1000 investment to amount to \$1500 if compounded continuously at 7%? $A = Pe^{rt}$
2. How long will it take an investment to double if compounded continuously at 6%?
3. If a bacteria population is 5000 and grows with a proportionality constant of 0.6 per hour, how long will it take for the population to reach 20,000?
4. If a bacteria population is 10,000 and grows with a proportionality constant of 0.5 per hour, how long will it take the population to double?

5. Find the half-life of a substance which satsifies $Q = Q_0e^{-0.01t}$, where t is in years.
6. Find the half-life of a substance which satsifies $Q = Q_0e^{-0.04t}$, where t is in years.
7. If a \$1000 investment is compounded continuously and amounts to \$1200 after 2.8 years, what is the interest rate?
8. If a \$2000 investment is compounded continuously and doubles in 12.5 years, what is the interest rate?
9. If a bacteria population grows from 1000 to 3000 in 6 hours, what is the constant of proportionality?
10. If a bacteria population doubles in 2 hours, what is its constant of proportionality?
11. If the half-life of a radioactive substance is 50 years, how long will it take 10 grams of the substance to disintegrate to 2?
12. If the half-life of a radioactive substance is 200 years, how long will it take 5 grams of the substance to disintegrate to 4 grams?
13. If a bacteria population of 50,000 grows to 75,000 in 20 hours, how much longer will it take for the population to reach 100,000?
14. If fluid friction is used to slow down a flywheel, then V, the number of revolutions per minute, after t seconds, is given by the formula $V = V_0e^{-kt}$, where V_0 is the value of V when $t = 0$ and k is a constant. If $k = 0.2$, how long must friction be applied to reduce the number of revolutions from 500 to 100 per minute?
15. The atmospheric pressure P, in pounds per square inch, at a height of z feet is approximately $P = P_0e^{-kz}$, where P_0 is the pressure at sea level and k is a constant. If $P_0 = 15$ and $P = 14.5$ at a height of 1100 feet, at what height is the pressure 14?
16. The temperature T (in degrees Celsius) of a body surrounded by cooler air, after t minutes, is given by $T = T_0 + (T_1 - T_0)e^{-kt}$, where T_1 is the initial temperature of the body, T_0 is the temperature of the air, and k is a constant. If $k = 0.13$, $T_1 = 60$, and $T_0 = 20$, how long will it take the body to cool to a temperature of 30°C?
17. If the unhealed area A (in cm^2) of a skin wound after n days is given by the formula

$$A = A_0e^{-n/10},$$

where A_0 is the initial area of the wound, how many days will it take for the wound to reduce in area by 75%?
18. The rapid compression or expansion of a gas is very nearly adiabatic. That is, in such a system $P_1V_1k = p_2V_2k$, where P_i and V_i represent the pressure and volume, and k is a constant which depends on the gas. Assume that the compression of a diesel engine is adiabatic. If the compression ratio, V_1/V_2, is 15, and the pressure at the beginning of the stroke is 15 lb/in.^{2} and 663 lb/in.^{2} at the end of the stroke, what is the value of k?

Chapter 5 Review

In Problems 1–8, sketch the graph.

1. $y = 3^{-x}$
2. $y = 2^{x+1}$
3. $y = -4^x$
4. $y = 1 - e^{x/2}$
5. $y = \log_3 x$
6. $y = -\log_2 x$
7. $y = \log_{1/4} x$
8. $y = \ln(-x)$

9. Convert the equation $16^{3/4} = 8$ into an equivalent logarithmic equation.
10. Convert the equation $\log_8 \frac{1}{32} = -\frac{5}{3}$ into an equivalent exponential equation.

11. If $\log x = 0.2$, $\log y = 0.6$, and $\log z = 0.8$, find $\log \dfrac{xy^2}{\sqrt{z}}$.

12. Write $\log \dfrac{x^2\sqrt{y}}{z^3}$ in terms of $\log x$, $\log y$, and $\log z$.

13. Write $\log x - \frac{1}{3} \log y + 2 \log z$ as a single logarithm.

In Problems 14–27, find the solution set (with exact solutions).

14. $5^x = 20$

15. $\left(\dfrac{1}{2}\right)^x = 10$

16. $5^{-2x} = 8^{x+1}$

17. $5^{-x^2} = \dfrac{1}{20}$

18. $2^{x^2} = 20^{2x^2+1}$

19. $\log 5x = 2$

20. $\log x = 1 + \log x^3$

21. $\log x - \log \dfrac{2}{x} = -2$

22. $\log (x + 21) = 2 - \log x$

23. $1 - \log \dfrac{3x}{2} = \log 5$

24. $\log \sqrt{1 - x^2} = -1$

25. $\log x^3 = (\log x)^3$

26. $\ln 2x - \ln x^2 = 5$

27. $\ln (-x) + \ln \left(\dfrac{-x}{4}\right) = 2$

In Problems 28–31, find the solution set (with approximate solutions to the nearest hundredth).

28. $3^{-x} = 17$

29. $19^{2x-1} = 7$

30. $\ln (3 - 2x) - \ln (1 - x) = .75$

31. $2 \ln x = 2 + \ln(x - 1)$

32. If fluid friction is used to slow down a flywheel, then V, the number of revolutions per minute, after t seconds, is given by the formula $V = V_0e^{-kt}$, where V_0 is the initial value of V, and k is a constant. If it takes 5 seconds to slow a flywheel from 200 to 100 revolutions per minute, what is the value of k?
33. A body of temperature 50° Celsius is placed in a room of temperature 20°. The temperature T of the body after t minutes is given by $T = 20 + 30e^{-kt}$, for some constant k. If it takes 5 minutes for the body to cool to 40°, how long will it take for it to cool from 40° to 30°?
34. If the half-life of a radioactive substance is 500 years, how long will it take 5 grams of the substance to disintegrate to 2?
35. If a bacteria population grows with a proportionality constant of 0.4 per hour, how long will it take the population to double?
36. If an investment is compounded continuously and doubles in 10 years, what is the interest rate?

Systems of Equations

6

In this chapter we consider sets of equations in two or more variables and study techniques for determining the common solutions.

6.1 Systems of Linear Equations in Two Variables

If p_{xy}, q_{xy}, r_{xy}, and s_{xy} are algebraic expressions in the variables x and y, then the pair of equations

$$\begin{cases} p_{xy} = q_{xy} \\ r_{xy} = s_{xy} \end{cases}$$

is called a **system of equations in the variables x and y.** An ordered pair is called a **solution** of the system provided that it is a solution of each of the equations in the system. (Thus a system of equations is the same as an "and" statement which includes both equations.) For example, we see that (3, 2) is a solution of the system

$$\begin{cases} x + y = 5 \\ x - y = 1 \end{cases}$$

because $3 + 2 = 5$ *and* $3 - 2 = 1$. The **solution set** of the system is the set of all solutions of the system. Of course, the solution set of a system is the intersection of the solution sets of the separate equations. Thus the intersection of the graphs of the two equations is the graph of the solution set of the system.

Two systems are said to be **equivalent** provided that they have the same solution set. We shall find equivalent systems by employing the following theorem.

Theorem 6.1

Let p_{xy}, q_{xy}, r_{xy}, and s_{xy} be algebraic expressions in the variables x and y, and let k_1 and k_2 be nonzero constants. Then the system

$$\begin{cases} p_{xy} = q_{xy} \\ r_{xy} = s_{xy} \end{cases}$$

is equivalent to

$$\begin{cases} p_{xy} = q_{xy} \\ k_1 p_{xy} + k_2 r_{xy} = k_1 q_{xy} + k_2 s_{xy}. \end{cases}$$

Theorem 6.1 tells us that we can multiply each equation by a nonzero constant, add the resulting equations to derive another, and replace either of the original equations with the one derived to obtain an equivalent system.

For some systems, we may choose multipliers which lead to a derived equation in one variable only. We can then solve this equation and substitute the value obtained in one of the original equations to solve for the remaining variable. This procedure is called the **elimination method** for determining the solution set of a system of equations. This technique is illustrated in Example 1.

Example 1

Use the elimination method to find the solution set of

$$\begin{cases} x - 3y = 10 \\ -3x + 2y = -2. \end{cases}$$

Solution. If we multiply the first equation by 3, we get

$$3x - 9y = 30.$$

Adding this equation to the second will give us an equation without the variable x.

$$\begin{array}{r} 3x - 9y = 30 \\ -3x + 2y = -2 \\ \hline -7y = 28. \end{array}$$

Replacing the second equation of the given system with the derived equation, we have the equivalent system

$$\begin{cases} x - 3y = 10 \\ \quad - 7y = 28. \end{cases}$$

After solving the second for y, and getting $y = -4$, we substitute -4 for y in the first and get

$$\begin{aligned} x + 12 &= 10 \\ x &= -2. \end{aligned}$$

Hence the solution set is $\{(-2, -4)\}$. ■

In Example 2 we see that it is sometimes necessary to multiply each equation in order to eliminate one of the variables.

Example 2
Use the elimination method to find the solution set of

$$\begin{cases} 5x + 3y = 1 \\ 3x - 2y = -7. \end{cases}$$

Solution. If we multiply the first equation by 2, we get

$$10x + 6y = 2.$$

If we multiply the second by 3, we have

$$9x - 6y = -21.$$

We now add these two to get an equation without the variable y.

$$\begin{array}{r} 10x + 6y = 2 \\ 9x - 6y = -21 \\ \hline 19x \quad\quad = -19. \end{array}$$

Replacing the first equation of the given system with the derived equation, we have the equivalent system

$$\begin{cases} 19x \quad\quad = -19 \\ 3x - 2y = -7. \end{cases}$$

After solving the first for x and getting $x = -1$, we substitute -1 for x in the second and get

$$\begin{aligned} -3 - 2y &= -7 \\ y &= 2. \end{aligned}$$

Hence our solution set is $\{(-1, 2)\}$. ■

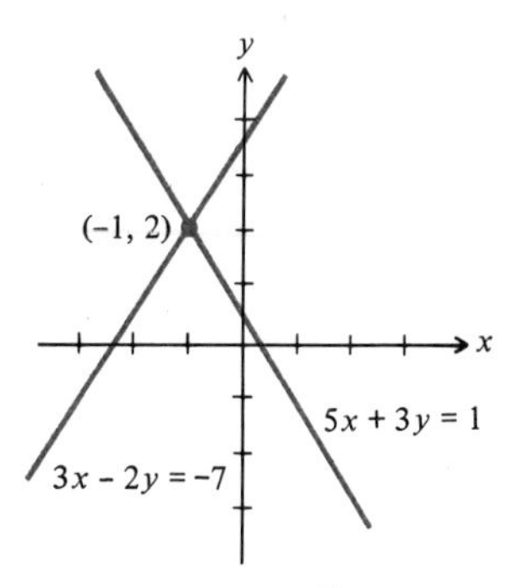

Figure 6.1

In the preceding example, we could have eliminated the variable x by multiplying the first equation by 3 and the second by -5 and adding. And we could have replaced the second equation instead of the first. The solution set would be the same.

Each system in the preceding examples is called a **system of linear equations** because each equation is linear. In such a system, since the graph of each equation is a line, one of three situations will exist. The two lines intersect at exactly one point (and the system has exactly one solution); the two lines are distinct but parallel (and the system has no solutions); or the two lines are really the same line (and the system has infinitely many solutions).

If we graphed each of the equations of Example 2, we would find two lines intersecting at $(-1, 2)$, as indicated in Figure 6.1.

Example 3
Use the elimination method to find the solution set of

$$\begin{cases} 2x = 2 - y \\ 4x + y = -2x + 8. \end{cases}$$

Solution. Our goal is to combine the two equations so that the derived equation has only one variable. Before choosing the multipliers, it is best to get all the x and y terms on the same side in each equation. (In the present form it might appear that multiplying the first equation by -2 and adding it to the second will eliminate x, but it will not.) Our system is

$$\begin{cases} 2x + y = 2 \\ 6x + y = 8. \end{cases}$$

To eliminate x, we can multiply the first by -3 and add to the second. However, it may be easier to simply subtract the second from the first and eliminate y. Doing the latter, we get

$$-4x = -6.$$

Upon solving for x, we have $x = \frac{3}{2}$. Substituting this value for x in the first or second and solving for y, we get

$$y = -1.$$

Hence the solution set is $\{(\frac{3}{2}, -1)\}$. ■

Example 4
Use the elimination method to find the solution set of

$$\begin{cases} -4x + 5y = -2 \\ 6x - 7y = 4. \end{cases}$$

Solution. To eliminate x, we may multiply the first by 3 and the second by 2 and add. To eliminate y, we may multiply the first by 7 and the second by 5 and add. We'll eliminate x because the multipliers are smaller. We get

$$\begin{array}{r} -12x + 15y = -6 \\ 12x - 14y = 8 \\ \hline y = 2. \end{array}$$

Substituting 2 for y in one of the original equations and solving for x, we get

$$x = 3.$$

The solution set is $\{(3, 2)\}$. ■

Exercise
Use the elimination method to determine the solution set.

(a) $\begin{cases} 5x + 2y = -4 \\ 10x + y = 1 \end{cases}$ (b) $\begin{cases} 5x - 3y = 8 \\ 9x + 2y = 7 \end{cases}$

Answers. (a) $\left\{\left(\frac{2}{5}, -3\right)\right\}$ (b) $\{(1, -1)\}$ ■

As indicated previously, the solution set does not have to consist of a single point. Consider Examples 5 and 6.

Example 5
Find the solution set of

$$\begin{cases} 2x - y = 12 \\ 4x - 2y = 12. \end{cases}$$

Solution. Upon multiplying the first equation by -2 and adding to the second, we get

$$\begin{array}{r} -4x + 2y = -24 \\ 4x - 2y = 12 \\ \hline 0 = -12. \end{array}$$

What does this mean? Recall that we can replace either of the original equations with the derived one and maintain an equivalent system. Thus

$$\begin{cases} 2x - y = -12 \\ \quad\;\; 0 = -12 \end{cases}$$

is equivalent, and we are looking for the solution set of this system. But since the second equation is never true, the system has no solutions. The solution set is $\varnothing$. (If we graph each of the equations, we find that we have two parallel lines.) ■

Example 6
Find the solution set of

$$\begin{cases} -2x + y = -2 \\ 6x - 3y = 6. \end{cases}$$

Solution. Multiplying the first equation by 3 and adding to the second, we get

$$\begin{array}{r} -6x + 3y = -6 \\ 6x - 3y = \;\;\, 6 \\ \hline 0 = \;\;\, 0. \end{array}$$

What does this mean? The given system is equivalent to

$$\begin{cases} -2x + y = -2 \\ \qquad\;\; 0 = 0 \end{cases}$$

Since the second equation is always true, any solution to the first will be a solution of the system. Hence there are infinitely many solutions, and the solution set is $\{(x, y) \mid -2x + y = -2\}$. (If we graph each of the equations, we will see that they describe the same line. The two equations are equivalent.) ■

Thus it sometimes happens that in eliminating one variable, *both* are eliminated. If this occurs and the derived equation is *always false*, then the system has no solutions. If this occurs and the derived equation is *always true*, then the system has infinitely many solutions.

Exercise
Determine the solution set of each of the following.

(a) $\begin{cases} 3x - 2y = 1 \\ -6x + 4y = -2 \end{cases}$ (b) $\begin{cases} 2x + 4y = -6 \\ -3x - 6y = -10 \end{cases}$

Answers. (a) $\{(x, y) \mid 3x - 2y = 1\}$ (b) $\varnothing$ ■

Problem Set 6.1 Extra credit quiz problem Thursday

Use the elimination method to determine the solution set.

1. $\begin{cases} 2x - y = 4 \\ x + y = 5 \end{cases}$

2. $\begin{cases} x + 3y = 3 \\ x - 5y = 11 \end{cases}$

3. $\begin{cases} 3x + y = 6 \\ 5x + y = 3 \end{cases}$

4. $\begin{cases} -x + 5y = -2 \\ x - y = 5 \end{cases}$

5. $\begin{cases} 2x - y = 8 \\ x + 3y = -10 \end{cases}$

6. $\begin{cases} 3x - 5y = 6 \\ x + y = -5 \end{cases}$

7. $\begin{cases} x + 5y = 20 \\ 3x + 2y = 8 \end{cases}$

8. $\begin{cases} 3x + 2y = 6 \\ 4x - y = -4 \end{cases}$

9. $\begin{cases} 3x + 5y = 9 \\ 2x + 7y = 17 \end{cases}$

10. $\begin{cases} 5x - 4y = -16 \\ 8x - 3y = -29 \end{cases}$

11. $\begin{cases} \dfrac{7x}{2} + y = \dfrac{11}{2} \\ \dfrac{11x}{2} - \dfrac{5y}{2} = 29 \end{cases}$

12. $\begin{cases} -4x + 8y = 72 \\ 2x + \dfrac{7y}{3} = \dfrac{25}{3} \end{cases}$

13. $\begin{cases} 2x - 5y = 5 \\ \dfrac{x}{2} + \dfrac{y}{3} = \dfrac{1}{2} \end{cases}$

14. $\begin{cases} 4x + 3y = 0 \\ 11x - 5y = \dfrac{53}{4} \end{cases}$

15. $\begin{cases} 6x + 5y = \dfrac{-35}{12} \\ 7x + 4y = \dfrac{-25}{6} \end{cases}$

16. $\begin{cases} 3x + 8y = -5 \\ 2x - 5y = \dfrac{53}{12} \end{cases}$

17. $\begin{cases} 2x - 4y = 5 \\ -3x + 6y = -4 \end{cases}$

18. $\begin{cases} x + 3y = -3 \\ 2x + 6y = -6 \end{cases}$

19. $\begin{cases} \sqrt{2}x + y = 3 \\ \sqrt{3}x + \sqrt{6}y = 2\sqrt{6} \end{cases}$

20. $\begin{cases} x - \sqrt{2}y = -\sqrt{3} - 6 \\ \sqrt{3}x + y = 3\sqrt{2} - 3 \end{cases}$

6.2 The Method of Substitution

The following theorem tells us of another method which may be used to obtain an equivalent system of equations.

Theorem 6.2

Let p_{xy} and q_{xy} be algebraic expressions in the variables x and y, and let f be a function. Then the system

$$\begin{cases} p_{xy} = q_{xy} \\ y = f(x) \end{cases}$$

is equivalent to

$$\begin{cases} p^*_{xy} = q^*_{xy} \\ y = f(x), \end{cases}$$

where the equation $p^*_{xy} = q^*_{xy}$ *is the one obtained when we substitute* $f(x)$ *for* y *in the equation* $p_{xy} = q_{xy}$.

Theorem 6.2 tells us that if one of the equations in a system is solved for y in terms of x, then we may replace y in the other equation by that expression in x, and the resulting system is equivalent to the original. Of course, after the substitution is made we will have an equation in the variable x only. We can then solve this equation for x. To determine y, we use the equation $y = f(x)$. This technique of finding the solution set to a system of equations is called the **substitution method** and is illustrated in Examples 1 and 2.

Example 1
Use the substitution method to determine the solution set of

$$\begin{cases} 2x - 5y = 6 \\ y = \dfrac{3x - 8}{5}. \end{cases}$$

Solution. Substituting the expression $\dfrac{3x - 8}{5}$ for y in the first equation, we get

$$2x - 5 \cdot \frac{3x - 8}{5} = 6.$$

Hence the following system is equivalent to the original.

$$\begin{cases} 2x - 5 \cdot \dfrac{3x - 8}{5} = 6 \\ y = \dfrac{3x - 8}{5}. \end{cases}$$

Solving the first equation for x, we get

$$\begin{aligned} 2x - (3x - 8) &= 6 \\ 2x - 3x + 8 &= 6 \\ -x &= -2 \\ x &= 2. \end{aligned}$$

Substituting 2 for x in the second equation of the system, we get

$$y = \frac{3(2) - 8}{5} = \frac{-2}{5}.$$

Hence the solution set is $\{(2, -\frac{2}{5})\}$. ■

Of course, it is not necessary that one of the equations in the original system be explicitly solved for y in terms of x. We may have to put it in that form ourselves, as in Example 2.

Example 2
Use the substitution method to find the solution set of

$$\begin{cases} 6x + 5y = -9 \\ 8x + 7y = -15. \end{cases}$$

Solution. To use the substitution method, we will solve one of the equations for y in terms of x. Considering the first, we have

$$y = \frac{-9 - 6x}{5}. \tag{6-1}$$

Substituting for y in the second equation, we get

$$8x + 7 \cdot \frac{-9 - 6x}{5} = -15$$

$$\begin{aligned} 40x + 7(-9 - 6x) &= -75 \\ 40x - 63 - 42x &= -75 \\ -2x &= -12 \\ x &= 6. \end{aligned}$$

Substituting 6 for x in equation (6-1), we get

$$y = \frac{-9 - 6(6)}{5} = -9.$$

The solution set is $\{(6, -9)\}$. ■

Exercise
Use the substitution method to find the solution set.

(a) $\begin{cases} x + 9y = 11 \\ y = \dfrac{26 - 4x}{9} \end{cases}$ (b) $\begin{cases} 6x + 8y = 5 \\ 13x + 6y = 8 \end{cases}$

Answers. (a) $\left\{\left(5, \frac{2}{3}\right)\right\}$ (b) $\left\{\left(\frac{1}{2}, \frac{1}{4}\right)\right\}$ ■

The solution set to each of the systems in Examples 1 and 2 can also be found by using the elimination method. Note that regardless of the method used, elimination or substitution, there is a common objective: to derive an equation that has a single variable. We then solve for it, substitute the obtained value in one of the original equations, and solve for the remaining variable.

Example 3
Find the solution set of

$$\begin{cases} 8x + 3y = 6 \\ \quad\quad\; y = x^2 + 1. \end{cases}$$

Solution. Substituting $x^2 + 1$ for y in the first equation gives the equivalent system

$$\begin{cases} 8x + 3(x^2 + 1) = 6 \\ \quad\quad\quad\quad\;\; y = x^2 + 1. \end{cases}$$

Solving the first equation for x, we have

$$3x^2 + 8x - 3 = 0$$
$$(3x - 1)(x + 3) = 0$$
$$x = \frac{1}{3} \quad \text{or} \quad x = -3.$$

Using the second equation of the system, we find that when $x = \frac{1}{3}$, $y = \frac{10}{9}$. When $x = -3$, $y = 10$. Hence the solution set is $\{(\frac{1}{3}, \frac{10}{9}), (-3, 10)\}$. ■

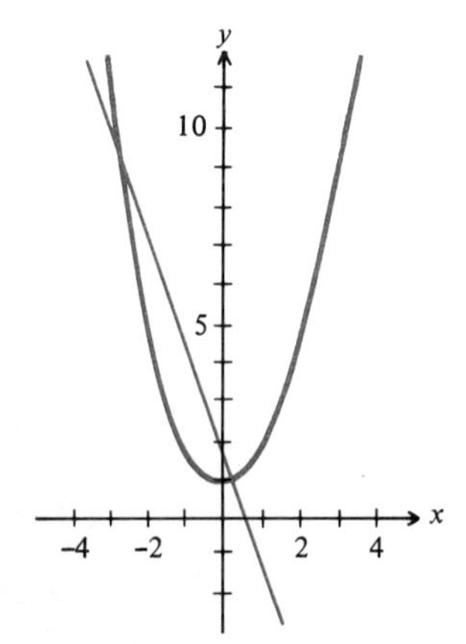

Figure 6.2

If we examine the graphs of the two equations in Example 3, we find a line (first equation) intersecting a parabola (second equation) at two points, as seen in Figure 6.2.

It is evident that the substitution method may also be used when one of the equations is solved for x in terms of y. In such a case we simply replace x in the other equation by the expression in y. The resulting equation will be in the variable y only.

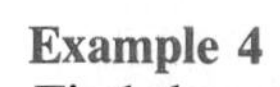

Example 4
Find the solution set of

$$\begin{cases} \;\; x + y^2 = 2 \\ 2x + 3y = 2. \end{cases}$$

Solution. We may use the substitution method, solving either the first or second equation for x or y. It appears that it will be easiest to solve the first for x. We get

$$x = 2 - y^2. \tag{6-2}$$

Replacing x in the second equation by $2 - y^2$, we have

$$2(2 - y^2) + 3y = 2.$$

Solving for y, we get

$$\begin{aligned} 2y^2 - 3y - 2 &= 0 \\ (2y + 1)(y - 2) &= 0 \end{aligned}$$

$$y = -\frac{1}{2} \quad \text{or} \quad y = 2.$$

Using equation (6-2), we find that when $y = -\frac{1}{2}$, $x = \frac{7}{4}$. And when $y = 2$, $x = -2$. Hence the solution set is $\{(\frac{7}{4}, -\frac{1}{2}), (-2, 2)\}$. ■

Exercise
Find the solution set.

(a) $\begin{cases} y = x^2 - 5 \\ 4x - y = 0 \end{cases}$ (b) $\begin{cases} x = y^2 - 2y + 1 \\ x + y = 3 \end{cases}$

Answers. (a) $\{(5, 20), (-1, -4)\}$ (b) $\{(1, 2), (4, -1)\}$ ■

Problem Set 6.2 Extra credit Quiz Problems Tuesday

Use substitution to determine the solution set.

1. $\begin{cases} 3x - 2y = 6 \\ y = \dfrac{2 - x}{2} \end{cases}$

2. $\begin{cases} x = \dfrac{y - 3}{4} \\ 3x - 2y = -6 \end{cases}$

3. $\begin{cases} x = \dfrac{2y + 3}{3} \\ 3x - 5y = 0 \end{cases}$

4. $\begin{cases} y = \dfrac{2 - x}{5} \\ 5x + 3y = -3 \end{cases}$

5. $\begin{cases} 2x - 3y = 7 \\ 3x - 5y = 6 \end{cases}$

6. $\begin{cases} 6x - 5y = -4 \\ 11x + y = 6 \end{cases}$

7. $\begin{cases} 2x - 5y^1 = 5 \\ -4x + 10y = -10 \end{cases}$

8. $\begin{cases} 3x - 6y = 1 \\ -5x + 10y = 2 \end{cases}$

9. $\begin{cases} 2x - y = 2 \\ y = x^2 - 1 \end{cases}$

10. $\begin{cases} x = 2y^2 - 3 \\ 2x - 3y = -5 \end{cases}$

11. $\begin{cases} 2x + y^2 = \dfrac{3}{5} \\ 5x + 3y = 2 \end{cases}$

12. $\begin{cases} 5x - 3y = -2 \\ -2x^2 + y = 1 \end{cases}$

13. $\begin{cases} y = x^2 - 3x + 2 \\ 2x - 3y = 5 \end{cases}$

14. $\begin{cases} 2x - y = -9 \\ x = 2y^2 + y - 5 \end{cases}$

6.3 Systems of Nonlinear Equations

In some systems both equations are nonlinear. But elimination or substitution may still be used to determine the solution set, as indicated in Examples 1 and 2.

Example 1
Find the solution set of

$$\begin{cases} \dfrac{-1}{x} + \dfrac{2}{y} = 5 \\ \dfrac{3}{x} - \dfrac{1}{y} = 3. \end{cases}$$

Solution. With the proper substitution we can get a system of linear equations. Letting $u = 1/x$ and $v = 1/y$, we have

$$\begin{cases} -u + 2v = 5 \\ 3u - v = 3. \end{cases} \tag{6-3}$$

We may now use substitution or elimination to solve this system. Multiplying the second equation by 2 and adding it to the first, we get

$$\begin{array}{rcr} -u + 2v &=& 5 \\ 6u - 2v &=& 6 \\ \hline 5u \quad &=& 11. \end{array}$$

Solving for u, we get $u = \frac{11}{5}$. When we substitute $\frac{11}{5}$ for u in one of the equations of system (6-3), we get $v = \frac{18}{5}$. Therefore,

$$x = \frac{5}{11} \quad \text{and} \quad y = \frac{5}{18},$$

and the solution set is $\{(\frac{5}{11}, \frac{5}{18})\}$. ■

Example 2
Find the solution set of

$$\begin{cases} x^2 - y^2 = 1 \\ \quad\;\; y^2 = \dfrac{3x^2 - 4}{2}. \end{cases}$$

Solution. Substituting $(3x^2 - 4)/2$ for y^2 in the first equation, we get

$$x^2 - \frac{3x^2 - 4}{2} = 1.$$

Solving this equation for x, we have

$$\begin{aligned} x^2 &= 2 \\ x &= \pm\sqrt{2}. \end{aligned}$$

Now when $x = \pm\sqrt{2}$, we get

$$\begin{aligned} y^2 &= \frac{6 - 4}{2} = 1 \\ y &= \pm 1. \end{aligned}$$

Hence the solution set is $\{(\sqrt{2}, 1), (\sqrt{2}, -1), (-\sqrt{2}, 1), (-\sqrt{2}, -1)\}$. ■

Exercise
Find the solution set.

(a) $\begin{cases} \dfrac{5}{x} - \dfrac{1}{y} = -29 \\ \dfrac{2}{x} + \dfrac{3}{y} = \quad 2 \end{cases}$ (b) $\begin{cases} x^2 + 4y^2 = 25 \\ \quad\;\; y^2 = 25 - 4x^2 \end{cases}$

Answers. (a) $\left\{\left(-\frac{1}{5}, \frac{1}{4}\right)\right\}$
(b) $\{(\sqrt{5}, \sqrt{5}), (\sqrt{5}, -\sqrt{5}), (-\sqrt{5}, \sqrt{5}), (-\sqrt{5}, -\sqrt{5})\}$ ■

Example 3
Find the solution set of

$$\begin{cases} x^2 + y^2 = 25 \\ 4x - 3y = 0. \end{cases}$$

Solution. Solving the second equation for x, we get

$$x = \frac{3y}{4}. \tag{6-4}$$

Substitution in the first equation gives

$$\begin{aligned}
\left(\frac{3y}{4}\right)^2 + y^2 &= 25 \\
\frac{9y^2}{16} + y^2 &= 25 \\
9y^2 + 16y^2 &= 400 \\
y^2 &= 16 \\
y &= \pm 4.
\end{aligned}$$

Using equation (6-4), we see that when $y = 4$, $x = 3$, and when $y = -4$, $x = -3$. Hence the solution set is $\{(3, 4), (-3, -4)\}$. ■

Now, in Example 3, we used equation (6-4) to find the corresponding x-values for the calculated y-values. What if we had used the first equation from the system instead? We would have noted that:

$$\begin{aligned}
\text{when } y = 4,\ x^2 + 16 &= 25 \\
x &= \pm 3; \\
\text{when } y = -4,\ x^2 + 16 &= 25 \\
x &= \pm 3;
\end{aligned}$$

and the solution set apparently would have been $\{(3, 4), (-3, 4), (3, -4), (-3, -4)\}$, a set containing an extra two solutions. But note that the graph of the first equation in the system is a circle, and the graph of the second is a line. Since a line cannot intersect a circle in more than two points, this apparent solution set must be incorrect. The lesson to be learned is that upon finding the values for one variable, we must use the equation of substitution to calculate the corresponding values for the remaining variable.

As seen in Example 4, we must sometimes use both the elimination and the substitution method.

Example 4

Find the solution set of

$$\begin{cases} x^2 + y^2 = 5 \\ x^2 - 2xy + y^2 = 1. \end{cases}$$

Solution. We must first use elimination. Subtracting the second equation from the first, we get

$$2xy = 4.$$

Thus the given system is equivalent to

$$\begin{cases} x^2 + y^2 = 5 \\ \quad\; 2xy = 4. \end{cases}$$

Now we want to solve the second equation for x or y. Solving for y, we have

$$y = \frac{2}{x}. \tag{6-5}$$

(Of course, this is valid only when $x \neq 0$. When $x = 0$, there is no value of y which makes the second equation true. Consequently, $2xy = 4$ is equivalent to $y = 2/x$.) Replacing y in the first equation by $2/x$, we get

$$x^2 + \left(\frac{2}{x}\right)^2 = 5.$$

Solving for x, we have

$$\begin{aligned} x^2 + \frac{4}{x^2} &= 5 \\ x^4 + 4 &= 5x^2 \\ x^4 - 5x^2 + 4 &= 0 \\ (x^2 - 1)(x^2 - 4) &= 0 \\ x = \pm 1 \quad \text{or} \quad x &= \pm 2. \end{aligned}$$

Using equation (6-5) to calculate the four corresponding values of y, we get the solution set $\{(1, 2), (-1, -2), (2, 1), (-2, -1)\}$. ■

Exercise
Find the solution set

(a) $\begin{cases} x^2 + y^2 = 169 \\ 12x + 5y = 0 \end{cases}$ (b) $\begin{cases} x^2 - xy + y^2 = 13 \\ x^2 + y^2 = 10 \end{cases}$

Answers. (a) $\{(5, -12), (-5, 12)\}$
(b) $\{(-1, 3), (1, -3), (3, -1), (-3, 1)\}$ ■

Problem Set 6.3

Find the solution set.

1. $\begin{cases} \dfrac{2}{x} + \dfrac{1}{y} = -2 \\ \dfrac{3}{x} - \dfrac{1}{y} = 1 \end{cases}$

2. $\begin{cases} \dfrac{1}{x} - \dfrac{3}{y} = 2 \\ \dfrac{1}{x} + \dfrac{5}{y} = 3 \end{cases}$

3. $\begin{cases} \dfrac{2}{x} - \dfrac{5}{y} = 0 \\ \dfrac{3}{x} - \dfrac{7}{y} = 1 \end{cases}$

4. $\begin{cases} \dfrac{3}{x} + \dfrac{2}{y} = 9 \\ \dfrac{8}{x} - \dfrac{5}{y} = \dfrac{17}{2} \end{cases}$

5. $\begin{cases} \dfrac{4}{y} - \dfrac{3}{x^2} = 0 \\ \dfrac{1}{y} = \dfrac{3}{x} - 3 \end{cases}$

6. $\begin{cases} \dfrac{1}{y} = \dfrac{1}{x^2} + \dfrac{4}{x} - 2 \\ \dfrac{1}{x} - \dfrac{1}{y} + 8 = 0 \end{cases}$

7. $\begin{cases} \dfrac{1}{x} + \dfrac{2}{3y} = 0 \\ \dfrac{-3}{5x} + \dfrac{1}{y} = \dfrac{7}{10} \end{cases}$

8. $\begin{cases} \dfrac{2}{x} - \dfrac{1}{2y} = \dfrac{16}{5} \\ \dfrac{1}{3x} + \dfrac{2}{y} = \dfrac{-3}{10} \end{cases}$

9. $\begin{cases} x^2 + y^2 = 25 \\ y = \dfrac{x+5}{3} \end{cases}$

10. $\begin{cases} x^2 + y^2 = 4 \\ x + y = 3 \end{cases}$

11. $\begin{cases} x^2 + y^2 = 7 \\ x^2 - y^2 = 1 \end{cases}$

12. $\begin{cases} x^2 + 3y^2 = 30 \\ x^2 - 7y^2 = -10 \end{cases}$

13. $\begin{cases} x^2 + y^2 = 8 \\ xy = 4 \end{cases}$

14. $\begin{cases} xy = 1 \\ x^2 - y^2 = 0 \end{cases}$

15. $\begin{cases} x^2 + y^2 = 9 \\ x^2 + 3x + y^2 = 6 \end{cases}$

16. $\begin{cases} x^2 - y^2 = 16 \\ x^2 - y^2 - 2y = 2 \end{cases}$

17. $\begin{cases} x^2 - y^2 = 3 \\ x^2 - x - y^2 + 3y = 4 \end{cases}$

18. $\begin{cases} x^2 + 2x + y^2 - y = 2 \\ x^2 + y^2 = 5 \end{cases}$

6.4 Word Problems

We now return to the subject of word problems. In Section 1.6 six steps were listed to be used in solving such problems. Their generalization, to be used with systems, is as follows:

1. Represent the quantities asked for with algebraic symbols.
2. Find two statements of equality.

3. If necessary, represent other unknown quantities in terms of the variables already introduced.
4. Translate the statements of equality.
5. Solve the system and state your answers.
6. Make certain that your answers satisfy the conditions stated in the problem.

Just as before, it is essential that you *understand the problem.* Read the problem as many times as necessary to comprehend the information being given and the information being requested. Then proceed through the six steps, as illustrated in Examples 1 and 2.

Example 1
The sum of two numbers is 21, and the square of the larger minus the square of the smaller is 105. What are the two numbers?

Solution. Let

$$\begin{aligned} x &= \text{larger number} \\ y &= \text{smaller number.} \end{aligned}$$

The *two* statements of equality are "the sum of two numbers is 21" and "the square of the larger minus the square of the smaller is 105." Translating them into equations, we get the system

$$\begin{cases} x + y = 21 \\ x^2 - y^2 = 105. \end{cases}$$

We solve this system by substitution. From the first equation we get $y = 21 - x$. Upon substituting into the second, we have

$$x^2 - (21 - x)^2 = 105.$$

Solving for x, we get

$$\begin{aligned} x^2 - (441 - 42x + x^2) &= 105 \\ -441 + 42x &= 105 \\ 42x &= 546 \\ x &= 13. \end{aligned}$$

Thus $y = 21 - x = 8$, and the numbers are 13 and 8. To check, we note that $13 + 8 = 21$ and $13^2 - 8^2 = 105$. ■

Example 2
Two cars start together and travel in the same direction down an interstate, one going 50% faster than the other. At the end of 3 hours, they are 54 miles apart. How fast is each traveling?

Solution. We begin by letting

$$x = \text{speed of slower car (in mph)}$$
$$y = \text{speed of faster car (in mph).}$$

The two statements of equality are "the faster car is 50% faster than the slower," and "after 3 hours, the distance traveled by the faster car is 54 miles greater than the distance traveled by the slower car." Translating, we get

$$\begin{cases} y = \dfrac{3x}{2} \\ 3y = 54 + 3x. \end{cases}$$

Upon solving this system, we get

$$x = 36$$
$$y = 54.$$

Hence our answer is 36 mph and 54 mph. ■

Example 3

A rectangular box with a square base and open top has a surface area of 85 square feet. If the base material costs 20 cents/ft², the side material costs 10 cents/ft², and the total cost is $11, what are the dimensions?

Solution. Let

$$x = \text{base dimension (in feet)}$$
$$y = \text{height (in feet).}$$

The two statements of equality are "the surface area is 85 square feet" and "the total cost is $11." Translating the first statement, we have

$$x^2 + 4xy = 85,$$

where x^2 is the area of the square base and $4xy$ is the area of the four rectangular sides. To translate the second statement, we must multiply the area of the base by its cost and add the area of the sides times that cost. And, of course, we must use only one unit of cost, either dollars or cents. Using dollars, we have

$$0.2x^2 + 0.1(4xy) = 11.$$

Hence our system is

$$\begin{cases} x^2 + 4xy = 85 \\ 0.2x^2 + 0.1(4xy) = 11. \end{cases}$$

Solving this system gives

$$x = 5$$
$$y = 3.$$

Thus the box has a 5-foot base and a 3-foot height. ■

Problem Set 6.4

1. Find two numbers whose sum is 27 and whose product is 180.
2. Find two numbers whose sum is 5 and whose product is −24.
3. Find two consecutive positive integers such that the difference of their squares is 371.
4. Find two consecutive negative integers such that the difference of their squares is 197.
5. A man made two investments totaling $4000. After 1 year, one investment has earned 6% and the other 8%. If the total income from the two investments was $270, how much was invested at each rate?
6. A sum of $3000 was invested, part at 6% and the rest at 9%. How much was invested at each rate if the interest on each investment is the same?
7. The sum of two distinct numbers is 26. If the smaller number is reduced by 1 and the larger by 3, the product of the two resulting numbers is 120. What are the numbers?
8. Two different routes between two cities differ by 15 miles. If a person averages 55 mph on the longer route, he can make the trip in the same length of time as traveling 50 mph on the shorter route. How long is each route?
9. An airplane traveled 1170 miles in the same time that a car traveled 330 miles. If the rate of the plane was 140 mph greater than the rate of the car, what was the speed of each?
10. Two cars start together and travel in the same direction down an interstate, one going 50% faster than the other. At the end of 3 hours, they are 54 miles apart. How fast is each traveling?
11. How many liters of 40% acid solution must be added to 15% acid solution to obtain a 20-liter solution that is 20% acid?
12. How many grams of 60–40 solder (60% tin, 40% lead) and 40–60 solder must be mixed together to produce 500 grams of 45–55 solder?
13. A real estate dealer owned two apartments, one of which rented for $25 per month more than the other. If the dealer received a total of $4210 for a 12-month period, and the more expensive apartment was vacant for 2 months, what was the monthly rental on each?
14. The Newtons spent $1710 on carpeting in their new house. The carpeting installed in the living room and hall cost $20 per square yard, and that installed in the bedrooms cost $22 per square yard. If the bedrooms required 9 square yards more than the living room/hall, how much was spent on each type?
15. The sum of the perimeters of two squares is 100 cm. What are the dimensions of

the smaller square if the perimeter of the larger is three times the perimeter of the smaller?

16. The sum of the perimeters of two equilateral triangles is 33 inches. What is the length of each side of the larger triangle if each side of the larger is 7 inches longer than each side of the smaller?
17. The perimeter of a rectangle is 34 inches. If the width is doubled and the length increased by 3 inches, the perimeter increases by 16 inches. What are the dimensions of the original rectangle?
18. A piece of wire 8 inches long is cut into two pieces. Each piece is bent into a square. If the combined area is 4 in.^{2}, what are the dimensions of each square?
19. The legs of a right triangle are 8 and 12 inches. From a point on the hypotenuse, perpendiculars are drawn to the legs, forming a rectangle whose area is 18 in.^{2}. What are the dimensions of the rectangle?
20. The hypotenuse of a right triangle is 10 feet. If the shorter leg is decreased by 1 foot and the longer leg is increased by 4 feet, the hypotenuse is increased by 3 feet. Find the dimensions of the original triangle.
21. Each of two rectangles has an area of 24 ft^2. If one rectangle is 2 feet longer in length and 1 foot shorter in width, what is the greater perimeter?
22. Square A has a perimeter of k inches, and square B has a perimeter of $3k + 10$ inches. If the sum of the two perimeters is $7\frac{1}{2}$ feet, what is the area of square A?
23. A boat traveled 24 miles downstream and 24 miles back in a total of 3 hours and 20 minutes. If its speed downstream was 50% greater than its speed upstream, what is the speed of the current and the speed of the boat in still water?
24. At a constant temperature, the product of the pressure and volume of a gas is constant. For a certain gas, this constant is 42 when the pressure is measured in lb/in.^{2} and the volume is in in.^{3}. Find the pressure and volume of a gas whose pressure increases by 3 lb/in.^{2} when the volume decreases 7/15 in.^{3}.
25. A rectangular piece of tin of area 600 in.^{2} is made into a rectangular box of 1000 in.^{3} by cutting a 5-inch square from each corner and folding up the sides. Find the dimensions of the original rectangle.
26. Two pipes, A and B, empty a tank. Working together, it takes 1 hour and 12 minutes to empty the tank. If pipe A is used alone for 1 hour and then pipe B is opened, the tank is empty 48 minutes later. How long would it take each alone to empty the tank?
27. Going into the Houston series, a Cincinnati baseball player had a .280 batting average. During the series, he batted 15 times and raised his average to .300. How many hits did he have in the series?
28. In a given two-digit number, the tens digit is 3 more than the ones digit, and the number is 7 times the sum of its digits. What is the number?
29. Find the points of intersection of the graphs of f and f^{-1} if $f(x) = \sqrt{2x - 3} + 1$.
30. Show that the rational function $y = \dfrac{2x^2 + 3x - 2}{x - 1}$ doesn't cross its slant asymptote.
31. Find the equation of the parabola which passes through $(3, \frac{11}{2})$ and $(-3, 7)$ and has axis $x = 1$.
32. Find the equation of the hyperbola whose asymptotes are $y = \pm\frac{3}{5}x$ and whose foci are $(\pm 2, 0)$.
33. Find the equation of the ellipse with center at the origin which passes through the points $(2, \sqrt{15})$ and $(4/\sqrt{5}, 4)$.
34. Two hyperbolas are called confocal if they have the same foci. Show that distinct confocal hyperbolas cannot intersect each other.

6.5 Systems of Linear Equations in Three Variables

An equation in variables x, y, and z is called **linear** in x, y, and z provided that it is equivalent to

$$Ax + By + Cz + D = 0,$$

where A, B, C, and D are constants, with A, B, and C not all zero. Some examples include

$$2x - y + z - 3 = 0, \qquad z = 2x - y + 5, \qquad \text{and} \qquad 3x - y = 0.$$

On the other hand,

$$x^2 - y + z = 0 \qquad \text{and} \qquad 5x + y + \sqrt{z} = 1$$

are *not* linear.

A real **solution** of an equation in x, y, and z is an ordered triple (a, b, c) which makes the equation true when x is replaced by a, y is replaced by b, and z is replaced by c. The set of all solutions is called the **solution set,** and equations which have the same solution set are called **equivalent.**

The graphs of equations in three variables involve the use of a three-dimensional coordinate system and will not be discussed in this book. However, it is worth noting that the graph of a linear equation in three variables is a plane.

In this section we learn how to find the solution set of a system of three linear equations in three variables. A solution of the system is an ordered triple that is a solution of each of the equations in the system.

The technique for solving such systems is analogous to that used for systems of two equations in two variables. Theorems 6.1 (which tells us how to get equivalent systems by elimination) and 6.2 (which tells us how to get equivalent systems by substitution) may be naturally extended. We use elimination or substitution or both to eventually obtain an equation in a single variable. Upon solving this derived equation, we substitute to find the corresponding values of the remaining variables. This technique is illustrated in Examples 1 and 2.

Example 1
Find the solution set of

$$\begin{cases} 3x + y + z = 8 \\ 2x - y + 3z = 5 \\ 3x + 3y - 2z = 8. \end{cases}$$

Solution. Our first task is to decide which variable to eliminate first. We must combine two pairs of equations (from the three pairs: first and second, first and third, and second and third) to derive two equations in two variables. Although it appears to make little difference in this system, elimination of the y variable looks easiest. Adding the first equation to the second, we have

$$\begin{array}{rcrcrcr} 3x & + & y & + & z & = & 8 \\ 2x & - & y & + & 3z & = & 5 \\ \hline 5x & & & + & 4z & = & 13. \end{array}$$

Multiplying the first by -3 and adding it to the third, we get

$$\begin{array}{rcrcrcr} -9x & - & 3y & - & 3z & = & -24 \\ 3x & + & 3y & - & 2z & = & 8 \\ \hline -6x & & & - & 5z & = & -16. \end{array}$$

Replacing the second and third equations of the given system with the derived equations, we have the equivalent system

$$\begin{cases} 3x + y + z = 8 \\ 5x \qquad + 4z = 13 \\ -6x \qquad - 5z = -16. \end{cases}$$

Next we want to combine the two equations in x and z to eliminate either x or z. Multiplying the second equation by 5, the third by 4, and adding, we get

$$\begin{array}{rcrcr} 25x & + & 20z & = & 65 \\ -24x & - & 20z & = & -64 \\ \hline x & & & = & 1. \end{array}$$

Replacing the third equation with the derived one, we have the equivalent system

$$\begin{cases} 3x + y + z = 8 \\ 5x \qquad + 4z = 13 \\ x \qquad\qquad = 1. \end{cases}$$

Substituting 1 for x in the second equation enables us to solve for z. We get

$$\begin{aligned} 5 + 4z &= 13 \\ z &= 2. \end{aligned}$$

Substituting 1 for x and 2 for z in the first equation allows us to solve for y. We get

$$\begin{aligned} 3 + y + 2 &= 8 \\ y &= 3. \end{aligned}$$

Hence the solution set is $\{(1, 3, 2)\}$. ■

In the preceding example, upon obtaining two equations in x and z, we substituted them for the second and third equations of the given system. Note that we could just as well have replaced the first and second or first and third. We would still get the proper solution set.

Example 2
Find the solution set of

$$\begin{cases} x + y + z = 2 \\ x - y + 2z = -4 \\ z = 2x + y - 1. \end{cases}$$

Solution. Our first goal is to derive two equations in two variables. This may be done by substituting the expression $2x + y - 1$ for z in the first and second equations. We get

$$x + y + (2x + y - 1) = 2 \quad \text{and} \quad x - y + 2(2x + y - 1) = -4.$$

Simplified, these become

$$3x + 2y = 3 \quad \text{and} \quad 5x + y = -2.$$

Thus the original system is equivalent to

$$\begin{cases} 3x + 2y = 3 \\ 5x + y = -2 \\ z = 2x + y - 1. \end{cases}$$

To solve the pair of equations in x and y, let's use elimination. Multiplying the second by -2 and adding to the first, we get

$$\begin{array}{rl} 3x + 2y &= 3 \\ -10x - 2y &= 4 \\ \hline -7x \quad\quad &= 7. \end{array}$$

Thus $x = -1$. Upon replacing x by -1 in either of the equations in x and y only, we get $y = 3$. Finally, using the third equation of the system, we get $z = 2(-1) + 3 - 1 = 0$. Therefore, the solution set is $\{(-1, 3, 0)\}$. ■

Exercise

Find the solution set.

(a) $\begin{cases} -3x + 2y + 3z = 15 \\ 2x + y + z = -1 \\ 4x - 3y + 5z = 7 \end{cases}$ (b) $\begin{cases} x = 2y - z + 3 \\ 2x - 5y - z = 7 \\ -x - 2y + 5z = -17 \end{cases}$

Answers. (a) $\{(-2, 0, 3)\}$ (b) $\{(8, 2, -1)\}$ ■

As with systems of equations in two variables, it is possible for a system in three variables to have no solutions. Whenever the process of elimination or substitution results in an equation that is always false, the solution set is $\varnothing$.

It is also possible to have an infinite solution set, as seen in Example 3.

Example 3

Find the solution set of

$$\begin{cases} x - 2y + z = 0 \\ x - y - z = 2 \\ 5x - 8y + z = 4. \end{cases}$$

Solution. We may eliminate the variable z by adding the first and second and the second and third equations. We get

$$\begin{array}{r} x - 2y + z = 0 \\ x - y - z = 2 \\ \hline 2x - 3y \quad = 2 \end{array} \qquad \begin{array}{r} x - y - z = 2 \\ 5x - 8y + z = 4 \\ \hline 6x - 9y \quad = 6. \end{array}$$

Upon multiplying the first equation from this pair by -3 and adding it to the second, we have

$$\begin{array}{r} -6x + 9y = -6 \\ 6x - 9y = 6 \\ \hline 0 = 0. \end{array}$$

Hence the given system is equivalent to

$$\begin{cases} x - 2y + z = 0 \\ 2x - 3y = 2 \\ 0 = 0. \end{cases}$$

Since the third equation is always true, the solution set consists of all solutions to

$$x - 2y + z = 0 \quad \text{and} \quad 2x - 3y = 2,$$

and there are infinitely many solutions to this statement. (Given any values for x and y which satisfy the second equation, we could calculate the appropriate value for z so that the first equation would also be satisfied.) The solution set may be written $\{(x, y, z)|x - 2y + z = 0 \text{ and } 2x - 3y = 2\}$. An alternative way of writing the solution set results when we solve two of the variables in terms of the other. If we take the second equation and solve for y, we get

$$y = \frac{2x - 2}{3}.$$

Substituting this expression for y in the first equation and solving for z, we have

$$x - 2\left(\frac{2x - 2}{3}\right) + z = 0$$

$$z = \frac{4x - 4}{3} - x$$

$$z = \frac{x - 4}{3}.$$

Hence the solution set may be written $\left\{\left(x, \frac{2x - 2}{3}, \frac{x - 4}{3}\right)\middle| x \in \mathrm{R}\right\}$. ■

Exercise
Find the solution set.

(a) $\begin{cases} x - 2y + z = 2 \\ -x + 3y + z = 0 \\ 2x - 3y + 4z = 1 \end{cases}$ (b) $\begin{cases} x + y - 3z = 3 \\ x - y + 2z = -2 \\ 3x - y + z = -1 \end{cases}$

Answers. (a) $\varnothing$ (b) $\{(x, 5x, 2x - 1)|x \in \mathrm{R}\}$ ■

Example 4
Three technicians, A, B, and C, can accomplish a certain task in 1 hour and 30 minutes when all three work together. If A and B only work together, then the job takes 2 hours and 24 minutes. If A and C only work together, the task takes 2 hours, How long does it take each technician alone to do the job?
Solution. Let

x = time required by A alone (in hours)
y = time required by B alone (in hours)
z = time required by C alone (in hours).

The *three* statements of equality are "A, B, and C can do the job in 1 hour and 30 minutes," "A and B can do the job in 2 hours and 24 minutes," and "A

and C can do the job in 2 hours.'' Translating, we get the system

$$\begin{cases} \dfrac{\frac{3}{2}}{x} + \dfrac{\frac{3}{2}}{y} + \dfrac{\frac{3}{2}}{z} = 1 \\ \dfrac{\frac{12}{5}}{x} + \dfrac{\frac{12}{5}}{y} = 1 \\ \dfrac{2}{x} + \dfrac{2}{z} = 1. \end{cases}$$

By letting $u = 1/x$, $v = 1/y$, $w = 1/z$, and simplifying, we get the system

$$\begin{cases} u + v + w = \dfrac{2}{3} \\ u + v = \dfrac{5}{12} \\ 2u + 2w = 1. \end{cases}$$

Solving this system gives $u = \frac{1}{4}$, $v = \frac{1}{6}$, and $w = \frac{1}{4}$. Thus $x = 4$, $y = 6$, and $z = 4$. So A can do the job alone in 4 hours, B in 6 hours, and C in 4 hours. ■

Problem Set 6.5

In Problems 1–18, find the solution set.

~ 1. $\begin{cases} 2x + y - 3z = -4 \\ 2x + 3y + z = 10 \\ -x + 2y + z = 3 \end{cases}$

2. $\begin{cases} x - 2y + z = -3 \\ 3x + 2y + 3z = -1 \\ -5x - y - 2z = 10 \end{cases}$

— 3. $\begin{cases} 2x + y - 2z = -5 \\ -2x + 3y + 4z = -4 \\ x - 2y + z = 8 \end{cases}$

4. $\begin{cases} 5x + 10y - 2z = -4 \\ x + 2y + 3z = 6 \\ -x + 3y - 5z = -9 \end{cases}$

5. $\begin{cases} x + y + z = 2 \\ x - y - z = 0 \\ 2x + y - z = -1 \end{cases}$

6. $\begin{cases} x + y + z = 1 \\ x + y - 2z = 3 \\ x + 2y + z = 2 \end{cases}$

7. $\begin{cases} x = y - 3z + 1 \\ 4x + 2y - z = 3 \\ 3x - 2y + 4z = 0 \end{cases}$

8. $\begin{cases} 3x - 2y - z = 7 \\ z = x + 2y - 5 \\ -x + 4y + 2z = -4 \end{cases}$

9. $\begin{cases} 3x - y + 2z = -5 \\ 2x - y + 3z = 14 \\ y + z = -4 \end{cases}$

10. $\begin{cases} 2x - y = 6 \\ 4x + y - 3z = -4 \\ -2y + 3z = 11 \end{cases}$

11. $\begin{cases} -x + y - z = 1 \\ -x + 3y + z = 3 \\ x + 2y + 4z = 2 \end{cases}$

12. $\begin{cases} x = 2y - z + 1 \\ 2x - 3y + z = 5 \\ -x - 2y + 3z = -13 \end{cases}$

13. $\begin{cases} 2x + y + 3z = 6 \\ 6x + y + z = 1 \\ -4x - y - 2z = 3 \end{cases}$

14. $\begin{cases} x + y + z = 0 \\ y = \dfrac{2x - z + 1}{5} \\ 3x - 5y + 2z = \dfrac{4}{5} \end{cases}$

15. $\begin{cases} 2x + y - 2z = 3 \\ 3x - 2y + 5z = 7 \\ -4x + 3y + 10z = -27 \end{cases}$

16. $\begin{cases} 2x - 3y + 2z = 7 \\ x + 2y + 5z = -3 \\ -3x + 3y + 4z = 1 \end{cases}$

17. $\begin{cases} \dfrac{1}{x} + \dfrac{2}{y} + \dfrac{1}{z} = 2 \\ \dfrac{2}{x} - \dfrac{3}{y} - \dfrac{2}{z} = 2 \\ \dfrac{4}{x} + \dfrac{1}{y} + \dfrac{3}{z} = -3 \end{cases}$

18. $\begin{cases} \dfrac{2}{x} + \dfrac{3}{y} - \dfrac{1}{z} = -2 \\ \dfrac{1}{x} + \dfrac{2}{y} + \dfrac{1}{z} = 3 \\ \dfrac{3}{x} - \dfrac{1}{y} - \dfrac{2}{z} = 1 \end{cases}$

19. Last year a man invested a total of \$20,000 in three different ways. One of the investments was in bonds that paid 8% annually; one was in a savings account that paid 5%; and the other was in stocks that paid 10%. The total yearly income from the three investments was \$1600. If the income from the bonds was equal to the sum of incomes from the stocks and the savings account, how much was invested each way?
20. The sum of the digits of a certain three-digit number is 18. The tens digit is equal to the sum of the hundreds digit and the ones digit. And if the ones and tens digits are reversed, the resulting number is 18 less than the original. What is the number?
21. Find the equation of the circle which passes through $(0, 2)$, $(6, -6)$, and $(-1, -5)$. (Recall that the general equation of a circle is $x^2 + y^2 + ax + by + c = 0$.)
22. Find the equation of the parabola with vertical axis which passes through $(1, -2)$, $(-1, -4)$, and $(2, 5)$. (Recall that the general equation of such a parabola is $y = ax^2 + bx + c$.)
23. Pipe A and pipe B may be used to fill a certain tank. If the pipes are used simultaneously and the drain is open, it takes 2 hours and 24 minutes to fill the tank. If the drain is closed and both pipes are used, it takes only 1 hour and 20 minutes. If pipe A only is used and the drain is open, it takes 6 hours to fill the tank. How long does it take each pipe alone to fill the tank when the drain is closed, and how long does it take to drain a full tank?

6.6 Matrix Methods

Consider the following system of equations:

$$\begin{cases} x + y - 3z = -10 \\ -x - 2y + z = 3 \\ -2x + y + 3z = 14. \end{cases} \tag{6-6}$$

Our first task is to combine pairs of equations to eliminate one of the variables. We may eliminate x by adding the first to the second and by adding two times the first to the third. We get the equivalent system

$$\begin{cases} x + y - 3z = -10 \\ - y - 2z = -7 \\ 3y - 3z = -6. \end{cases}$$

We may eliminate y by multiplying the second equation of the preceding system by three and adding it to the third. We have the system

$$\begin{cases} x + y - 3z = -10 \\ - y - 2z = -7 \\ - 9z = -27. \end{cases} \tag{6-7}$$

From this last system, we determine the value of z and upon substituting, the corresponding values for y and x.

Now the solution of a system of linear equations is determined by the coefficients and the constants. So if each equation is written in an orderly fashion, such as

$$ax + by + cz = d,$$

we can simplify our calculations by omitting the variables. A convenient way of doing this is to use a matrix. A **matrix** is an array of numbers arranged in horizontal rows and vertical columns. For example,

$$\begin{pmatrix} 2 & 3 \\ -1 & 4 \end{pmatrix}, \quad \begin{pmatrix} 2 & 1 & 3 \\ -1 & -1 & 2 \end{pmatrix}, \quad \text{and} \quad \begin{pmatrix} 0 & 6 & 5 \\ 1 & -5 & 2 \\ 7 & 4 & -8 \end{pmatrix}$$

are matrices. The numbers in the matrix are called the elements. If a matrix has m rows and n columns, then we have an $m \times n$ (read "m by n") matrix. The matrices just listed are 2×2, 2×3, and 3×3, respectively.

Double-subscript notation is convenient for representing the elements of a matrix. The symbol a_{ij} refers to the element in the ith row and jth column. Thus an $m \times n$ matrix may be denoted

$$\begin{pmatrix} a_{11} & a_{12} & a_{13} & \cdots & a_{1n} \\ a_{21} & a_{22} & a_{23} & \cdots & a_{2n} \\ a_{31} & a_{32} & a_{33} & \cdots & a_{3n} \\ \cdot & & & & \\ \cdot & & & & \\ \cdot & & & & \\ a_{m1} & a_{m2} & a_{m3} & \cdots & a_{mn} \end{pmatrix}.$$

The elements a_{11}, a_{22}, a_{33}, . . . form the **main diagonal** of the matrix.

Returning to system (6-6), we will represent the system with the following matrix, called the **augmented matrix** of the system.

$$\begin{pmatrix} 1 & 1 & -3 & -10 \\ -1 & -2 & 1 & 3 \\ -2 & 1 & 3 & 14 \end{pmatrix}.$$

And since system (6-7) may be written

$$\begin{cases} x + y - 3z = -10 \\ 0x - y - 2z = -7 \\ 0x + 0y - 9z = -27 \end{cases}$$

its augmented matrix is

$$\begin{pmatrix} 1 & 1 & -3 & -10 \\ 0 & -1 & -2 & -7 \\ 0 & 0 & -9 & -27 \end{pmatrix}.$$

To solve a system of linear equations with matrices, each equation must be put in the form $ax + by + cz = d$. The augmented matrix is formed with this assumption. Once the augmented matrix is determined, we combine the rows in such a way to obtain a matrix which represents an equivalent system and which has zeros below the main diagonal. The rules for combining rows are listed in the following theorem.

Theorem 6.3

Given an augmented matrix of a system of linear equations, each of the following produces a matrix that represents an equivalent system.
(i) *Interchanging any two rows.*
(ii) *Replacing a row by a nonzero multiple of itself.*
(iii) *Replacing a row by the sum of itself and the multiple of another row.*

These operations, called **elementary row operations,** are possible because they represent manipulations with equations which produce equivalent systems. Taken together they mean that we may multiply any row by a nonzero constant, add it to the nonzero multiple of another row, and replace either of the two original rows with the result. The matrix method is illustrated in Example 1.

Example 1
Use the matrix method to find the solution set of

$$\begin{cases} 2x + y + z = 3 \\ x + 4y - 2z = -3 \\ -5x - 2y + 3z = -14. \end{cases}$$

Solution. The augmented matrix is

$$\begin{pmatrix} 2 & 1 & 1 & 3 \\ 1 & 4 & -2 & -3 \\ -5 & -2 & 3 & -14 \end{pmatrix}.$$

We first want to get two zeros in the first column. If we multiply the second row by -2, add it to the first, and replace the second, we get

$$\begin{pmatrix} 2 & 1 & 1 & 3 \\ 0 & -7 & 5 & 9 \\ -5 & -2 & 3 & -14 \end{pmatrix}.$$

Next we multiply the first row by 5, the third row by 2, add them together, and replace the third row. We get

$$\begin{pmatrix} 2 & 1 & 1 & 3 \\ 0 & -7 & 5 & 9 \\ 0 & 1 & 11 & -13 \end{pmatrix}.$$

To get the final zero below the main diagonal, we multiply the third row by 7, add it to the second, and replace the third. We get

$$\begin{pmatrix} 2 & 1 & 1 & 3 \\ 0 & -7 & 5 & 9 \\ 0 & 0 & 82 & -82 \end{pmatrix}.$$

This matrix represents the system

$$\begin{cases} 2x + y + z = 3 \\ \quad - 7y + 5z = 9 \\ \qquad 82z = -82. \end{cases}$$

From this last system we see that $z = -1$. After substituting, we get $y = -2$ and $x = 3$. Hence the solution set is $\{(3, -2, -1)\}$. ■

Example 2

Use the matrix method to find the solution set of

$$\begin{cases} 2y - w = 2 \\ 2x - 3y + z - 2w = -5 \\ x + y + 2z = -3 \\ - z - 3w = 5. \end{cases}$$

Solution. The augmented matrix is

$$\begin{pmatrix} 0 & 2 & 0 & -1 & 2 \\ 2 & -3 & 1 & -2 & -5 \\ 1 & 1 & 2 & 0 & -3 \\ 0 & 0 & -1 & -3 & 5 \end{pmatrix}.$$

After interchanging the first and third rows, we multiply the (new) first row by -2, add it to the second row, and replace the second row. We get

$$\begin{pmatrix} 1 & 1 & 2 & 0 & -3 \\ 0 & -5 & -3 & -2 & 1 \\ 0 & 2 & 0 & -1 & 2 \\ 0 & 0 & -1 & -3 & 5 \end{pmatrix}.$$

Multiplying the second row by 2, the third row by 5, adding the two, and replacing the third row, we get

$$\begin{pmatrix} 1 & 1 & 2 & 0 & -3 \\ 0 & -5 & -3 & -2 & 1 \\ 0 & 0 & -6 & -9 & 12 \\ 0 & 0 & -1 & -3 & 5 \end{pmatrix}.$$

Finally, we multiply the last row by -6, add it to the third row, and replace the last row. We have

$$\begin{pmatrix} 1 & 1 & 2 & 0 & -3 \\ 0 & -5 & -3 & -2 & 1 \\ 0 & 0 & -6 & -9 & 12 \\ 0 & 0 & 0 & 9 & -18 \end{pmatrix}.$$

This matrix represents the system

$$\begin{cases} x + y + 2z = -3 \\ \quad -5y - 3z - 2w = 1 \\ \quad\quad -6z - 9w = 12 \\ \quad\quad\quad 9w = -18. \end{cases}$$

We get $w = -2$, $z = 1$, $y = 0$, and $x = -5$. Hence the solution set, containing an ordered 4-tuple, is $\{(-5, 0, 1, -2)\}$. ■

Exercise

Use the matrix method to find the solution set of

(a) $\begin{cases} -3x + 2y + 2z = -5 \\ x + y + 5z = 2 \\ 5x - 3y - z = 6 \end{cases}$ (b) $\begin{cases} 2y - z = 4 \\ x - 2y + 3z - w = -10 \\ -2x + y + w = 9 \\ -2z + 3w = 15 \end{cases}$

Answers. (a) $\{(5, 7, -2)\}$ (b) $\{(-1, 2, 0, 5)\}$ ■

Problem Set 6.6

Use the matrix method to find the solution set.

1. $\begin{cases} x + 2y - 2z = 16 \\ -x + y + 3z = -11 \\ x + 5y + 2z = 9 \end{cases}$

2. $\begin{cases} x + 3y + z = 7 \\ x + y + 4z = 18 \\ -x - y + z = 7 \end{cases}$

3. $\begin{cases} x + 2y + 3z = 2 \\ -x + 3y + 5z = 20 \\ 5x + 10y + 2z = 27 \end{cases}$

4. $\begin{cases} x + y + z = 4 \\ x + y - 2z = -11 \\ 2x + 3y + z = 5 \end{cases}$

5. $\begin{cases} x + y - z = 9 \\ x - y - z = 5 \\ 2x + y - z = 12 \end{cases}$

6. $\begin{cases} x + 3y - 2z = 15 \\ -2x - 4y + 5z = -32 \\ x + 4y - 3z = 23 \end{cases}$

7. $\begin{cases} x + y + z = 6 \\ -x + 3y + 2z = 20 \\ x - 4y - z = -23 \end{cases}$

8. $\begin{cases} x + 2y - z = -5 \\ 3x + 2y + 3z = -7 \\ 5x - y - 2z = -30 \end{cases}$

9. $\begin{cases} 2x + y - 2z = 5 \\ -2x + 3y + 4z = 4 \\ x + 2y + z = 4 \end{cases}$

10. $\begin{cases} 2x + y + 2z = -1 \\ -3x + 2y - 5z = -7 \\ -4x + 3y + 10z = -27 \end{cases}$

11. $\begin{cases} x + y = 2z + 1 \\ x + z = y + 4 \\ 6z = 6x + 2y - 11 \end{cases}$

12. $\begin{cases} y + z = 1 - x \\ x = 5 + z + y \\ 2x + z = 4 - y \end{cases}$

13. $\begin{cases} 2y + z = -5 \\ x - y + 2z - w = 0 \\ 2x + y + w = 13 \\ 3z - w = -8 \end{cases}$

14. $\begin{cases} x - 2y + z - 3w = 12 \\ x + 2y - z + 2w = -13 \\ -2x - y - 2z + w = -9 \\ y + z - w = 8 \end{cases}$

6.7 Determinants

A *square* matrix is one which has the same number of columns as rows. In this section we assign a value to each such matrix. To do this, we must first consider a 2×2 matrix. Let

$$A = \begin{pmatrix} a & b \\ c & d \end{pmatrix}.$$

Then the **determinant** of A, denoted det A or $\begin{vmatrix} a & b \\ c & d \end{vmatrix}$, is the number $ad - bc$.

That is,

$$\det A = \begin{vmatrix} a & b \\ c & d \end{vmatrix} = ad - bc.$$

Thus the determinant of a matrix is a number (and *not* a matrix). As indicated in Figure 6.3, the determinant of a 2×2 matrix is obtained by taking the product of the elements in the upper left–lower right diagonal and subtracting the product of the elements in the lower left–upper right diagonal.

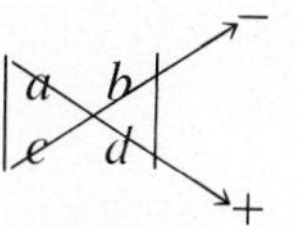

Figure 6.3

Example 1
Evaluate the determinants.

(a) $\begin{vmatrix} 2 & 3 \\ 1 & 5 \end{vmatrix}$ (b) $\begin{vmatrix} -3 & 1 \\ -5 & 3 \end{vmatrix}$

Solution. By the definition we have

(a) $\begin{vmatrix} 2 & 3 \\ 1 & 5 \end{vmatrix} = 10 - 3 = 7$

(b) $\begin{vmatrix} -3 & 1 \\ -5 & 3 \end{vmatrix} = -9 - (-5) = -9 + 5 = -4$ ■

Exercise
Evaluate the determinants.

(a) $\begin{vmatrix} 6 & 1 \\ 4 & 2 \end{vmatrix}$ (b) $\begin{vmatrix} 2 & -1 \\ 4 & -5 \end{vmatrix}$

Answers. (a) 8 (b) -6 ■

A third-order determinant (the determinant of a 3×3 matrix) will be defined in terms of second-order determinants (2×2 matrices). To do this, we must first consider some additional concepts. Let

$$A = \begin{pmatrix} a_{11} & a_{12} & a_{13} \\ a_{21} & a_{22} & a_{23} \\ a_{31} & a_{32} & a_{33} \end{pmatrix}.$$

The **minor** of a_{ij} (the element in the ith row and jth column), denoted M_{ij}, is the determinant of the matrix derived by deleting the ith row and jth column. Thus M_{12}, the minor of a_{12}, is the determinant of the matrix derived by deleting the first row and second column.

$$\begin{pmatrix} \not{a}_{11} & \not{a}_{12} & \not{a}_{13} \\ a_{21} & \not{a}_{22} & a_{23} \\ a_{31} & \not{a}_{32} & a_{33} \end{pmatrix}.$$

Hence,

$$M_{12} = \begin{vmatrix} a_{21} & a_{23} \\ a_{31} & a_{33} \end{vmatrix}.$$

The **cofactor** of a_{ij}, denoted A_{ij}, is defined by

$$A_{ij} = (-1)^{i+j} M_{ij}.$$

So

$$A_{12} = (-1)^{1+2} M_{12} = -\begin{vmatrix} a_{21} & a_{23} \\ a_{31} & a_{33} \end{vmatrix}.$$

Note that the cofactor of an element is either the same as its minor or the negative of its minor. It depends on the element's position. If the sum of an element's row and column numbers is even, then $(-1)^{i+j} M_{ij} = M_{ij}$. If the sum of its row and column numbers is odd, then $(-1)^{i+j} M_{ij} = -M_{ij}$.

Example 2

For the matrix $\begin{pmatrix} -3 & 1 & 0 \\ 5 & 2 & -4 \\ 1 & -2 & -1 \end{pmatrix}$ determine the cofactors (a) A_{13}; (b) A_{21}.

Solution. (a) Deleting the first row and third column, we get the matrix

$$\begin{pmatrix} \not{-3} & \not{1} & \not{0} \\ 5 & 2 & \not{-4} \\ 1 & -2 & \not{-1} \end{pmatrix}.$$

Thus, using the definition, we get

$$A_{13} = (-1)^{1+3}\begin{vmatrix} 5 & 2 \\ 1 & -2 \end{vmatrix} = -10 - 2 = -12.$$

(b) Deleting the second row and first column, we have the matrix

$$\begin{pmatrix} -3 & 1 & 0 \\ 5 & 2 & 4 \\ 1 & -2 & -1 \end{pmatrix}.$$

Hence we have

$$A_{21} = (-1)^{2+1}\begin{vmatrix} 1 & 0 \\ -2 & -1 \end{vmatrix} = -1(-1 - 0) = 1.$$

We are now in a position to define third-order determinants. Let

$$A = \begin{pmatrix} a_{11} & a_{12} & a_{13} \\ a_{21} & a_{22} & a_{23} \\ a_{31} & a_{32} & a_{33} \end{pmatrix}.$$

Then $\det A = a_{11}A_{11} + a_{12}A_{12} + a_{13}A_{13}$. That is, the determinant is the result obtained by multiplying each element of the first row by its cofactor and adding. ■

Example 3

Evaluate $\begin{vmatrix} -2 & 1 & 5 \\ 1 & 2 & 3 \\ 4 & -3 & 0 \end{vmatrix}$.

Solution. Using the definition, we have

$$\begin{vmatrix} -2 & 1 & 5 \\ 1 & 2 & 3 \\ 4 & -3 & 0 \end{vmatrix} = -2(-1)^{1+1}\begin{vmatrix} 2 & 3 \\ -3 & 0 \end{vmatrix} + 1(-1)^{1+2}\begin{vmatrix} 1 & 3 \\ 4 & 0 \end{vmatrix}$$

$$+ 5(-1)^{1+3}\begin{vmatrix} 1 & 2 \\ 4 & -3 \end{vmatrix}$$

$$= -2(9) - (-12) + 5(-11) = -61.$$ ■

Exercise

Evaluate $\begin{vmatrix} 5 & 4 & -1 \\ -1 & 0 & 2 \\ 3 & 5 & 1 \end{vmatrix}$.

Answer. -17 ■

With third-order determinants defined, we can extend the definitions of minor and cofactor to 4×4 matrices, where minors and cofactors will be third-order determinants. We may then extend the definition of determinant to 4×4 matrices. If

$$A = \begin{pmatrix} a_{11} & a_{12} & a_{13} & a_{14} \\ a_{21} & a_{22} & a_{23} & a_{24} \\ a_{31} & a_{32} & a_{33} & a_{34} \\ a_{41} & a_{42} & a_{43} & a_{44} \end{pmatrix},$$

then $\det A = a_{11}A_{11} + a_{12}A_{12} + a_{13}A_{13} + a_{14}A_{14}$.

Example 4

Find $\det A$ if $A = \begin{pmatrix} 2 & 0 & 0 & 1 \\ -3 & 1 & 0 & 5 \\ 1 & 2 & -1 & 0 \\ 4 & -1 & 0 & 3 \end{pmatrix}$.

Solution. From the definition we have

$$\det A = 2(-1)^2 \begin{vmatrix} 1 & 0 & 5 \\ 2 & -1 & 0 \\ -1 & 0 & 3 \end{vmatrix} + 0(-1)^3 \begin{vmatrix} -3 & 0 & 5 \\ 1 & -1 & 0 \\ 4 & 0 & 3 \end{vmatrix}$$
$$+ 0(-1)^4 \begin{vmatrix} -3 & 1 & 5 \\ 1 & 2 & 0 \\ 4 & -1 & 3 \end{vmatrix} + 1(-1)^5 \begin{vmatrix} -3 & 1 & 0 \\ 1 & 2 & -1 \\ 4 & -1 & 0 \end{vmatrix}.$$

Fortunately, two of the first row elements are zero. Hence we need not calculate their cofactors because the products will be zero regardless. Thus

$$\det A = 2\left(1 \begin{vmatrix} -1 & 0 \\ 0 & 3 \end{vmatrix} + 5 \begin{vmatrix} 2 & -1 \\ -1 & 0 \end{vmatrix}\right)$$
$$- 1\left(-3 \begin{vmatrix} 2 & -1 \\ -1 & 0 \end{vmatrix} - 1 \begin{vmatrix} 1 & -1 \\ 4 & 0 \end{vmatrix}\right)$$
$$= 2(-3 - 5) - (3 - 4) = -15.$$

■

With fourth-order determinants defined, we may extend the definitions of minor and cofactor to 5×5 matrices and consequently define fifth-order determinants. Continuing in this fashion, we will be able to define determinants of any order greater than two. The determinant of an $n \times n$ matrix (where $n > 2$) is the result obtained by multiplying each element of the first row by its cofactor (a determinant of order $n - 1$) and adding.

The following theorem serves as an extremely valuable tool in evaluating determinants.

Theorem 6.4

Let A be an $n \times n$ matrix. Then
(i) $\det A = a_{i1}A_{i1} + a_{i2}A_{i2} + \cdots + a_{in}A_{in}$ *for any i*
(ii) $\det A = a_{1j}A_{1j} + a_{2j}A_{2j} + \cdots + a_{nj}A_{nj}$ *for any j*

Part (i) of this theorem says that we don't have to use the first row to evaluate the determinant. We may use any row! We simply multiply each element of one of the rows by its cofactors and add the results. Part (ii) says that we can even use a column to evaluate the determinant!

Let's look again at the determinant of Example 4.

$$A = \begin{pmatrix} 2 & 0 & 0 & 1 \\ -3 & 1 & 0 & 5 \\ 1 & 2 & -1 & 0 \\ 4 & -1 & 0 & 3 \end{pmatrix}.$$

By selecting the row or column which has the most zeros, we will minimize the number of determinants to be evaluated in the calculation. (The product of 0 and its cofactor is 0.) Thus if we choose the third column for matrix A, we get

$$\begin{aligned} \det A &= (-1)(-1)^{3+3} \begin{vmatrix} 2 & 0 & 1 \\ -3 & 1 & 5 \\ 4 & -1 & 3 \end{vmatrix} \\ &= -\left(2 \begin{vmatrix} 1 & 5 \\ -1 & 3 \end{vmatrix} + 1 \begin{vmatrix} -3 & 1 \\ 4 & -1 \end{vmatrix}\right) \\ &= -(16 - 1) = -15. \end{aligned}$$

Note also that if any matrix has an all-zero row or all-zero column, its determinant is zero.

A device may be used to evaluate third-order determinants. We write the first two columns of the matrix to the immediate right. Then we determine six diagonal products, adding or subtracting as indicated.

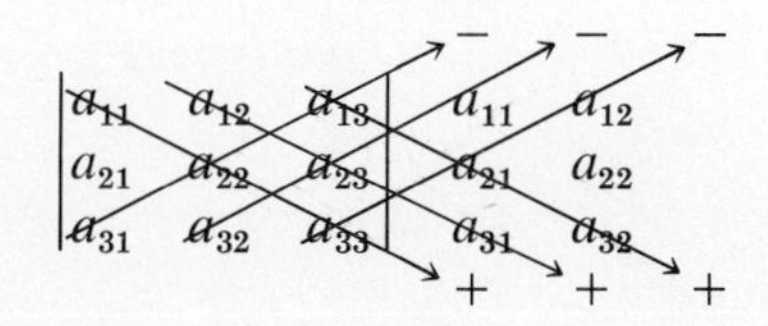

There is no such similar method for evaluating higher-order determinants. The cofactors must be used.

Example 5

Evaluate $\begin{vmatrix} -1 & 4 & 3 \\ 2 & -1 & 5 \\ -5 & 1 & 6 \end{vmatrix}$.

Solution. Using the diagonal method just explained, we get

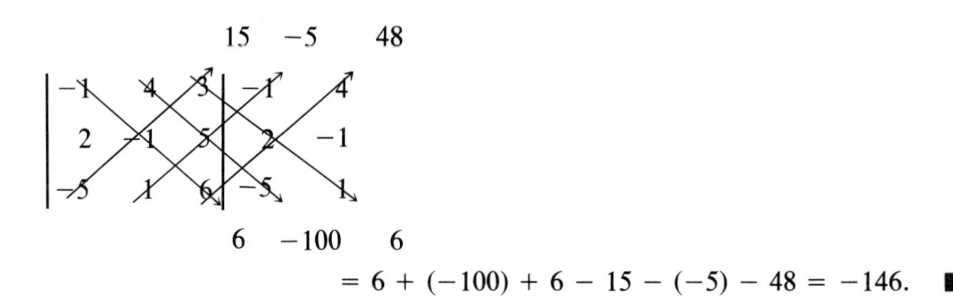

$= 6 + (-100) + 6 - 15 - (-5) - 48 = -146.$ ■

Exercise

Evaluate $\begin{vmatrix} 3 & -1 & 1 \\ 2 & 1 & 5 \\ 3 & -2 & 4 \end{vmatrix}$.

Answer. 28 ■

Problem Set 6.7

In Problems 1–6, evaluate the determinant.

1. $\begin{vmatrix} 3 & 1 \\ 2 & 5 \end{vmatrix}$

2. $\begin{vmatrix} 2 & 6 \\ 3 & 4 \end{vmatrix}$

3. $\begin{vmatrix} -3 & 5 \\ -2 & 1 \end{vmatrix}$

4. $\begin{vmatrix} 6 & 2 \\ -3 & -4 \end{vmatrix}$

5. $\begin{vmatrix} 0 & 5 \\ 6 & 2 \end{vmatrix}$

6. $\begin{vmatrix} -2 & 0 \\ 3 & -2 \end{vmatrix}$

For Problems 7 and 8, let

$$A = \begin{vmatrix} -1 & 2 & 5 \\ 6 & 1 & 3 \\ -2 & 1 & 4 \end{vmatrix}.$$

7. Determine the minors: M_{12} and M_{23}.

8. Determine the cofactors: A_{13} and A_{32}.

In Problems 9–16, evaluate the determinant.

9. $\begin{vmatrix} 2 & 1 & -4 \\ -3 & 1 & -2 \\ 1 & 2 & 3 \end{vmatrix}$ **10.** $\begin{vmatrix} 1 & -1 & 2 \\ 4 & -1 & 6 \\ -5 & 2 & 3 \end{vmatrix}$ **11.** $\begin{vmatrix} 6 & 1 & -3 \\ 4 & -5 & 1 \\ 7 & 2 & 1 \end{vmatrix}$

12. $\begin{vmatrix} -7 & 1 & 9 \\ 2 & 1 & 3 \\ 2 & 2 & -4 \end{vmatrix}$ **13.** $\begin{vmatrix} 7 & 0 & -5 \\ -3 & 2 & 1 \\ 6 & 1 & -3 \end{vmatrix}$ **14.** $\begin{vmatrix} 5 & -1 & 1 \\ 2 & 2 & -3 \\ -1 & 0 & 9 \end{vmatrix}$

15. $\begin{vmatrix} -2 & 1 & 5 & 2 \\ 1 & 0 & -2 & 0 \\ 3 & 1 & 2 & -2 \\ 8 & 1 & -1 & 4 \end{vmatrix}$ **16.** $\begin{vmatrix} 1 & 5 & 0 & 6 \\ -1 & 2 & -2 & -3 \\ 6 & 1 & 3 & 8 \\ 5 & 2 & 0 & -4 \end{vmatrix}$

In Problems 17–26, verify the equation by evaluating the determinants.

17. $\begin{vmatrix} a & b \\ c & d \end{vmatrix} = -\begin{vmatrix} c & d \\ a & b \end{vmatrix}$

18. $\begin{vmatrix} a & b \\ c & d \end{vmatrix} = -\begin{vmatrix} b & a \\ d & c \end{vmatrix}$

19. $\begin{vmatrix} ka & kb \\ c & d \end{vmatrix} = k\begin{vmatrix} a & b \\ c & d \end{vmatrix}$

20. $\begin{vmatrix} ka & b \\ kc & d \end{vmatrix} = k\begin{vmatrix} a & b \\ c & d \end{vmatrix}$

21. $\begin{vmatrix} a & b \\ c & d \end{vmatrix} = \begin{vmatrix} a & b \\ ka + c & kb + d \end{vmatrix}$

22. $\begin{vmatrix} a & b \\ c & d \end{vmatrix} = \begin{vmatrix} a & ka + b \\ c & kc + d \end{vmatrix}$

23. $\begin{vmatrix} a & b & c \\ a & b & c \\ d & e & f \end{vmatrix} = 0$

24. $\begin{vmatrix} a & a & d \\ b & b & e \\ c & c & f \end{vmatrix} = 0$

25. $\begin{vmatrix} a & b & c \\ ka & kb & kc \\ d & e & f \end{vmatrix} = 0$

26. $\begin{vmatrix} a & ka & d \\ b & kb & e \\ c & kc & f \end{vmatrix} = 0$

27. Show that the equation of the line through the points (x_1, y_1) and (x_2, y_2) $(x_1 \neq x_2)$ is

$$\begin{vmatrix} x & y & 1 \\ x_1 & y_1 & 1 \\ x_2 & y_2 & 1 \end{vmatrix} = 0.$$

6.8 Solving Linear Systems Using Determinants

In this section we learn how to use determinants to solve a system of n linear equations in n variables. We begin with the simplest of such systems: two linear equations in two unknowns. Consider the following system, written in **standard form,**

$$\begin{cases} ax + by = c \\ dx + ey = f, \end{cases}$$

where a, b, c, d, e, and f are constants, with a and d not both zero, and b and e not both zero. Without loss of generality, we may assume that $a \neq 0$. Let's solve this system by substitution. Solving for x in the first equation, we get

$$x = \frac{c - by}{a}. \tag{6-8}$$

Substituting for x in the second equation enables us to solve for y. We have

$$\frac{cd - bdy}{a} + ey = f$$

$$cd - bdy + aey = af$$

$$y(ae - bd) = af - cd$$

$$y = \frac{af - cd}{ae - bd} \qquad \text{if } ae - bd \neq 0.$$

Replacing y in equation (6-8) by this quotient and solving for x, we get

$$x = \frac{c - b \cdot \dfrac{af - cd}{ae - ba}}{a}$$

$$= \frac{c(ae - bd) - b(af - cd)}{a(ae - bd)}$$

$$= \frac{ace - abf}{a(ae - bd)}$$

$$= \frac{ce - bf}{ae - bd}.$$

We may write this result in terms of determinants. If we let

$$D = \begin{vmatrix} a & b \\ d & e \end{vmatrix}, \qquad D_x = \begin{vmatrix} c & b \\ f & e \end{vmatrix}, \qquad \text{and} \qquad D_y = \begin{vmatrix} a & c \\ d & f \end{vmatrix},$$

then

$$x = \frac{D_x}{D} \qquad \text{and} \qquad y = \frac{D_y}{D}.$$

Now the **coefficient matrix** of a system is the matrix consisting of the coefficients only. Thus D is the determinant of the coefficient matrix. And D_x is the determinant of the matrix obtained by replacing the first column of the coefficient matrix with the column of constants. Similarly, D_y is the determinant of the matrix obtained by replacing the second column of the coefficient matrix by the constants. In this way, determinants may be used to solve a system of linear equations, as illustrated in Example 1.

Example 1

Use determinants to find the solution set.

(a) $\begin{cases} 2x - 7y = -20 \\ -5x + 3y = 21 \end{cases}$ (b) $\begin{cases} 2x + 4y + 9 = 0 \\ x = 3y + 8 \end{cases}$

Solution. (a) We first evaluate D, the determinant of the coefficient matrix. We get

$$D = \begin{vmatrix} 2 & -7 \\ -5 & 3 \end{vmatrix} = -29.$$

Replacing the first column with the constants, we have

$$D_x = \begin{vmatrix} -20 & -7 \\ 21 & 3 \end{vmatrix} = 87.$$

And replacing the second column with the constants gives

$$D_y = \begin{vmatrix} 2 & -20 \\ -5 & 21 \end{vmatrix} = -58.$$

Therefore,

$$x = \frac{D_x}{D} = \frac{87}{-29} = -3 \quad \text{and} \quad y = \frac{D_y}{D} = \frac{-58}{-29} = 2.$$

Hence the solution set is $\{(-3, 2)\}$.

(b) Our first task is to put the system in standard form. We get

$$\begin{cases} 2x + 4y = -9 \\ x - 3y = 8. \end{cases}$$

Thus

$$D = \begin{vmatrix} 2 & 4 \\ 1 & -3 \end{vmatrix} = -10,$$

$$D_x = \begin{vmatrix} -9 & 4 \\ 8 & -3 \end{vmatrix} = -5,$$

$$D_y = \begin{vmatrix} 2 & -9 \\ 1 & 8 \end{vmatrix} = 25.$$

Therefore,

$$x = \frac{D_x}{D} = \frac{-5}{-10} = \frac{1}{2},$$

$$y = \frac{D_y}{D} = \frac{25}{-10} = \frac{-5}{2},$$

and the solution set is $\{(\frac{1}{2}, -\frac{5}{2})\}$. ■

Exercise

Use determinants to find the solution set.

(a) $\begin{cases} 2x - 3y = -16 \\ 5x + 7y = -11 \end{cases}$ (b) $\begin{cases} 2x + 3y - 4 = 0 \\ \qquad 9y = 11 - 8x \end{cases}$

Answers. (a) $\{(-5, 2)\}$ (b) $\left\{\left(\frac{-1}{2}, \frac{5}{3}\right)\right\}$ ■

As indicated in the following theorem, this method of determinants may be extended to n equations in n unknowns.

Theorem 6.5 (Cramer's Rule)

Consider a system of n linear equations in n variables $x_1, x_2, \ldots, x_n$, written in the following standard form:

$$\begin{cases} a_{11}x_1 + a_{12}x_2 + \cdots + a_{1n}x_n = k_1 \\ a_{21}x_1 + a_{22}x_2 + \cdots + a_{2n}x_n = k_2 \\ \quad\vdots \qquad\quad \vdots \qquad\qquad\quad \vdots \qquad \vdots \\ a_{n1}x_1 + a_{n2}x_2 + \cdots + a_{nn}x_n = k_n. \end{cases}$$

Let D be the determinant of the coefficient matrix:

$$D = \begin{vmatrix} a_{11} & a_{12} & \cdots & a_{1n} \\ a_{21} & a_{22} & \cdots & a_{2n} \\ \vdots & \vdots & & \vdots \\ a_{n1} & a_{n2} & \cdots & a_{nn} \end{vmatrix}.$$

And let D_{x_i} be the determinant of the matrix obtained by replacing the ith *column of the coefficient matrix by the column of constants:*

$$D_{x_i} = \begin{vmatrix} a_{11} & a_{12} & \cdots & k_1 & \cdots & a_{1n} \\ a_{21} & a_{22} & \cdots & k_2 & \cdots & a_{2n} \\ \cdot & \cdot & & \cdot & & \cdot \\ \cdot & \cdot & & \cdot & & \cdot \\ \cdot & \cdot & & \cdot & & \cdot \\ a_{n1} & a_{n2} & & k_n & \cdots & a_{nn} \end{vmatrix},$$

where the ith *column is the one containing $k_1, k_2, \ldots, k_n$. Then if $D \neq 0$,*

$$x_i = \frac{D_{x_i}}{D}, \qquad i = 1, 2, \ldots, n.$$

Example 1 was the application of Cramer's Rule to a system of two equations. In Example 2 we apply it to a system of three equations in three variables.

Example 2
Use Cramer's Rule to find the solution set of

$$\begin{cases} 2x + y - 2z = 0 \\ x - 3y + 2z = 6 \\ 5x - 2y + z = 7. \end{cases}$$

Solution. We have

$$D = \begin{vmatrix} 2 & 1 & -2 \\ 1 & -3 & 2 \\ 5 & -2 & 1 \end{vmatrix} = -15$$

$$D_x = \begin{vmatrix} 0 & 1 & -2 \\ 6 & -3 & 2 \\ 7 & -2 & 1 \end{vmatrix} = -10$$

$$D_y = \begin{vmatrix} 2 & 0 & -2 \\ 1 & 6 & 2 \\ 5 & 7 & 1 \end{vmatrix} = 30$$

$$D_z = \begin{vmatrix} 2 & 1 & 0 \\ 1 & -3 & 6 \\ 5 & -2 & 7 \end{vmatrix} = 5.$$

Therefore,

$$x = \frac{D_x}{D} = \frac{-10}{-15} = \frac{2}{3}$$

$$y = \frac{D_y}{D} = \frac{30}{-15} = -2$$

$$z = \frac{D_z}{D} = \frac{5}{-15} = \frac{-1}{3}.$$

Hence the solution set is $\{(\frac{2}{3}, -2, -\frac{1}{3})\}$. ■

Exercise
Use Cramer's Rule to find the solution set of

$$\begin{cases} 2x + y + 4z = 9 \\ x - y + z = -1 \\ -5x + 8y - 7z = 6. \end{cases}$$

Answer. $\left\{\left(\frac{-3}{2}, 2, \frac{5}{2}\right)\right\}$ ■

Note that Cramer's Rule may only be used on systems of *linear* equations, and the number of equations must equal the number of variables. Further, the determinant of the coefficient matrix must be nonzero. If $D = 0$, then the solution set is either empty or infinite. We must use methods previously described when $D = 0$.

Problem Set 6.8

In Problems 1–16, use Cramer's Rule to find the solution set.

1. $\begin{cases} 3x + y = 3 \\ -5x + y = 11 \end{cases}$

2. $\begin{cases} -x + 2y = 8 \\ 3x + y = -10 \end{cases}$

3. $\begin{cases} 2x + 3y = 5 \\ -x + 2y = 1 \end{cases}$

4. $\begin{cases} 5x + 3y = 9 \\ 7x + 2y = 17 \end{cases}$

5. $\begin{cases} 5x = 2y - 5 \\ 2x + 3y = 3 \end{cases}$

6. $\begin{cases} 3x = -4y \\ 5x - 11y = \dfrac{-53}{4} \end{cases}$

7. $\begin{cases} 2x - 4y = 5 \\ -3x + 6y = -3 \end{cases}$

8. $\begin{cases} 3x + y = -3 \\ 6x + 2y = -6 \end{cases}$

9. $\begin{cases} 2x - y + z = -1 \\ x - y - z = 0 \\ x + y + z = 2 \end{cases}$

10. $\begin{cases} x + y + z = -3 \\ x - 2y - z = 1 \\ x - y + 2z = 2 \end{cases}$

11. $\begin{cases} x + 2y = 3z - 4 \\ 3x + 2y + z = 10 \\ y = 2x + z - 3 \end{cases}$

12. $\begin{cases} x = 2y - z - 3 \\ 10 + 2x = -y - 5z \\ 3x + 2y + 3z = -1 \end{cases}$

13. $\begin{cases} y + z = -4 \\ 2x - y + 3z = 14 \\ 3x - y + 2z = -5 \end{cases}$

14. $\begin{cases} 2x - z = 6 \\ 4x - 3y + z = -4 \\ 3y - 2z = 11 \end{cases}$

15. $\begin{cases} -2x + 2y + z = -5 \\ 4x - 2y + 3z = -4 \\ x + y - 2z = 8 \end{cases}$

16. $\begin{cases} x + 3y + 2z = 6 \\ 5x - 2y + 10z = -4 \\ -x - 5y + 3z = -9 \end{cases}$

17. Use Cramer's Rule to solve for w.

$$\begin{cases} x - 2y + z - w = -10 \\ 3x + 2z + w = -1 \\ -2x + y - 2z - 2w = -6 \\ x - z + 3w = 8 \end{cases}$$

18. Use Cramer's Rule to solve for y.

$$\begin{cases} x + y + w = -3 \\ 3x + y - 2z + w = -4 \\ 4x + 6y - 3w = -1 \\ 5x - 3y + 3z + 2w = 6 \end{cases}$$

Chapter 6 Review

In Problems 1–12, find the solution set.

1. $\begin{cases} 2x + y = 1 \\ 3x - y = -11 \end{cases}$

2. $\begin{cases} 3x + 10y = 2 \\ 5x + 8y = 12 \end{cases}$

3. $\begin{cases} 4x + 3y = \dfrac{3}{4} \\ 4x - 2y = \dfrac{-11}{2} \end{cases}$

4. $\begin{cases} 5x - 4y = -13 \\ 7x + 3y = \dfrac{33}{10} \end{cases}$

5. $\begin{cases} x = \dfrac{5 - 2y}{3} \\ 6x + 5y = 4 \end{cases}$

6. $\begin{cases} y = 2x^2 + x \\ 3x + 2y = 0 \end{cases}$

7. $\begin{cases} x - 4y^2 + 3 = 0 \\ x + 3y + 1 = 0 \end{cases}$

8. $\begin{cases} \dfrac{2}{x} + \dfrac{3}{y} = \dfrac{1}{2} \\ \dfrac{1}{x} - \dfrac{2}{y} = \dfrac{-11}{12} \end{cases}$

9. $\begin{cases} \dfrac{1}{2x} - \dfrac{3}{y} = \dfrac{-23}{4} \\ \dfrac{1}{x} + \dfrac{1}{2y} = \dfrac{-2}{3} \end{cases}$

10. $\begin{cases} x^2 - 2y^2 = 1 \\ x + 2y = -1 \end{cases}$

11. $\begin{cases} x^2 - 2y^2 = 2 \\ x^2 + y^2 = 8 \end{cases}$

12. $\begin{cases} x^2 - 12x + y^2 = -32 \\ x^2 + y^2 = 9 \end{cases}$

13. A rectangle of perimeter 26 inches has length 5 inches greater than its width. What are its dimensions?

14. A rectangular box with a square base and open top has a surface area of 156 ft^2. If the base material cost $\$0.24/\text{ft}^2$, the side material cost $\$0.16/\text{ft}^2$, and the total cost was \$27.84, what are the dimensions?

15. For a certain two-digit number, twice the sum of the digits is 23 less than the number. If the digits are reversed, the resulting number is 45 more than the number. What is the number?

16. A civic club approved a project that would cost \$960. Before the project was completed, 16 new members joined and agreed to pay their share. If this resulted in a \$2 refund for the original members, what was the initial cost per member, and how many original members were there?

17. Find the equation of the hyperbola which passes through the points $\left(1, \frac{\sqrt{45}}{4}\right)$ and $\left(\sqrt{2}, \frac{3\sqrt{6}}{4}\right)$ and has foci on the y-axis.

In Problems 18 and 19, find the solution set.

18. $\begin{cases} 3x + 4y + 2z = 4 \\ 6x - 2y + z = 4 \\ 3x - 2y - z = 1 \end{cases}$

19. $\begin{cases} 2x - y + 3z = -10 \\ 5x + 4y + z = 1 \\ 3x + 2y - 5z = -7 \end{cases}$

In Problems 20 and 21, use the matrix method to find the solution set.

20. $\begin{cases} 4x + 2y - z = 3 \\ x - y + 3z = 1 \\ 3x - 2y + 4z = 0 \end{cases}$

21. $\begin{cases} 5x + 5y - z = 0 \\ x + 2y - 3z = 5 \\ -7x + y + 2z = -6 \end{cases}$

22. Use Cramer's Rule to find the solution set.

$$\begin{cases} 3x + 5y = 11 \\ -2x - y = 2 \end{cases}$$

23. Use Cramer's Rule to solve for y.

$$\begin{cases} 2x + 4y + 3z = -1 \\ 3x + 2y + 3z = -1 \\ x + 6y + 6z = 0 \end{cases}$$

Mathematical Induction, Binomial Theorem, and Sequences

7

This chapter is an introduction to several important concepts and theorems connected with the set of counting numbers.

7.1 Mathematical Induction

In order to establish many formulas, we must rely on an important principle known as the Principle of Mathematical Induction. This principle is based on the following axiom and theorem.

Axiom

Every (nonempty) set of positive integers has a smallest positive integer.

Theorem 7.1

Let S *be a set of positive integers. If*

(i) $1 \in S$, *and*

(ii) $k \in S$ *implies that* $k + 1 \in S$ *for any positive integer* k,

then $S = Z^+$.

Proof. Suppose that $S \neq Z^+$. Then the set $Z^+ - S$ is a nonempty set of positive integers. By the preceding axiom it has a smallest positive integer r. And since $1 \in S$, $1 \notin Z^+ - S$, and $r \geq 2$. Because r is the smallest integer in $Z^+ - S$, $r - 1 \in Z^+ - S$. And since $r - 1$ is a positive integer ($r \geq 2$), $r - 1 \in S$. By property (ii), $r - 1 + 1 = r \in S$. But this contradicts the statement that $r \in Z^+ - S$. Hence our supposition that $S \neq Z^+$ is incorrect, and $S = Z^+$. ■

Theorem 7.2 (Principle of Mathematical Induction)

For each positive integer n, let P_n be an associated statement. If
(i) *P_1 is true, and*
(ii) *the truth of P_k implies the truth of P_{k+1} for any positive integer k,*
then P_n is true for all positive integers n.

Proof. Let $S = \{n \in Z^+ | P_n \text{ is true}\}$. Because P_1 is true, $1 \in S$. Because the truth of P_k implies the truth of P_{k+1}, $k \in S$ implies that $k + 1 \in S$. By Theorem 7.1, $S = Z^+$. Hence P_n is true for all positive integers n. ■

The Principle of Mathematical Induction may be compared to an infinite set of dominoes. Let's suppose that these dominoes are lined up and are numbered 1, 2, 3, . . . If

1. the first domino falls, and
2. the falling of the kth domino causes the $k + 1$st domino to fall for any positive integer k,

then every domino will fall. That is, for all positive integers n, the nth domino will fall.

Let's see how this principle is applied.

Example 1
Let a and b be any real numbers. Prove that $(ab)^n = a^n b^n$ for all positive integers n.

Solution. Let S_n be the statement: $(ab)^n = a^n b^n$. We must show that S_1 is true and that the truth of S_k implies the truth of S_{k+1}. Since

$$(ab)^1 = ab = a^1 b^1,$$

S_1 is true. Now suppose that S_k is true. That is, suppose that

$$(ab)^k = a^k b^k. \tag{7-1}$$

We shall show that this implies that S_{k+1} is true. That is, we shall show that

$$(ab)^{k+1} = a^{k+1} b^{k+1}.$$

Now

$$(ab)^{k+1} = (ab)^k (ab).$$

Multiplying both sides of equation (7-1) by ab, we have

$$(ab)^k (ab) = a^k b^k (ab).$$

By the commutativity and associativity of multiplication, we get

$$a^k b^k (ab) = a^k a b^k b = a^{k+1} b^{k+1}.$$

From the three preceding equations we have

$$(ab)^{k+1} = a^{k+1} b^{k+1}.$$

By the Principle of Mathematical Induction, S_n is true for all positive integers n. That is, $(ab)^n = a^n b^n$ for all positive integers n. ■

Example 2
Prove that $n^2 + n$ is an even integer for all positive integers n.

Solution. Let S_n be the statement: $n^2 + n$ is an even integer. We must first show that S_1 is true. This is clear because

$$1^2 + 1 = 2.$$

Now suppose that S_k is true. That is, assume that $k^2 + k$ is an even integer. Hence there exists some integer a such that

$$k^2 + k = 2a.$$

We want to show that this implies that S_{k+1} is true. That is, we want to prove that

$$(k + 1)^2 + (k + 1) = 2b$$

for some integer b. We have

$$\begin{aligned}(k + 1)^2 + (k + 1) &= k^2 + 2k + 1 + k + 1 \\ &= k^2 + k + 2k + 2 \\ &= 2a + 2k + 2 \\ &= 2(a + k + 1).\end{aligned}$$

Since $a + k + 1$ is an integer, S_{k+1} is true. By mathematical induction, S_n is true for all positive integers n. That is, $n^2 + n$ is an even integer for all positive integers n. ■

Example 3
Prove that $n < 2^n$ for all positive integers n.

Solution. Let S_n be the statement: $n < 2^n$. Since

$$1 < 2^1,$$

S_1 is true. Now suppose that S_k is true. That is, assume that

$$k < 2^k.$$

We want to show that this implies that S_{k+1} is true. That is, we want to prove that

$$k + 1 < 2^{k+1}.$$

Because $k < 2^k$ and $k \geq 1$,

$$k + 1 \leq k + k = 2k < 2 \cdot 2^k = 2^{k+1}.$$

Hence S_{k+1} is true. By mathematical induction then, $n < 2^n$ for all positive integers n. ■

Example 4

Prove that $1 + 2 + 3 + \cdots + n = \dfrac{n(n + 1)}{2}$ for all positive integers n.

Solution. Let S_n be the statement: $1 + 2 + 3 + \cdots + n = \dfrac{n(n + 1)}{2}$. S_1 is true because

$$1 = \frac{1(1 + 1)}{2}.$$

Now suppose that S_k is true. That is, assume that

$$1 + 2 + 3 + \cdots + k = \frac{k(k + 1)}{2}.$$

We want to show that this implies that S_{k+1} is true. That is, we want to prove that

$$1 + 2 + 3 + \cdots + (k + 1) = \frac{(k + 1)(k + 1 + 1)}{2}.$$

We have

$$\begin{aligned} 1 + 2 + 3 + \cdots + (k + 1) &= (1 + 2 + 3 + \cdots + k) + (k + 1) \\ &= \frac{k(k + 1)}{2} + k + 1 \\ &= \frac{k(k + 1) + 2(k + 1)}{2} \end{aligned}$$

$$= \frac{(k + 1)(k + 2)}{2}$$

$$= \frac{(k + 1)(k + 1 + 1)}{2}.$$

Hence S_{k+1} is true. Therefore, $1 + 2 + 3 + \cdots + n = \dfrac{n(n + 1)}{2}$ for all positive integers n. ■

Problem Set 7.1

Prove that the statement is true for all positive integers n.

1. $2 + 4 + 6 + \cdots + 2n = n(n + 1)$
2. $1 + 3 + 5 + \cdots + (2n - 1) = n^2$
3. $1^3 + 2^3 + 3^3 + \cdots + n^3 = \dfrac{n^2(n + 1)^2}{4}$
4. $1^2 + 2^2 + 3^2 + \cdots + n^2 = \dfrac{n(n + 1)(2n + 1)}{6}$
5. $\left(\dfrac{a}{b}\right)^n = \dfrac{a^n}{b^n}$
6. $(a^x)^n = a^{xn}$
7. $\log x^n = n \log x,\ x > 0$
8. $x^n > 1$ for all $x > 1$
9. $n^3 + 2n$ is divisible by 3
10. $5^n - 1$ is divisible by 4
11. $2n \le 2^n$
12. The number of subsets of a set of n elements is 2^n.

7.2 Binomial Theorem

By multiplying we can see that

$$(a + b)^2 = a^2 + 2ab + b^2$$
$$(a + b)^3 = a^3 + 3a^2b + 3ab^2 + b^3$$
$$(a + b)^4 = a^4 + 4a^3b + 6a^2b^2 + 4ab^3 + b^4.$$

The following theorem tells us how to expand $(a + b)^n$ for any positive integer n.

Theorem 7.3 (Binomial Theorem) *For any positive integer n,*

$$(a + b)^n = a^n + na^{n-1}b + \frac{n(n-1)}{2}a^{n-2}b^2 + \frac{n(n-1)(n-2)}{3 \cdot 2}a^{n-3}b^3 + \cdots + b^n.$$

Proof. We shall use mathematical induction. Let S_n be the statement of the theorem. S_1 is certainly true. Now suppose that S_k is true. That is, assume that

$$(a + b)^k = a^k + ka^{k-1}b + \frac{k(k-1)}{2}a^{k-2}b^2 + \frac{k(k-1)(k-2)}{3 \cdot 2}a^{k-3}b^3 + \cdots + b^k. \quad (7\text{-}2)$$

To find $(a + b)^{k+1}$, let's note that

$$(a + b)^{k+1} = (a + b)(a + b)^k = a(a + b)^k + b(a + b)^k.$$

Multiplying both sides of equation (7-2) by a, we get

$$a(a + b)^k = a^{k+1} + ka^k b + \frac{k(k-1)}{2}a^{k-1}b^2 + \frac{k(k-1)(k-2)}{3 \cdot 2}a^{k-2}b^3 + \cdots + ab^k. \quad (7\text{-}3)$$

Multiplying both sides of equation (7-2) by b, we have

$$b(a + b)^k = a^k b + ka^{k-1}b^2 + \frac{k(k-1)}{2}a^{k-2}b^3 + \frac{k(k-1)(k-2)}{3 \cdot 2}a^{k-3}b^4 + \cdots + b^{k+1}. \quad (7\text{-}4)$$

Upon adding equations (7-3) and (7-4) and grouping like terms, we have

$$\begin{aligned} a(a + b)^k + b(a + b)^k &= a^{k+1} + ka^k b + a^k b \\ &\quad + \frac{k(k-1)}{2}a^{k-1}b^2 + ka^{k-1}b^2 \\ &\quad + \frac{k(k-1)(k-2)}{3 \cdot 2}a^{k-2}b^3 + \frac{k(k-1)}{2}a^{k-2}b^3 \\ &\quad + \cdots + b^{k+1} \end{aligned}$$

$$= a^{k+1} + a^k b(k + 1)$$
$$+ a^{k-1}b^2 \left[\frac{k(k-1)}{2} + k\right]$$
$$+ a^{k-2}b^3 \left[\frac{k(k-1)(k-2)}{3 \cdot 2} + \frac{k(k-1)}{2}\right]$$
$$+ \cdots + b^{k+1}.$$

Let's examine the third term of this expression. Its coefficient is

$$\frac{k(k-1)}{2} + k = \frac{k(k-1) + 2k}{2}$$
$$= \frac{k(k-1+2)}{2}$$
$$= \frac{k(k+1)}{2}$$
$$= \frac{(k+1)k}{2}.$$

Similarly, the coefficient of the fourth term is

$$\frac{(k+1)k(k-1)}{3 \cdot 2}.$$

Therefore,

$$(a + b)^{k+1} = a^{k+1} + (k + 1)a^k b + \frac{(k+1)k}{2} a^{k-1}b^2$$
$$+ \frac{(k+1)k(k-1)}{3 \cdot 2} a^{k-2}b^3 + \cdots + b^{k+1}.$$

Hence S_{k+1} is true. By mathematical induction, S_n is true for all positive integers n. ■

Let's note several important features of this expansion.

1. There are $n + 1$ terms.
2. The first term is a^n, and the last term is b^n.
3. The exponent of a decreases by one from term to term, and the exponent of b increases by one from term to term.
4. The sum of the exponents of a and b in any term is n.
5. If we multiply the coefficient of any term by the exponent of a and divide by the number of that term, we get the coefficient of the next term.

Example 1
Expand $(x + y)^5$.

Solution. Using the Binomial Theorem, we have

$$\begin{aligned}(x+y)^5 &= x^5 + 5x^4y + \frac{5\cdot 4}{2}x^3y^2 + \frac{5\cdot 4\cdot 3}{3\cdot 2}x^2y^3 \\ &\quad + \frac{5\cdot 4\cdot 3\cdot 2}{4\cdot 3\cdot 2}xy^4 + y^5 \\ &= x^5 + 5x^4y + 10x^3y^2 + 10x^2y^3 + 5xy^4 + y^5.\end{aligned}$$ ■

Example 2
Expand $(x - y)^6$.

Solution. We use the Binomial Theorem with $b = -y$. We get

$$\begin{aligned}(x-y)^6 &= x^6 + 6x^5(-y) + 15x^4(-y)^2 + 20x^3(-y)^3 \\ &\quad + 15x^2(-y)^4 + 6x(-y)^5 + (-y)^6 \\ &= x^6 - 6x^5y + 15x^4y^2 - 20x^3y^3 + 15x^2y^4 \\ &\quad - 6xy^5 + y^6.\end{aligned}$$ ■

Exercise
Expand $(x - y)^7$.

Answer. $(x - y)^7 = x^7 - 7x^6y + 21x^5y^2 - 35x^4y^3 + 35x^3y^4 - 21x^2y^5 + 7xy^6 - y^7$ ■

Example 3
Expand $(2x - y)^4$.

Solution. Using the Binomial Theorem with $a = 2x$ and $b = -y$, we get

$$\begin{aligned}(2x-y)^4 &= (2x)^4 + 4(2x)^3(-y) + 6(2x)^2(-y)^2 \\ &\quad + 4(2x)(-y)^3 + (-y)^4 \\ &= 16x^4 - 32x^3y + 24x^2y^2 - 8xy^3 + y^4.\end{aligned}$$ ■

Exercise
Expand $(x + 2y)^4$.

Answer. $(x + 2y)^4 = x^4 + 8x^3y + 24x^2y^2 + 32xy^3 + 16y^4$ ■

Sometimes we may be interested in only one term of the expansion. Instead of calculating the entire expansion, we need only note that the rth term of $(a + b)^n$ is

$$\frac{n(n-1)(n-2)\cdots(n-r+2)}{(r-1)(r-2)\cdots 2}a^{n-(r-1)}b^{r-1}.$$

Note that for the rth term, the exponent of b is $r - 1$. The exponent of a is simply the difference between n and the exponent of b. The factors of the denominator begin at $r - 1$ and decline to 2. And the factors of the numerator begin at n and decline until the *number* of factors is $r - 1$.

Example 4
Find the fifth term of $(2x + y)^7$.

Solution. The exponent for y will be 4, so the exponent for $2x$ will be $7 - 4 = 3$. We have

$$\frac{7 \cdot 6 \cdot 5 \cdot 4}{4 \cdot 3 \cdot 2}(2x)^3y^4 = 35(8x^3)y^4 = 280x^3y^4. \qquad \blacksquare$$

Exercise
Find the fourth term of $(x + 3y)^6$.

Answer. $540x^3y^3$ ■

Problem Set 7.2

In Problems 1–14, expand.

1. $(x + y)^6$
2. $(x + y)^7$
3. $(x - y)^5$
4. $(x - y)^8$
5. $(x - 2y)^4$
6. $(3x + y)^5$
7. $(x - 5y)^4$
8. $(2x + y)^5$
9. $(2x + 3y)^3$
10. $(4x - 2y)^3$
11. $(1 + x)^8$
12. $(1 - x)^7$
13. $(xy + 2)^6$
14. $\left(x - \frac{y}{x}\right)^5$

15. Find the fourth term in the expansion of $(x + y)^{10}$.
16. Find the fifth term in the expansion of $(x - 2y)^9$.
17. Find the third term in the expansion of $(2x - 3y)^{12}$.

7.3 Sequences and Sums

In this section we study ordered arrays of numbers and their sums. We begin with the following definition. A **sequence** is a function whose domain is Z^+, the set of positive integers, or is $\{1, 2, \ldots, n\}$, the set of the first n positive integers for some positive integer n. A sequence whose domain is Z^+ is called infinite, while one whose domain is $\{1, 2, \ldots, n\}$ for some n is called finite. For example, let f be the infinite sequence defined by

$$f(n) = \frac{n}{n + 1}. \qquad (7\text{-}5)$$

Since f is an infinite sequence its domain is Z^+. Thus we may calculate

$$f(1) = \frac{1}{1 + 1} = \frac{1}{2},$$

$$f(2) = \frac{2}{2 + 1} = \frac{2}{3},$$

$$f(3) = \frac{3}{3 + 1} = \frac{3}{4}.$$

Now $f(1)$ is called the first term of the sequence, $f(2)$ is called the second term, $f(3)$ is the third term, and so on. Hence the formula of definition for f, equation (7-5), gives us the nth term of the sequence. For sequences it is customary to denote the nth term by a_n instead of $f(n)$. Hence the preceding sequence is more commonly defined by

$$a_n = \frac{n}{n + 1},$$

and the first three terms are written

$$a_1 = \frac{1}{2},$$

$$a_2 = \frac{2}{3},$$

$$a_3 = \frac{3}{4}.$$

Exercise

Find the first three terms of the infinite sequence $a_n = \dfrac{n + 1}{n}$.

Answer. $a_1 = 2,\ a_2 = \dfrac{3}{2},\ a_3 = \dfrac{4}{3}$ ■

Our study of sequences will be restricted to infinite sequences. Thus when we use the word "sequence" without a qualifier, we are referring to an infinite sequence.

Another way of defining a sequence is to give a_1, the first term, and give a formula for a_{k+1}, the $k + 1$st term, in terms of a_k, the kth term. Such a definition is called **recursive.** (By mathematical induction all terms of the sequence are defined.)

For example, suppose that $a_1 = 1$ and $a_{k+1} = 2a_k + 1$. The first four terms of this sequence are

$$\begin{aligned} a_1 &= 1, \\ a_2 &= 2a_1 + 1 = 2(1) + 1 = 3, \\ a_3 &= 2a_2 + 1 = 2(3) + 1 = 7, \\ a_4 &= 2a_3 + 1 = 2(7) + 1 = 15. \end{aligned}$$

Note that to find a_{10}, the tenth term, we would have to find the first nine terms of the sequence. In this sense a sequence defined recursively is more difficult to work with than one which gives a formula for the nth term. However, some sequences defined recursively may also be defined in the more general sense. The first four terms of the preceding sequence may be obtained with the formula $a_n = 2^n - 1$. Let's prove that this formula is valid for the entire sequence by using mathematical induction.

We have already noted that the formula is valid for $n = 1$ (as well as for $n = 2, 3$, and 4). Hence we now assume that the formula is valid for $n = k$. That is, assume that

$$a_k = 2^k - 1.$$

We must show that the formula is valid for $n = k + 1$. Thus we must show that

$$a_{k+1} = 2^{k+1} - 1.$$

By the recursive definition we have

$$a_{k+1} = 2a_k + 1.$$

Since $a_k = 2^k - 1$, we get

$$\begin{aligned} a_{k+1} &= 2(2^k - 1) + 1 \\ &= 2^{k+1} - 2 + 1 \\ &= 2^{k+1} - 1. \end{aligned}$$

By mathematical induction, the formula is valid for all positive integers n. That is, $a_n = 2^n - 1$.

Exercise
Find the first four terms of the sequence defined by $a_1 = 1$ and $a_{k+1} = 3a_k + 2$.

Answer. $a_1 = 1$, $a_2 = 5$, $a_3 = 17$, $a_4 = 53$ ■

Note that a sequence is not defined by listing its first few terms. For example, suppose that

$$a_1 = 2,$$
$$a_2 = 4,$$
$$a_3 = 6,$$

and we want to calculate a_4. We might guess that $a_n = 2n$ and thus $a_4 = 8$. However, the sequence

$$a_n = 2(n - 2)^3 + 4$$

has the same first three terms, but $a_4 = 20$. Hence a sequence must be defined with a formula for its nth term or must be defined recursively. But it is proper to define a sequence with a list as long as we indicate the nth term. For example, the infinite sequence $a_n = 2n$ may be written.

$$2, 4, 6, \ldots, 2n, \ldots.$$

And the infinite sequence $a_n = 2(n - 2)^3 + 4$ may be written

$$2, 4, 6, \ldots, 2(n - 2)^3 + 4, \ldots.$$

A **series** is the indicated sum of the terms of a sequence. If the sequence is finite, then the corresponding series is called finite. And if the sequence is infinite, then the corresponding series is called infinite.

For example,

$$1 + 3 + 5 + 7 + 9$$

is the finite series that corresponds to the finite sequence 1, 3, 5, 7, 9. And

$$2 + 4 + 6 + \cdots + 2n + \cdots$$

is the infinite series that corresponds to the infinite sequence $a_n = 2n$.

In this text we confine our study to finite series because, at present, we can find the sum of finitely many terms only. An infinite sum requires concepts that will be developed in calculus.

A special notation, the capital Greek letter Σ (sigma), is frequently used to denote a series. We define

$$\sum_{i=1}^{n} a_i = a_1 + a_2 + \cdots + a_n.$$

Hence this sigma notation refers to the sum of the first n terms of a sequence. This notation is read "the sum of the a_i's from $i = 1$ to n." The variable i is called the **index of summation.** We could have used another variable as the index without altering the series. For example,

$$\sum_{j=1}^{n} a_j = a_1 + a_2 + \cdots + a_n$$

and

$$\sum_{k=1}^{n} a_k = a_1 + a_2 + \cdots + a_n$$

are the same series as $\sum_{i=1}^{n} a_i$.

Example 1

Find the value of the series $\sum_{i=1}^{5} (2i - 1)$.

Solution. We will expand the series and add. To find the terms in the sum, we successively substitute the values 1, 2, 3, 4, and 5 for i in the formula $2i - 1$. We get

$$\begin{aligned}\sum_{i=1}^{5} (2i - 1) &= (2(1) - 1) + (2(2) - 1) + (2(3) - 1) + (2(4) - 1) \\ &\quad + (2(5) - 1) \\ &= 1 + 3 + 5 + 7 + 9 = 25.\end{aligned}$$

■

Exercise

Find the value of the series $\sum_{i=1}^{5} (3i + 1)$.

Answer. 50 ■

Example 2
Write the series $2 - 5 + 8 - 11 + 14$ with sigma notation.

Solution. We need to determine the formula a_i for the first five terms of the sequence. We have

$$\begin{aligned} a_1 &= 2, \\ a_2 &= -5, \\ a_3 &= 8, \\ a_4 &= -11, \\ a_5 &= 14. \end{aligned}$$

Ignoring the sign, the ith term may be written $3i - 1$. But we want the odd terms to be positive and the even terms to be negative. This may be accomplished by multiplying by $(-1)^{i+1}$. Thus

$$a_i = (-1)^{i+1}(3i - 1) \qquad \text{for } i = 1, 2, \ldots, 5.$$

Hence we have

$$2 - 5 + 8 - 11 + 14 = \sum_{i=1}^{5} (-1)^{i+1}(3i - 1). \qquad ■$$

Exercise

Write the series $5 - 11 + 17 - 23$ with sigma notation.

Answer. $\sum_{i=1}^{4} (-1)^{i+1}(6i - 1)$ ■

Problem Set 7.3

In Problems 1–12, find the first three terms of the sequence.

1. $a_n = \dfrac{2n}{3} - 1$

2. $a_n = \dfrac{3n}{2} + 1$

3. $a_n = (n - 1)^2$

4. $a_n = 2n^2 + 1$

5. $a_n = \dfrac{1}{n} + \dfrac{1}{n + 1}$

6. $a_n = \dfrac{1 - n}{n^2}$

7. $a_n = \dfrac{\log n}{n}$

8. $a_n = \dfrac{n}{e^n}$

9. $a_n = (-1)^n \dfrac{n}{2^n}$

10. $a_n = (-1)^{n+1} \dfrac{n^2}{2^n}$

11. $a_n = (-1)^{n+1} x^n$

12. $a_n = \dfrac{(-1)^n}{x^n}$

In Problems 13–20, find the first four terms of the sequence defined.

13. $a_1 = 5$, $a_{k+1} = 2a_k - 1$

14. $a_1 = 3$, $a_{k+1} = 1 - 4a_k$

15. $a_1 = 1$, $a_{k+1} = 1 - (a_k)^2$

16. $a_1 = 0$, $a_{k+1} = (a_k)^2 + 1$

17. $a_1 = -1$, $a_{k+1} = ka_k$

18. $a_1 = 1$, $a_{k+1} = \dfrac{k}{a_k}$

19. $a_1 = x$, $a_{k+1} = (-1)^k (a_k)^{k+1}$

20. $a_1 = -x^2$, $a_{k+1} = (-1)^{k-1}(a_k)^k$

In Problems 21–26, find the value.

21. $\sum_{i=1}^{4} (-2i)$

22. $\sum_{i=1}^{3} \dfrac{i}{i + 1}$

23. $\sum_{j=1}^{4} (2^j - 1)$

24. $\sum_{k=1}^{3} \dfrac{3^k}{k}$

25. $\sum_{i=0}^{3} i^2$

26. $\sum_{j=2}^{4} \dfrac{2^j}{2^{j-1} + 1}$

In Problems 27 and 28, expand the series and add.

27. $\sum_{i=0}^{2} \frac{(-1)^{i+1}}{x^i}$ **28.** $\sum_{i=1}^{3} (-1)^i x$

In Problems 29–36, write the series in sigma notation.

29. $1 + \frac{1}{4} + \frac{1}{9} + \frac{1}{16} + \frac{1}{25}$ **30.** $4 + 9 + 16 + 25$

31. $1 - 2 + 3 - 4 + 5$ **32.** $-1 + \frac{1}{8} - \frac{1}{27} + \frac{1}{64}$

33. $1 + e + e^2$ **34.** $\frac{\log 2}{3} + \frac{\log 4}{5} + \frac{\log 6}{7}$

35. $x - x^3 + x^5 - x^7 + x^9$ **36.** $\sqrt{x} + x + x\sqrt{x} + x^2$

37. The Fibonacci sequence is defined as $a_1 = 1$, $a_2 = 1$, and $a_n = a_{n-1} + a_{n-2}$ for $n > 2$. Find the first seven terms of this sequence.

38. Let a_n be the sequence defined by a_n = the number of primes less than n. Find the first six terms of this sequence.

39. Find the first four terms of the sequence

$$a_n = \begin{cases} n^2 & \text{when } n \text{ is odd} \\ \frac{1}{n} & \text{when } n \text{ is even.} \end{cases}$$

40. Find the first four terms of the sequence

$$a_n = \begin{cases} -n & \text{when } n \text{ is odd} \\ \frac{-1}{n^2} & \text{when } n \text{ is even.} \end{cases}$$

41. Prove that $\sum_{i=1}^{n} (a_i + b_i) = \sum_{i=1}^{n} a_i + \sum_{i=1}^{n} b_i$.

42. Prove that $\sum_{i=1}^{n} (a_i - b_i) = \sum_{i=1}^{n} a_i - \sum_{i=1}^{n} b_i$.

43. Prove that $\sum_{i=1}^{n} ca_i = c \sum_{i=1}^{n} a_i$, where c is a constant.

44. Prove that $\sum_{i=1}^{n} c = nc$, where c is a constant.

7.4 Arithmetic and Geometric Progressions

In this section we study two very special types of sequences called progressions. The first type, an **arithmetic progression,** is a sequence in which successive terms differ by the same constant. That is, given the first term, every other term can be obtained by adding the same constant to the preceding term. Thus arithmetic progressions are defined recursively. Given a_1, we have

$$a_{k+1} = a_k + d$$

for some constant d, called the **common difference.** Hence if the first term of an arithmetic progression is 5 and the common difference is 4, the first five terms of the progression are

$$5, 9, 13, 17, 21.$$

Examining the first few terms of an arithmetic progression, we have

$$\begin{aligned} a_1 &= a_1, \\ a_2 &= a_1 + d, \\ a_3 &= a_2 + d = a_1 + 2d, \\ a_4 &= a_3 + d = a_1 + 3d. \end{aligned}$$

We speculate that

$$a_n = a_1 + (n - 1)d.$$

Indeed this is the case. It can be shown by mathematical induction.

We may thus calculate any term of an arithmetic progression without calculating all of the preceding terms. For example, the tenth term of an arithmetic progression whose first term is 5 and whose common difference is 2 is 23 because

$$\begin{aligned} a_n &= 5 + (n - 1)2, \\ a_{10} &= 5 + (9)2 = 23. \end{aligned}$$

Exercise

Write (a) the first five terms and (b) the 20th term of the arithmetic progression whose first term is -5 and whose common difference is 3.

Answers. (a) $-5, -2, 1, 4, 7$ (b) 52 ■

We now focus our attention on arithmetic series, the sum of terms from an arithmetic progression.

Theorem 7.4

Let $a_1, a_2, \ldots$ be an arithmetic progression. Then

$$\sum_{i=1}^{n} a_i = n\left(\frac{a_1 + a_n}{2}\right).$$

Proof. We will use mathematical induction. Let S_n be the statement of the theorem. Now S_1 is true because

$$\sum_{i=1}^{1} a_i = a_1 = 1\left(\frac{a_1 + a_1}{2}\right).$$

Assuming that S_k is true, we have

$$\sum_{i=1}^{k} a_i = k\left(\frac{a_1 + a_k}{2}\right).$$

We must use this fact to show that S_{k+1} is true. That is, show that

$$\sum_{i=1}^{k+1} a_i = (k + 1)\left(\frac{a_1 + a_{k+1}}{2}\right).$$

We have

$$\begin{aligned}
\sum_{i=1}^{k+1} a_i &= \left(\sum_{i=1}^{k} a_i\right) + a_{k+1} \\
&= k\left(\frac{a_1 + a_k}{2}\right) + a_{k+1} \\
&= k\left(\frac{a_1 + a_{k+1} - d}{2}\right) + a_1 + kd \\
&= \frac{ka_1 + ka_{k+1} - kd + 2a_1 + 2kd}{2} \\
&= \frac{ka_1 + ka_{k+1} + a_1 + a_1 + kd}{2} \\
&= \frac{k(a_1 + a_{k+1}) + a_1 + a_{k+1}}{2} \\
&= \frac{(a_1 + a_{k+1})(k + 1)}{2} \\
&= (k + 1)\left(\frac{a_1 + a_{k+1}}{2}\right).
\end{aligned}$$

By mathematical induction, S_n is true for all positive integers n. That is,

$$\sum_{i=1}^{n} a_i = n\left(\frac{a_1 + a_n}{2}\right).$$

■

This theorem says that to get the sum of the first n terms of an arithmetic progression, we multiply the average of the first and nth terms by n.

Example 1
Find the sum of the first ten terms of the arithmetic progression whose first term is 2 and whose common difference is -3.

Solution. To do this, we must first calculate the tenth term of the progression. Since

$$a_n = a_1 + (n - 1)d = 2 - 3(n - 1),$$

we have

$$a_{10} = 2 - 3(9) = -25.$$

Therefore,

$$\sum_{i=1}^{10} a_i = 10\left(\frac{2 - 25}{2}\right) = -115.$$

■

Exercise
Find the sum of the first ten terms of the arithmetic progression whose first term is 3 and whose common difference is 2.

Answer. 120 ■

A **geometric progression** is a sequence in which the ratio of successive terms is constant. That is, given the first term, every other term can be obtained by multiplying the preceding term by the same nonzero constant. Thus geometric progressions are defined recursively. Given a_1, we have

$$a_{k+1} = ra_k$$

for some constant r, called the **common ratio.** Hence if the first term of a geometric progression is 3 and the common ratio is 2, the first five terms of the progression are

$$3, 6, 12, 24, 48.$$

Let's examine the first few terms of a geometric progression. We have

$$\begin{aligned} a_1 &= a_1, \\ a_2 &= ra_1, \\ a_3 &= ra_2 = r^2a_1, \\ a_4 &= ra_3 = r^3a_1. \end{aligned}$$

We speculate that

$$a_n = r^{(n-1)}a_1.$$

By mathematical induction it can be shown that this is the case.

We may thus calculate any term of a geometric progression without calculating all of the preceding terms. For example, the sixth term of the geometric progression whose first term is 5 and whose common ratio is $\frac{1}{2}$ is $\frac{5}{32}$ because

$$a_n = \left(\frac{1}{2}\right)^{n-1} \cdot 5,$$

$$a_6 = \left(\frac{1}{2}\right)^{5} \cdot 5 = \frac{5}{32}.$$

Exercise

(a) Write the first three terms of the geometric progression whose first term is -5 and whose common ratio is 3.

(b) Write the sixth term of the geometric progression whose first term is 6 and whose common ratio is $\frac{1}{3}$.

Answers. (a) $-5, -15, -45$ (b) $\frac{2}{81}$ ■

Let's now turn our attention to geometric series, the sum of terms from a geometric progression.

Theorem 7.5

Let $a_1, a_2, \ldots$ be a geometric progression with common ratio $r \neq 1$. Then

$$\sum_{i=1}^{n} a_i = a_1\left(\frac{1-r^n}{1-r}\right).$$

Proof. Although this may be proven by mathematical induction, we'll use a different approach. Let

$$S_n = \sum_{i=1}^{n} a_i.$$

Then

$$S_n = a_1 + a_2 + \cdots + a_n$$

$$S_n = a_1 + ra_1 + \cdots + r^{n-1}a_1. \qquad (7\text{-}6)$$

After multiplying both sides of equation (7-6) by r, we have

$$rS_n = ra_1 + r^2a_1 + \cdots + r^na. \qquad (7\text{-}7)$$

Subtracting equation (7-7) from equation (7-6), we get

$$S_n - rS_n = a_1 - r^n a_1.$$

Solving for S_n, we have

$$S_n(1 - r) = a_1(1 - r^n)$$

$$S_n = a_1\left(\frac{1 - r^n}{1 - r}\right). \quad \blacksquare$$

Example 2

Find the sum of the first seven terms of the geometric progression whose first term is 3 and whose common ratio is $-\frac{1}{2}$.

Solution.

$$\sum_{i=1}^{7} a_i = a_1\left(\frac{1 - r^n}{1 - r}\right)$$

$$= 3\left(\frac{1 - \left(\frac{-1}{2}\right)^7}{1 - \left(\frac{-1}{2}\right)}\right)$$

$$= 3\left(\frac{1 + \frac{1}{128}}{1 + \frac{1}{2}}\right)$$

$$= 3\left(\frac{128 + 1}{128 + 64}\right) = \frac{129}{64} \quad \blacksquare$$

Exercise

Find the sum of the first four terms of the geometric progression whose first term is 2 and whose common ratio is $\frac{1}{3}$.

Answer. $\frac{80}{27}$ ■

Problem Set 7.4

In Problems 1–4, (a) write the first four terms of the arithmetic progression (b) find the tenth term, and (c) calculate the sum of the first ten terms.

1. $a_1 = 1, d = 5$ **2.** $a_1 = -3, d = 2$ **3.** $a_1 = -2, d = -6$
4. $a_1 = 5, d = -4$

In Problems 5–12, write the first four terms of the geometric progression.

5. $a_1 = 1, r = 3$ **6.** $a_1 = -3, r = 2$ **7.** $a_1 = -5, r = -2$

8. $a_1 = 6, r = -3$ **9.** $a_1 = -1, r = \frac{1}{3}$ **10.** $a_1 = 4, r = \frac{1}{2}$

11. $a_1 = \sqrt{2}, r = -\sqrt{3}$ **12.** $a_1 = -\sqrt{3}, r = -\sqrt{5}$

In Problems 13–16, find the sum of the first four terms of the geometric series.

13. $a_1 = 5, r = \frac{1}{3}$ **14.** $a_1 = -5, r = \frac{-1}{2}$ **15.** $a_1 = -2, r = 2$

16. $a_1 = 3, r = \sqrt{2}$

17. For a geometric series, $r = \frac{1}{2}$ and $a_6 = \frac{1}{8}$. Find a_1.

18. For a geometric series, $r = \frac{1}{3}$ and $\sum_{i=1}^{4} a_i = \frac{80}{27}$. Find a_1.

19. For what value of x are $4x - 1$, $2x$, $3x + 4$ consecutive terms of an arithmetic progression?

20. For what values of x are $2x$, $2 - 7x$, $18x$ consecutive terms of a geometric progression?

21. If log 3 and log 9 are the first two terms of a geometric progression, what is the sum of the first four terms?

22. A harmonic progression is a sequence in which the reciprocals form an arithmetic progression. Show that $a_n = \frac{1}{2n + 1}$ is harmonic.

23. A ball rolling down a hill covers 5 feet during the first second, and during each subsequent second it covers 3 more feet than in the preceding second. How far has the ball traveled after 10 seconds?

24. A man is paid \$50 on the first day and receives a \$10 increase on each successive day. How much will he receive for 30 days?

25. A city with a population of 200,000 has a 20% increase every decade. What will the population be in 30 years?

26. If a car depreciates 20% per year, what is its value after 3 years if the original cost was \$8000?

27. A ball, dropped from a height of 10 feet, rebounds one-half of the distance it has fallen. How far has the ball traveled when it strikes the ground for the sixth time?

28. Let a_1, b_1, c_1 and a_2, b_2, c_2 be consecutive terms from arithmetic progressions. Find the solution set of

$$\begin{cases} a_1x + b_1y = c_1 \\ a_2x + b_2y = c_2. \end{cases}$$

29. Prove that the nth term of an arithmetic progression is

$$a_1 + (n - 1)d,$$

where a_1 is the first term and d is the common difference.

30. Prove that the nth term of a geometric progression is

$$r^{(n-1)}a_1,$$

where a_1 is the first term and r is the common ratio.

Chapter 7 Review

1. Prove that if $0 < a < b$, then $a^n < b^n$ for every positive integer n.
2. Prove that $2^n \geq 1 + n$ for every positive integer n.
3. Prove that $x^n - 1$ is divisible by $x - 1$ ($x \neq 1$) for every positive integer n.
4. Prove that $\sum_{i=1}^{n} \frac{1}{i(i+1)} = \frac{n}{n+1}$ for all positive integers n.
5. Prove that $\sum_{i=1}^{n} i(i+1) = \frac{n(n+1)(n+2)}{3}$ for all positive integers n.
6. Expand $(x - 1)^5$.
7. Expand $(4x + y)^4$.
8. Expand $(x - 3y)^6$.
9. Find the third term in the expansion of $(x + 2y)^{10}$.

In Problems 10–17, write the first three terms of the sequence defined.

10. $a_n = 3 - 2n^2$
11. $a_n = \frac{1 - 2n}{n^3}$
12. $a_n = \frac{(-1)^n 3^n}{n}$
13. $a_n = \frac{(-1)^{n+1}}{n2^n}$
14. $a_n = (-1)^{n-1} n x^n$
15. $a_1 = x,\ a_{k+1} = k a_k$
16. $a_1 = -3,\ a_{k+1} = 2 - 5a_k$
17. $a_1 = -2,\ a_{k+1} = (-1)^k (a_k)^2$

In Problems 18–21, find the value.

18. $\sum_{i=1}^{3} \frac{i - 1}{i}$
19. $\sum_{j=1}^{3} (1 + 3^j)$
20. $\sum_{i=0}^{3} (-i^2)$
21. $\sum_{i=1}^{3} (-1)^{i+1} x^2$

In Problems 22 and 23, write the series in sigma notation.

22. $1 - 1 + 1 - 1 + 1$
23. $1 + \frac{2}{3} + \frac{3}{5} + \frac{4}{7} + \frac{5}{9}$

In Problems 24 and 25, (a) write the first four terms of the arithmetic progression and (b) find the sum of the first ten terms.

24. $a_1 = -3,\ d = -5$
25. $a_1 = 6,\ d = 4$

In Problems 26 and 27, write the first four terms of the geometric progression.

26. $a_1 = \frac{-1}{2}, r = 3$

27. $a_1 = 5, r = \sqrt{2}$

In Problems 28 and 29, find the sum of the first four terms of the geometric series.

28. $a_1 = -5, r = \frac{3}{2}$

29. $a_1 = 4, r = -\sqrt{2}$

30. A stack of cans has 2 cans on the top row and 10 on the bottom row. If each row has one less can than the row below, how many cans are there in the stack?

31. A man is paid 1 cent on the first day and his pay is doubled on each successive day. How much will he receive for 30 days?

Selected Topics

8

This chapter contains an assortment of interesting topics which, although independent of the mainstream of ideas in the preceding chapters, are nonetheless important.

8.1 Variation

Many relationships between quantities are described in terms of **variation.** Given two variables x and y, we say that **y varies directly as x** provided that

$$y = kx$$

for some constant $k \neq 0$. This constant is called the **constant of variation.** For example, for a motorist averaging 50 mph, we might say that his distance traveled varies directly as his time since $d = 50t$. The constant of variation is 50.

In variation problems we first use given information to determine the constant of variation. Then we use this to answer a question concerning some other set of conditions. See Examples 1 and 2.

Example 1
Suppose that y varies directly as x and that $y = -8$ when $x = 2$. What is the value of y when $x = 6$?

Solution. Since y varies directly as x,

$$y = kx$$

for some constant $k \neq 0$. Because $y = -8$ when $x = 2$, we have

$$-8 = 2k$$
$$-4 = k.$$

Thus

$$y = -4x.$$

When $x = 6$, $y = -24$. ∎

Exercise
Suppose that y varies directly as x and that $y = -9$ when $x = -3$. What is the value of y when $x = -5$?

Answer. $y = -15$ ∎

Example 2
The pressure at a point in a liquid varies directly as its distance beneath the surface. In a certain liquid the pressure at a depth of 2 feet is 100 pounds/ft². What is the pressure at a depth of 5 feet?

Solution. Let

$$p = \text{pressure (in pounds/ft}^2\text{)}$$
$$d = \text{distance beneath surface (in feet)}$$

Because p varies directly as d,

$$p = kd$$

for some constant $k \neq 0$. Since $p = 100$ when $d = 2$, we have

$$100 = 2k$$
$$50 = k.$$

Thus

$$p = 50d.$$

When $d = 5$, $p = 250$. So our answer is 250 pounds/ft². ∎

To say that **y varies inversely as x** means that

$$y = \frac{k}{x}$$

for a constant $k \neq 0$. This type of variation is illustrated in Example 3.

Example 3
The illumination from a light source varies inversely as the square of the distance from the source. If the illumination is 100 candlepower 2 feet from the source, what is the illumination 5 feet from the source?

Solution. Let

$$\begin{aligned} i &= \text{illumination (in candlepower)} \\ d &= \text{distance from source (in feet)} \end{aligned}$$

Since i varies inversely as d^2,

$$i = \frac{k}{d^2}$$

for some constant $k \neq 0$. Because $i = 100$ when $d = 2$,

$$100 = \frac{k}{2^2}$$

$$400 = k.$$

Thus

$$i = \frac{400}{d^2}.$$

When $d = 5$, $i = 16$. Our answer is 16 candlepower. ■

As seen in Example 4, the variation is sometimes a combination of both direct and inverse.

Example 4
Suppose that z varies directly as x and inversely as y and that $z = \frac{5}{2}$ when $x = -2$ and $y = 4$. What is the value of z when $x = 6$ and $y = 3$?

Solution. Since z varies directly as x and inversely as y,

$$z = \frac{kx}{y}$$

for some constant $k \neq 0$. Because $z = \frac{5}{2}$ when $x = -2$ and $y = 4$,

$$\frac{5}{2} = \frac{-2k}{4}$$

$$k = -5.$$

Thus

$$z = \frac{-5x}{y}.$$

When $x = 6$ and $y = 3$, $z = -10$. ■

Exercise
Suppose that z varies directly as x and inversely as y and that $z = -\frac{3}{4}$ when $x = -2$ and $y = 8$. What is the value of z when $x = 4$ and $y = -5$?

Answer. $y = -\dfrac{12}{5}$. ■

Problem Set 8.1

1. Suppose that y varies directly as x and that $y = 12$ when $x = -2$. What is the value of y when $x = 5$?
2. Suppose that y varies directly as x and that $y = 4$ when $x = 8$. What is the value of y when $x = -4$?
3. Suppose that y varies inversely as x and that $y = \frac{1}{6}$ when $x = 2$. What is the value of y when $x = -\frac{1}{2}$?
4. Suppose that y varies inversely as x and that $y = 2$ when $x = -\frac{1}{2}$. What is the value of y when $x = 5$?
5. Suppose that y varies directly as x and that $y = \frac{1}{3}$ when $x = \frac{1}{2}$. What is the value of x when $y = -4$?
6. Suppose that y varies inversely as x and that $y = -\frac{5}{4}$ when $x = 2$. What is the value of x when $y = 2$?
7. Suppose that z varies directly as x and inversely as y and that $z = -30$ when $x = -3$ and $y = \frac{1}{2}$. What is the value of z when $x = \frac{1}{2}$ and $y = 8$?
8. Suppose that z varies directly as x and inversely as y and that $z = -\frac{1}{10}$ when $x = \frac{1}{2}$ and $y = 2$. What is the value of z when $x = -3$ and $y = \frac{1}{2}$?
9. The surface area of a sphere varies directly as the square of its radius. What is the constant of variation?
10. The area of a circle varies directly as the square of its diameter. What is the constant of variation?
11. The horsepower required to propel a ship at a desired speed varies directly as the cube of the speed. If the horsepower required for 12 knots is 6048, what horsepower is required for 20 knots?
12. The acceleration a of an object varies directly as the force F applied to it. If an object accelerates 30 ft/sec^2 when a force of 5 newtons is applied, what will the acceleration of the object be if a force of 12 newtons is applied?
13. According to Hooke's Law, the distance a vertical spring stretches varies directly as the weight placed on it. If a 50-kilogram weight stretches a vertical spring 6 centimeters, how much weight is needed to stretch the spring 15 centimeters?
14. The period of a pendulum (time of one complete oscillation) varies directly as the square root of its length. If an 8-foot pendulum has a period of 3 seconds, what length would be necessary for a period of 2 seconds?
15. The strength of a radio signal varies inversely as the square of its distance from the transmitter. If a signal is 60 kilowatts when 100 meters from the transmitter, what is its strength 150 meters from the transmitter?
16. The weight of a body varies inversely as the square of its distance from the center of the earth. How much will a 200-pound object weigh 400 miles above the earth's surface? (The earth's radius is 4000 miles.)
17. The resistance of a wire varies directly as its length and inversely as the square of its radius. If a wire 100 meters long, with a radius of 2 millimeters, has a resis-

tance of 15 ohms, what will be the resistance of a similar wire which is 125 meters long and which has a radius of 2.5 millimeters?

18. The volume V of a gas varies directly as its temperature T and inversely as its pressure P. If the volume of a gas is 150 ft^3 when the temperature is 180° and the pressure is 18 pounds/$in.^2$, what is the volume when the temperature is decreased by $\frac{1}{3}$ and the pressure is increased by $\frac{1}{3}$?
19. The safe load of a horizontal beam supported at each end varies directly as the product of the width and the squared depth of the cross section and inversely as its length. If a 2-inch by 8-inch beam 10 feet long has a safe load of 1920 pounds when supported on the 2-inch edge, what is its safe load when supported on the 8-inch edge?
20. The force of attraction between spheres varies directly as the product of their masses and inversely as the square of the distance between centers. What is the effect on a force between two spheres if the distance between them is cut in half?
21. According to Kepler's third law, the square of the time it takes a planet to revolve around the sun varies directly as the cube of its mean distance from the sun. If it takes Jupiter 12 years to revolve around the sun, what is the ratio (to the nearest tenth) of its mean distance to that of the earth?

8.2 Literal Equations

In many applications of mathematics, problems (such as max-min and related rates problems in calculus) are stated in words, and one must produce a correct equation as well as solve it. In this section we concentrate on writing the equation only. We will not be concerned with a solution.

In each problem of this section, we will be told which algebraic symbols will represent some of the unknown quantities. Upon finding a statement of equality, we may wish to introduce some additional symbols. Then, using the symbols, we translate the statement of equality into an equation. If the statement is correctly translated, we should be able to identify what each term (or group of terms) represents in the problem. Writing the statement of equality in English before attempting to write the equation will help produce a correct equation. Finally, if necessary, we must substitute for any of the extra variables that we introduced.

It is essential that you *understand the problem*. Read the problem as many times as necessary to comprehend the information given. If you introduce a variable not specified in the problem, define it before using it. If the variable represents a measurable quantity, such as distance, length, or time, identify the units of measurement.

Examples 1 through 3 illustrate the procedure to be used.

Example 1

A right circular cylinder with open top has a capacity of 24π cubic inches. The cost of the material for the bottom is $\frac{3}{4}$ of a cent per square inch, and the cost of the side material is $\frac{1}{4}$ of a cent per square inch. Let r represent the radius (in inches), and let C represent the total cost of the material (in cents). Write an equation involving r and C only.

Solution. The statement of equality is "The cost of the cylinder equals the cost of the side plus the cost of the bottom." To find the cost of the side material, we must calculate the lateral surface area of the right circular cylinder. Thus we must have some expression to represent height. Let's introduce another variable by letting

$$h = \text{height (in inches)}.$$

Therefore, the lateral surface area is $2\pi rh$, and its cost is

$$(0.25)2\pi rh.$$

Now the area of the bottom is πr^2. Thus its cost is

$$(0.75)\pi r^2.$$

Consequently, the total cost may be written

$$C = (0.25)2\pi rh + (0.75)\pi r^2.$$

Since the equation required is to involve C and r only, we must substitute for h. Hence we must find a relationship between h and r (or between h and C). Because the volume of this cylinder is 24π in.^{3}, we have

$$\pi r^2 h = 24\pi.$$

Solving this equation for h, we get

$$h = \frac{24}{r^2}.$$

Therefore, the equation involving C and r only is

$$C = (0.25)\ 2\pi r \frac{24}{r^2} + (0.75)\pi r^2.$$

That is,

$$C = \frac{12\pi}{r} + (0.75)\pi r^2.$$

■

Example 2

A man 6 feet tall is standing near a lamppost 24 feet high. Let x represent the distance (measured along the ground in feet) from the man to the end of his shadow, and let y represent his distance (in feet) from the lamppost. Write an equation involving x and y only.

Solution. Let's begin with a diagram, Figure 8.1. The statement of equality is implied. It depends on the fact that the smaller triangle is similar to the larger one (Angle–Angle Similarity Theorem). That is, the ratios of corresponding sides are equal. Noting that the base of the larger triangle is $x + y$, we have

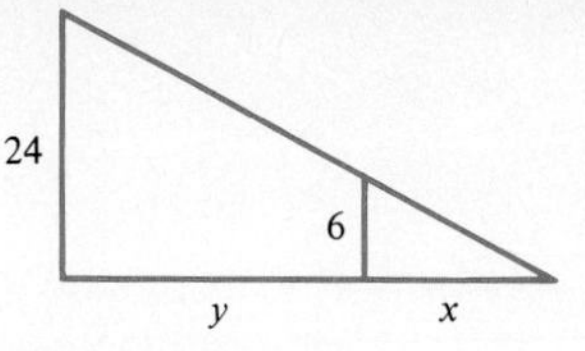

Figure 8.1

$$\frac{x}{6} = \frac{x + y}{24}. \qquad \blacksquare$$

Example 3

A man is in a boat 4 miles off a straight coast and wants to reach a point 10 miles down and on the coast. Let x represent the distance (in miles) from his landing point to his destination, and let T represent the time (in hours) of the entire trip. Write an equation involving x and T only if he rows 4 mph and runs along the beach 5 mph. (Assume that he lands at some point between his present position and destination.)

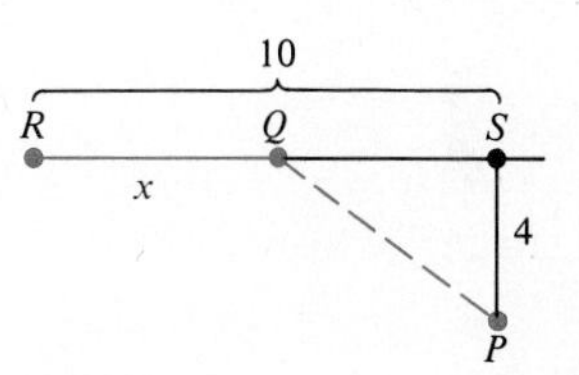

Figure 8.2

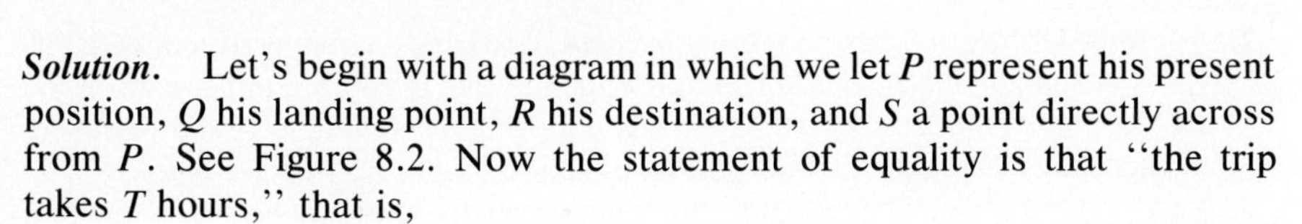

Solution. Let's begin with a diagram in which we let P represent his present position, Q his landing point, R his destination, and S a point directly across from P. See Figure 8.2. Now the statement of equality is that "the trip takes T hours," that is,

$$(\text{time from } P \text{ to } Q) + (\text{time from } Q \text{ to } R) = T.$$

To calculate these times, we use the formula $d = rt$, where d represents distance, r represents rate (average speed), and t represents time. Solving this equation for t, we find that $t = d/r$. Therefore,

$$\text{time from } P \text{ to } Q = \frac{\text{distance from } P \text{ to } Q}{4}$$

$$\text{time from } Q \text{ to } R = \frac{\text{distance from } Q \text{ to } R}{5}.$$

Now the distance from Q to R is x, but what is the distance from P to Q? Let's introduce another variable by letting

$$y = \text{distance from } P \text{ to } Q \text{ (in miles)}.$$

Thus our equation is

$$\frac{y}{4} + \frac{x}{5} = T.$$

To write this equation in terms of x and T only, we must substitute for y. Since $RQ = x$ and $RS = 10$, we have $QS = 10 - x$. And because $PS = 4$, we have, by the Pythagorean Theorem,

$$y = \sqrt{16 + (10 - x)^2}.$$

Therefore, the desired equation is

$$\frac{\sqrt{16 + (10 - x)^2}}{4} + \frac{x}{5} = T.$$

■

Problem Set 8.2

1. A long rectangular sheet of metal 10 inches wide is bent into a rain gutter by turning up the sides at right angles. Let x represent the height (in inches), and let A represent the cross-sectional area (in in.^{2}). Write an equation involving x and A only.
2. A farmer wishes to fence a rectangular field bordering a straight stream with 500 yards of fencing material. He will not fence the side bordering the stream. Let x represent the length (in yards) of the side opposite the stream, and let A represent the area (in yd^2) of the field. Write an equation involving x and A only.
3. Two vertical poles, 20 feet apart, standing on level ground, are 6 feet and 8 feet tall. A cable reaches straight from the top of one pole to some point on the ground in between and then goes straight to the top of the other pole. Let l represent the length (in feet) of the cable, and let x represent the distance (in feet) from the base of the 6-foot pole to the point where the cable touches the ground. Write an equation involving l and x only.
4. A rectangle of height 2 inches is inscribed in a circle. Let r represent the radius (in inches), and let A represent the area (in in.^{2}) of the rectangle. Write an equation involving r and A only.
5. Ship A is steaming north at 10 miles per hour, and ship B is steaming east at 15 mph. At 2:00 ship B is 5 miles due west of ship A. Let d represent the distance (in miles) between ships, and let t represent the lapsed time (in hours) since 2:00. Write an equation involving d and t only.
6. A certain baserunner runs at a top speed of 32 ft/sec. The opposing catcher can throw the ball at a speed of 130 ft/sec. The runner attempts to steal third base. The catcher (who is 90 feet from third base) throws the ball toward the third baseman when the runner is 30 feet from third base. Let d represent the distance (in feet) between the ball and the runner, and let t represent lapsed time (in seconds) after the catcher's throw. Write an equation involving d and t only.
7. A farmer wants to fence in 60,000 square feet of land in a rectangular plot along a straight highway. The fence he plans to use along the highway costs \$2 per foot, and the fence for the other three sides costs \$1 per foot. Let x represent the length (in feet) of fence along the highway, and let C represent the total cost (in dollars) of the fence. Write an equation involving x and C only.
8. A rectangular box, open at the top, with a square base, has a volume of 4000 cubic inches. Let h represent its height (in inches) and S represent its surface area (in in.^{2}). Write an equation involving h and S only.
9. A page of a book has an area of 90 square inches. It has a 1-inch margin at the top and 1/2-inch margins at the bottom and sides. Let x represent the width (in

inches) of the page, and let A represent the area (in in.^{2}) of the printed section. Write an equation involving x and A only.

10. A printed page has 1-inch margins at the top and bottom and 3/4-inch margins on the sides. The area of the printed portion is 44 square inches. Let x represent the width (in inches) of the paper, and let A represent the area of the entire page (in in.^{2}). Write an equation involving x and A only.

11. A trough with triangular ends is 12 feet long, 2 feet high, and 3 feet wide at the top. Let d represent the depth (in feet) of the water it contains, and let V represent the volume (in ft^3) of the water. Write an equation involving d and V only.

12. A fence 8 feet tall runs parallel to a tall building and is 2 feet from the building. A ladder leaning against the building over the fence touches the top of the fence. Let y represent the distance (in feet) from the base of the building to the top of the ladder, and let x represent the distance (in feet) from the base of the ladder to the base of the fence. Write an equation involving y and x only.

13. A 10-inch wire is cut into two pieces. One is bent into a square and the other an equilateral triangle. Let x represent the length (in inches) of the piece bent into the triangle, and let S represent the sum of the areas (in in.^{2}). Write an equation involving x and S only.

14. A wire 20 centimeters long is cut into two pieces. One piece is bent into the shape of a square and the other into the shape of a circle. Let x represent the length (in centimeters) of the piece bent into the circle, and let S represent the sum (in cm^2) of the areas. Write an equation involving x and S only.

15. A conical tent has a volume of 27π cubic feet and has no canvas bottom. Let r represent the radius (in feet), and let S represent the surface area (in ft^2). Write an equation involving r and S only.

16. A right cylindrical barrel has a volume of 32π cubic meters. The cost of the material on the sides is \$0.25 per square meter, and the cost of the top and bottom is \$0.50 per square meter. Let r represent the radius (in meters), and let C represent the total cost (in dollars). Write an equation involving r and C only.

17. A Norman window (a rectangular piece of glass topped by a semicircular sheet of glass) has a perimeter of 30 feet. Let r represent the radius (in feet) of the semicircle, and let A represent the area (in ft^2) of the window. Find an equation involving r and A only.

18. A steel storage tank for propane gas is in the shape of a right circular cylinder with a hemisphere at each end. Its volume is 200 cubic feet. Let r represent its radius (in feet), and let S represent its surface area (in ft^2). Write an equation involving r and S only.

19. A clear rectangle of glass is inserted in a colored semicircular glass window (see figure below). The radius of the window is 3 feet. The colored glass admits 50 lumens of light per square foot, and the clear glass passes 100 lumens per square foot. Let x represent the width (in feet), of the rectangle, and let L represent the amount of light (in lumens) which passes through the entire window. Write an equation involving x and L only.

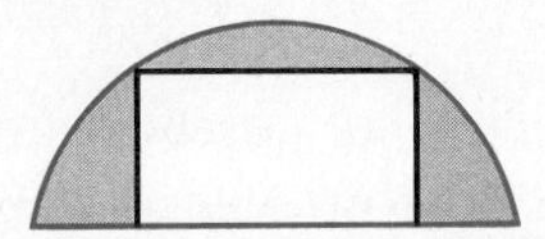

20. An isosceles triangle of height 4 inches is inscribed in a circle. Let r represent the radius (in inches), and let A represent the area (in in.^{2}) of the triangle. Write an equation involving r and A only.

8.3 Complex Numbers

So far all variables considered have been *real* variables. That is, they have represented real numbers only. Hence the solution set of $x^2 = -7$ is $\varnothing$ because the square of a real number is always nonnegative. Similarly, if we use the quadratic formula to solve a quadratic equation and the discriminant $b^2 - 4ac$ is negative, then there are no real solutions. For some applications we need a system of numbers in which such equations have solutions.

To get such a system, we need a number whose square is negative. We thus define i to be the number such that

$$i^2 = -1.$$

(Of course, this number is not real, since its square is negative.) In view of the above we will represent i as

$$i = \sqrt{-1}.$$

Then the set of all numbers of the form

$$a + bi,$$

where a and b are real numbers, is called the set of **complex numbers** and is denoted C. If $z = a + bi$, then a is called the **real part** of z and bi is called the **imaginary part** of z. A complex number like $a + (-b)i$ will be written as $a - bi$. Also, we will agree that

$$0i = 0 \text{ and } a + 0i = a.$$

Hence every real number will also be a complex number.

Two complex numbers $a + bi$ and $c + di$ are said to be **equal** provided that $a = c$ and $b = d$. In other words, equal complex numbers must have equal real parts and equal imaginary parts.

For complex numbers the operations of addition and multiplication will be defined so that when the complex numbers are also real we will get the sum and product expected. In this way, the set of complex numbers, with these operations, is just an extension of the real number system.

The sum of two complex numbers is defined as follows:

$$(a + bi) + (c + di) = (a + c) + (b + d)i.$$

Thus to add complex numbers we merely add the real parts and add the imaginary parts. Note that this is the same as viewing each complex number as a binomial and adding like terms.

The product of two complex numbers is defined as

$$(a + bi)(c + di) = (ac - bd) + (ad + bc)i.$$

Note that this product may be achieved by considering each complex number as a binomial and using the fact that $i^2 = -1$.

$$\begin{aligned}(a + bi)(c + di) &= ac + adi + bci + bdi^2 \\ &= ac + adi + bci + bd(-1) \\ &= (ac - bd) + (ad + bc)i.\end{aligned}$$

Example 1

Perform the indicated operation, expressing the result in the form $a + bi$.
(a) $(2 + 3i) + (5 + 4i)$ (b) $(4 + 2i) + (-2 - 3i)$ (c) $(5 + i)(3 + 2i)$
(d) $(3 - 2i)(4 + 3i)$

Solution. (a) $(2 + 3i) + (5 + 4i) = (2 + 5) + (3i + 4i) = 7 + 7i$
(b) $(4 + 2i) + (-2 - 3i) = (4 - 2) + (2i - 3i) = 2 - i$
(c) $(5 + i)(3 + 2i) = 15 + 10i + 3i + 2i^2 = 15 + 13i + 2(-1) = 13 + 13i$
(d) $(3 - 2i)(4 + 3i) = 12 + 9i - 8i - 6i^2 = 12 + i - 6(-1) = 18 + i$ ■

Exercise

Perform the indicated operation, expressing the result in the form $a + bi$.
(a) $(1 + 2i) + (5 + 3i)$ (b) $(3 + 2i) + (6 - 5i)$ (c) $(1 + 3i)(2 - 4i)$
(d) $(2 - 3i)(4 + 5i)$

Answers. (a) $6 + 5i$ (b) $9 - 3i$ (c) $14 + 2i$ (d) $23 - 2i$ ■

If z_1, z_2, and z_3 are complex numbers, we have the following properties.

Commutative Laws

$$z_1 + z_2 = z_2 + z_1 \quad \text{and} \quad z_1z_2 = z_2z_1$$

Associative Laws

$$(z_1 + z_2) + z_3 = z_1 + (z_2 + z_3) \quad \text{and} \quad (z_1z_2)z_3 = z_1(z_2z_3)$$

Distributive Laws

$$z_i(z_2 + z_3) = z_1z_2 + z_1z_3 \quad \text{and} \quad (z_1 + z_2)z_3 = z_1z_3 + z_2z_3$$

Identity Elements

There exist two complex numbers, 0 and 1 ($0 + 0i$ and $1 + 0i$, respectively) such that

$$z_1 + 0 = 0 + z_1 = z_1 \quad \text{and} \quad z_1 \cdot 1 = 1 \cdot z_1 = z_1.$$

Inverse Elements

For every complex number z, there exists a complex number denoted $-z$ such that

$$z + (-z) = (-z) + z = 0,$$

and for every $z \neq 0$, there exists a complex number denoted $1/z$ such that

$$z\left(\frac{1}{z}\right) = \left(\frac{1}{z}\right)z = 1.$$

The proofs of the first four properties are straightforward. With respect to inverse elements, let $z = a + bi$. Then since

$$(a + bi) + (-a - bi) = 0 + 0i = 0,$$

$-z = -a - bi$. Also, if $z \neq 0$, then

$$\begin{aligned}\left(\frac{a}{a^2 + b^2} - \frac{b}{a^2 + b^2}i\right)\left(a + bi\right) &= \frac{a^2}{a^2 + b^2} + \frac{ab}{a^2 + b^2}i - \frac{ab}{a^2 + b^2}i - \frac{b^2}{a^2 + b^2}i^2 \\ &= \frac{a^2}{a^2 + b^2} + \frac{b^2}{a^2 + b^2} = 1.\end{aligned}$$

Thus

$$\frac{1}{z} = \frac{a}{a^2 + b^2} - \frac{b}{a^2 + b^2}i.$$

The complex number $-z$ is called the additive inverse of z. And the complex number $1/z$ is called the multiplicative inverse of z.

The multiplicative inverse is perhaps easiest to describe with the use of conjugates. Let $z = a + bi$. Then the **conjugate** of z, denoted $\bar{z}$ (read "z bar") is defined by

$$\bar{z} = a - bi.$$

For example, the conjugates of $2 + 3i$ and $5 - 4i$ are $2 - 3i$ and $5 + 4i$, respectively. (Of course, the conjugate of a real number is just itself.)

Note that

$$z\bar{z} = (a + bi)(a - bi) = (a^2 + b^2) + (ab - ab)i = a^2 + b^2.$$

Thus the multiplicative inverse of z may be written:

$$\frac{1}{z} = \frac{a}{a^2 + b^2} - \frac{b}{a^2 + b^2}i = \frac{a - bi}{a^2 + b^2} = \frac{\bar{z}}{z\bar{z}}.$$

The operations of subtraction and division are defined in terms of addition and multiplication, respectively. If z and w are complex numbers, then $z - w = z + (-w)$, and $z \div w = z(1/w)$. As with addition, we may view each complex number as a binomial when subtracting and combine like terms. When dividing, we may simply multiply both the numerator and the denominator by the conjugate of the denominator. See Example 2.

Example 2
Perform the indicated operation, expressing the result in the form $a + bi$.
(a) $(5 + 2i) - (3 + i)$ (b) $(6 - 3i) - (2 - i)$ (c) $(5 + i) \div (3 + 2i)$
(d) $(3 - 2i) \div (4 - 3i)$

Solution. For parts (a) and (b) we merely use the definition and combine like terms.
(a) $(5 + 2i) - (3 + i) = 5 - 3 + 2i - i = 2 + i$
(b) $(6 - 3i) - (2 - i) = 6 - 2 - 3i + i = 4 - 2i$
(c) We have

$$(5 + i) \div (3 + 2i) = \frac{5 + i}{3 + 2i}.$$

Multiplying both the numerator and the denominator by the conjugate of the denominator, we get

$$\begin{aligned}\frac{5 + i}{3 + 2i} &= \frac{5 + i}{3 + 2i} \cdot \frac{3 - 2i}{3 - 2i} \\ &= \frac{15 - 7i - 2i^2}{9 - 4i^2} \\ &= \frac{17 - 7i}{13} \\ &= \frac{17}{13} - \frac{7}{13} i.\end{aligned}$$

(d) We have

$$\begin{aligned}(3 - 2i) \div (4 - 3i) &= \frac{3 - 2i}{4 - 3i} \cdot \frac{4 + 3i}{4 + 3i} \\ &= \frac{12 + i - 6i^2}{16 - 9i^2} \\ &= \frac{18 + i}{25} \\ &= \frac{18}{25} + \frac{1}{25} i.\end{aligned}$$

■

Exercise
Perform the indicated operation, expressing the result in the form $a + bi$.
(a) $(7 + 2i) - (3 + 5i)$ (b) $(11 - 3i) - (7 - 2i)$ (c) $(3 + 2i) \div (4 + i)$
(d) $(5 - 4i) \div (2 - 5i)$

Answers. (a) $4 - 3i$ (b) $4 - i$ (c) $\frac{14}{17} + \frac{5}{17}i$ (d) $\frac{30}{29} + \frac{17}{29}i$ ■

Just as there are two real numbers whose square is 4, there are two complex numbers whose square is -4.

$$(2i)^2 = 2^2i^2 = 4(-1) = -4$$
$$(-2i)^2 = (-2)^2i^2 = 4(-1) = -4.$$

We want the radical symbol to refer to only one of these. Thus, for any positive number a, we define

$$\sqrt{-a} = \sqrt{a}i.$$

Hence

$$\sqrt{-4} = \sqrt{4}i = 2i.$$

Example 3
Find the complex solution set of $x^2 = -9$.

Solution. Using the method of extraction, we have

$$x^2 = -9$$
$$x = \pm\sqrt{-9}$$
$$x = \pm 3i.$$

Therefore, the solution set is $\{3i, -3i\}$. ■

Example 4
Find the complex solution set of $x^2 + x + 1 = 0$.

Solution. Using the quadratic formula, we get

$$x = \frac{-1 \pm \sqrt{1 - 4}}{2}$$
$$= \frac{-1 \pm \sqrt{-3}}{2}$$
$$= \frac{-1 \pm \sqrt{3}i}{2}.$$

Hence the solution set is $\left\{\frac{-1 + \sqrt{3}i}{2}, \frac{-1 - \sqrt{3}i}{2}\right\}$. ■

Example 5
Find the complex solution set of $x^3 = -8$.

Solution. This problem is equivalent to finding the zeros of the polynomial

$$P(x) = x^3 + 8.$$

Clearly, -2 is a zero. Therefore, $x + 2$ is a factor of $P(x)$ by the Factor Theorem. By long division

$$\begin{array}{r|l} & x^2 - 2x + 4 \\ \hline x + 2 & x^3 \qquad\qquad + 8 \\ & x^3 + 2x^2 \\ \hline & -2x^2 \\ & -2x^2 - 4x \\ \hline & 4x + 8 \\ & 4x + 8 \\ \hline & 0 \end{array}$$

Thus $P(x) = (x + 2)(x^2 - 2x + 4)$. To find the solutions of $x^2 - 2x + 4 = 0$, we use the quadratic formula.

$$\begin{aligned} x &= \frac{2 \pm \sqrt{4 - 16}}{2} \\ &= \frac{2 \pm \sqrt{-12}}{2} \\ &= \frac{2 \pm 2\sqrt{3}i}{2} \\ &= 1 \pm \sqrt{3}i. \end{aligned}$$

Hence the solution set is $\{-2, 1 + \sqrt{3}i, 1 - \sqrt{3}i\}$. ■

Exercise
In each of the following, find the complex solution set.
(a) $x^2 = -36$ (b) $x^2 - x + 2 = 0$ (c) $x^3 = -1$

Answers. (a) $\{6i, -6i\}$ (b) $\left\{\frac{1 - \sqrt{7}i}{2}, \frac{1 - \sqrt{7}i}{2}\right\}$
(c) $\left\{-1, \frac{1 + \sqrt{3}i}{2}, \frac{1 - \sqrt{3}i}{2}\right\}$ ■

Note that the equations of Examples 3 and 4 have no real solutions. But with the allowance of complex solutions, the solution set is nonempty. In fact, given any quadratic equation, we can see that the quadratic formula will always give us at least one complex solution. What about other equations of the type

$$P(x) = 0,$$

where P is a polynomial of positive degree? Will these equations also have at least one complex zero? The following theorem (whose proof requires advanced theory) provides the answer.

Theorem 8.1 (Fundamental Theorem of Algebra)

Every polynomial of positive degree has at least one complex zero.

Let's also note that in Examples 3, 4, and 5, the conjugate of each solution is also a solution. Will this always occur for equations of this type? The answer is "yes."

Theorem 8.2 (Conjugate Roots Theorem)

If z is a complex root of $P(x) = 0$, where P is a polynomial of positive degree, then $\bar{z}$ is a complex root of $P(x) = 0$.

Problem Set 8.3

In Problems 1–23, perform the indicated operation, expressing the result in the form $a + bi$.

1. $(6 + 4i) + (5 + 2i)$
2. $(12 + i) + (6 + 7i)$
3. $(4 - 7i) + (-3 + 2i)$
4. $(-5 - 6i) + (-2 - 5i)$
5. $(5 + 7i) - (3 + 2i)$
6. $(6 + 2i) - (7 + 4i)$
7. $(-3 + 2i) - (1 - i)$
8. $(11 - i) - (-5 + 3i)$
9. $(2 + 3i)(5 + 4i)$
10. $(5 + 2i)(7 + i)$
11. $(5 - 2i)(-6 + 3i)$
12. $(-4 - 3i)(2 + 3i)$
13. $(-4 + 5i) \div (3 + 5i)$
14. $(5 + 3i) \div (4 + 2i)$
15. $\dfrac{6 - i}{-3 + 2i}$
16. $\dfrac{-4 + 3i}{5 - 4i}$
17. $(2 + i)(2 - i)$
18. $5(2 - 3i)$
19. $(2 + 3i)^2$
20. $(1 - 2i)^2$
21. i^{36}
22. i^{50}
23. i^{27}

24. Prove that $i^n = 1, i, -1$, or $-i$ if the remainder upon dividing n by 4 is 0, 1, 2, or 3, respectively.
25. Prove that for any complex number z, $\bar{\bar{z}} = z$.
26. Prove that for any complex numbers z and w, $\overline{z \cdot w} = \bar{z} \cdot \bar{w}$.
27. Prove that for any complex number z, $\overline{z^n} = \bar{z}^n$ for all positive integers n.
28. Let f be a function with domain C (the set of complex numbers) defined by $f(x) = x^2 + 2x + 5$. Evaluate **(a)** $f(i)$ **(b)** $f(1 + i)$ **(c)** $f(2 - i)$

In Problems 29–40, find the complex solution set.

29. $x^2 = -16$

30. $x^2 = -25$

31. $(x + 3)^2 = -4$

32. $(x - 2)^2 = -9$

33. $x^2 + 3x + 5 = 0$

34. $x^2 - 5x + 7 = 0$

35. $2x^2 - 4x + 3 = 0$

36. $3x^2 + 2x + 2 = 0$

37. $x^3 = 27$

38. $(x - 1)^3 = -8$

39. $2x + 1 + 3i = 6 - i$

40. $3x - 2 - 5i = 7 + 3i$

Chapter 8 Review

1. The force needed to loosen a bolt with a wrench varies inversely as the length of the wrench. Suppose that for a certain bolt, a 9-inch wrench requires a force of 200 pounds. What size wrench would require only 150 pounds?
2. According to Kepler's third law, the time it takes a planet to revolve around the sun varies directly as the 3/2 power of the maximum radius of its orbit. Using 94 million miles as the maximum radius of the earth's orbit and 68 million miles as the maximum radius of Venus's orbit, how many days does it take Venus to make one revolution?
3. The current I in a wire varies directly as the electromotive force E and inversely as the resistance R. If a resistance of 50 ohms and a force of 120 volts produces a current of 2.4 amperes in a given wire, how much current is produced if the length of the wire is doubled and the radius halved? (Resistance in a wire varies directly as the length and inversely as the square of its radius.)
4. A tank has the shape of a cone with height 10 feet and radius 4 feet. Let d represent the depth (in feet) of the water contained, and let V represent the volume (in ft^3) of the water. Write an equation involving d and V only.
5. A trapezoid has three sides of length 10 feet each and is not a parallelogram. Let x represent the length (in feet) of the fourth side, and let A represent the area (in ft^2). Write an equation involving x and A only.
6. A wire 36 centimeters long is cut into two pieces. One piece is bent into an equilateral triangle and the other into a rectangle whose length is twice its width. Let x represent the length (in centimeters) of the piece bent into the triangle, and let S represent the sum (in cm^2) of the areas. Write an equation involving x and S only.
7. A trough with trapezoidal ends is 12 feet long, 2 feet high, 3 feet wide at the top, and 2 feet wide at the bottom. Let d represent the depth (in feet) of water it contains, and let V represent the volume (in ft^3) of the water. Write an equation involving d and V only.

In Problems 8–15, perform the indicated operation, expressing the result in the form $a + bi$.

8. $(5 + 2i) + (-3 + 5i)$

9. $(7 - i) - (2 + 6i)$

10. $(5 + i)(3 - 2i)$

11. $(-3 - 4i)(1 + i)$

12. $(4 + 2i) \div (3 - 2i)$

13. $\dfrac{6 + 5i}{5 + 2i}$

14. $(2 - 3i)^2$

15. i^{51}

In Problems 16–20, find the complex solution set.

16. $x^2 = -49$

17. $(2x - 1)^2 = -1$

18. $x^2 - 3x + 3 = 0$

19. $2x^2 - 6x + 7 = 0$

20. $x^3 = 8$

21. Prove that for any complex numbers z and w, $\overline{z + w} = \bar{z} + \bar{w}$.

Appendix

Irrationality of $\sqrt{2}$

We will prove that $\sqrt{2}$ is an irrational number by contradiction. That is, we will show that the assumption that it is not irrational leads to a contradiction.

Suppose that $\sqrt{2}$ is *not* irrational. Then it must be rational. Hence there exist integers a and b, relatively prime (no common factors except 1 and -1), such that

$$\sqrt{2} = \frac{a}{b}.$$

The fact that a and b are relatively prime means that the quotient a/b is in reduced form. Upon squaring both sides, we have

$$2 = \frac{a^2}{b^2}.$$

Therefore,

$$a^2 = 2b^2. \tag{A-1}$$

Hence a^2 must be even. And because the square of an odd integer is always odd, a must be even. So there exists an integer k such that $a = 2k$. Substituting for a in equation (A-1), we get

$$
\begin{aligned}
(2k)^2 &= 2b^2 \\
4k^2 &= 2b^2 \\
2k^2 &= b^2.
\end{aligned}
$$

Thus b^2 must be even. It follows that b must also be even. Consequently, both a and b are even. But this contradicts the fact that a and b are relatively prime. Therefore, the supposition that $\sqrt{2}$ is not irrational is false, and we conclude that $\sqrt{2}$ *is* irrational.

Angles

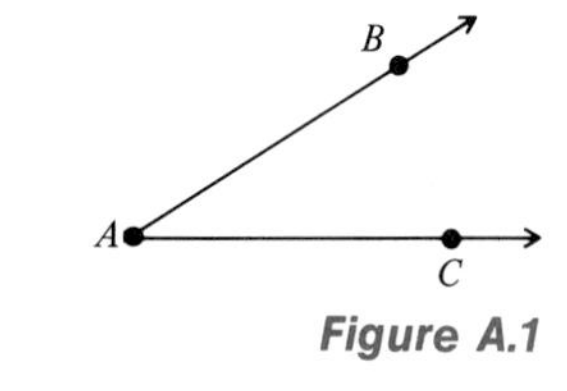

Figure A.1

An **angle** is the union of two rays emanating from a common point. In Figure A.1 we see the angle formed by the two rays $\overrightarrow{AB}$ and $\overrightarrow{AC}$.

Each ray is called a **side**, and the common point is called the **vertex**. For the angle in Figure A.1, the sides are $\overrightarrow{AB}$ and $\overrightarrow{AC}$, and the vertex is A.

The notation for angle is $\angle$. In specifying an angle, we generally use three points: the vertex and one from each side (different from the vertex). With respect to the notation, the vertex appears in the middle. Thus the angle in Figure A.1 is labeled $\angle BAC$ or $\angle CAB$. When the vertex belongs to one angle only (and hence there is no possibility for confusion), we sometimes refer to an angle by its vertex only, like $\angle A$ of Figure A.1.

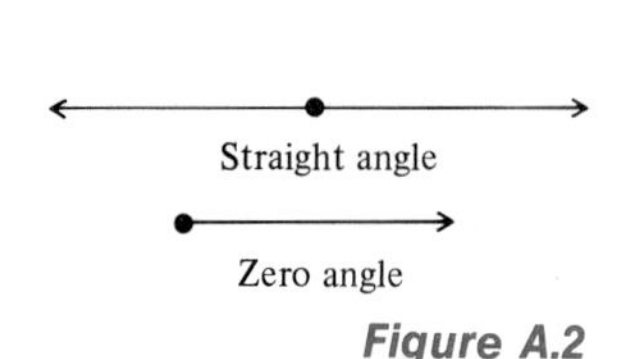

Figure A.2

A **straight angle** is one in which the two rays lie along the same line and point in different directions. A **zero angle** is one in which the two rays are coincident. See Figure A.2.

Postulate

For every angle there is a unique number between 0 and 180 inclusive called the degree measure of the angle.

Protractor Postulate

The rays in a half-plane can be placed in a one-to-one correspondence with the real numbers from 0 to 180 inclusive so that number differences measure angles.

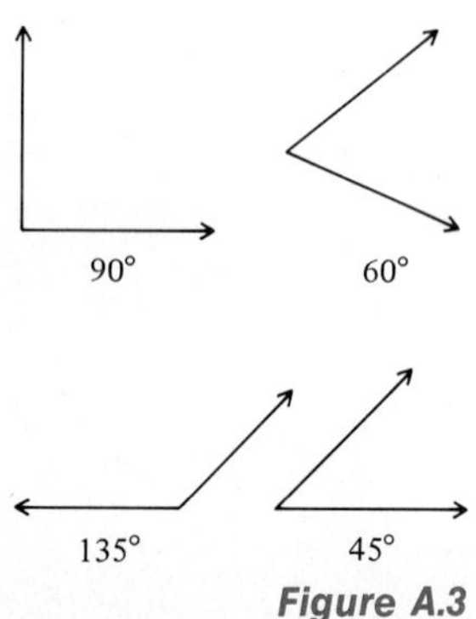

Figure A.3

The measure of $\angle BAC$ is denoted $m\angle BAC$. The measurement system described (*degree* measure) is denoted with the symbol °. The measure of a straight angle is defined to be 180°, and the measure of a zero angle is 0°. For the angle in Figure A.1, we have $m\angle A = 35°$. Some other angles and their measures are shown in Figure A.3.

An angle whose measure is 90° is called a **right angle.** An angle whose measure is between 0° and 90° is called **acute.** And an angle is called **obtuse** if its measure is between 90° and 180°.

Note: If we say that two different angles are equal, we mean that their *measures* are equal (technically, such angles are *congruent,* not equal). And if α represents the measure of an angle, we sometimes refer to the angle as "angle α" when there is no possibility for confusion.

Triangles

A **triangle** is the union of the three line segments connecting three noncollinear points. Each of the line segments is called a side, and each of the points is called a vertex.

Two triangles are said to be **similar** provided that there exists a correspondence between vertices such that the measures of corresponding angles are equal and the lengths of corresponding sides are proportional. Thus we write $\triangle ABC \sim \triangle DEF$ ("triangle ABC is similar to triangle DEF") provided that

$$m\angle A = m\angle D, \qquad m\angle B = m\angle E, \qquad m\angle C = m\angle F,$$

and

$$\frac{AB}{DE} = \frac{AC}{DF} = \frac{BC}{EF}.$$

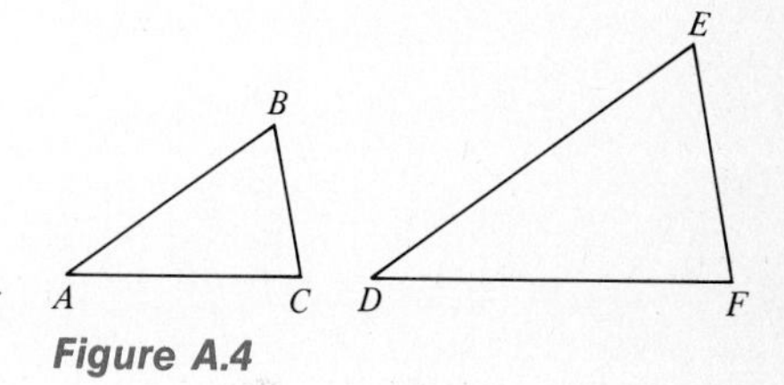

Figure A.4

This definition means that two triangles are similar when they have exactly the same shape, as seen in Figure A.4.

To determine that two triangles are similar, we may use any of the following Similarity Theorems:

Theorem A.1 (Similarity Theorems)

SSS (side–side–side): If there exists a correspondence between the vertices of two triangles such that corresponding sides are proportional, then the triangles are similar.

AA (angle–angle): If there exists a correspondence between the vertices of two triangles such that two pairs of corresponding angles have equal measures, then the triangles are similar.

SAS (side–angle–side): If there exists a correspondence between the vertices of two triangles such that two sides of one triangle are proportional to the corresponding sides of the other, and the measures of the included angles are equal, then the triangles are similar.

Note that the angle–angle theorem requires only *two* pairs of corresponding angles to have equal measures. However, it follows that the third pair must also have equal measures, since the sum of the measures of the three angles of a triangle is 180°.

Further note that although the definition of similar triangles requires two conditions: equal angles and proportional sides, the first two Similarity Theorems indicate that each condition implies the other.

A **right triangle** is a triangle which has a right angle. The side opposite the right angle is called the **hypotenuse**, and the other two sides are called **legs**. An important characteristic of right triangles is the following:

Theorem A.2 (Pythagorean Theorem)

In a right triangle, the sum of the squares of the lengths of the two legs is equal to the square of the length of the hypotenuse.

Thus if we label a right triangle as in Figure A.5, the Pythagorean Theorem says that $a^2 + b^2 = c^2$.

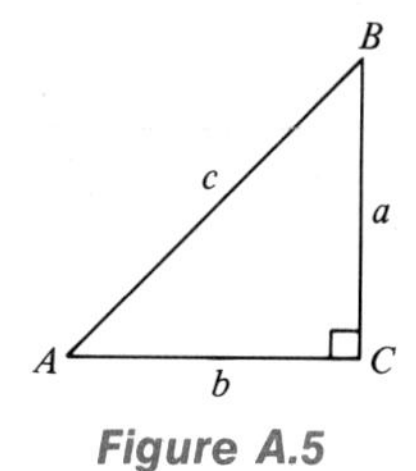

Figure A.5

Table I. Exponentials

x	e^x	e^{-x}	x	e^x	e^{-x}
0.00	1.0000	1.0000	2.5	12.182	0.0821
0.05	1.0513	0.9512	2.6	13.464	0.0743
0.10	1.1052	0.9048	2.7	14.880	0.0672
0.15	1.1618	0.8607	2.8	16.445	0.0608
0.20	1.2214	0.8187	2.9	18.174	0.0550
0.25	1.2840	0.7788	3.0	20.086	0.0498
0.30	1.3499	0.7408	3.1	22.198	0.0450
0.35	1.4191	0.7047	3.2	24.533	0.0408
0.40	1.4918	0.6703	3.3	27.113	0.0369
0.45	1.5683	0.6376	3.4	29.964	0.0334
0.50	1.6487	0.6065	3.5	33.115	0.0302
0.55	1.7333	0.5769	3.6	36.598	0.0273
0.60	1.8221	0.5488	3.7	40.447	0.0247
0.65	1.9155	0.5220	3.8	44.701	0.0224
0.70	2.0138	0.4966	3.9	49.402	0.0202
0.75	2.1170	0.4724	4.0	54.598	0.0183
0.80	2.2255	0.4493	4.1	60.340	0.0166
0.85	2.3396	0.4274	4.2	66.686	0.0150
0.90	2.4596	0.4066	4.3	73.700	0.0136
0.95	2.5857	0.3867	4.4	81.451	0.0123
1.0	2.7183	0.3679	4.5	90.017	0.0111
1.1	3.0042	0.3329	4.6	99.484	0.0101
1.2	3.3201	0.3012	4.7	109.95	0.0091
1.3	3.6693	0.2725	4.8	121.51	0.0082
1.4	4.0552	0.2466	4.9	134.29	0.0074
1.5	4.4817	0.2231	5	148.41	0.0067
1.6	4.9530	0.2019	6	403.43	0.0025
1.7	5.4739	0.1827	7	1096.6	0.0009
1.8	6.0496	0.1653	8	2981.0	0.0003
1.9	6.6859	0.1496	9	8103.1	0.0001
2.0	7.3891	0.1353	10	22026	0.00005
2.1	8.1662	0.1225			
2.2	9.0250	0.1108			
2.3	9.9742	0.1003			
2.4	11.023	0.0907			

Table II. Natural Logarithms

N	ln N	N	ln N	N	ln N
		4.5	1.5041	9.0	2.1972
0.1	−2.3026	4.6	1.5261	9.1	2.2083
0.2	−1.6094	4.7	1.5476	9.2	2.2192
0.3	−1.2040	4.8	1.5686	9.3	2.2300
0.4	−0.9163	4.9	1.5892	9.4	2.2407
0.5	−0.6931	5.0	1.6094	9.5	2.2513
0.6	−0.5108	5.1	1.6292	9.6	2.2618
0.7	−0.3567	5.2	1.6487	9.7	2.2721
0.8	−0.2231	5.3	1.6677	9.8	2.2824
0.9	−0.1054	5.4	1.6864	9.9	2.2925
1.0	0.0000	5.5	1.7047	10	2.3026
1.1	0.0953	5.6	1.7228	11	2.3979
1.2	0.1823	5.7	1.7405	12	2.4849
1.3	0.2624	5.8	1.7579	13	2.5649
1.4	0.3365	5.9	1.7750	14	2.6391
1.5	0.4055	6.0	1.7918	15	2.7081
1.6	0.4700	6.1	1.8083	16	2.7726
1.7	0.5306	6.2	1.8245	17	2.8332
1.8	0.5878	6.3	1.8405	18	2.8904
1.9	0.6419	6.4	1.8563	19	2.9444
2.0	0.6931	6.5	1.8718	20	2.9957
2.1	0.7419	6.6	1.8871	25	3.2189
2.2	0.7885	6.7	1.9021	30	3.4012
2.3	0.8329	6.8	1.9169	35	3.5553
2.4	0.8755	6.9	1.9315	40	3.6889
2.5	0.9163	7.0	1.9459	45	3.8067
2.6	0.9555	7.1	1.9601	50	3.9120
2.7	0.9933	7.2	1.9741	55	4.0073
2.8	1.0296	7.3	1.9879	60	4.0943
2.9	1.0647	7.4	2.0015	65	4.1744
3.0	1.0986	7.5	2.0149	70	4.2485
3.1	1.1314	7.6	2.0281	75	4.3175
3.2	1.1632	7.7	2.0412	80	4.3820
3.3	1.1939	7.8	2.0541	85	4.4427
3.4	1.2238	7.9	2.0669	90	4.4998
3.5	1.2528	8.0	2.0794	95	4.5539
3.6	1.2809	8.1	2.0919	100	4.6052
3.7	1.3083	8.2	2.1041		
3.8	1.3350	8.3	2.1163		
3.9	1.3610	8.4	2.1282		
4.0	1.3863	8.5	2.1401		
4.1	1.4110	8.6	2.1518		
4.2	1.4351	8.7	2.1633		
4.3	1.4586	8.8	2.1748		
4.4	1.4816	8.9	2.1861		

Using the Common Logarithm Table

The table may be used quite easily to find the common logarithm of a number greater than or equal to 1 and less than 10, written to no more than two decimal places. For example, to determine the log of 3.94, we proceed down the leftmost column (labeled N) until we get to 3.9. We then move across until we reach the column labeled 4. The entry here, 0.5955, at the intersection of the row labeled 3.9 and the column 4, is the logarithm of 3.94. Thus

$$\log 3.94 = 0.5955.$$

As another example, let us find log 7. We note that the entry in row 7.0 and column 0 is 0.8451. Thus

$$\log 7 = 0.8451.$$

Finding the logarithm of a positive number outside the interval $[1, 10)$ requires that we write the number in scientific notation. Any positive number x may be expressed as

$$x = k \times 10^n, \qquad \text{where } 1 \leq k < 10,\ n \in \mathsf{Z}.$$

A number expressed in this form, as a number between 1 and 10 times a power of 10, is said to be written in **scientific notation.** Therefore,

$$\begin{aligned} \log x &= \log (k \times 10^n) \\ &= \log k + \log 10^n \\ &= (\log k) + n. \end{aligned}$$

Now $\log k$ may be found from the table, since k is between 1 and 10. It is called the **mantissa** of $\log x$. The integer n is called the **characteristic** of $\log x$.

For example, to find the logarithm of 285 we note that

$$285 = 2.85 \times 10^2.$$

From the table we see that $\log 2.85 = 0.4548$. Thus

$$\log 285 = 0.4548 + 2 = 2.4548.$$

As another example, let us find log 0.00953. We have

$$0.00953 = 9.53 \times 10^{-3}.$$

From the table we get $\log 9.53 = 0.9791$. Hence

$$\log 0.00953 = 0.9791 - 3 = -2.0209.$$

Table III. Common Logarithms

N	0	1	2	3	4	5	6	7	8	9
1.0	.0000	.0043	.0086	.0128	.0170	.0212	.0253	.0294	.0334	.0374
1.1	.0414	.0453	.0492	.0531	.0569	.0607	.0645	.0682	.0719	.0755
1.2	.0792	.0828	.0864	.0899	.0934	.0969	.1004	.1038	.1072	.1106
1.3	.1139	.1173	.1206	.1239	.1271	.1303	.1335	.1367	.1399	.1430
1.4	.1461	.1492	.1523	.1553	.1584	.1614	.1644	.1673	.1703	.1732
1.5	.1761	.1790	.1818	.1847	.1875	.1903	.1931	.1959	.1987	.2014
1.6	.2041	.2068	.2095	.2122	.2148	.2175	.2201	.2227	.2253	.2279
1.7	.2304	.2330	.2355	.2380	.2405	.2430	.2455	.2480	.2504	.2529
1.8	.2553	.2577	.2601	.2625	.2648	.2672	.2695	.2718	.2742	.2765
1.9	.2788	.2810	.2833	.2856	.2878	.2900	.2923	.2945	.2967	.2989
2.0	.3010	.3032	.3054	.3075	.3096	.3118	.3139	.3160	.3181	.3201
2.1	.3222	.3243	.3263	.3284	.3304	.3324	.3345	.3365	.3385	.3404
2.2	.3424	.3444	.3464	.3483	.3502	.3522	.3541	.3560	.3579	.3598
2.3	.3617	.3636	.3655	.3674	.3692	.3711	.3729	.3747	.3766	.3784
2.4	.3802	.3820	.3838	.3856	.3874	.3892	.3909	.3927	.3945	.3962
2.5	.3979	.3997	.4014	.4031	.4048	.4065	.4082	.4099	.4116	.4133
2.6	.4150	.4166	.4183	.4200	.4216	.4232	.4249	.4265	.4281	.4298
2.7	.4314	.4330	.4346	.4362	.4378	.4393	.4409	.4425	.4440	.4456
2.8	.4472	.4487	.4502	.4518	.4533	.4548	.4564	.4579	.4594	.4609
2.9	.4624	.4639	.4654	.4669	.4683	.4698	.4713	.4728	.4742	.4757
3.0	.4771	.4786	.4800	.4814	.4829	.4843	.4857	.4871	.4886	.4900
3.1	.4914	.4928	.4942	.4955	.4969	.4983	.4997	.5011	.5024	.5038
3.2	.5051	.5065	.5079	.5092	.5105	.5119	.5132	.5145	.5159	.5172
3.3	.5185	.5198	.5211	.5224	.5237	.5250	.5263	.5276	.5289	.5302
3.4	.5315	.5328	.5340	.5353	.5366	.5378	.5391	.5403	.5416	.5428
3.5	.5441	.5453	.5465	.5478	.5490	.5502	.5514	.5527	.5539	.5551
3.6	.5563	.5575	.5587	.5599	.5611	.5623	.5635	.5647	.5658	.5670
3.7	.5682	.5694	.5705	.5717	.5729	.5740	.5752	.5763	.5775	.5786
3.8	.5798	.5809	.5821	.5832	.5843	.5855	.5866	.5877	.5888	.5899
3.9	.5911	.5922	.5933	.5944	.5955	.5966	.5977	.5988	.5999	.6010
4.0	.6021	.6031	.6042	.6053	.6064	.6075	.6085	.6096	.6107	.6117
4.1	.6128	.6138	.6149	.6160	.6170	.6180	.6191	.6201	.6212	.6222
4.2	.6232	.6243	.6253	.6263	.6274	.6284	.6294	.6304	.6314	.6325
4.3	.6335	.6345	.6355	.6365	.6375	.6385	.6395	.6405	.6415	.6425
4.4	.6435	.6444	.6454	.6464	.6474	.6484	.6493	.6503	.6513	.6522
4.5	.6532	.6542	.6551	.6561	.6571	.6580	.6590	.6599	.6609	.6618
4.6	.6628	.6637	.6646	.6656	.6665	.6675	.6684	.6693	.6702	.6712
4.7	.6721	.6730	.6739	.6749	.6758	.6767	.6776	.6785	.6794	.6803
4.8	.6812	.6821	.6830	.6839	.6848	.6857	.6866	.6875	.6884	.6893
4.9	.6902	.6911	.6920	.6928	.6937	.6946	.6955	.6964	.6972	.6981
5.0	.6990	.6998	.7007	.7016	.7024	.7033	.7042	.7050	.7059	.7067
5.1	.7076	.7084	.7093	.7101	.7110	.7118	.7126	.7135	.7143	.7152
5.2	.7160	.7168	.7177	.7185	.7193	.7202	.7210	.7218	.7226	.7235
5.3	.7243	.7251	.7259	.7267	.7275	.7284	.7292	.7300	.7308	.7316
5.4	.7324	.7332	.7340	.7348	.7356	.7364	.7372	.7380	.7388	.7396

Table III. *(continued)*

N	0	1	2	3	4	5	6	7	8	9
5.5	.7404	.7412	.7419	.7427	.7435	.7443	.7451	.7459	.7466	.7474
5.6	.7482	.7490	.7497	.7505	.7513	.7520	.7528	.7536	.7543	.7551
5.7	.7559	.7566	.7574	.7582	.7589	.7597	.7604	.7612	.7619	.7627
5.8	.7634	.7642	.7649	.7657	.7664	.7672	.7679	.7686	.7694	.7701
5.9	.7709	.7716	.7723	.7731	.7738	.7745	.7752	.7760	.7767	.7774
6.0	.7782	.7789	.7796	.7803	.7810	.7818	.7825	.7832	.7839	.7846
6.1	.7853	.7860	.7868	.7875	.7882	.7889	.7896	.7903	.7910	.7917
6.2	.7924	.7931	.7938	.7945	.7952	.7959	.7966	.7973	.7980	.7987
6.3	.7993	.8000	.8007	.8014	.8021	.8028	.8035	.8041	.8048	.8055
6.4	.8062	.8069	.8075	.8082	.8089	.8096	.8102	.8109	.8116	.8122
6.5	.8129	.8136	.8142	.8149	.8156	.8162	.8169	.8176	.8182	.8189
6.6	.8195	.8202	.8209	.8215	.8222	.8228	.8235	.8241	.8248	.8254
6.7	.8261	.8267	.8274	.8280	.8287	.8293	.8299	.8306	.8312	.8319
6.8	.8325	.8331	.8338	.8344	.8351	.8357	.8363	.8370	.8376	.8382
6.9	.8388	.8395	.8401	.8407	.8414	.8420	.8426	.8432	.8439	.8445
7.0	.8451	.8457	.8463	.8470	.8476	.8482	.8488	.8494	.8500	.8506
7.1	.8513	.8519	.8525	.8531	.8537	.8543	.8549	,8555	.8561	.8567
7.2	.8573	.8579	.8585	.8591	.8597	.8603	.8609	.8615	.8621	.8627
7.3	.8633	.8639	.8645	.8651	.8657	.8663	.8669	.8675	.8681	.8686
7.4	.8692	.8698	.8704	.8710	.8716	.8722	.8727	.8733	.8739	.8745
7.5	.8751	.8756	.8762	.8768	.8774	.8779	.8785	.8791	.8797	.8802
7.6	.8808	.8814	.8820	.8825	.8831	.8837	.8842	.8848	.8854	.8859
7.7	.8865	.8871	.8876	.8882	.8887	.8893	.8899	.8904	.8910	.8915
7.8	.8921	.8927	.8932	.8938	.8943	.8949	.8954	.8960	.8965	.8971
7.9	.8976	.8982	.8987	.8993	.8998	.9004	.9009	.9015	.9020	.9025
8.0	.9031	.9036	.9042	.9047	.9053	.9058	.9063	.9069	.9074	.9079
8.1	.9085	.9090	.9096	.9101	.9106	.9112	.9117	.9122	.9128	.9133
8.2	.9138	.9143	.9149	.9154	.9159	.9165	.9170	.9175	.9180	.9186
8.3	.9191	.9196	.9201	.9206	.9212	.9217	.9222	.9227	.9232	.9238
8.4	.9243	.9248	.9253	.9258	.9263	.9269	.9274	.9279	.9284	.9289
8.5	.9294	.9299	.9304	.9309	.9315	.9320	.9325	.9330	.9335	.9340
8.6	.9345	.9350	.9355	.9360	.9365	.9370	.9375	.9380	.9385	.9390
8.7	.9395	.9400	.9405	.9410	.9415	.9420	.9425	.9430	.9435	.9440
8.8	.9445	.9450	.9455	.9460	.9465	.9469	.9474	.9479	.9484	.9489
8.9	.9494	.9499	.9504	.9509	.9513	.9518	.9523	.9528	.9533	.9538
9.0	.9542	.9547	.9552	.9557	.9562	.9566	.9571	.9576	.9581	.9586
9.1	.9590	.9595	.9600	.9605	.9609	.9614	.9619	.9624	.9628	.9633
9.2	.9638	.9643	.9647	.9652	.9657	.9661	.9666	.9671	.9675	.9680
9.3	.9685	.9689	.9694	.9699	.9703	.9708	.9713	.9717	.9722	.9727
9.4	.9731	.9736	.9741	.9745	.9750	.9754	.9759	.9763	.9768	.9773
9.5	.9777	.9782	.9786	.9791	.9795	.9800	.9805	.9809	.9814	.9818
9.6	.9823	.9827	.9832	.9836	.9841	.9845	.9850	.9854	.9859	.9863
9.7	.9868	.9872	.9877	.9881	.9886	.9890	.9894	.9899	.9903	.9908
9.8	.9912	.9917	.9921	.9926	.9930	.9934	.9939	.9943	.9948	.9952
9.9	.9956	.9961	.9965	.9969	.9974	.9978	.9983	.9987	.9991	.9996

Table IV. Compound Interest

	$i = 0.005$				$i = 0.01$				$i = 0.015$		
k	$(1+i)^k$	k	$(1+i)^k$	k	$(1+i)^k$	k	$(1+i)^k$	k	$(1+i)^k$	k	$(1+i)^k$
1	1.0050	51	1.2896	1	1.0100	51	1.6611	1	1.0150	51	2.1368
2	1.0100	52	1.2961	2	1.0201	52	1.6777	2	1.0302	52	2.1689
3	1.0151	53	1.3026	3	1.0303	53	1.6945	3	1.0457	53	2.2014
4	1.0202	54	1.3091	4	1.0406	54	1.7114	4	1.0614	54	2.2344
5	1.0253	55	1.3156	5	1.0510	55	1.7285	5	1.0773	55	2.2679
6	1.0304	56	1.3222	6	1.0615	56	1.7458	6	1.0934	56	2.3020
7	1.0355	57	1.3288	7	1.0721	57	1.7633	7	1.1098	57	2.3365
8	1.0407	58	1.3355	8	1.0829	58	1.7809	8	1.1265	58	2.3715
9	1.0459	59	1.3421	9	1.0937	59	1.7987	9	1.1434	59	2.4071
10	1.0511	60	1.3489	10	1.1046	60	1.8167	10	1.1605	60	2.4432
11	1.0564	61	1.3556	11	1.1157	61	1.8349	11	1.1779	61	2.4799
12	1.0617	62	1.3624	12	1.1268	62	1.8532	12	1.1956	62	2.5171
13	1.0670	63	1.3692	13	1.1381	63	1.8717	13	1.2136	63	2.5548
14	1.0723	64	1.3760	14	1.1495	64	1.8905	14	1.2318	64	2.5931
15	1.0777	65	1.3829	15	1.1610	65	1.9094	15	1.2502	65	2.6320
16	1.0831	66	1.3898	16	1.1726	66	1.9285	16	1.2690	66	2.6715
17	1.0885	67	1.3968	17	1.1843	67	1.9477	17	1.2880	67	2.7116
18	1.0939	68	1.4038	18	1.1961	68	1.9672	18	1.3073	68	2.7523
19	1.0994	69	1.4108	19	1.2081	69	1.9869	19	1.3270	69	2.7936
20	1.1049	70	1.4178	20	1.2202	70	2.0068	20	1.3469	70	2.8355
21	1.1104	71	1.4249	21	1.2324	71	2.0268	21	1.3671	71	2.8780
22	1.1160	72	1.4320	22	1.2447	72	2.0471	22	1.3876	72	2.9212
23	1.1216	73	1.4392	23	1.2572	73	2.0676	23	1.4084	73	2.9650
24	1.1272	74	1.4464	24	1.2697	74	2.0882	24	1.4295	74	3.0094
25	1.1328	75	1.4536	25	1.2824	75	2.1091	25	1.4509	75	3.0546
26	1.1385	76	1.4609	26	1.2953	76	2.1302	26	1.4727	76	3.1004
27	1.1442	77	1.4682	27	1.3082	77	2.1515	27	1.4948	77	3.1469
28	1.1499	78	1.4755	28	1.3213	78	2.1730	28	1.5172	78	3.1941
29	1.1556	79	1.4829	29	1.3345	79	2.1948	29	1.5400	79	3.2420
30	1.1614	80	1.4903	30	1.3478	80	2.2167	30	1.5631	80	3.2907
31	1.1672	81	1.4978	31	1.3613	81	2.2389	31	1.5865	81	3.3400
32	1.1730	82	1.5053	32	1.3749	82	2.2613	32	1.6103	82	3.3901
33	1.1789	83	1.5128	33	1.3887	83	2.2839	33	1.6345	83	3.4410
34	1.1848	84	1.5204	34	1.4026	84	2.3067	34	1.6590	84	3.4926
35	1.1907	85	1.5280	35	1.4166	85	2.3298	35	1.6839	85	3.5450
36	1.1967	86	1.5356	36	1.4308	86	2.3531	36	1.7091	86	3.5982
37	1.2027	87	1.5433	37	1.4451	87	2.3766	37	1.7348	87	3.6521
38	1.2087	88	1.5510	38	1.4595	88	2.4004	38	1.7608	88	3.7069
39	1.2147	89	1.5588	39	1.4741	89	2.4244	39	1.7872	89	3.7625
40	1.2208	90	1.5666	40	1.4889	90	2.4486	40	1.8140	90	3.8189
41	1.2269	91	1.5744	41	1.5038	91	2.4731	41	1.8412	91	3.8762
42	1.2330	92	1.5823	42	1.5188	92	2.4979	42	1.8688	92	3.9344
43	1.2392	93	1.5902	43	1.5340	93	2.5228	43	1.8969	93	3.9934
44	1.2454	94	1.5981	44	1.5493	94	2.5481	44	1.9253	94	4.0533
45	1.2516	95	1.6061	45	1.5648	95	2.5735	45	1.9542	95	4.1141
46	1.2579	96	1.6141	46	1.5805	96	2.5993	46	1.9835	96	4.1758
47	1.2642	97	1.6222	47	1.5963	97	2.6253	47	2.0133	97	4.2384
48	1.2705	98	1.6303	48	1.6122	98	2.6515	48	2.0435	98	4.3020
49	1.2768	99	1.6385	49	1.6283	99	2.6780	49	2.0741	99	4.3665
50	1.2832	100	1.6467	50	1.6446	100	2.7048	50	2.1052	100	4.4320

Table IV. *(continued)*

$i = 0.02$				$i = 0.025$				$i = 0.03$	
k	$(1+i)^k$	k	$(1+i)^k$	k	$(1+i)^k$	k	$(1+i)^k$	k	$(1+i)^k$
1	1.0200	51	2.7454	1	1.0250	51	3.5230	1	1.0300
2	1.0404	52	2.8003	2	1.0506	52	3.6111	2	1.0609
3	1.0612	53	2.8563	3	1.0769	53	3.7014	3	1.0927
4	1.0824	54	2.9135	4	1.1038	54	3.7939	4	1.1255
5	1.1041	55	2.9717	5	1.1314	55	3.8888	5	1.1593
6	1.1262	56	3.0312	6	1.1597	56	3.9860	6	1.1941
7	1.1487	57	3.0918	7	1.1887	57	4.0856	7	1.2299
8	1.1717	58	3.1536	8	1.2184	58	4.1878	8	1.2668
9	1.1951	59	3.2167	9	1.2489	59	4.2925	9	1.3048
10	1.2190	60	3.2810	10	1.2801	60	4.3998	10	1.3439
11	1.2434	61	3.3467	11	1.3121	61	4.5098	11	1.3842
12	1.2682	62	3.4136	12	1.3449	62	4.6225	12	1.4258
13	1.2936	63	3.4819	13	1.3785	63	4.7381	13	1.4685
14	1.3195	64	3.5515	14	1.4130	64	4.8565	14	1.5126
15	1.3459	65	3.6225	15	1.4483	65	4.9780	15	1.5580
16	1.3728	66	3.6950	16	1.4845	66	5.1024	16	1.6047
17	1.4002	67	3.7689	17	1.5216	67	5.2300	17	1.6528
18	1.4282	68	3.8443	18	1.5597	68	5.3607	18	1.7024
19	1.4568	69	3.9211	19	1.5987	69	5.4947	19	1.7535
20	1.4859	70	3.9996	20	1.6386	70	5.6321	20	1.8061
21	1.5157	71	4.0795	21	1.6796	71	5.7729	21	1.8603
22	1.5460	72	4.1611	22	1.7216	72	5.9172	22	1.9161
23	1.5769	73	4.2444	23	1.7646	73	6.0652	23	1.9736
24	1.6084	74	4.3293	24	1.8087	74	6.2168	24	2.0328
25	1.6406	75	4.4158	25	1.8539	75	6.3722	25	2.0938
26	1.6734	76	4.5042	26	1.9003	76	6.5315	26	2.1566
27	1.7069	77	4.5942	27	1.9478	77	6.6948	27	2.2213
28	1.7410	78	4.6861	28	1.9965	78	6.8622	28	2.2879
29	1.7758	79	4.7798	29	2.0464	79	7.0337	29	2.3566
30	1.8114	80	4.8754	30	2.0976	80	7.2096	30	2.4273
31	1.8476	81	4.9729	31	2.1500	81	7.3898	31	2.5001
32	1.8845	82	5.0724	32	2.2038	82	7.5746	32	2.5751
33	1.9222	83	5.1739	33	2.2589	83	7.7639	33	2.6523
34	1.9607	84	5.2773	34	2.3153	84	7.9580	34	2.7319
35	1.9999	85	5.3829	35	2.3732	85	8.1570	35	2.8139
36	2.0399	86	5.4905	36	2.4325	86	8.3609	36	2.8983
37	2.0807	87	5.6003	37	2.4933	87	8.5699	37	2.9852
38	2.1223	88	5.7124	38	2.5557	88	8.7842	38	3.0748
39	2.1647	89	5.8266	39	2.6196	89	9.0038	39	3.1670
40	2.2080	90	5.9431	40	2.6851	90	9.2289	40	3.2620
41	2.2522	91	6.0620	41	2.7522	91	9.4596	41	3.3599
42	2.2972	92	6.1832	42	2.8210	92	9.6961	42	3.4607
43	2.3432	93	6.3069	43	2.8915	93	9.9385	43	3.5645
44	2.3901	94	6.4330	44	2.9638	94	10.1869	44	3.6715
45	2.4379	95	6.5617	45	3.0379	95	10.4416	45	3.7816
46	2.4866	96	6.6929	46	3.1139	96	10.7026	46	3.8950
47	2.5363	97	6.8268	47	3.1917	97	10.9702	47	4.0119
48	2.5871	98	6.9633	48	3.2715	98	11.2445	48	4.1323
49	2.6388	99	7.1026	49	3.3533	99	11.5256	49	4.2562
50	2.6916	100	7.2446	50	3.4371	100	11.8137	50	4.3839

Table IV. *(continued)*

$i = 0.04$		$i = 0.05$		$i = 0.06$		$i = 0.07$		$i = 0.08$	
k	$(1 + i)^k$	k	$(1 + i)^k$	k	$(1 + i)^k$	k	$(1 + i)^k$	k	$(1 + i)^k$
1	1.0400	1	1.0500	1	1.0600	1	1.0700	1	1.0800
2	1.0816	2	1.1025	2	1.1236	2	1.1449	2	1.1664
3	1.1249	3	1.1576	3	1.1910	3	1.2250	3	1.2597
4	1.1699	4	1.2155	4	1.2625	4	1.3108	4	1.3605
5	1.2167	5	1.2763	5	1.3382	5	1.4026	5	1.4693
6	1.2653	6	1.3401	6	1.4185	6	1.5007	6	1.5869
7	1.3159	7	1.4071	7	1.5036	7	1.6058	7	1.7138
8	1.3686	8	1.4775	8	1.5938	8	1.7182	8	1.8509
9	1.4233	9	1.5513	9	1.6895	9	1.8385	9	1.9990
10	1.4802	10	1.6289	10	1.7908	10	1.9672	10	2.1589
11	1.5395	11	1.7103	11	1.8983	11	2.1049	11	2.3316
12	1.6010	12	1.7959	12	2.0122	12	2.2522	12	2.5182
13	1.6651	13	1.8856	13	2.1329	13	2.4098	13	2.7196
14	1.7317	14	1.9799	14	2.2609	14	2.5785	14	2.9372
15	1.8009	15	2.0789	15	2.3966	15	2.7590	15	3.1722
16	1.8730	16	2.1829	16	2.5404	16	2.9522	16	3.4259
17	1.9479	17	2.2920	17	2.6928	17	3.1588	17	3.7000
18	2.0258	18	2.4066	18	2.8543	18	3.3799	18	3.9960
19	2.1068	19	2.5270	19	3.0256	19	2.6165	19	4.3157
20	2.1911	20	2.6533	20	3.2071	20	3.8697	20	4.6610
21	2.2788	21	2.7860	21	3.3996	21	4.1406	21	5.0338
22	2.3699	22	2.9253	22	3.6035	22	4.4304	22	5.4365
23	2.4647	23	3.0715	23	3.8197	23	4.7405	23	5.8715
24	2.5633	24	3.2251	24	4.0489	24	5.0724	24	6.3412
25	2.6658	25	3.3864	25	4.2919	25	5.4274	25	6.8485
26	2.7725	26	3.5557	26	4.5494	26	5.8074	26	7.3964
27	2.8834	27	3.7335	27	4.8223	27	6.2139	27	7.9881
28	2.9987	28	3.9201	28	5.1117	28	6.6488	28	8.6271
29	3.1187	29	4.1161	29	5.4184	29	7.1143	29	9.3173
30	3.2434	30	4.3219	30	5.7435	30	7.6123	30	10.0627
31	3.3731	31	4.5380	31	6.0881	31	8.1451	31	10.8677
32	3.5081	32	4.7649	32	6.4534	32	8.7153	32	11.7371
33	3.6484	33	5.0032	33	6.8406	33	9.3253	33	12.6760
34	3.7943	34	5.2533	34	7.2510	34	9.9781	34	13.6901
35	3.9461	35	5.5160	35	7.6861	35	10.6766	35	14.7853
36	4.1039	36	5.7918	36	8.1473	36	11.4239	36	15.9682
37	4.2681	37	6.0814	37	8.6361	37	12.2236	37	17.2456
38	4.4388	38	6.3855	38	9.1543	38	13.0793	38	18.6253
39	4.6164	39	6.7048	39	9.7035	39	13.9948	39	20.1153
40	4.8010	40	7.0400	40	10.2857	40	14.9745	40	21.7245
41	4.9931	41	7.3920	41	10.9029	41	16.0227	41	23.4625
42	5.1928	42	7.7616	42	11.5570	42	17.1443	42	25.3395
43	5.4005	43	8.1497	43	12.2505	43	18.3444	43	27.3666
44	5.6165	44	8.5572	44	12.9855	44	19.6285	44	29.5560
45	5.8412	45	8.9850	45	13.7646	45	21.0025	45	31.9204
46	6.0748	46	9.4343	46	14.5905	46	22.4726	46	34.4741
47	6.3178	47	9.9060	47	15.4659	47	24.0457	47	37.2320
48	6.5705	48	10.4013	48	16.3939	48	25.7289	48	40.2106
49	6.8333	49	10.9213	49	17.3775	49	27.5299	49	43.4274
50	7.1067	50	11.4674	50	18.4202	50	29.4570	50	46.9016

Answers to Selected Problems

Problem Set 0.1, page 4

1. false
2. true
3. false
4. true
5. false
6. false
7. {1, 2, 3, 4, 5}
8. {5}
9. {1, 3}
10. {2, 4}
11. {2, 4}
12. {1, 3, 5}
13. {1, 3, 5, 6}
14. {1, 2, 3, 4, 6}
15. {6}
16. {2, 4}
17. {2, 4, 5, 6}
18. {1, 3}
19. (a), (b), (d), (e), and (f) are always true

Problem Set 0.2, page 9

1. **(a)** {7} **(b)** {−5, 0, 7} **(c)** $\{-5, \frac{22}{7}, 7, \frac{-1}{3}, 0.87, 0, 0.1\overline{3}\}$

2. $\frac{2}{9} = 0.\overline{2}$

$\frac{3}{11} = 0.\overline{27}$

$\frac{5}{12} = 0.41\overline{6}$

3. $0.2\overline{9} = \frac{3}{10}$

$0.\overline{18} = \frac{2}{11}$

$0.0\overline{6} = \frac{1}{15}$

4. yes

5. yes

6. no

7. yes

8. no

9. $A = \frac{-9}{2}$; $B = -1$; $C = \frac{7}{2}$

10.

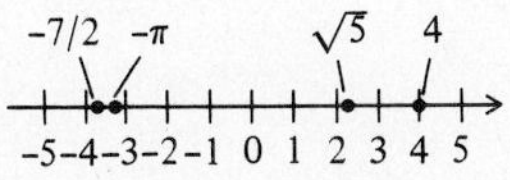

Problem Set 0.3, page 14

1. **(a)** -7 **(b)** 5 **(c)** 0 **(d)** $-\pi$ **(e)** $\frac{2}{5}$

2. **(a)** $\frac{-1}{3}$ **(b)** $\frac{1}{6}$ **(c)** no reciprocal **(d)** 3 **(e)** $-\pi$

7. **(a)** additive identity **(b)** multiplicative inverse **(c)** additive inverse **(d)** distributive law **(e)** associative law **(f)** commutative law **(g)** multiplicative identity

8. **(a)** true **(b)** true **(c)** false **(d)** true **(e)** false **(f)** false **(g)** true **(h)** true **(i)** true

9. 11
10. 16
11. 13
12. -25
13. -3
14. 72
15. 60
16. 70
17. 180
18. 300
19. $\frac{5}{12} = \frac{10}{24}$, $\frac{7}{8} = \frac{21}{24}$

20. $\frac{3}{10} = \frac{18}{60}$, $\frac{1}{4} = \frac{15}{60}$, $\frac{2}{3} = \frac{40}{60}$

21. $\frac{1}{6}$
22. $\frac{3}{4}$
23. $\frac{1}{2}$
24. $\frac{-1}{6}$
25. $\frac{1}{36}$
26. $\frac{2}{15}$

Problem Set 0.4, page 18

1. **(a)** $<$ **(b)** $>$ **(c)** $>$ **(d)** $<$
2. **(a)** $>$ **(b)** $<$ **(c)** $<$ **(d)** $>$
3. **(a)** $c \leq 4$ **(b)** $a > b$ **(c)** $a \geq 0$ **(d)** $3 < b \leq 5$ **(e)** $b < a$ **(f)** $-3 < a < -2$ **(g)** $a \geq -5$
4. **(a)** true **(b)** true **(c)** false **(d)** false **(e)** false **(f)** false
5. no
6. **(a)** -5 **(b)** 3 **(c)** π **(d)** -8
7. 7
8. 3
9. $\frac{2}{3}$
10. 12
11. -3
12. -5
13. 1
14. -1
15. $4 - \pi$
16. $\pi - 3$
17. 0
18. -1
19. no
20. yes
21. no
22. yes, always
23. when $a \leq 0$ and $b \leq 0$, or when $a \geq 0$ and $b \geq 0$

Problem Set 0.5, page 24

1. 81
2. 25
3. $\frac{1}{5}$
4. $\frac{1}{8}$
5. -25
6. 1
7. $\frac{1}{36}$
8. $\frac{3}{2}$
9. $\frac{25}{4}$
10. $\frac{27}{25}$
11. $\frac{82}{9}$
12. $\frac{26}{5}$
13. $\frac{1}{a^3}$
14. $3x$
15. x^{10}

16. $\frac{1}{a^6}$
17. a^7
18. $\frac{2}{x^5}$
19. $\frac{y^3}{x^6}$
20. $9a^2$
21. $9x^6y^4$
22. $\frac{x^6}{y^3}$
23. -4
24. -3
25. $2\sqrt[3]{2}$
26. $2\sqrt{3}$
27. $\frac{\sqrt[3]{5}}{2}$
28. $\frac{2}{\sqrt{3}}\left(\text{or } \frac{2\sqrt{3}}{3}\right)$
29. $2|x|$
30. $-x^2$
31. $\frac{-2\sqrt{5}}{5}$
32. $2\sqrt{3}$
33. $\frac{\sqrt{2} + \sqrt{6}}{4}$
34. $2(\sqrt{5} - 1)$
35. 8
36. -125
37. 4
38. $\frac{1}{9}$
39. $\frac{-1}{3}$
40. $\frac{1}{2}$
41. $\frac{1}{2}$
42. 8
43. x
44. $a^{1/2}$
45. $a^{2/3}$
46. $\frac{1}{x^2}$
47. $\frac{1}{2a^{1/3}}$
48. $\left(\frac{x}{y}\right)^{3/2}$
49. $x^{3/2}$
50. $a^{3/4}$
51. $a^{1/4}$
52. $x^{1/6}$

Problem Set 0.6, page 30

1. 22
2. 11
3. $\frac{-11}{3}$
4. $\frac{-8}{5}$
5. $5x^2 - x$
6. $4x^2 - 2xy + y^2$
7. $-y^2 - 5y + 3$
8. $4x^2 + 5x + 3$
9. $2x^2 - 3xy$
10. $2xy^4 - 2x^2y^3 + 10x^2y^2$
11. $6x^2 - 11x - 10$
12. $3x^2 - xy - 10y^2$
13. $2x^3 + x^2y - 7xy^2 + 3y^3$
14. $2x^4 + 4x^3 - 3x^2 + 17x - 5$
15. $3x - 2 + xy$
16. $\frac{2}{y} - y + 2x^2$
17. $2x + 8 + \frac{39}{2x - 5}$
18. $2x^2 + \frac{5}{3}x + \frac{10}{9} + \frac{-\frac{25}{9}x + 7}{3x^2 - 2x}$
19. $4x^2 + 20x + 25$
20. $9x^4y^2 + 6x^2y^3 + y^4$
21. $9y^2 - 12y + 4$
22. $25x^2y^4 - 10x^3y^2 + x^4$
23. $4x^2 - y^2$
24. $9y^4 - 16x^2y^2$
25. $x^2 + 8x + 15$
26. $x^2 - x - 12$

27. $x^2 + 3x - 10$

28. $x^2 - 10x + 24$

29. $8x^3 + 12x^2y + 6xy^2 + y^3$

30. $x^3 - 9x^2y + 27xy^2 - 27y^3$

Problem Set 0.7, page 37

1. $2(x + 4)$
2. $x(xy + 1)$
3. $xy(x - 3y + xy)$
4. $3xy^2(3x^2 + 1 - 2xy)$
5. $(y + 1)(x + y)$
6. $(2y - 3)(2x^2 - 1)$
7. $(x + 3)(x - 2y)$
8. $(y - 3)(2y - x^2)$
9. $(2y - 3)(2y + 3)$
10. $(x^2 + 4)(x + 2)(x - 2)$
11. $(x + 5)(x + 2)$
12. $(x + 4)(x + 5)$
13. $(x + 4)(x - 3)$
14. $(x - 5)(x + 3)$
15. $(x - 4)(x - 3)$
16. $(x - 2)(x - 7)$
17. $(2x + 1)(x + 5)$
18. $(x - 3)(3x - 4)$
19. $(x + 1)(3x - 8)$
20. $(x - 2)(5x + 3)$
21. $(3x + 4)(4x + 1)$
22. $(6x - 5)(x - 2)$
23. $(2x - 3)(5x + 4)$
24. $(12x - 1)(x + 2)$
25. $2x(x + y)^2$
26. $x^2(3x + y)(2x - 3y)$
27. $x(2x - y)(x + 4y)$
28. $(x + 2)(x - 2)(2x - 3)$

Problem Set 0.8, page 41

1. $-5x$
2. $\frac{1}{3xy^2}$
3. $\frac{-x - 1}{x}$
4. $\frac{1}{x + 1}$
5. $\frac{x + 4}{2x + 1}$
6. $\frac{x(x + 3)}{3(x - 3)}$
7. $\frac{x + 2}{x + y}$
8. $\frac{2x^2 + 5}{x^2 + 1}$
9. $\frac{x^2}{x^2 - 4}$
10. $\frac{x^2 - y^2}{4xy}$
11. $\frac{y^2}{x^2}$
12. $\frac{x(x + 2)}{2y}$
13. $\frac{2x + 5}{x + 5}$
14. $\frac{x + 2}{x + 3}$
15. $\frac{2x - 5}{(x + 2)(x + 3)(x - 1)}$
16. $\frac{x - y}{x(3x + y)(x - 2y)}$
17. $\frac{2x + 1}{7x}$
18. $\frac{2}{x}$
19. $\frac{1}{x - 5}$
20. $\frac{2x + 1}{(2x - 3)(x + 2)}$

21. $\frac{(x-1)(x+7)}{(x-2)(x+1)}$

22. $\frac{x^2-4x-9}{(x+5)(x-1)}$

23. $\frac{x(3x-y)}{2(x+y)(x-y)}$

24. $\frac{1}{5(x-2)}$

Chapter 0 Review, page 42

1. **(a)** {0, 1, 2, 3, 4} **(b)** {4} **(c)** {0, 1} **(d)** {2, 3}
2. **(a)** true **(b)** false **(c)** false **(d)** true **(e)** true **(f)** false
3. **(a)** $a \leq 0$ **(b)** $-10 < a < -8$ **(c)** $a \geq 1$

4. $\frac{\sqrt{5}}{3\sqrt{2}}\left(\text{or } \frac{\sqrt{10}}{6}\right)$

5. $3\sqrt{5}$

6. $-2\sqrt[3]{2}$

7. $\frac{3\sqrt{3}}{2}$

8. $2+\sqrt{3}$

9. $\frac{3\sqrt{2}+\sqrt{3}}{3}$

10. 32

11. 3

12. $\frac{1}{2}$

13. -8

14. $(6x+1)(2x+5)$

15. $a^2b^2(a+b)(a-b)(a^2+b^2)$

16. $(2x-1)(x+1)(x-1)$

17. $(c-d)(a-b)$

18. $\frac{a^2+3}{3-a^2}$

19. $1+2x$

20. $\frac{a-b}{ab}$

21. $\frac{1}{x+y}$

22. $\frac{1}{x-y}$

23. $\frac{y-x}{xy}$

24. $\frac{xy}{x+y}$

25. $\frac{x^2-y^2}{xy}$

26. $\frac{1}{y+x}$

27. $\frac{1+y^{1/2}}{(xy)^{1/2}}$

28. $\frac{y^{1/3}-x^{1/3}}{(xy)^{1/3}}$

29. $x+2(xy)^{1/2}+y$

30. $x-2(xy)^{1/2}+y$

31. $\frac{3}{x^2(x+3)}$

32. $\frac{y}{2x^2}$

33. $\frac{x-6}{x+5}$

34. $x(x-y)$

35. $\frac{x}{x-2y}$

36. $\frac{3}{2x^2}$

37. $\frac{1}{x + 4}$

38. $x^2 + 3x + 1 + \frac{8x - 1}{x^2 - 1}$

Problem Set 1.1, page 51

1. $\mathsf{R} - \{0\}$
3. $\mathsf{R} - \{2, -5\}$
5. all real numbers ≥ 1
7. all nonpositive real numbers
9. all positive real numbers
11. all real numbers ≥ -2, except 0
13. $\left\{\frac{-10}{3}\right\}$
15. $\{4\}$
17. $\left\{\frac{6}{7}\right\}$
19. $\left\{\frac{16}{5}\right\}$
21. $\{2\}$
23. $\left\{\frac{-5}{2}\right\}$
25. $\left\{\frac{-1}{2}\right\}$
27. $\left\{\frac{-20}{3}\right\}$
29. $\left\{\frac{-4}{13}\right\}$
31. $\left\{\frac{2}{3}\right\}$
33. $\{0\}$
35. $\{2\}$
37. $\varnothing$
39. $\{18\}$
41. $\{-\sqrt{5} - 1\}$
43. $x = 9a$
45. $r = \frac{d}{t}, t \neq 0$
47. $C = \frac{5}{9}(F - 32)$
49. $y = 6x - 6x_1 + y_1, x \neq x_1$
51. $y = \frac{6x}{3x - 5}, x \neq \frac{5}{3}$
53. $y' = \frac{x^2}{3 - 2x}, x \neq \frac{3}{2}$
55. $\left\{\frac{2a}{b - a}\right\}, \varnothing$
57. $k = \frac{19}{3}$
59. $k = -6$

Problem Set 1.2, page 55

1. $\{0, 3\}$
3. $\{2, 3\}$
5. $\left\{-3, \frac{7}{2}\right\}$
7. $\left\{\frac{1}{2}, \frac{-1}{5}\right\}$
9. $\{-5, 4\}$
11. $\left\{2, \frac{-3}{2}\right\}$
13. $\left\{-4, \frac{1}{4}\right\}$
15. $\{6, -3\}$
17. $\left\{\frac{-2}{3}, \frac{-3}{4}\right\}$

19. $\left\{-2, \frac{3}{5}\right\}$

21. $\{-4, 2\}$

23. $\{-1\}$

25. $\{2, -3\}$

27. $k = -4$ or $k = 3$

29. $k = -2$

Problem Set 1.3, page 61

1. $\{5, -5\}$

3. $\left\{\frac{7}{2}, \frac{-1}{2}\right\}$

5. $\{2 + \sqrt{7}, 2 - \sqrt{7}\}$

7. $\varnothing$

9. $\left\{\frac{1}{3}, \frac{-1}{7}\right\}$

11. $\left\{2, \frac{-1}{2}\right\}$

13. $\{3 + \sqrt{3}, 3 - \sqrt{3}\}$

15. $\left\{\frac{1 + \sqrt{41}}{4}, \frac{1 - \sqrt{41}}{4}\right\}$

17. $\left\{\frac{1 + \sqrt{13}}{3}, \frac{1 - \sqrt{13}}{3}\right\}$

19. $\varnothing$

21. $\left\{\frac{2 + \sqrt{15}}{2}, \frac{2 - \sqrt{15}}{2}\right\}$

23. $\{3\sqrt{2}, -2\sqrt{2}\}$

25. $k = 4$

Problem Set 1.4, page 64

1. $\{5\}$

3. $\left\{\frac{21}{2}\right\}$

5. $\left\{\frac{7}{6}\right\}$

7. $\varnothing$

9. $\left\{\frac{5}{13}\right\}$

11. $\{-1, 7\}$

13. $\{3\}$

15. $\{3\}$

17. $\{8\}$

19. $\left\{\frac{-3 + \sqrt{41}}{2}, \frac{-3 - \sqrt{41}}{2}\right\}$

21. $\varnothing$

23. $\{4, 8\}$

25. $\{9, 21\}$

27. $\left\{\frac{-1}{4}\right\}$

29. $\{-2, 1\}$

31. $\left\{\frac{3 - \sqrt{5}}{2}\right\}$

33. $\varnothing$

Problem Set 1.5, page 69

1. $\{0, 2, -2\}$

3. $\{0, \sqrt{5}, -\sqrt{5}\}$

5. $\{0\}$

7. $\{2\}$

9. $\left\{2, -2, \frac{1}{2}\right\}$

11. $\left\{\frac{-1}{2}\right\}$

13. $\left\{0, \frac{5}{2}, \sqrt{5}, -\sqrt{5}\right\}$

15. $\left\{0, \frac{3+\sqrt{17}}{4}, \frac{3-\sqrt{17}}{4}\right\}$

17. $\{\sqrt{3}, -\sqrt{3}\}$

19. $\{2, -2, 3, -3\}$

21. $\{\sqrt{3}, -\sqrt{3}\}$

23. $\varnothing$

25. $\left\{\frac{\sqrt{\sqrt{33}-5}}{2}, \frac{-\sqrt{\sqrt{33}-5}}{2}\right\}$

27. $\{-1, 1, -3\}$

29. $\left\{\frac{-1+\sqrt{13}}{2}, \frac{-1-\sqrt{13}}{2}\right\}$

31. $\left\{0, \frac{-1}{8}\right\}$

33. $\left\{\frac{1}{2}, \frac{-1}{2}, -1, 1\right\}$

35. $\{0, 3\}$

37. $\{1, 1+\sqrt{2}, 1-\sqrt{2}\}$

39. $\{3+\sqrt{11}, 3-\sqrt{11}, -3+\sqrt{7}, -3-\sqrt{7}\}$

41. $\varnothing$

43. $\{1\}$

45. $\left\{1, \frac{3}{2}\right\}$

47. $\left\{1, \frac{-5}{3}\right\}$

49. $\varnothing$

Problem Set 1.6, page 74

1. 80
3. \$50,000
5. 28 ft
7. $2\sqrt{2}$($\approx$2.83) in.
9. 60 mph
11. 165 miles
13. 1237.5 ft
15. 50 mph
17. 10 lb
19. 14 gal
21. 2.4 liters
23. 2 hr and 55 min
25. 5 days
27. 3 hr
29. 12 hr

Problem Set 1.7, page 79

1. $\left(-\infty, \frac{5}{3}\right)$

5/3

3. $[-2, \infty)$

-2

5. $(-\infty, -2)$

-2

7. $(-1, \infty)$

-1

9. $(-\infty, 0]$

11. $\left(-\infty, \frac{-5}{8}\right)$

13. $(-2, \infty)$

15. $\left[\frac{2}{3}, \infty\right)$

17. $\left(-\infty, \frac{-1}{6}\right]$

19. $\left(-\infty, \frac{18}{23}\right]$

21. $\left(-\infty, \frac{1}{2}\right]$

23. $[5, \infty)$

25. $\left[\frac{3}{2}, 3\right]$

27. $\left(\frac{13}{4}, \infty\right)$

Problem Set 1.8, page 85

1. $(-3, 1)$

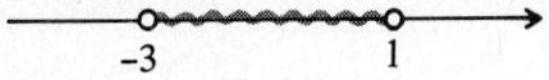

3. $(-\infty, -2] \cup \left[\frac{1}{2}, \infty\right)$

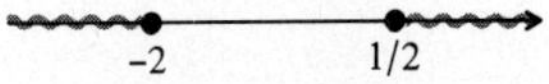

5. $[-3, -2]$

7. $(-\infty, -1) \cup \left(\frac{3}{2}, \infty\right)$

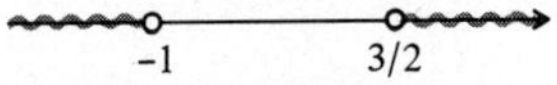

9. $\left[\frac{-1}{2}, \frac{1}{5}\right]$

11. $\left[-4, \frac{1}{2}\right]$

13. $(0, 2)$

15. $(-\sqrt{2}, \sqrt{2})$

17. $(-\infty, -1 - \sqrt{6}] \cup [-1 + \sqrt{6}, \infty)$

19. $\{-1\}$

21. R

23. $\left(\frac{-9}{2}, -1\right) \cup (1, \infty)$

25. $(-1, 2)$

27. $[-5, -3) \cup (-2, \infty)$

29. $(-\infty, 0) \cup (1, \infty)$

31. $(2, 3]$

33. $(-\infty, -2) \cup (1, 4)$

35. $(-\infty, -3) \cup [-1, 1) \cup [3, \infty)$

37. $(-5, -2) \cup (3, \infty)$

39. $(-3, 1) \cup (1, \infty)$

41. $\left(-\infty, \frac{4}{3}\right) \cup (4, \infty)$

Problem Set 1.9, page 89

1. $\left(\frac{-1}{2}, 3\right]$

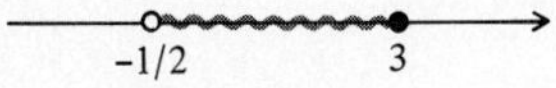

3. $\left(\frac{-7}{5}, \frac{1}{5}\right)$

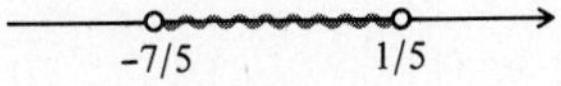

5. $\left(-1, \frac{2}{3}\right]$

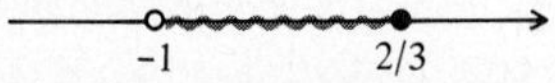

7. $(-4, 0] \cup [1, 5)$

9. $(-2, -1) \cup (2, 3)$

11. $\left[-2, \frac{1}{2}\right)$

13. $[1, 2)$

15. $\left(\frac{-5}{2}, \frac{5}{2}\right)$

17. $\left(-\infty, \frac{-7}{3}\right] \cup \left[\frac{7}{3}, \infty\right)$

19. $(-\infty, -3] \cup [2, \infty)$

21. $\left(\frac{-11}{3}, \frac{1}{3}\right)$

23. $\left(-\infty, \frac{1}{2}\right] \cup \left[\frac{5}{2}, \infty\right)$

25. $\left(\frac{-13}{2}, \frac{7}{2}\right)$

27. $\varnothing$

29. $\left(-3, \frac{-3}{2}\right) \cup \left(-1, \frac{1}{2}\right)$

31. $\left(\frac{1}{3}, \infty\right)$

33. $(-\infty, 2)$

Chapter 1 Review, page 90

1. $\{8\}$

2. $\{-1\}$

3. $\left\{\frac{-24}{5}\right\}$

4. $\left\{-1, \frac{-1}{2}\right\}$

5. $\left\{\frac{-3 + \sqrt{5}}{2}, \frac{-3 - \sqrt{5}}{2}\right\}$

6. $\left\{\frac{3}{2}, \frac{-2}{3}\right\}$

7. $\left\{\frac{1}{2}, -1\right\}$

8. $\left\{\frac{-1 + \sqrt{7}}{3}, \frac{-1 - \sqrt{7}}{3}\right\}$

9. $\left\{\sqrt{3}, \frac{-\sqrt{3}}{2}\right\}$

10. $\varnothing$

11. $\left\{-\frac{1}{3}\right\}$

12. $\{2\}$

13. $\left\{\frac{5}{3}\right\}$

14. $\left\{\frac{\sqrt{6}}{2}, \frac{-\sqrt{6}}{2}\right\}$

15. $\{2 + \sqrt{2}, 2 - \sqrt{2}, -2 + \sqrt{6}, -2 - \sqrt{6}\}$

16. $\left\{\frac{23}{2}, \frac{-17}{2}\right\}$

17. $(-\infty, -2]$

18. $\left(-1, \frac{7}{3}\right)$

19. $(-\infty, -2) \cup (5, \infty)$

20. $(1, 2]$

21. $\mathsf{R} - \{2\}$

22. $(-\infty, -3) \cup (1, \infty)$

23. $\left[\frac{-3}{2}, 2\right)$

24. $(-4, -3] \cup [0, 1)$

25. $\left(\frac{-5 - \sqrt{41}}{2}, -4\right) \cup \left(-1, \frac{-5 + \sqrt{41}}{2}\right)$

26. $[9, \infty)$

27. $(-\infty, 0] - \{-1, -7\}$

28. $(-\infty, -3] \cup [-2, \infty)$

29. $x = a + 3$

30. $y' = \dfrac{1 - 2y}{2x + 1}, x \neq \dfrac{-1}{2}$

31. $x = \dfrac{-y}{3}$ or $x = 3y$

32. $y = \dfrac{-x \pm \sqrt{x^2 + 24x}}{4}$

33. 4 runs

34. 460 miles

35. 8 miles

36. 3 gal

37. 3 hr and 36 min

Problem Set 2.1, page 99

1. $\{(1, 3), (1, 4), (2, 3), (2, 4)\}$

3. $\{(3, 3), (3, 4), (4, 3), (4, 4)\}$

5. $\{(a, b), (a, c), (b, b), (b, c)\}$

7. $\{(0, y) \mid y \in \mathsf{R}\}$

9. $\{(x, y) \mid x \in Z, y \in \mathsf{R}\}$

11. $A = (-3, 3), B = (3, 2), C = (5, 0)$
$D = (3, -4), E = (0, -3), F = (-5, -2)$

13. 5

15. 13

17. $\sqrt{10}$

19. $2\sqrt{5}$

21. 6

23. $\sqrt{74}$

25. $\dfrac{17}{10}$

27. $(2, 5)$

29. $\left(\dfrac{-1}{2}, \dfrac{17}{2}\right)$

31. $\left(2\sqrt{3}, \dfrac{3\sqrt{5}}{2}\right)$

33. $\left(\dfrac{7}{40}, \dfrac{-5}{12}\right)$

35. $\left(\dfrac{19}{10}, 0\right)$

37. $a = 11, b = 1$

Problem Set 2.2, page 103

1. domain $= \{5, -3, 6\}$, range $= \{7, 2, 4\}$

3. domain $= \{2, 3, -1\}$, range $= \{0\}$

5. domain $= \{0, \pi\}$, range $= \{0, \pi\}$

11.

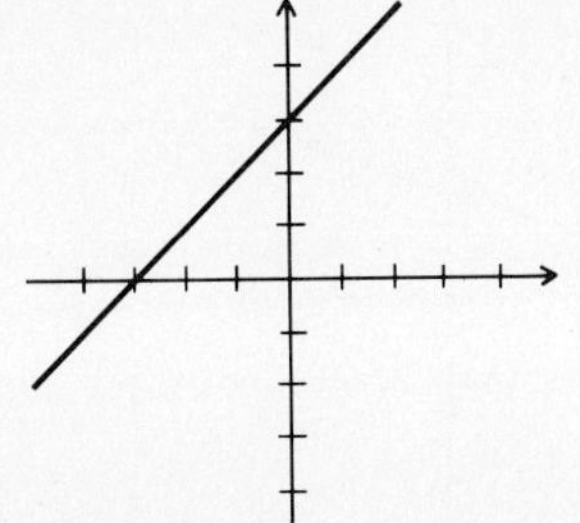

13.

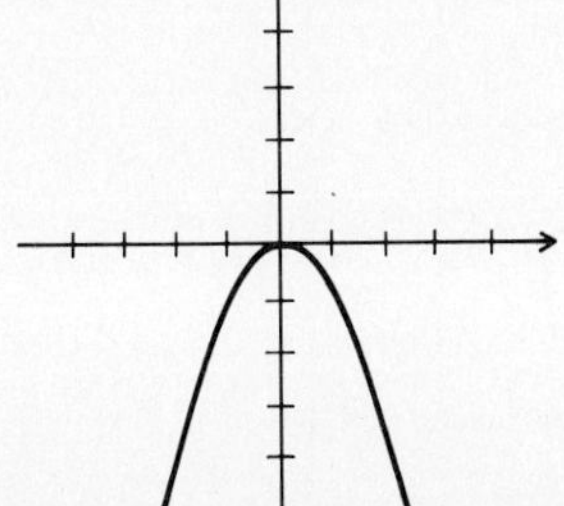

15.

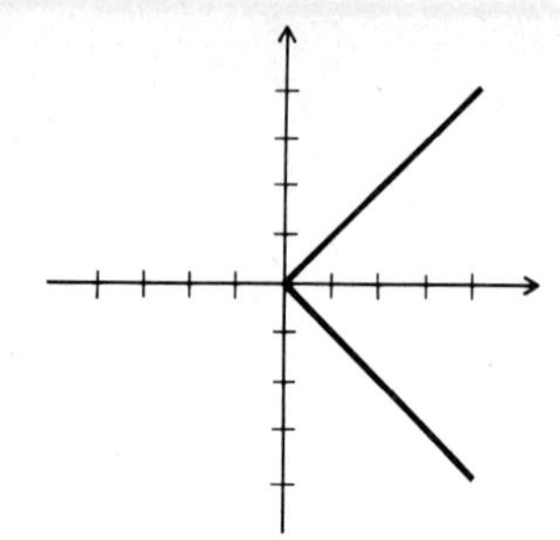

17.

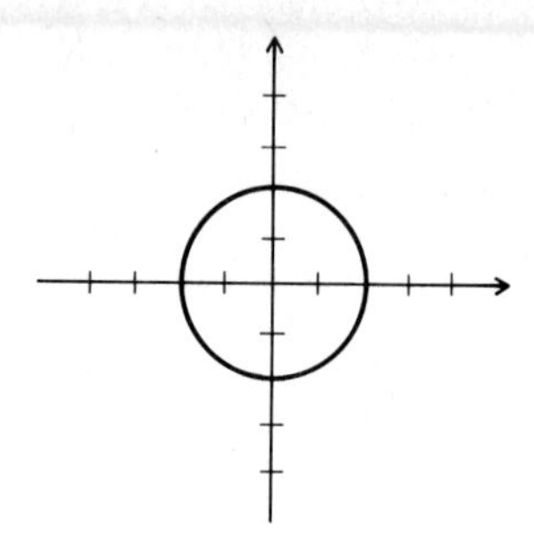

19. domain = R, range = R
21. domain = $\{-2\}$, range = R
23. domain = $[-4, 4]$, range = $[-3, 3]$
25. domain = R, range = $[-1, \infty)$
27. domain = $(-\infty, -1] \cup [1, \infty)$, range = R

Problem Set 2.3, page 114

1. x-intercept $= -5$, y-intercept $= 4$, $m = \frac{4}{5}$

3. x-intercept $= \frac{-3}{5}$, y-intercept $= \frac{3}{2}$, $m = \frac{5}{2}$

5. x-intercept $= 0$, no y-intercept, slope undefined

7.

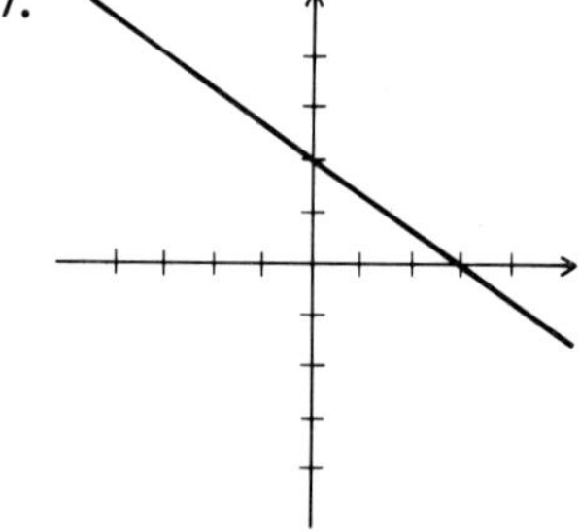

9.

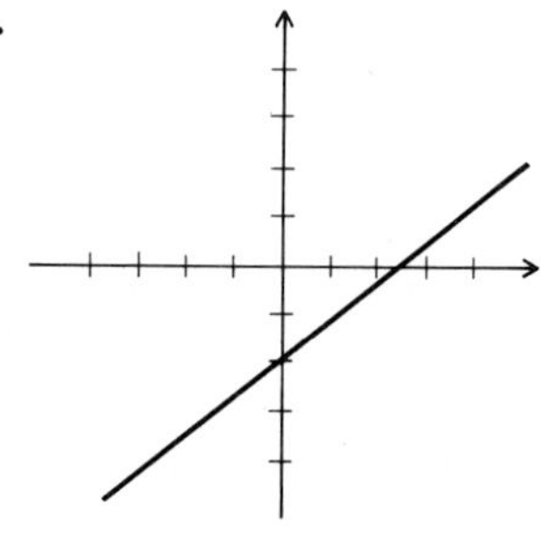

11.

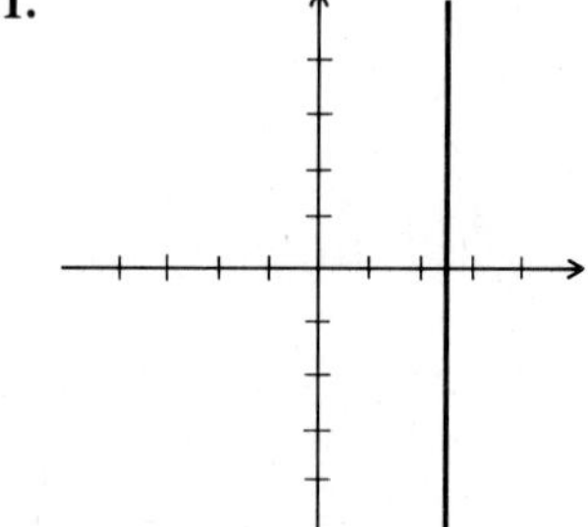

13.

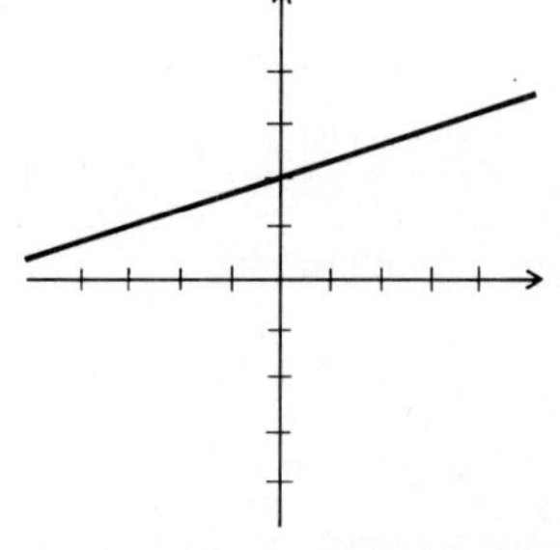

15.

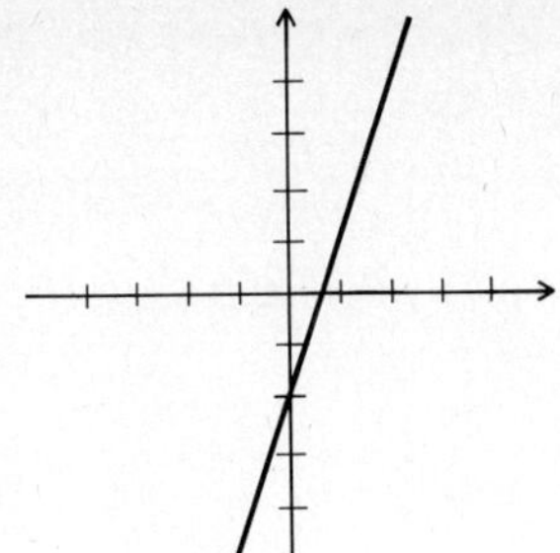

17. -2

19. $\frac{4}{5}$

21. $\frac{-5}{32}$

23. $\sqrt{10}$

25. 0

27. $m = 2$, y-intercept $= -5$

29. $m = \frac{-2}{5}$, y-intercept $= \frac{1}{5}$

31. $m = \frac{-5}{3}$, y-intercept $= \frac{2}{3}$

33. slope undefined, no y-intercept

35. 2

37. -5

39. 0

41. $\frac{-9}{4}$

43. $\frac{4}{3}$

Problem Set 2.4, page 118

1. $y = 3x - 2$

3. $y = \frac{-1}{2}x + \frac{3}{2}$

5. $y = -2$

7. $y = \frac{-3}{5}x + 3$

9. $y = \frac{3}{2}x + \frac{3}{5}$

11. $y = -2x - 1$

13. $y = \frac{2}{3}x - 4$

15. $y = -2x + \frac{11}{5}$

17. $y = \frac{1}{2}x + \frac{23}{40}$

19. $y = -5$

21. $y = -2x + 20$

23. $y = \frac{1}{3}x - \frac{7}{3}$

25. $y = \frac{-5}{3}x + \frac{16}{3}$

27. $y = \frac{25}{22}x - \frac{331}{132}$

29. $x = 1$

31. (a) $y = -2x + 1$ (b) $y = \frac{1}{2}x + 6$

33. (a) $y = \frac{3}{5}x + \frac{41}{30}$ (b) $y = \frac{-5}{3}x + \frac{5}{2}$

35. (a) $x = \frac{5}{2}$ (b) $y = -3$

37. $y = x - 2$

Problem Set 2.5, page 124

1. $(x - 2)^2 + (y + 5)^2 = 9$

3. $x^2 + y^2 = \dfrac{25}{4}$

5. $\left(x - \dfrac{5}{2}\right)^2 + \left(y - \dfrac{5}{3}\right)^2 = \dfrac{9}{4}$

7. $(x - 1)^2 + (y - 1)^2 = 41$

9. $x^2 + 6x + 9 = (x + 3)^2$

11. $x^2 - \dfrac{5x}{2} + \dfrac{25}{16} = \left(x - \dfrac{5}{4}\right)^2$

13. circle of radius 3, centered at (2, 3)

15. circle of radius $\sqrt{2}$, centered at $(-2, 1)$

17. circle of radius 2, centered at $\left(0, \dfrac{1}{2}\right)$

19. circle of radius $\dfrac{3}{2}$, centered at $\left(\dfrac{1}{2}, \dfrac{-7}{3}\right)$

21. point $(-5, 3)$

23. $y = \sqrt{3}x + 8$

25. $y = \dfrac{-9}{5}x + \dfrac{3}{5}$

Problem Set 2.6, page 135

1. $y = \dfrac{x^2}{12}$

3. $y = \dfrac{x^2}{-20}$

5. $y - 8 = \dfrac{(x - 2)^2}{-12}$

7. $x - 1 = \dfrac{3(y + 4)^2}{16}$

9. $x + \dfrac{1}{2} = \dfrac{(y + 2)^2}{-10}$

11. $y - 3 = \dfrac{(x - 3)^2}{-4}$

13. $y - \dfrac{23}{5} = \dfrac{-5(x + 1)^2}{32}$

15. $x + \dfrac{7}{2} = \dfrac{(y + 3)^2}{-6}$

17. $y = \dfrac{x^2}{12}$

19. $y - 3 = \dfrac{(x + 1)^2}{6}$

21. $x + 1 = \dfrac{(y + 3)^2}{-12}$

23. vertex $(2, -3)$

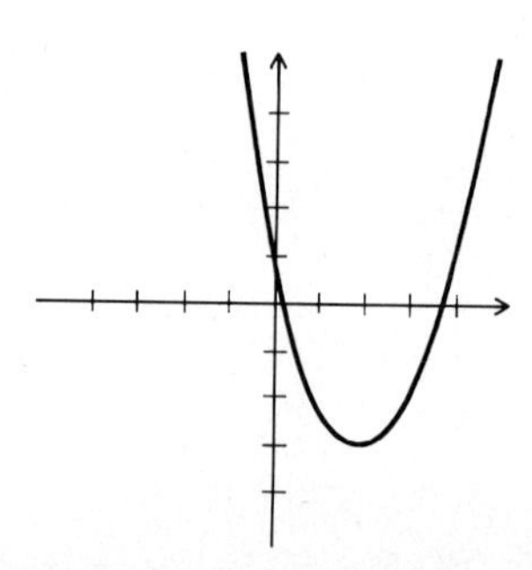

25. vertex $(-1, -4)$

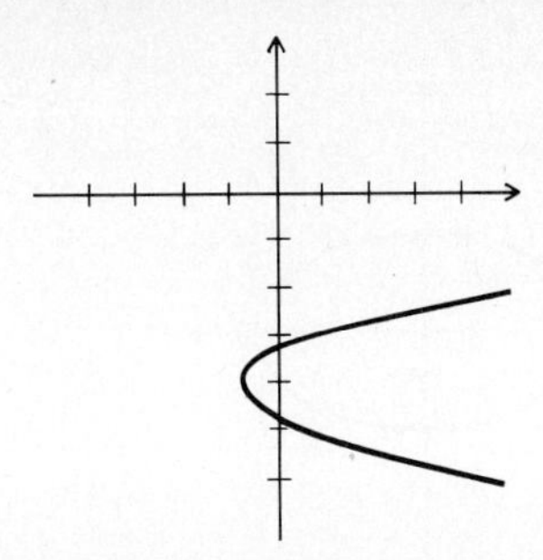

27. vertex $(4, 2)$

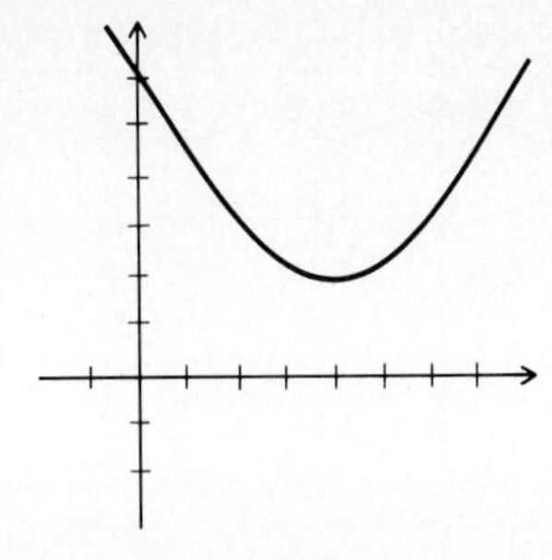

29. vertex $(3, -2)$

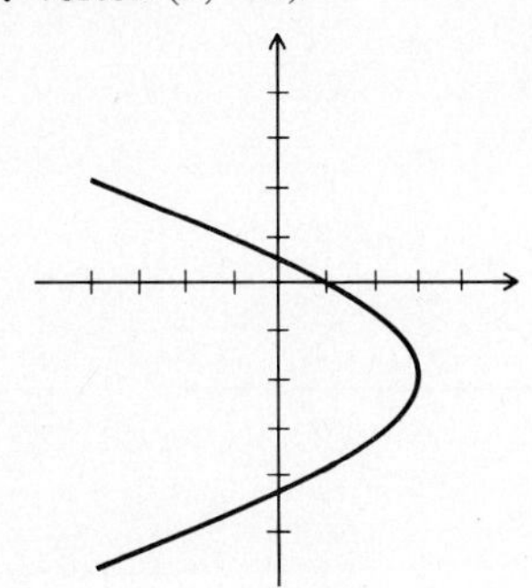

31. focus $\left(3, \frac{-15}{16}\right)$, directrix $y = \frac{-17}{16}$

33. $y + 3 = \frac{(x-2)^2}{10}$

35. 1250

37. \$3.50

39. 125,000 ft^2

41. 32 ft; 64 ft

43. $y = x - \frac{x^2}{288}$

45. 96 ft/sec

47. $\frac{1}{25}$ ft (0.48 in.)

49. 36 ft; 1.5 sec

51. 1.25 in.

53. 9 ft

55. 91 ft

57. $y = \frac{4bx^2}{a^2}$

Problem Set 2.7, page 143

1.

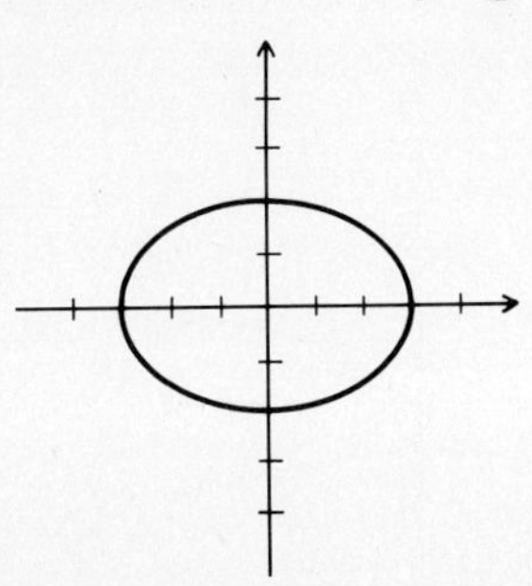

3.

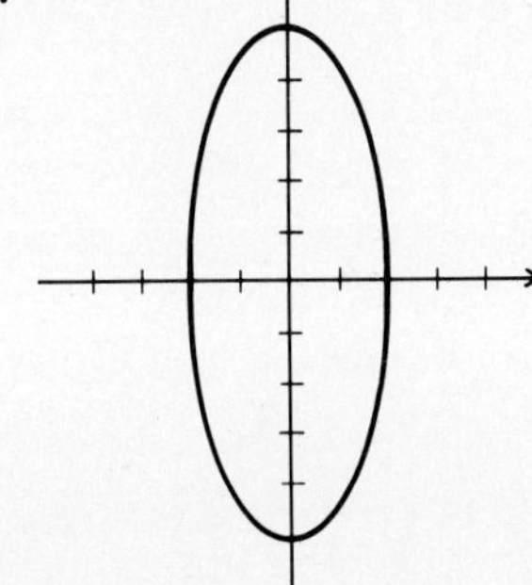

5.

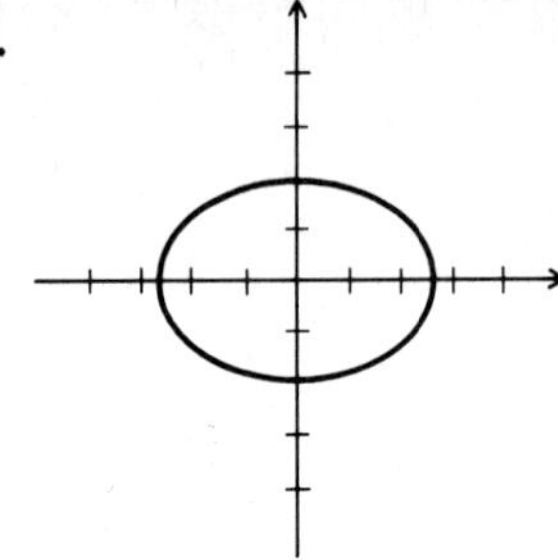

7. $\frac{x^2}{\frac{49}{4}} + \frac{y^2}{4} = 1$

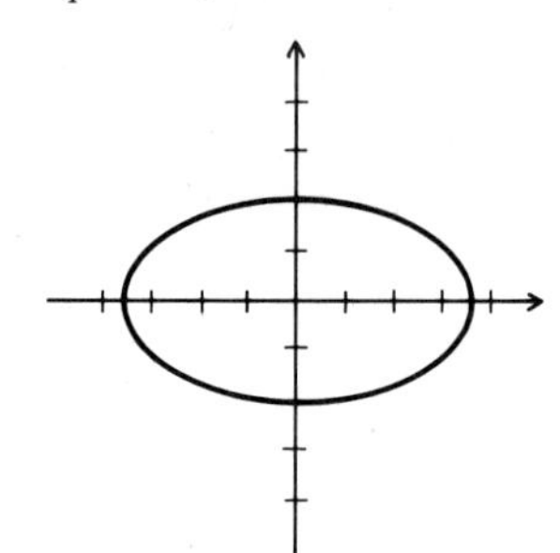

9. $\frac{x^2}{9} + \frac{y^2}{\frac{49}{4}} = 1$

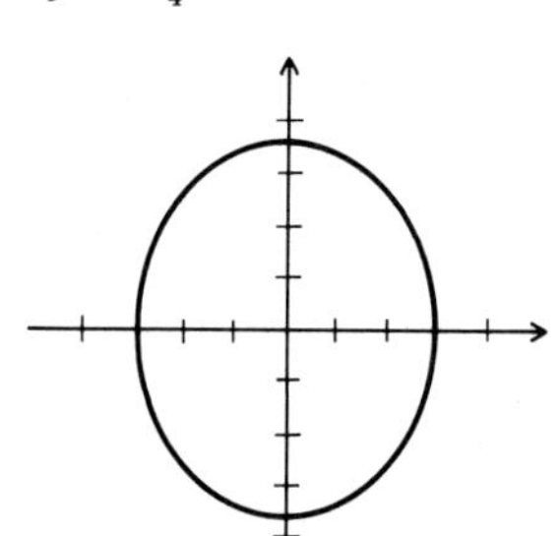

11. $\frac{x^2}{6} + \frac{y^2}{\frac{15}{4}} = 1$

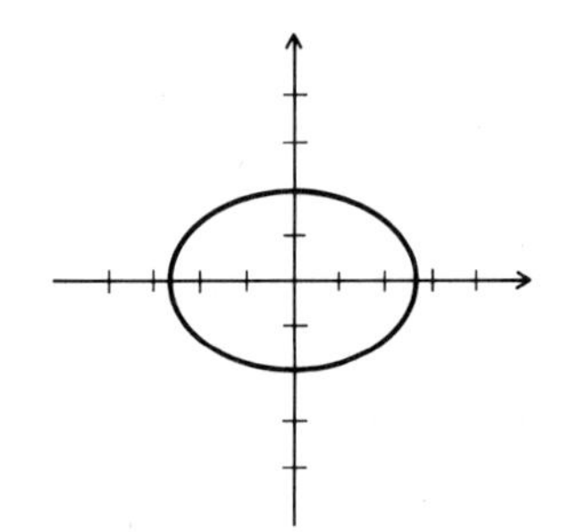

13. $\frac{x^2}{36} + \frac{y^2}{16} = 1$

15. $\frac{x^2}{16} + \frac{y^2}{4} = 1$

17. $\frac{x^2}{9} + \frac{y^2}{25} = 1$

19. $\frac{x^2}{5} + \frac{y^2}{16} = 1$

21. $\frac{x^2}{15} + \frac{y^2}{10} = 1$

23. $(4, 0)$ and $(-4, 0)$

25. Each focus should be 2 ft from the longer sides and $2\sqrt{3}$ (≈ 3.46) ft away from the center. String should be 8 ft.

27. $4\sqrt{15}$ (≈ 15.49) ft

29. ≈ 207 million km

31. $\frac{x^2}{(4400)^2} + \frac{y^2}{(4390)^2} = 1$

Problem Set 2.8, page 150

1.

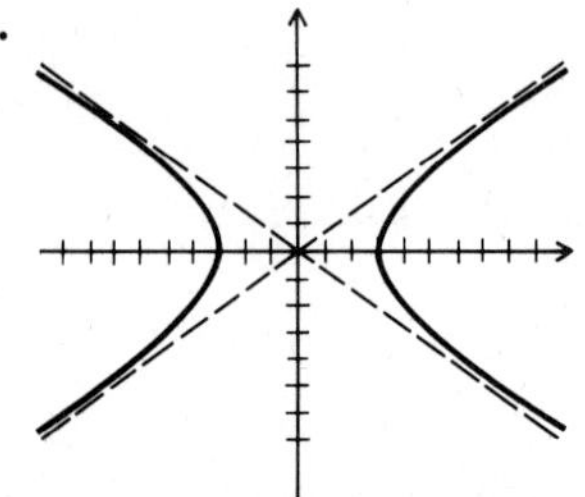

3.

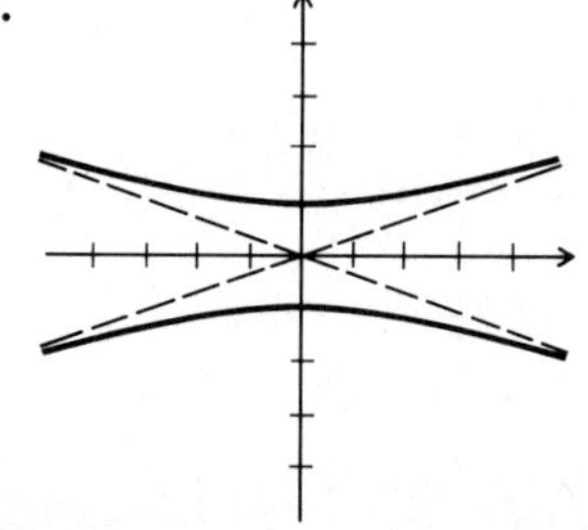

5.

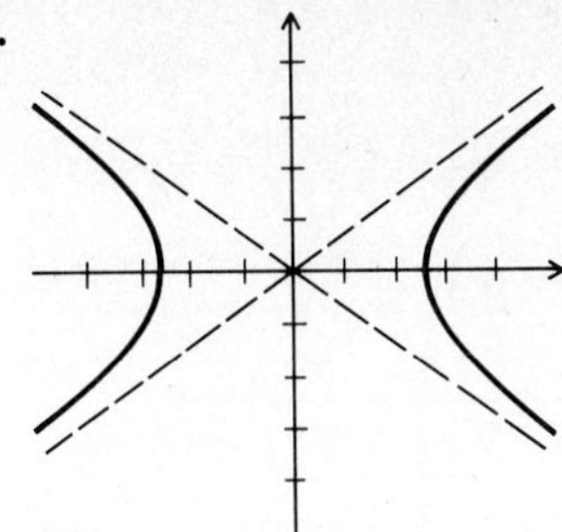

7. $\dfrac{x^2}{\frac{49}{4}} - \dfrac{y^2}{4} = 1$

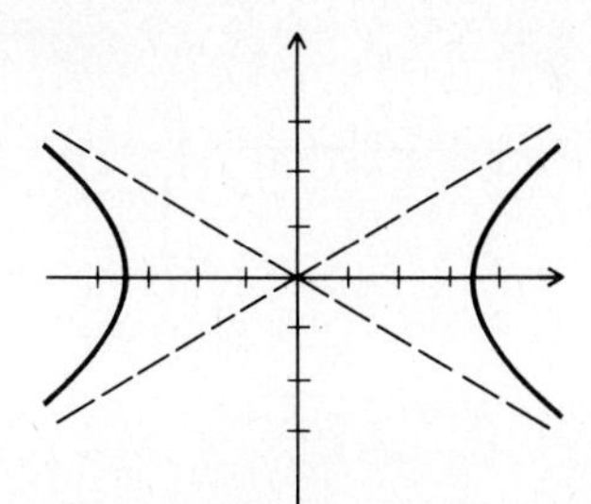

9. $\dfrac{y^2}{9} - \dfrac{x^2}{\frac{49}{4}} = 1$

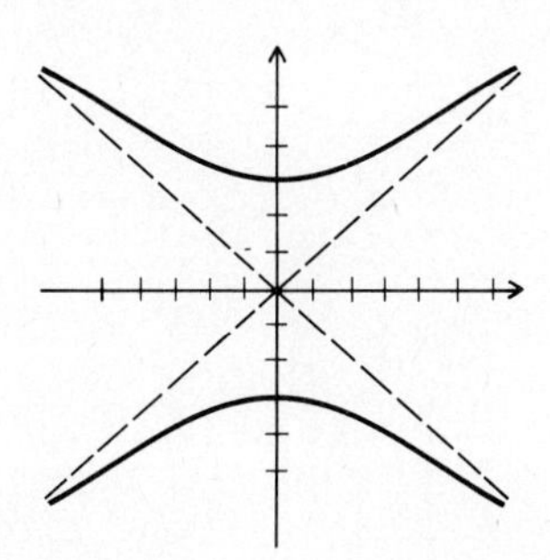

11. $\dfrac{y^2}{\frac{15}{4}} - \dfrac{x^2}{6} = 1$

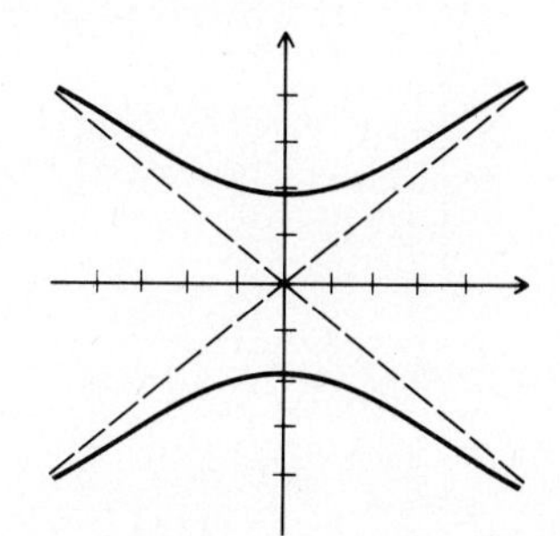

13. $\dfrac{x^2}{20} - \dfrac{y^2}{16} = 1$

15. $\dfrac{y^2}{9} - \dfrac{x^2}{16} = 1$

17. $\dfrac{x^2}{11} - \dfrac{y^2}{25} = 1$

19. $\dfrac{x^2}{6} - \dfrac{y^2}{4} = 1$

21. $\dfrac{x^2}{16} - \dfrac{y^2}{20} = 1$

23. $(\pm 2\sqrt{10}, 0)$

25. $\dfrac{x^2}{(2200)^2} - \dfrac{y^2}{3 \cdot (2200)^2} = 1,\ x \leq -2200,$

if Mike $= (-4400, 0)$, Jim $= (4400, 0)$

27. $\dfrac{x^2}{900} - \dfrac{y^2}{1600} = 1,\ x \geq 30,$

if $A = (-50, 0)$, $B = (50, 0)$

Chapter 2 Review, page 151

1. $PQ = \dfrac{\sqrt{181}}{6}$, midpoint $= \left(\dfrac{-13}{6}, \dfrac{7}{4}\right)$

2. **(a)** domain $= \mathsf{Z}^+$, range $= \{0\}$ **(b)** domain $= \mathsf{R}$, range $= \mathsf{R}$
(c) domain $= [0, \infty)$, range $= (-\infty, -2] \cup [2, \infty)$

3.

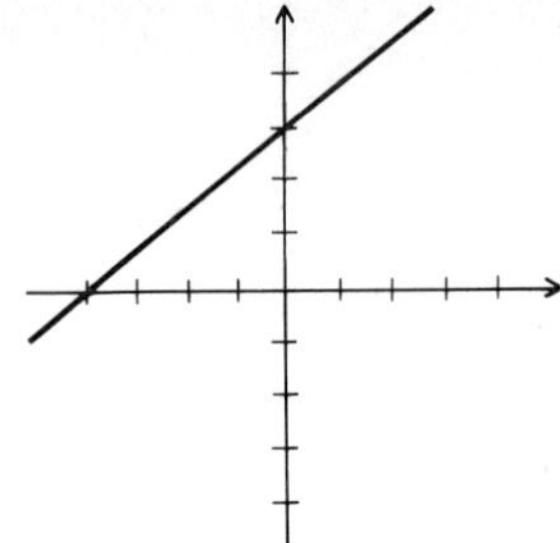

4.

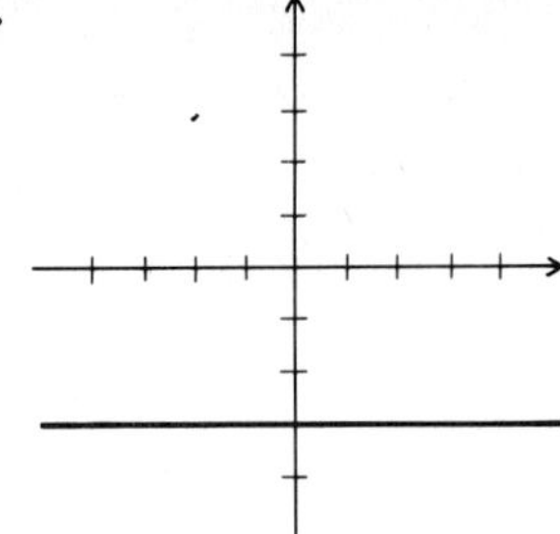

5.

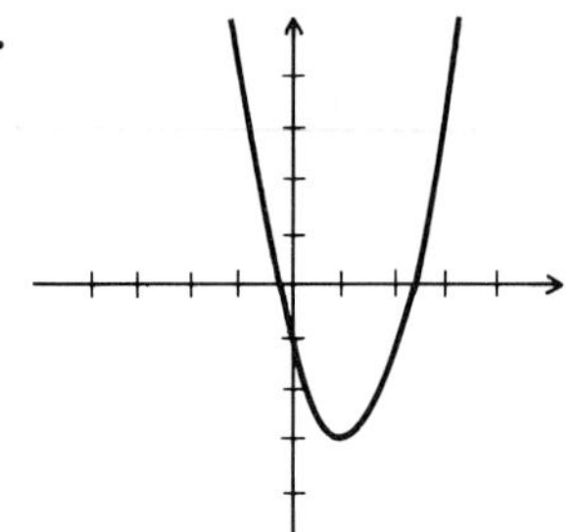

6.

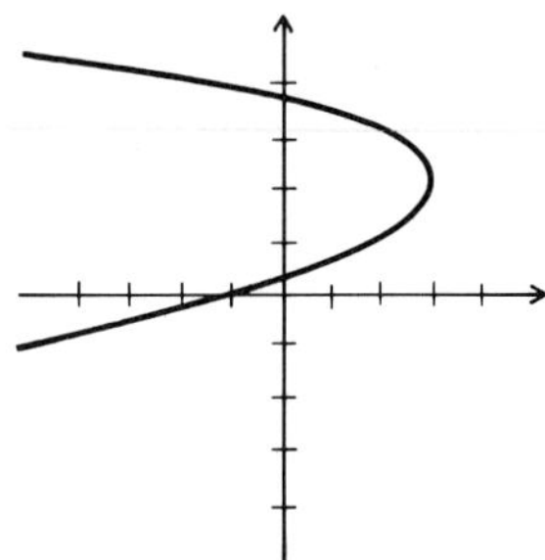

7.

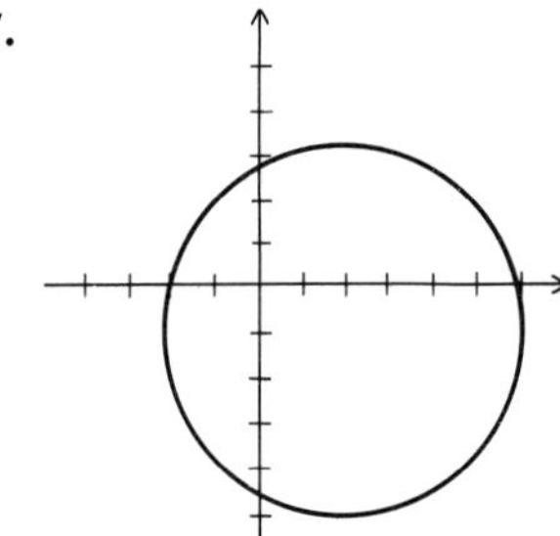

8.

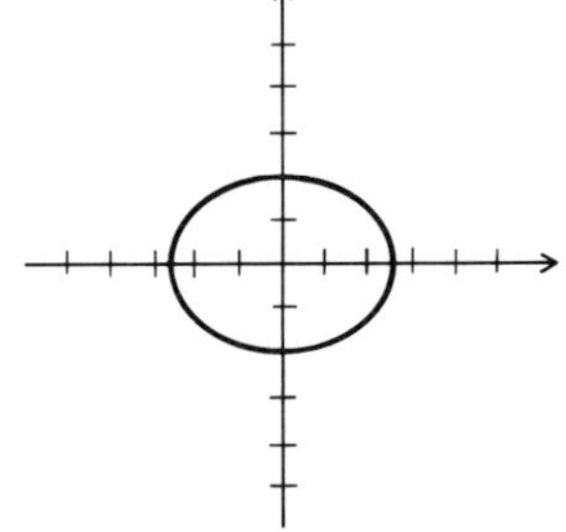

9.

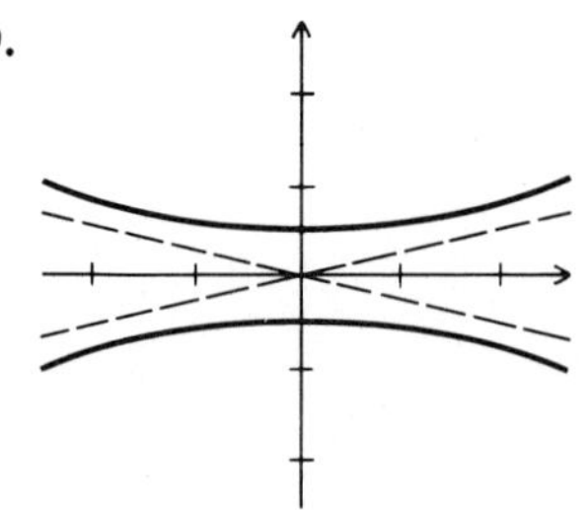

10. $y = \frac{1}{2}x - 2$

11. $y = 5x + 7$

12. $y = \frac{-1}{6}x + \frac{17}{6}$

13. $y = -2x + 5$

14. $y = \frac{3}{2}x - \frac{3}{2}$

15. $x = -2$

16. $y = \frac{3}{2}x - 3$

17. $(x + 3)^2 + y^2 = 4$

18. $(x - 2)^2 + (y + 3)^2 = 20$

19. $x + \frac{1}{2} = \frac{(y - 2)^2}{-10}$

20. $y - \frac{3}{2} = \frac{(x - 1)^2}{2}$

21. $y + 3 = \frac{(x + 2)^2}{-8}$

22. $x - 4 = \frac{3(y + 3)^2}{16}$

23. $\frac{x^2}{4} + \frac{y^2}{\frac{25}{16}} = 1$

24. $\frac{x^2}{\frac{25}{16}} + \frac{y^2}{\frac{225}{16}} = 1$

25. $\frac{y^2}{4} - \frac{x^2}{12} = 1$

26. $\frac{x^2}{8} - \frac{y^2}{12} = 1$

27. $\frac{y^2}{4} - \frac{x^2}{2} = 1$

28. focus $\left(\frac{-2}{3}, \frac{5}{2}\right)$,

directrix $y = \frac{-3}{2}$

29. $(\pm 2, 0)$

30. $\left(0, \frac{\pm\sqrt{15}}{2}\right)$

31. $y = \frac{1}{3}x + \frac{7}{3}$

32. $\left(x + \frac{3}{2}\right)^2 + (y - 3)^2 = 5$

33. 0.51 ft

34. 8.4 ft

35. $1.15

36. 286 ft

37. $6\sqrt{6}$ (≈ 14.70) ft

38. ≈ 22 million miles

39. $\frac{y^2}{4} - \frac{x^2}{96} = 1,\ x \geq 0,\ y \geq 2$

Problem Set 3.1, page 161

1. is
3. is
5. is
7. is not
9. is not
11. is
13. is
15. is
17. is not
19. is
21. is not
23. is
25. is
27. R
29. R $-$ {0}
31. $[0, \infty)$
33. R
35. $[-3, 1]$
37. (a) 5 (b) 6 (c) 3

39. (a) 3 (b) $2a^2 - a + 3$ (c) $7 - \sqrt{2}$ (d) 4 (e) 31 (f) $4t + 2h - 1$
41. (a) $2x + 2 + h$ (b) $x^4 + 2x^2 - 1$ (c) $x^4 + 4x^3 + 4x^2 - 2$
43. $f(x) = 2x + 5$

45. $f(x) = \frac{2x^2 + x - 7}{2}$

47. (a) $[-3, 6]$ (b) $[-6, 6]$ (c) -4 (d) 4 (e) -2 or 4

Problem Set 3.2, page 167

1. range = $\{-2\}$

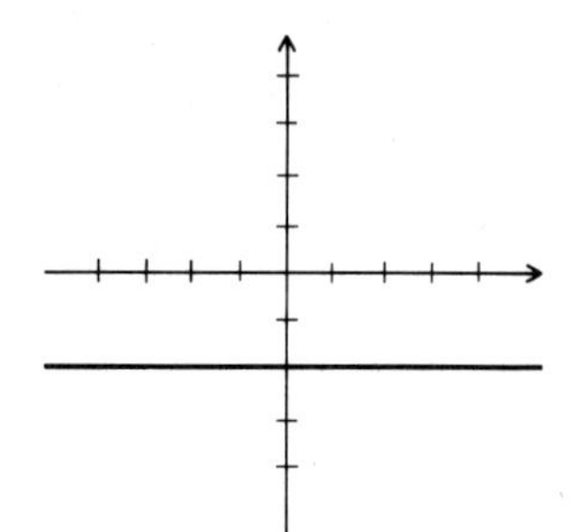

3. range = R

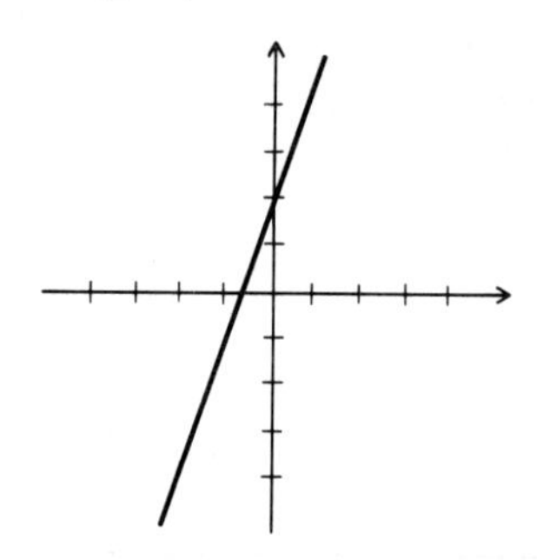

5. range = R

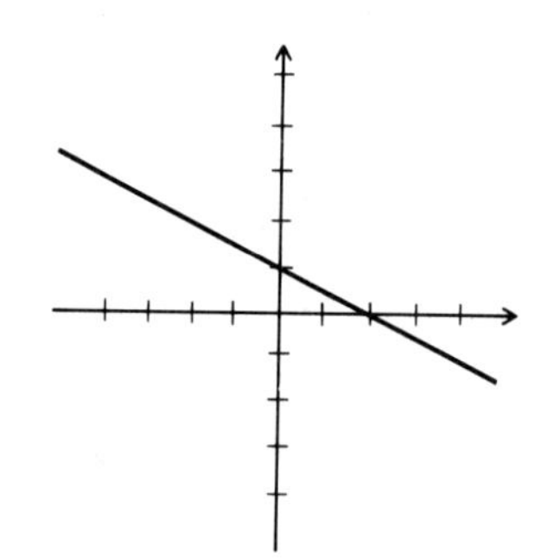

7. range = $[0, \infty)$

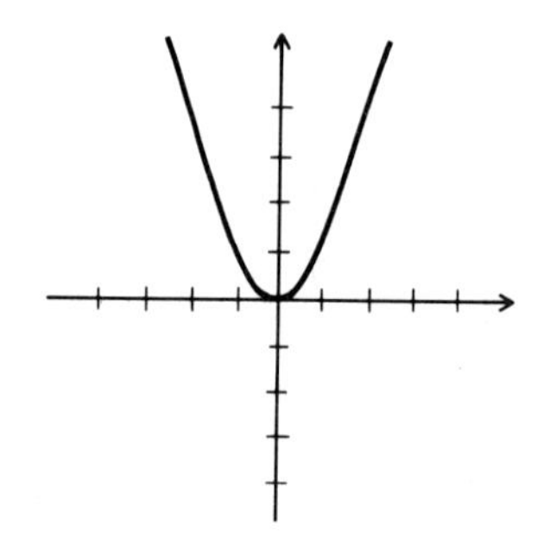

9. range = $[-1, \infty)$

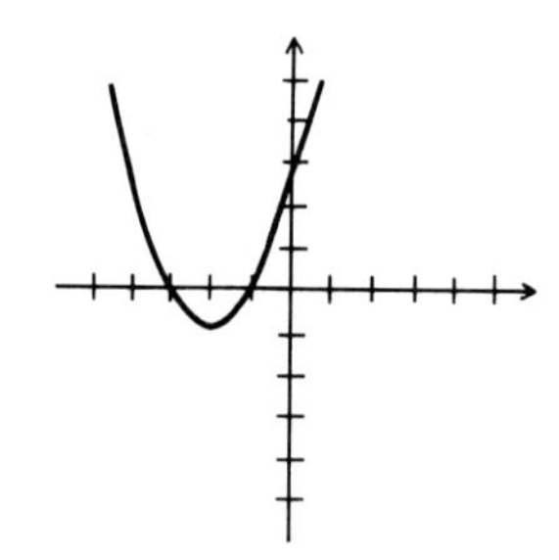

11. range = $(-\infty, 3]$

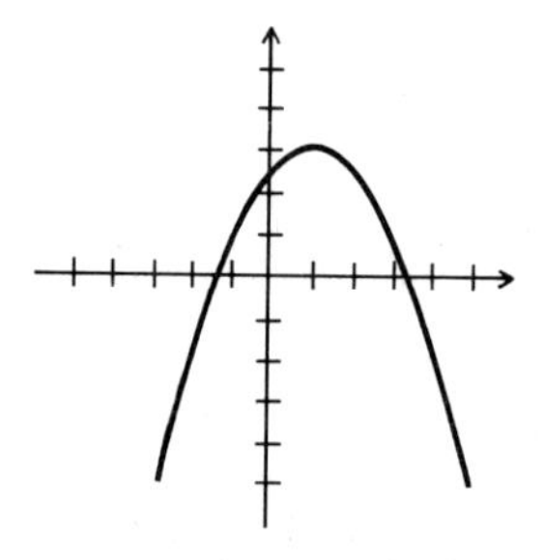

13.

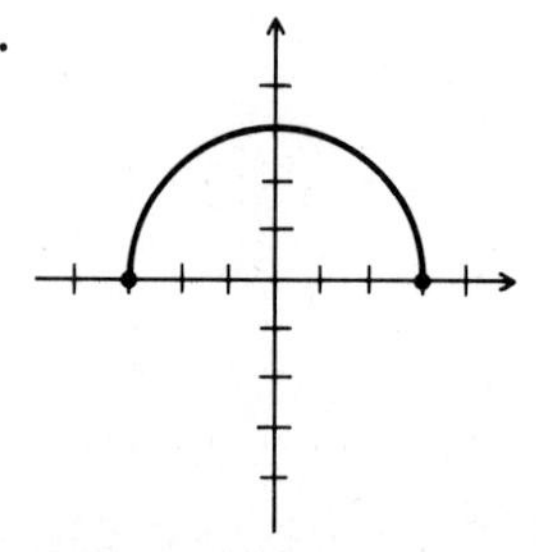

15.

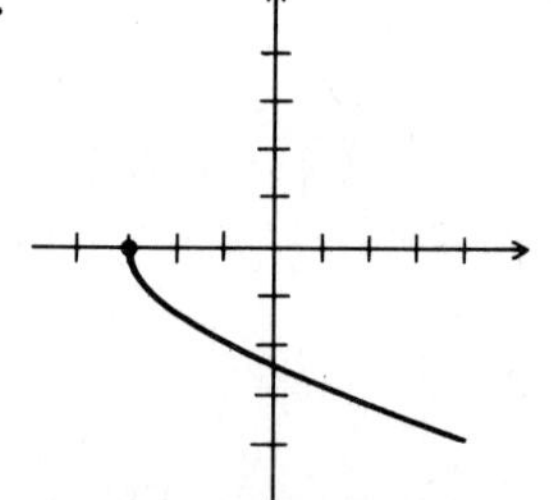

17.

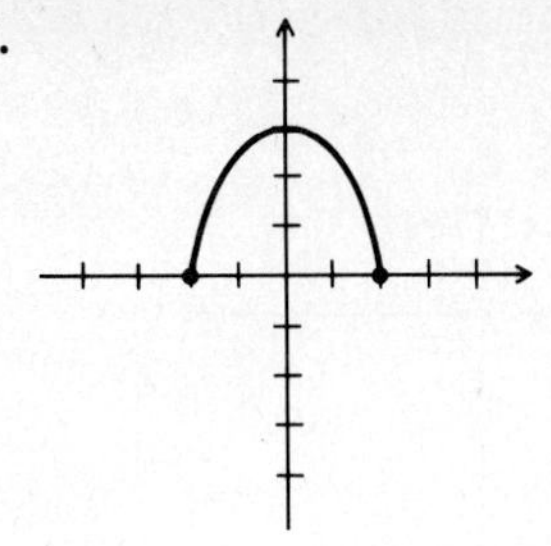

19.

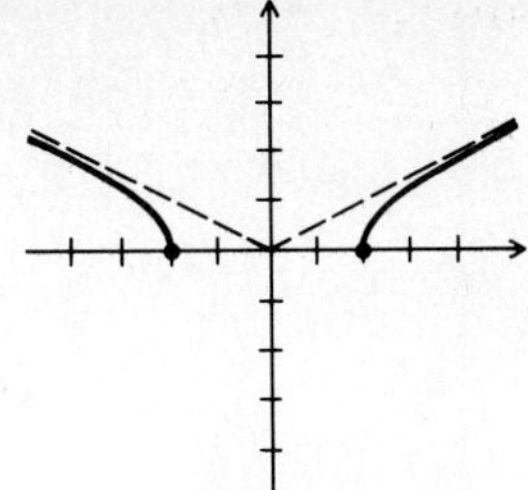

21.

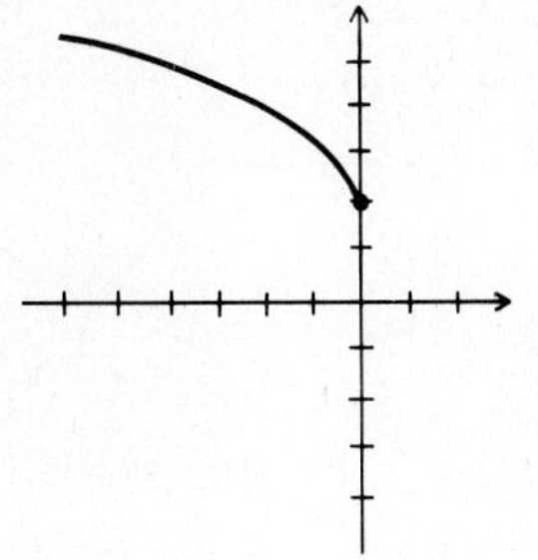

23.

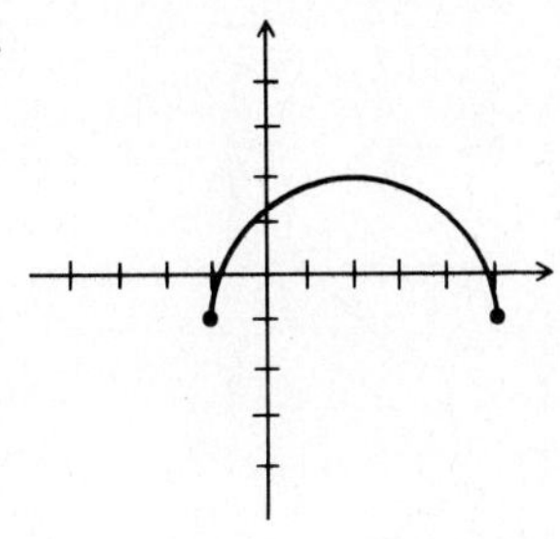

25.

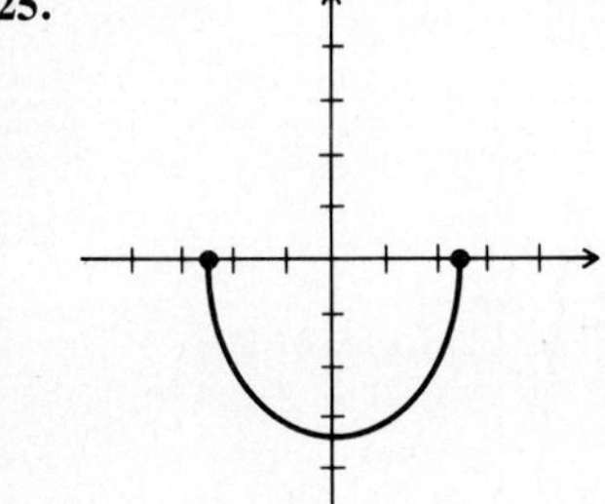

27.

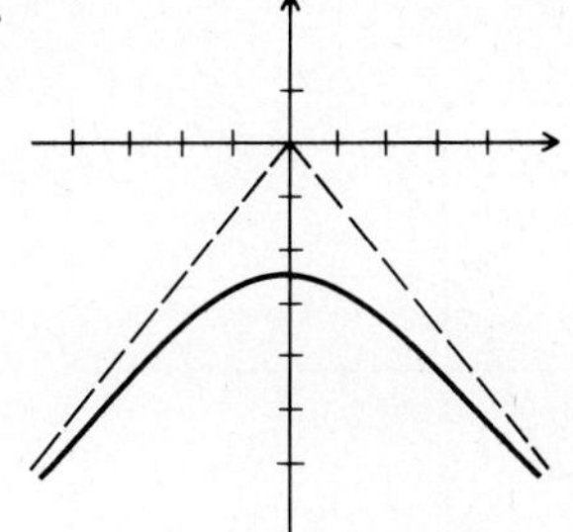

Problem Set 3.3, page 171

1.

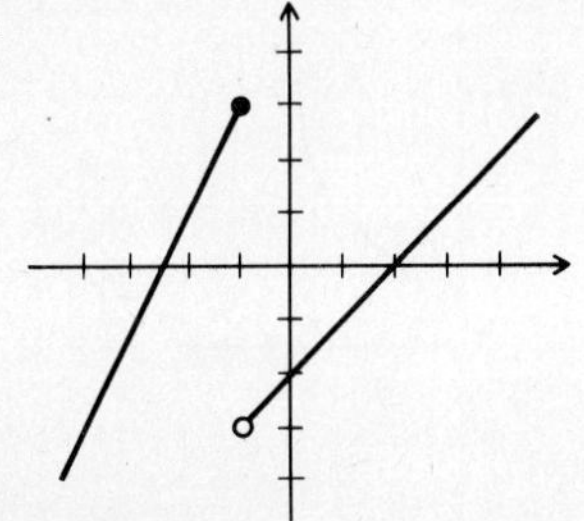

3.

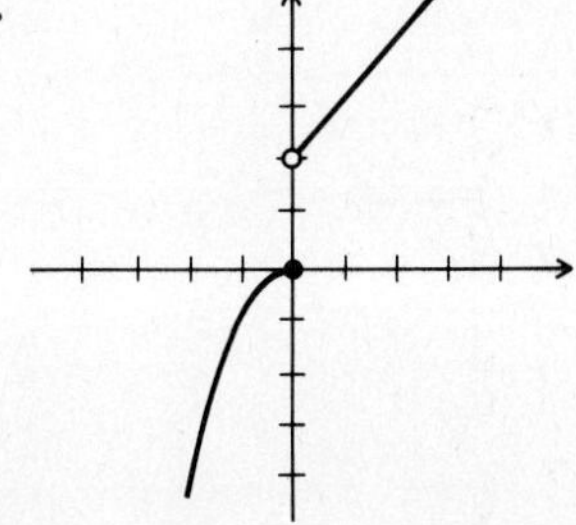

5.

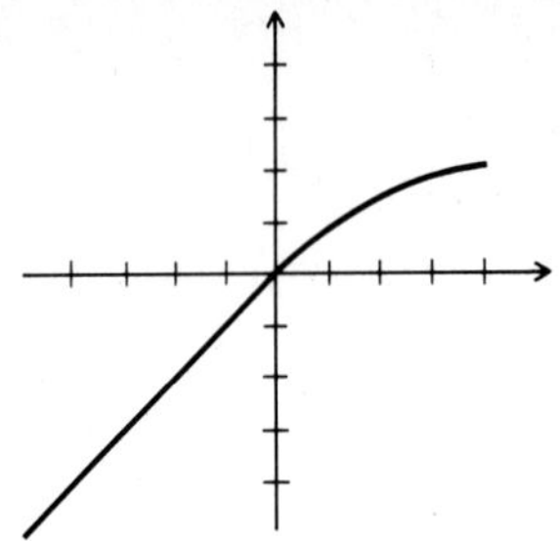

7.

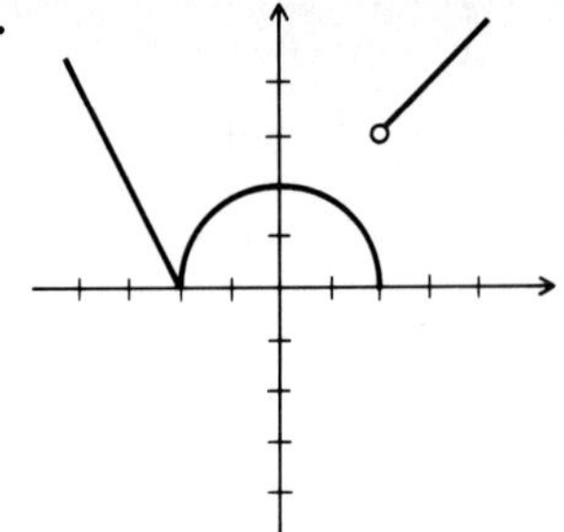

9.

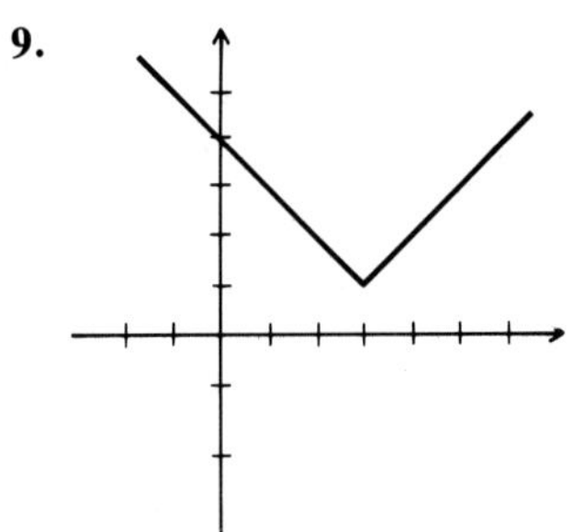

11.

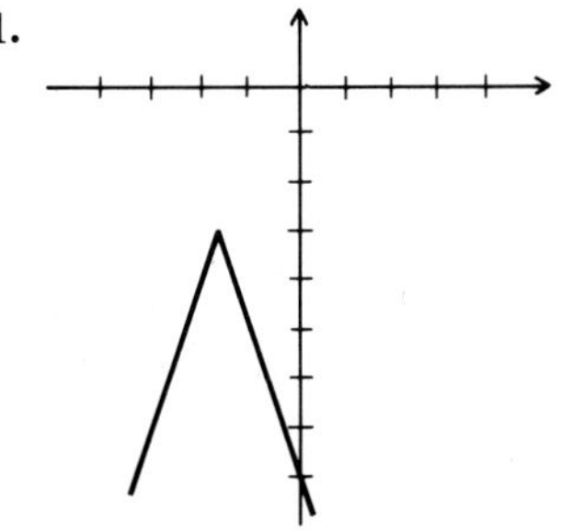

13.

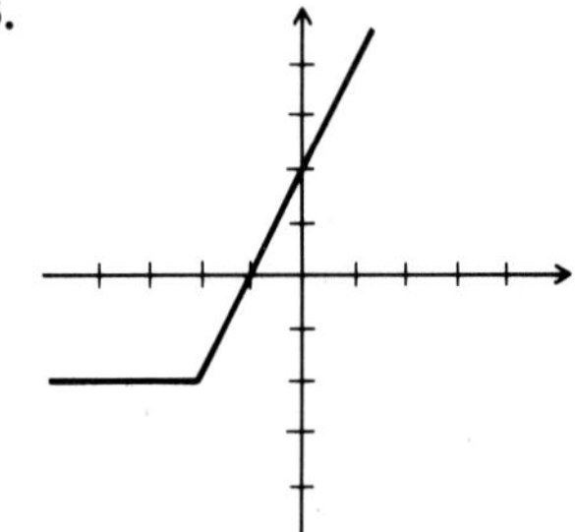

15.

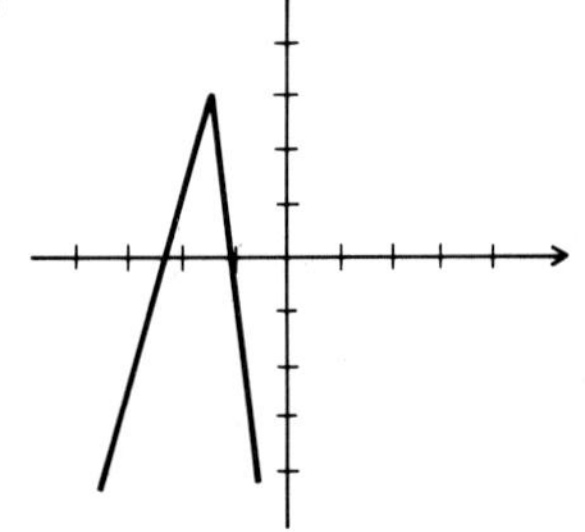

17.

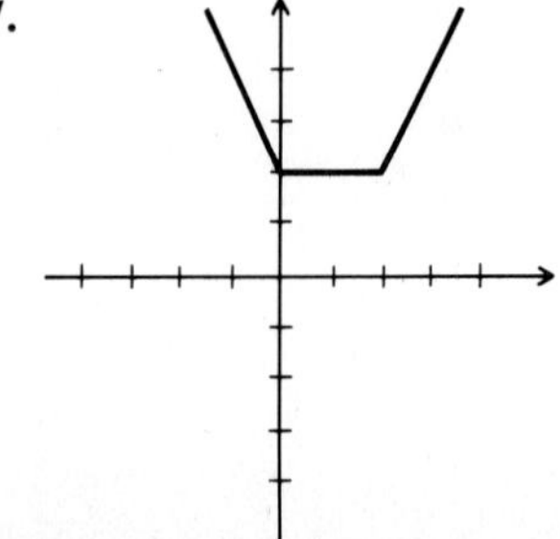

19.

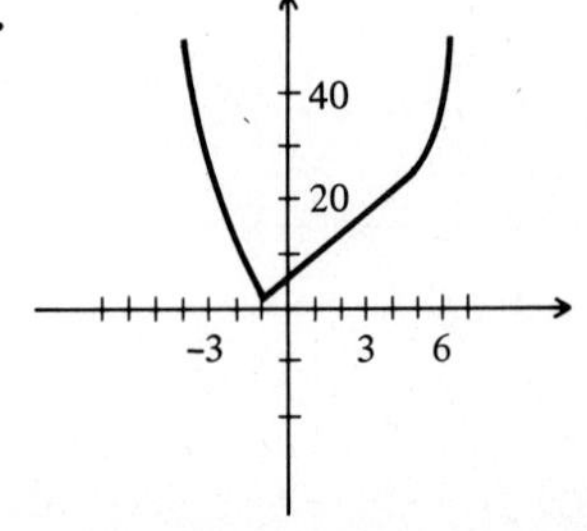

21.

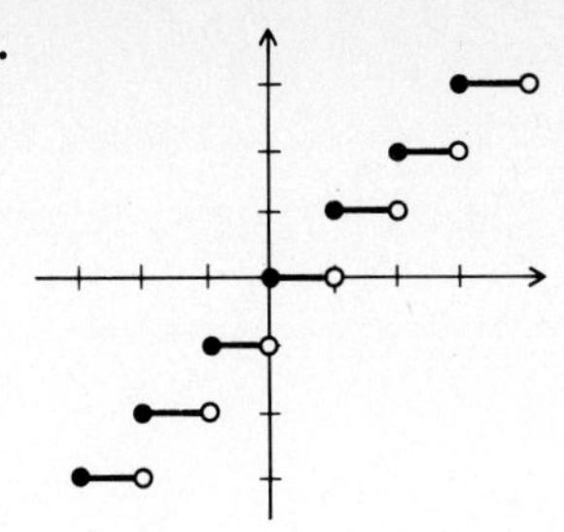

23.

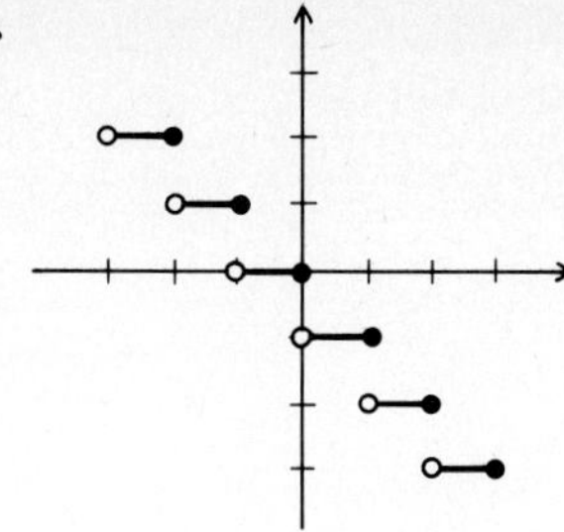

25.

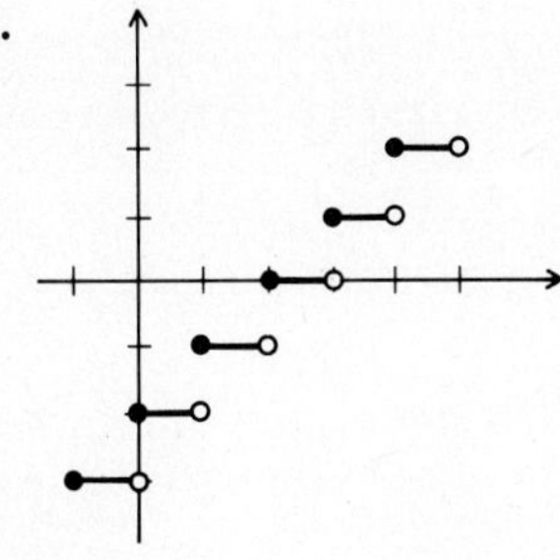

27. $f(x) = [x + 0.5]$, $g(x) = \dfrac{[10x + 0.5]}{10}$

Problem Set 3.4, page 176

1. $(f + g)(x) = x^2 - 2$

$(f - g)(x) = x^2 - 2x + 2$

$(fg)(x) = x^3 - 3x^2 + 2x$

$\left(\dfrac{f}{g}\right)(x) = \dfrac{x^2 - x}{x - 2}$

3. $(f + g)(x) = 3x^2 - x + 3$ for $x \in (0, \infty)$

$(f - g)(x) = -x^2 + x + 5$ for $x \in (0, \infty)$

$(fg)(x) = 2x^4 - x^3 + 7x^2 - 4x - 4$ for $x \in (0, \infty)$

$\left(\dfrac{f}{g}\right)(x) = \dfrac{x^2 + 4}{2x^2 - x - 1}$ for $x \in (0, 1) \cup (1, \infty)$

5. $(f + g)(x) = \dfrac{2x^3 - 7x^2 + 5x + 4}{x - 2}$

$(f - g)(x) = \dfrac{2x^3 - 7x^2 + 3x + 4}{x - 2}$

$(fg)(x) = 2x^2 + x$ for $x \in \mathsf{R} - \{2\}$

$\left(\dfrac{f}{g}\right)(x) = \dfrac{2x^3 - 7x^2 + 4x + 4}{x}$ for $x \in \mathsf{R} - \{2, 0\}$

7. $f + g = \{(1, 2), (2, -5)\}$
$f - g = \{(1, 10), (2, -5)\}$
$fg = \{(1, -24), (2, 0)\}$

$$\frac{f}{g} = \left\{\left(1, \frac{-3}{2}\right)\right\}$$

9. $(f \circ g)(x) = x^2 - 2x + 1$
$(g \circ f)(x) = -x^2 + 1$

11. $(f \circ g)(x) = 2x^2 - 7x + 6$ for $x > 3$

$(g \circ f)(x) = 2x^2 + x - 2$ for $x \in \left(-\infty, \frac{-3}{2}\right) \cup (1, \infty)$

13. $(f \circ g)(x) = 3x + 7$ for $x \geq -2$
$(g \circ f)(x) = \sqrt{3x^2 + 3}$

15. $f \circ g = \{(6, 7), (7, 5)\}$
$g \circ f = \{(-2, 7), (5, -2), (0, 5)\}$

17. $(f \circ f)(x) = 4x - 5$ for $x \in [1, 2]$

Problem Set 3.5, page 182

1. is
3. is
5. is not
7. is
9. is not
11. is not
13. is not
15. $f^{-1}(x) = \dfrac{x + 2}{3}$
17. $f^{-1}(x) = -\sqrt{2 - x}$
19. $f^{-1}(x) = \sqrt{x + 1} - 2$
21. $f^{-1}(x) = \dfrac{x^2 - 1}{2}, x \geq 0$
23. $f^{-1}(x) = \dfrac{x - 1}{x - 2}$
25. $f^{-1}(x) = \begin{cases} \dfrac{x - 1}{3} & \text{when } x < 1 \\ \dfrac{\sqrt{2x - 2}}{2} & \text{when } x \geq 1 \end{cases}$
27.

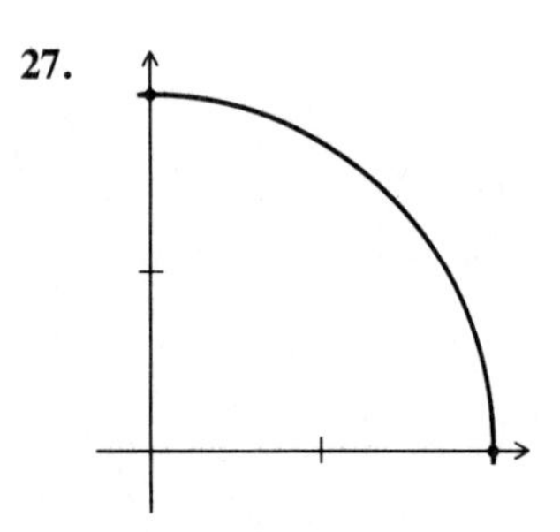

$f^{-1}(x) = \sqrt{4 - x^2}, 0 \leq x \leq 2$

29.

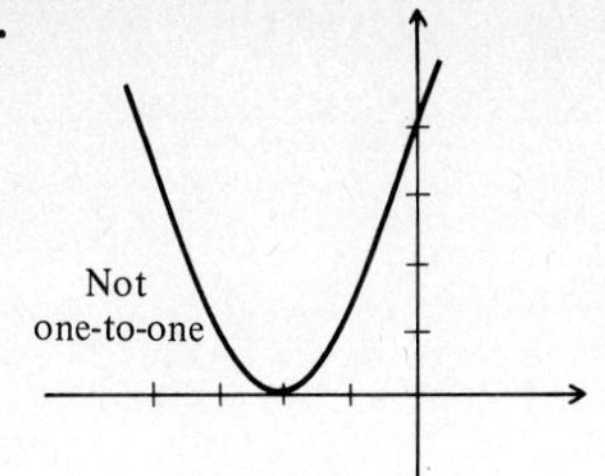

31. $\frac{3}{2}$

33.

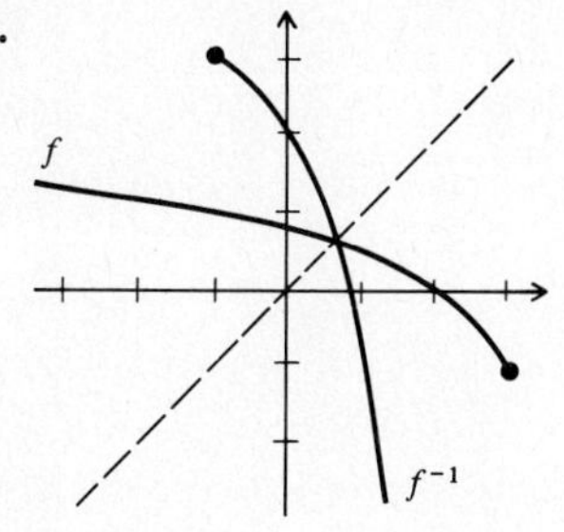

Chapter 3 Review, page 183

1. is
2. is not
3. is
4. is
5. is
6. is not
7. is not
8. is
9. $\mathsf{R} - \{1\}$
10. R
11. $(-\infty, -3] \cup (0, \infty)$
12. $[-1, 1) \cup (1, 2]$
13. **(a)** 0 **(b)** 3 **(c)** 3
14. **(a)** 1 **(b)** $\frac{1}{2}$ **(c)** $\frac{1}{|a|}$ **(d)** 2
15. $a = \frac{-5}{2}$ or $a = 3$
16. **(a)** $\mathsf{R} - \{0\}$ **(b)** $(-1, \infty)$ **(c)** 1 **(d)** not defined
17. range = R

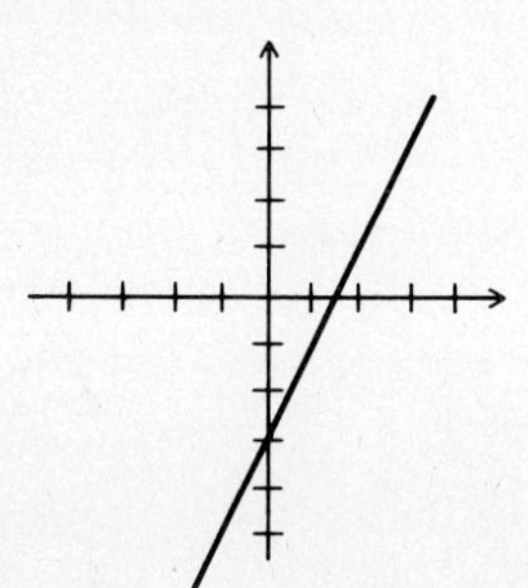

18. range = $\{2\}$

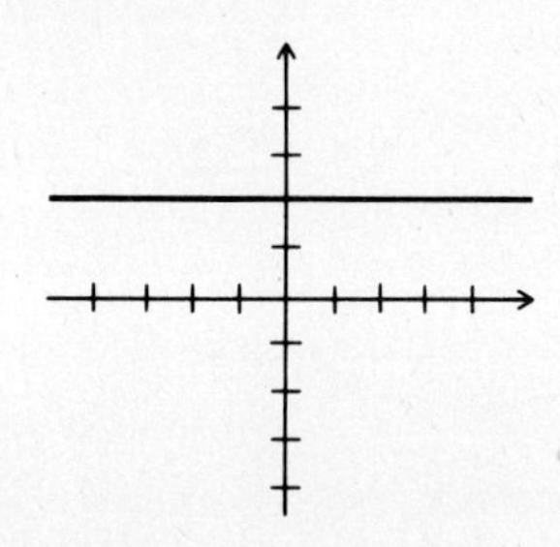

19. range = $[-6, \infty)$

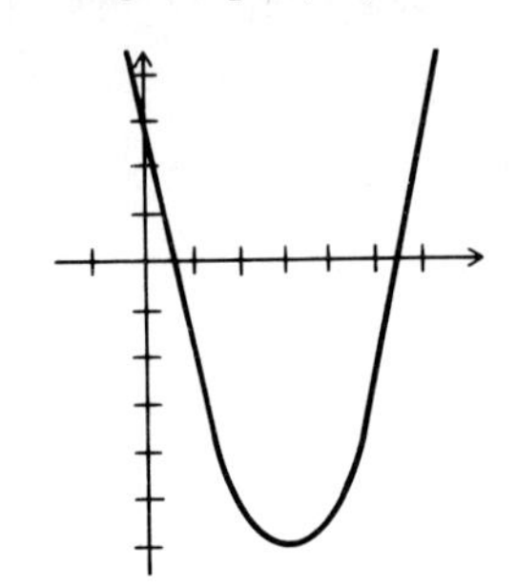

20. range = $(-\infty, 4]$

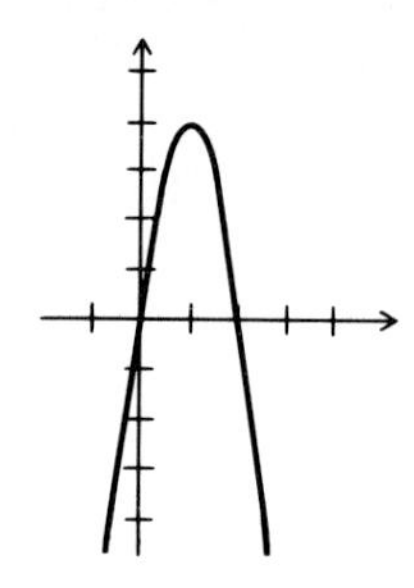

21.

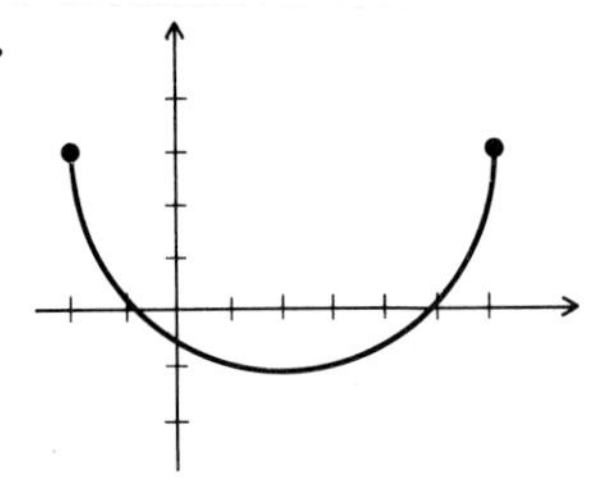

22.

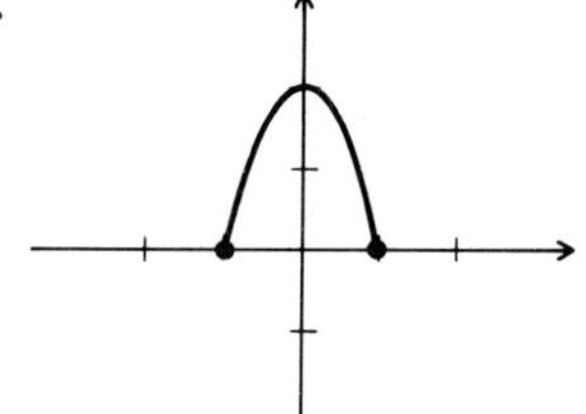

23.

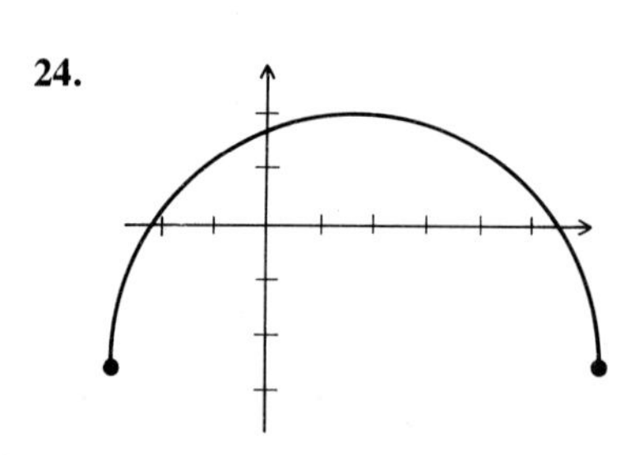

24.

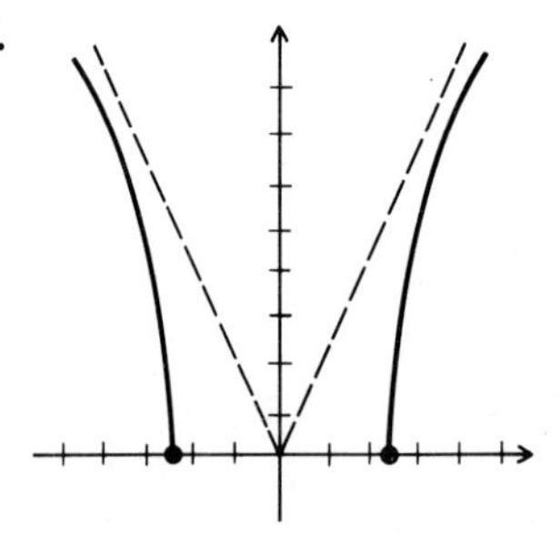

25.

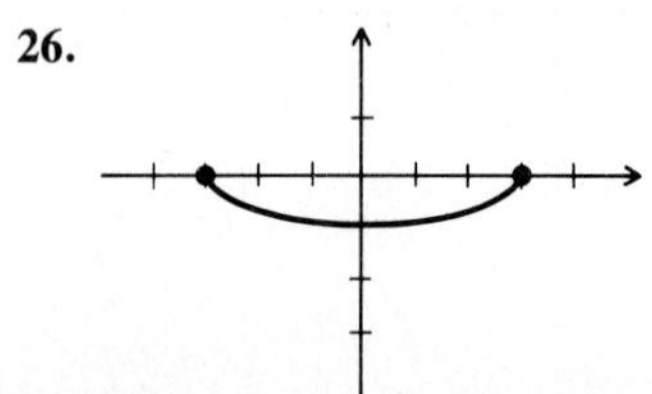

26.

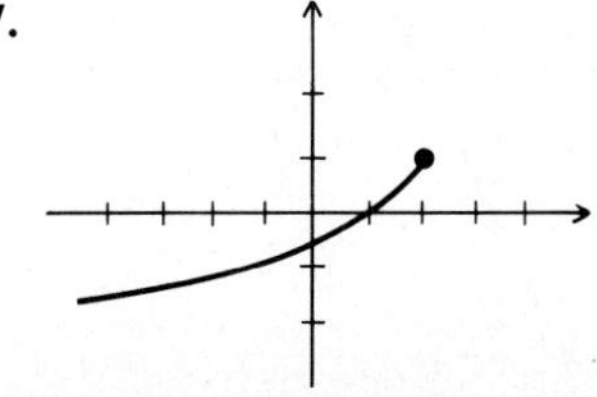

27.

28.

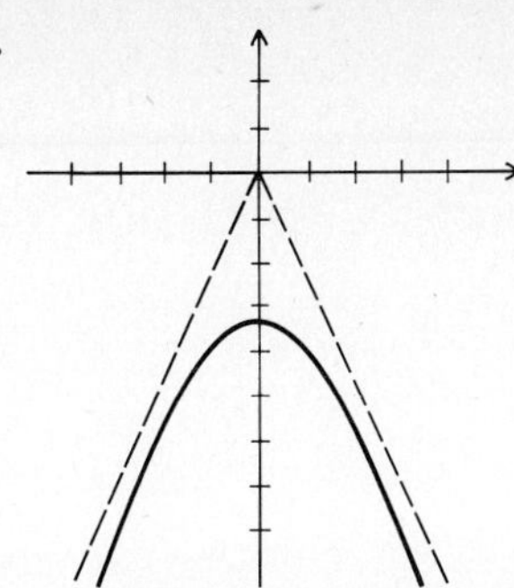

29.

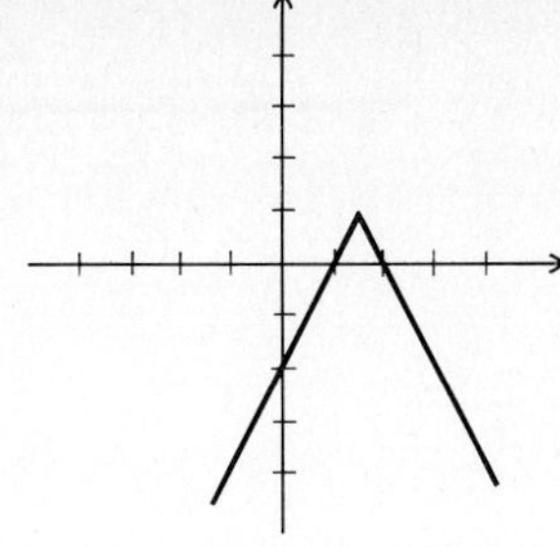

30.

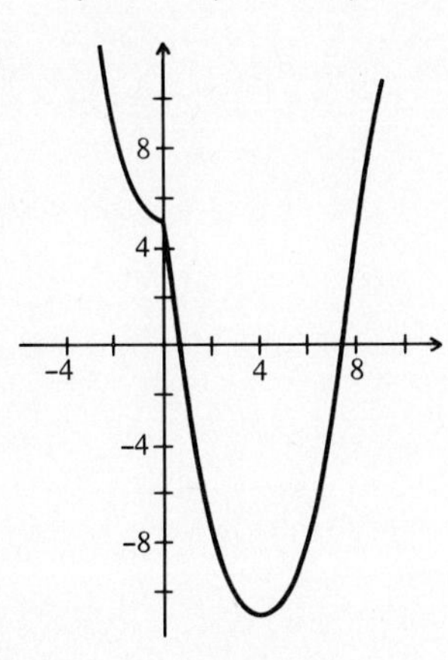

31.

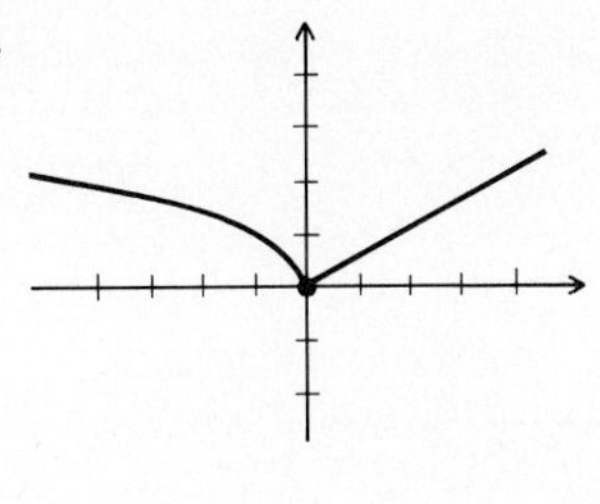

32.

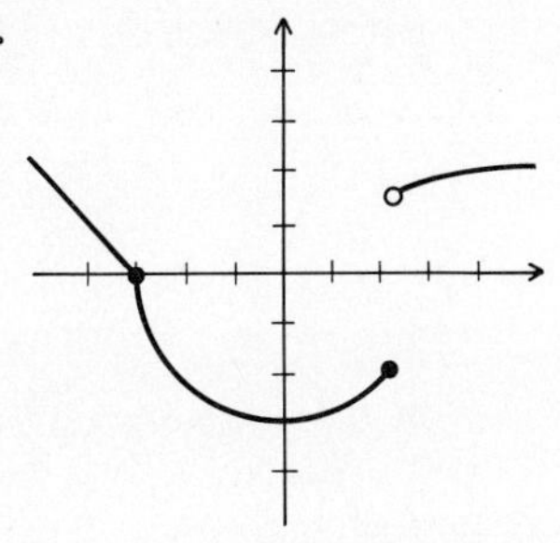

33.

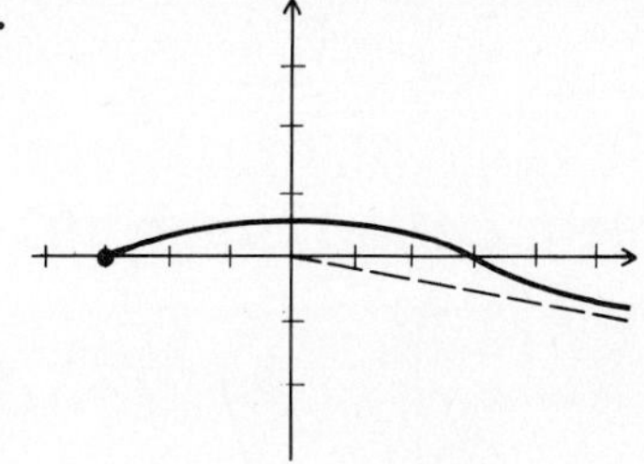

34. $(f+g)(x) = \dfrac{x^3 - x^2 + 2x + 1}{x(2x+1)}$

$(f-g)(x) = \dfrac{-x^3 + x^2 + 2x + 1}{x(2x+1)}$

$(fg)(x) = \dfrac{x-1}{2x+1}$ for $x \in \mathsf{R} - \left\{\dfrac{-1}{2}, 0\right\}$

$\left(\dfrac{f}{g}\right)(x) = \dfrac{2x+1}{x^2(x-1)}$ for $x \in \mathsf{R} - \left\{\dfrac{-1}{2}, 0, 1\right\}$

35. $(f \circ g)(x) = \dfrac{1-x}{3-x}$ for $x \leq 1$

$(g \circ f)(x) = \sqrt{\dfrac{2}{x^2+2}}$

36. is not

37. is

38. is not

39. $f^{-1}(x) = \dfrac{x - 5}{2}$

40. $f^{-1}(x) = \dfrac{x^2 + 2}{3}, \ x \geq 0$

41. $f^{-1}(x) = \dfrac{1 + 5x}{2 - x}$

42. $f^{-1}(x) = 2 - \sqrt{4 - x^2}, \ -\sqrt{3} \leq x \leq 0$

43. $f^{-1}(x) = \dfrac{x^2 - 4x + 5}{3}, \ x \geq 2$

44. $f^{-1}(x) = \sqrt{1 - x} + 3$

45.

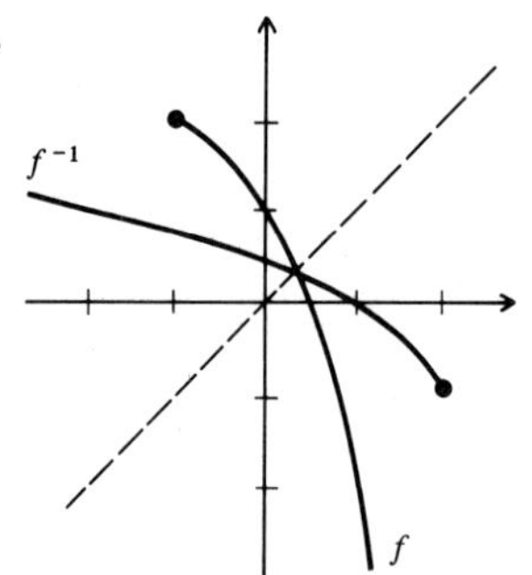

Problem Set 4.1, page 193

1.

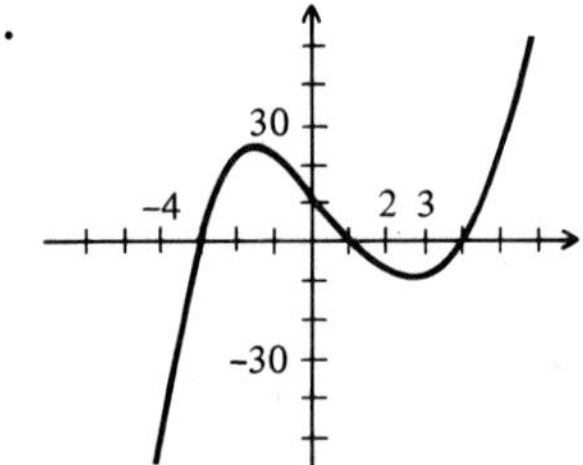

3.

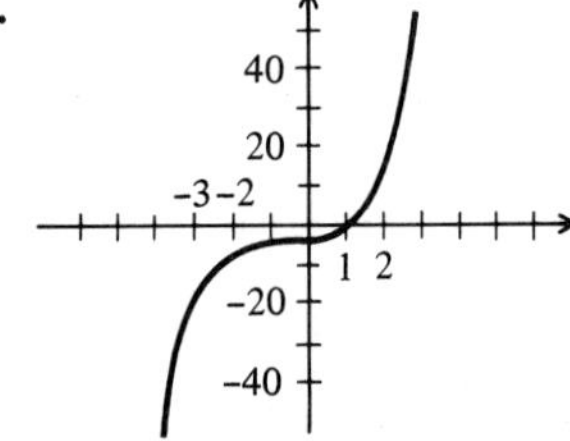

5.

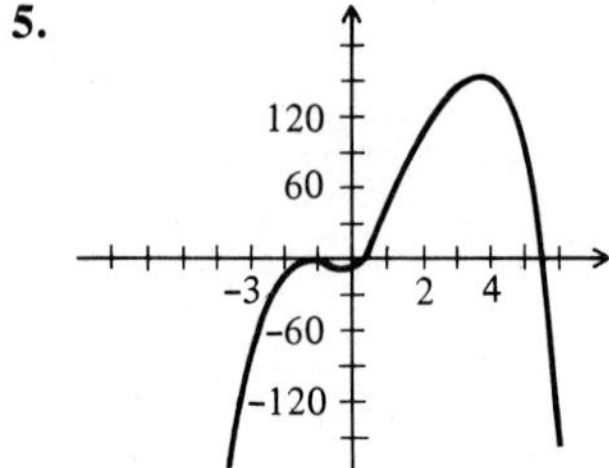

7.

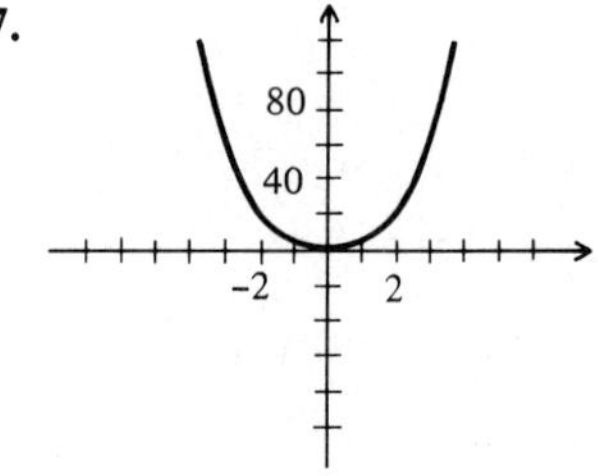

9.

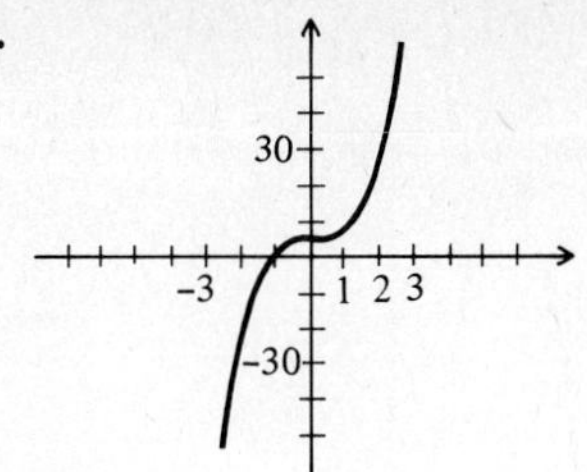

Problem Set 4.2, page 199

1.

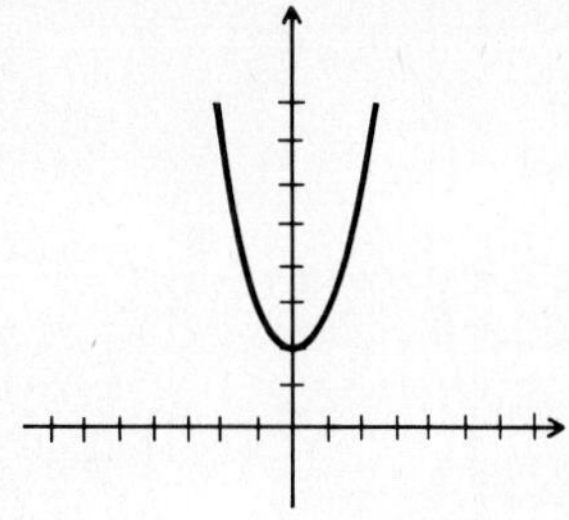

3.

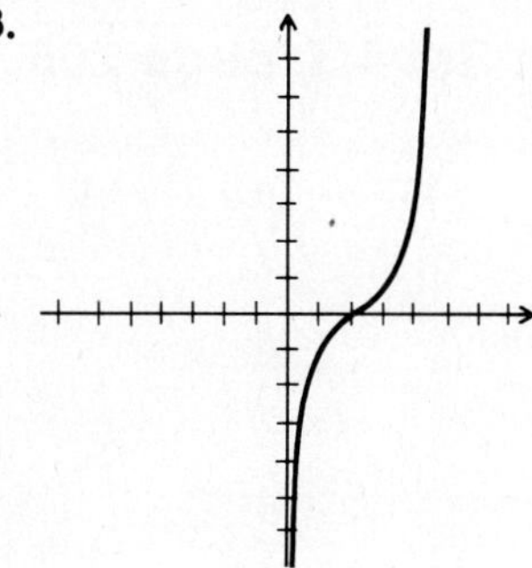

5.

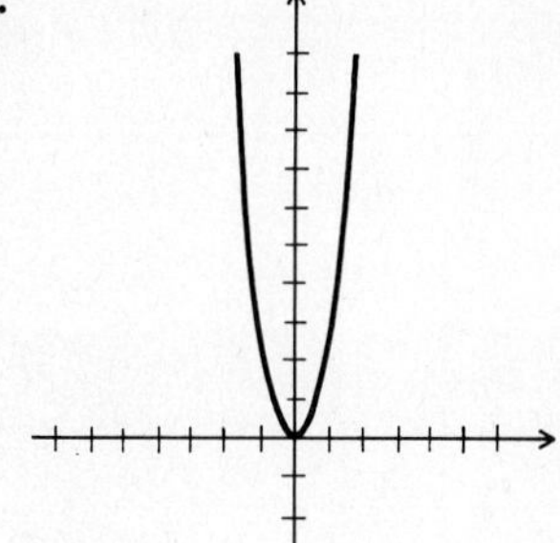

7.

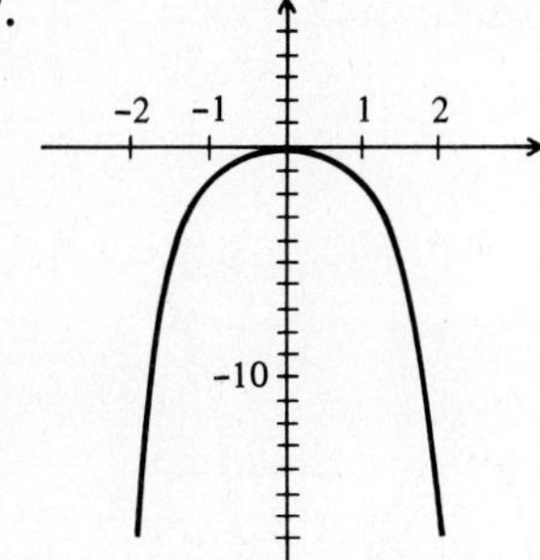

9.

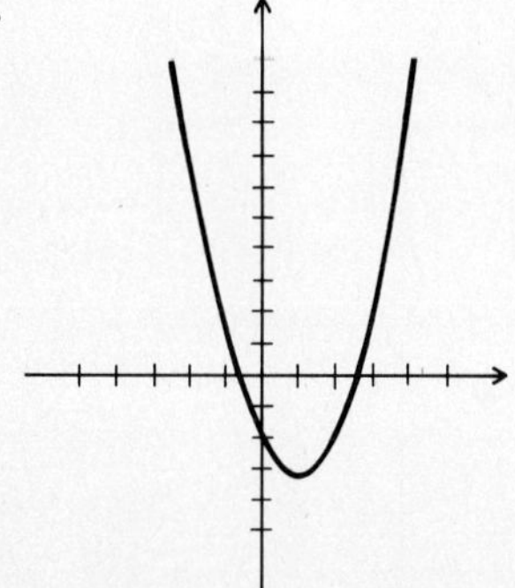

11.

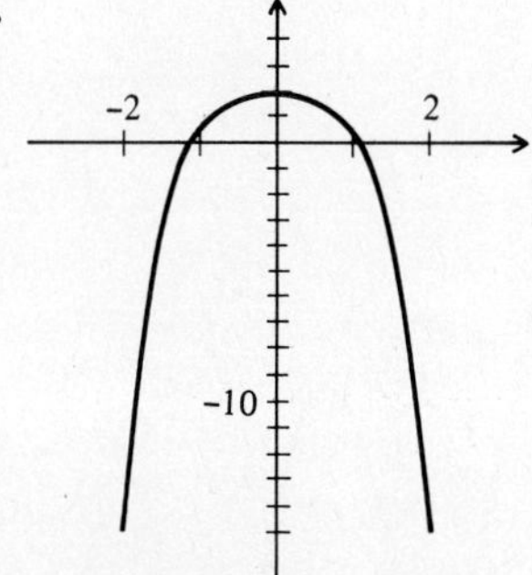

13.

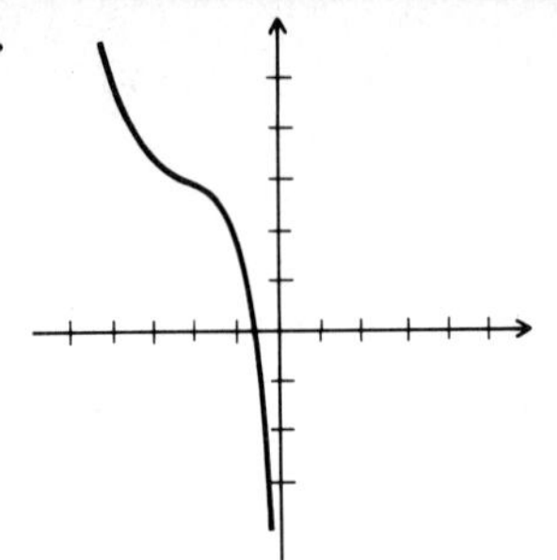

15.

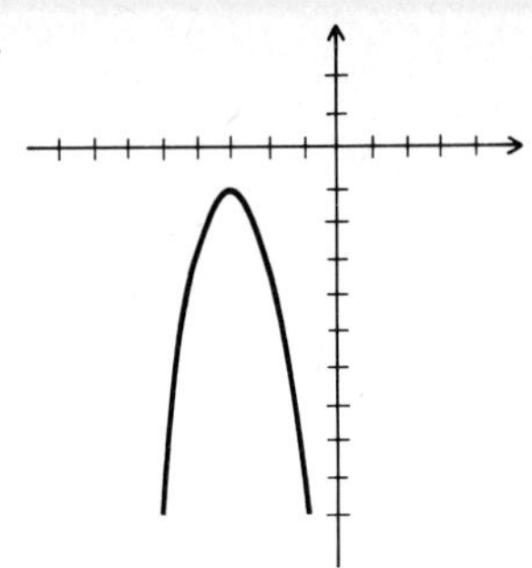

17. (c)

19. (b)

Problem Set 4.3, page 205

1. as $x \to \infty$, $y \to \infty$; and as $x \to -\infty$, $y \to -\infty$

3. as $x \to \infty$, $y \to -\infty$; and as $x \to -\infty$, $y \to -\infty$

5.

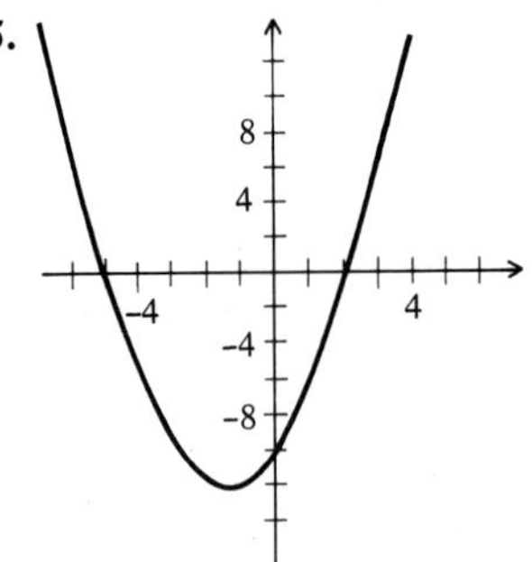

7.

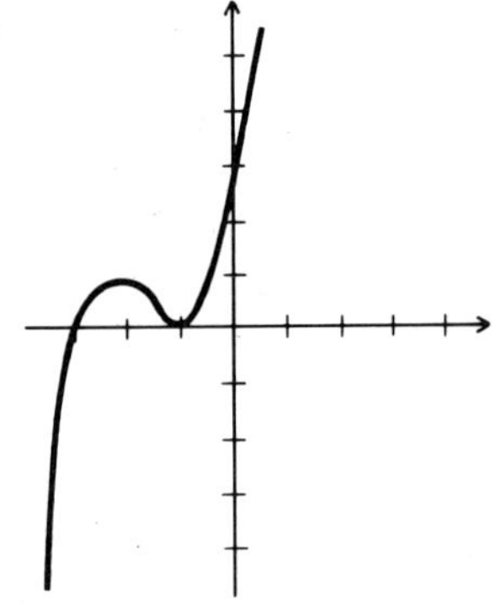

9.

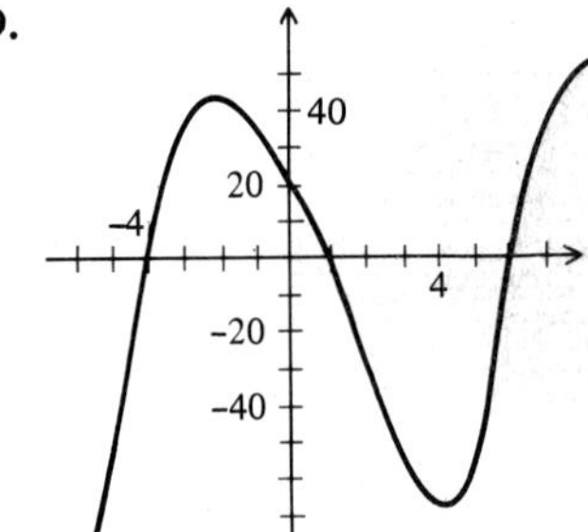

11.

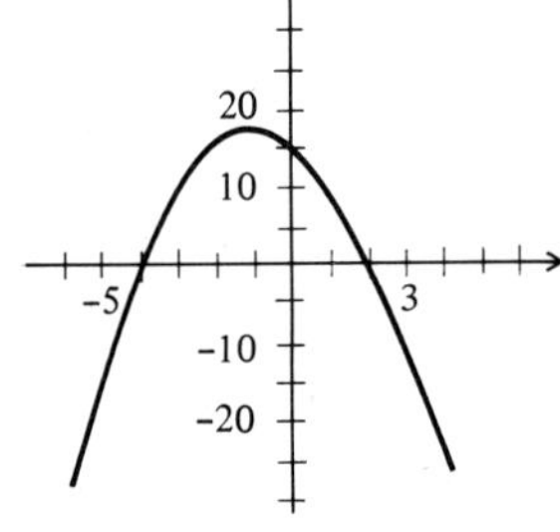

13.

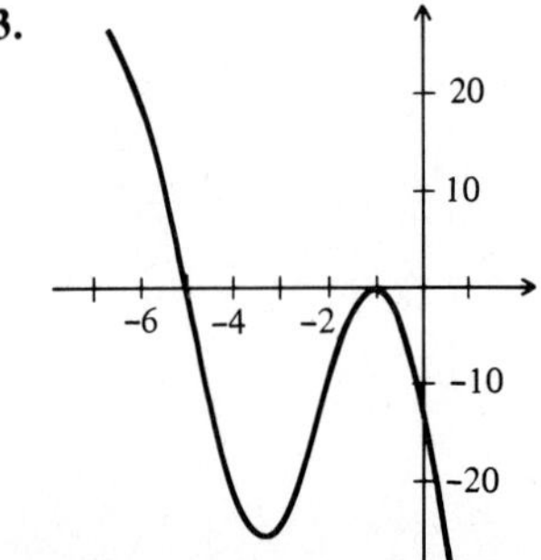

15.

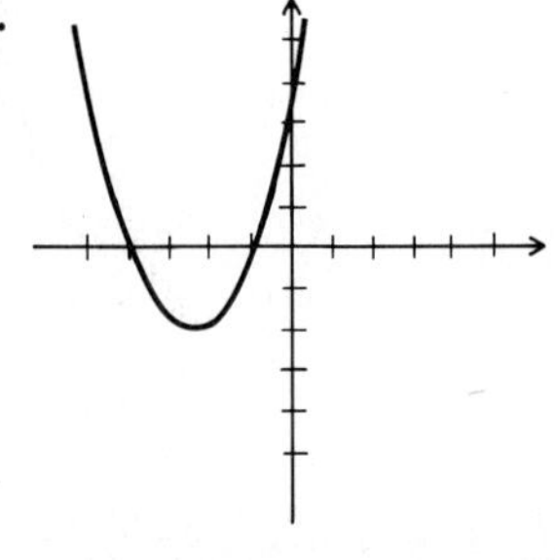

17.

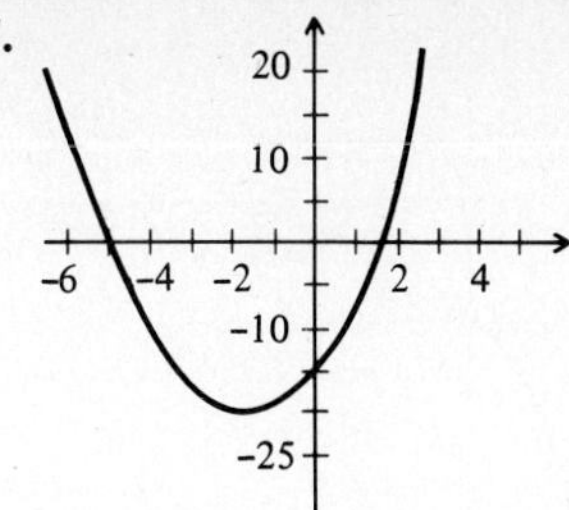

19.

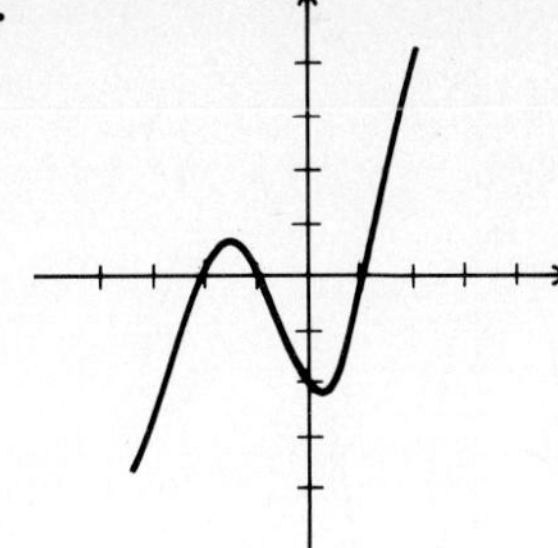

21.

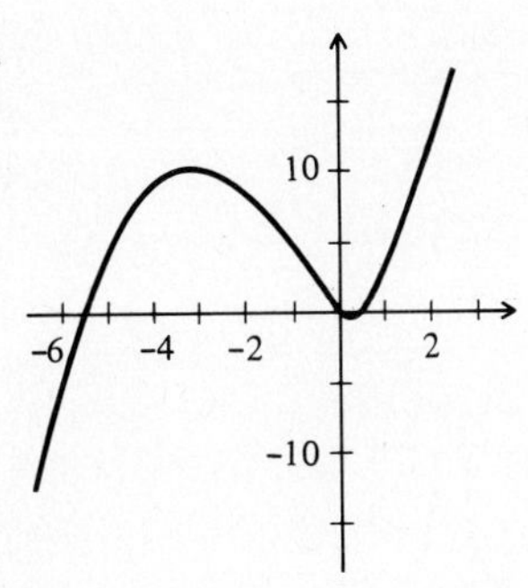

23. (b)

25. **(a)** is **(b)** is not

27.

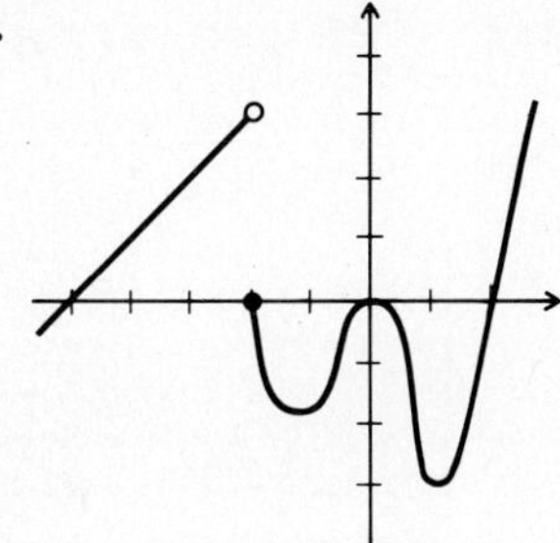

Problem Set 4.4, page 212

1. at most four; two or no positive zeros; two or no negative zeros
3. at most three; at least one; two or no positive zeros; one negative zero

5. one negative zero

7. $-2, 1, \frac{1}{2}$

9. 1 (multiplicity two), $-2, \frac{2}{3}$

11. $\frac{1}{2}, \frac{-2}{3}$

13. $\frac{2}{5}, \frac{-1+\sqrt{7}}{3}, \frac{-1-\sqrt{7}}{3}$

15.

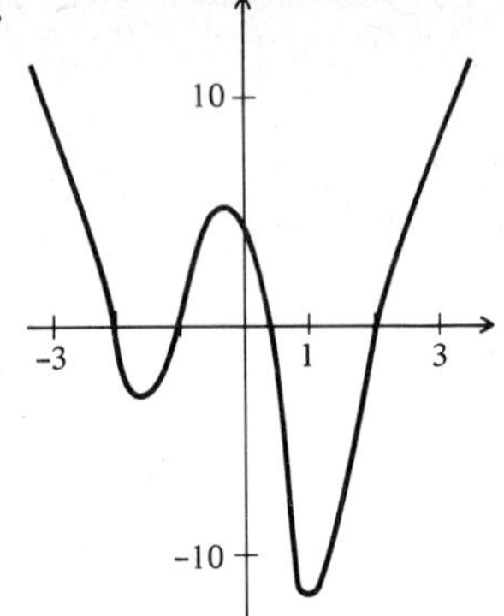

17.

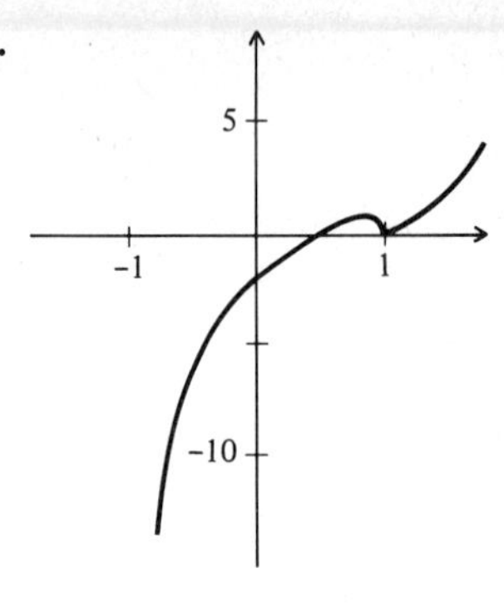

Problem Set 4.5, page 216

1.

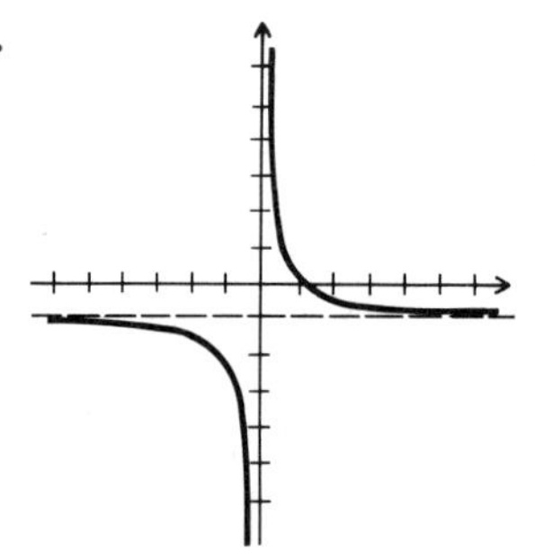

3.

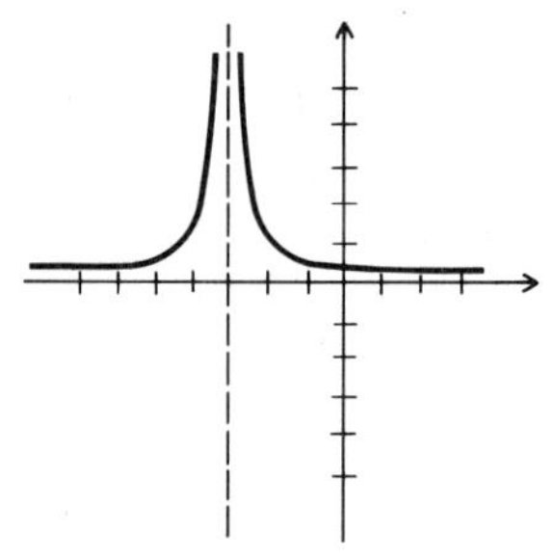

5.

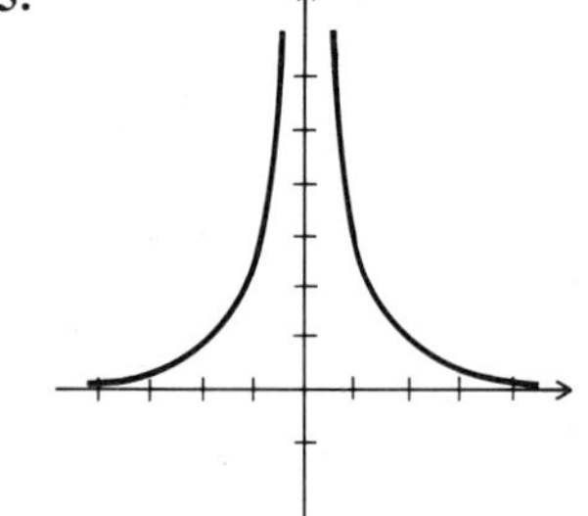

7.

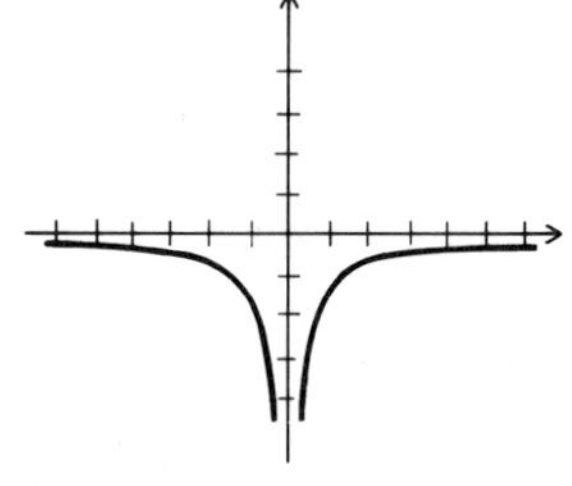

9.

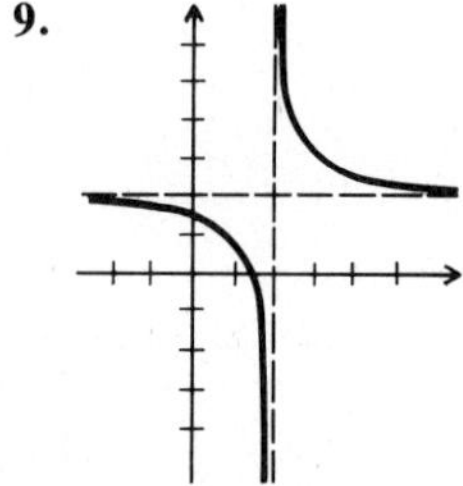

11.

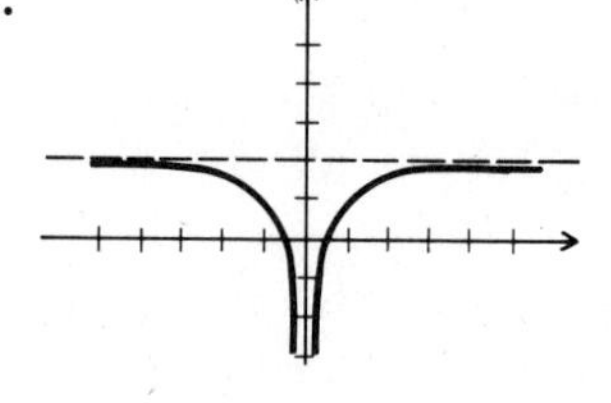

13.

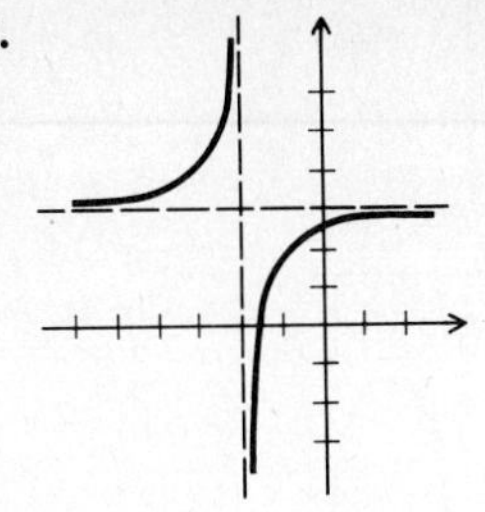

15.

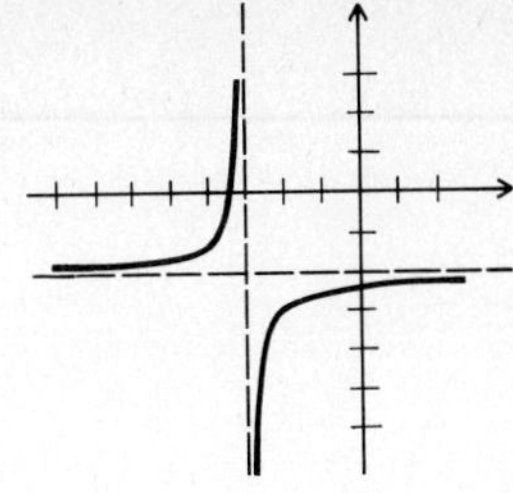

17. (d)

19. (e)

21. $y = \frac{2}{x - 1} + 3$

23.

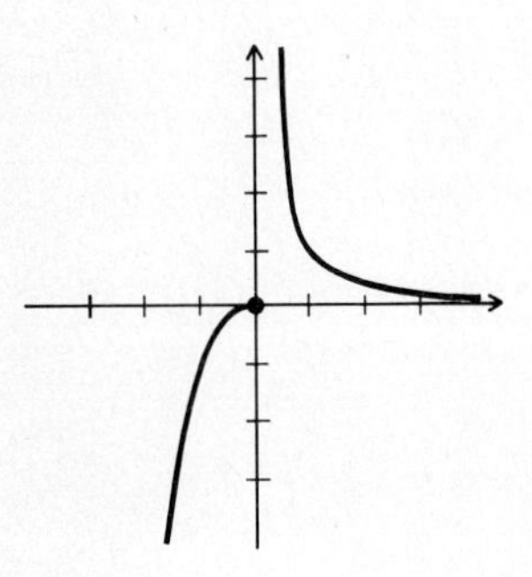

25.

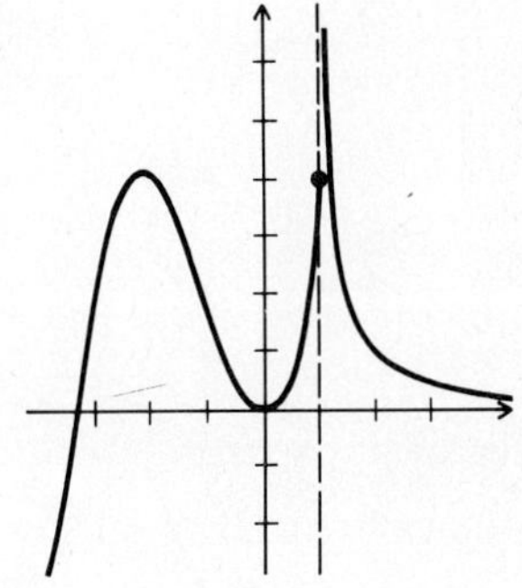

Problem Set 4.6, page 227

1. vertical: $x = -1$
horizontal: $y = 3$
slant: none

3. vertical: $x = \frac{1}{2}, x = -1$
horizontal: $y = \frac{-1}{2}$
slant: none

5. vertical: $x = 2, x = -1$
horizontal: $y = 0$
slant: none

7. vertical: $x = 5$
horizontal: none
slant: $y = x + 6$

9. vertical: $x = -2$
horizontal: none
slant: none

11. vertical: $x = 3 + 2\sqrt{2}$,
$x = 3 - 2\sqrt{2}$
horizontal: $y = 0$
slant: none

13.

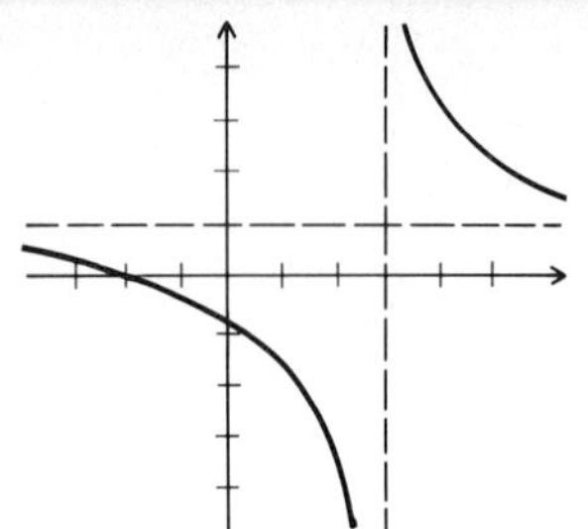

15.

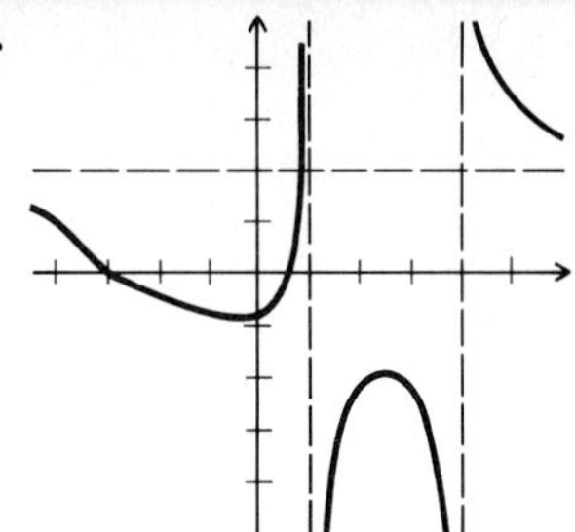

17.

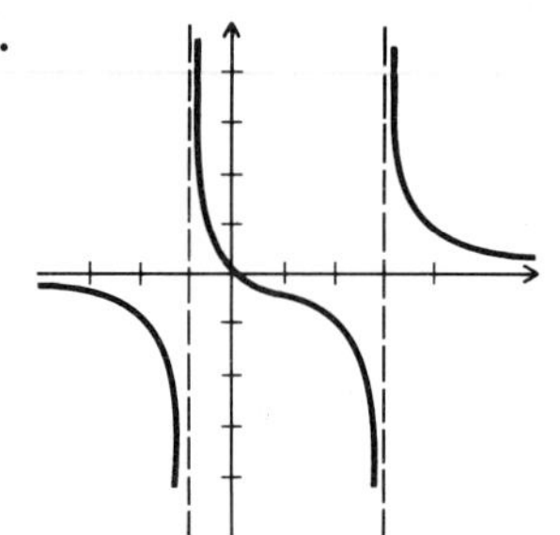

19.

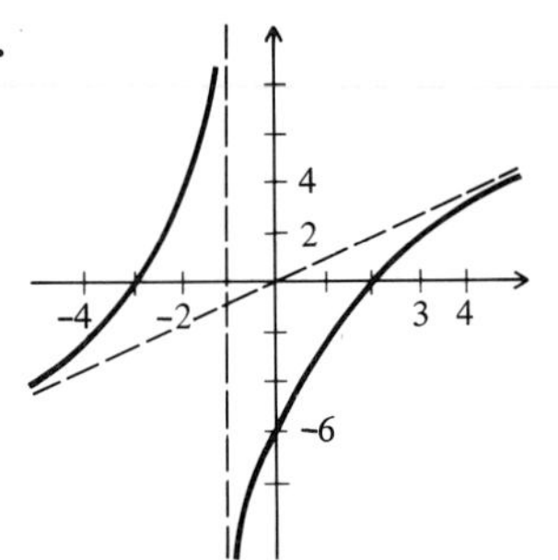

21.

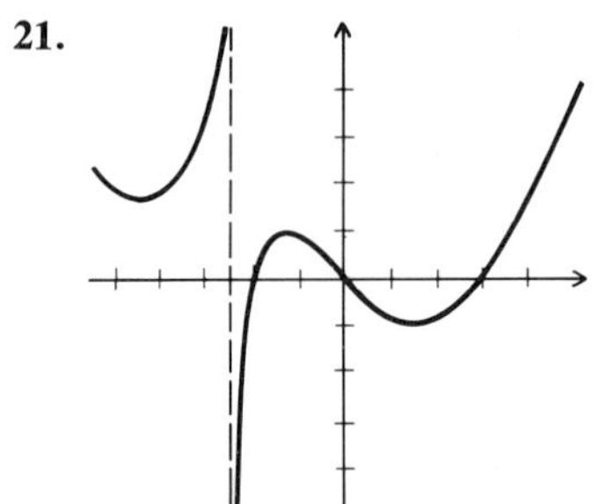

23.

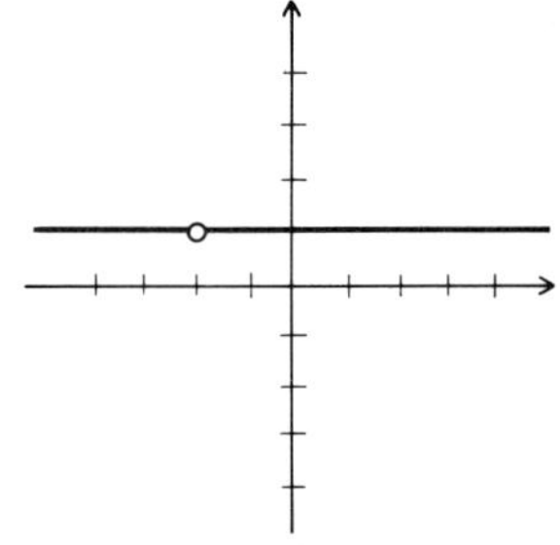

25.

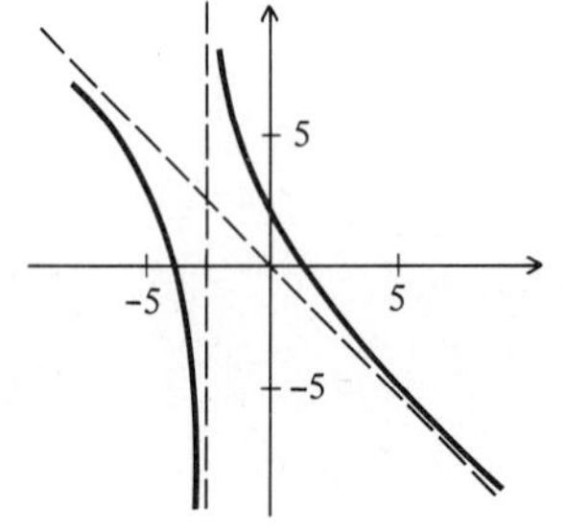

27.

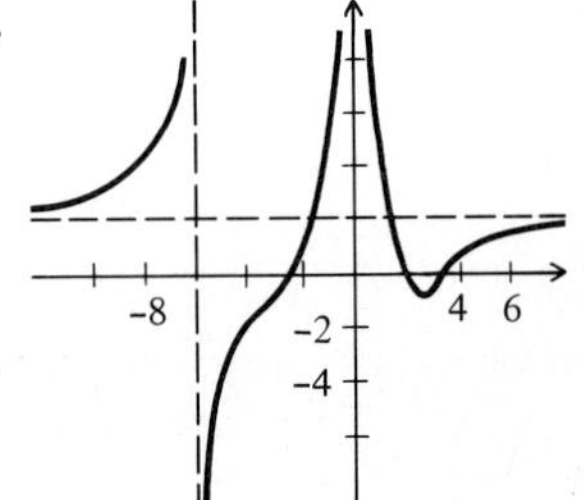

29. (c)

31. $\left(\frac{11}{15}, 2\right)$

35.

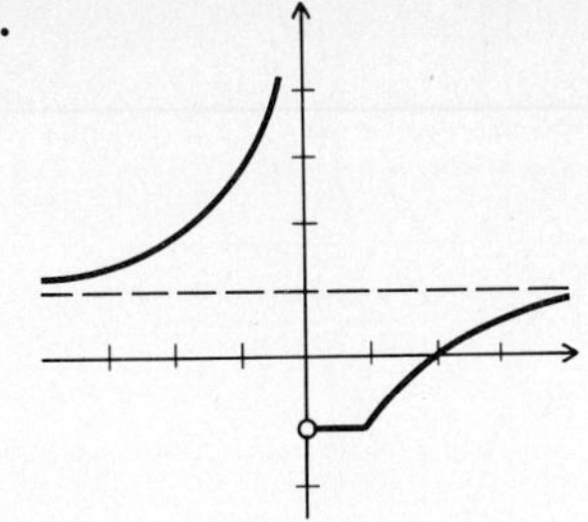

Chapter 4 Review, page 229

1.

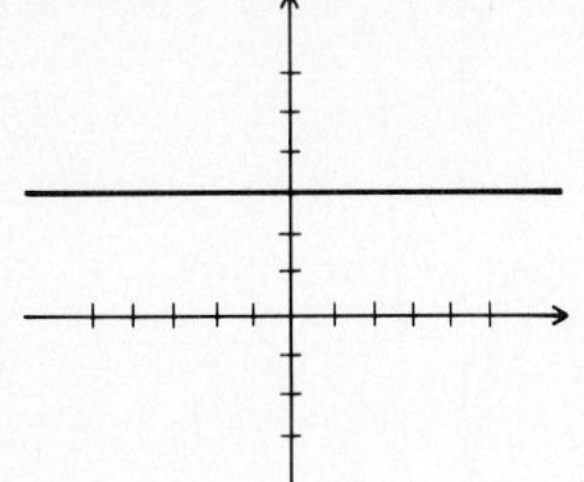

2.

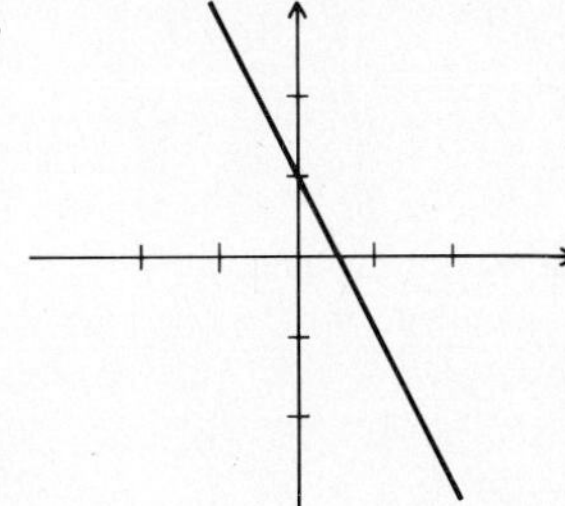

3.

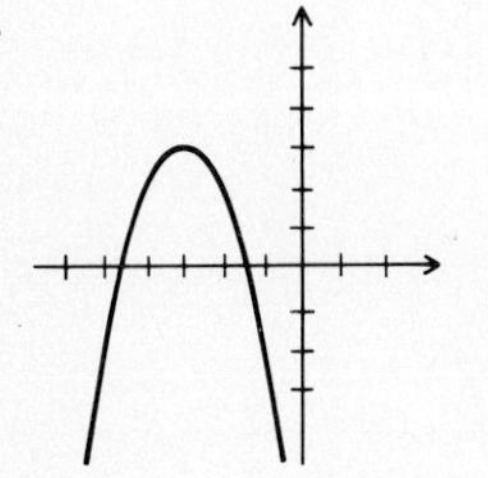

4.

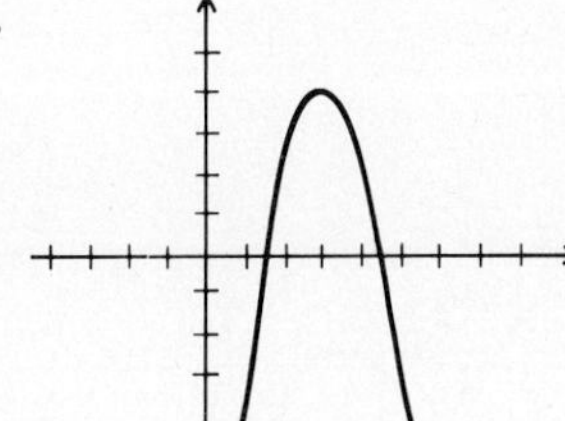

5.

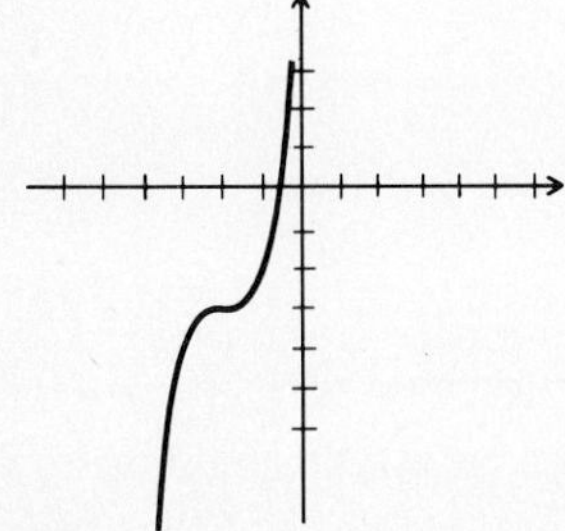

6.

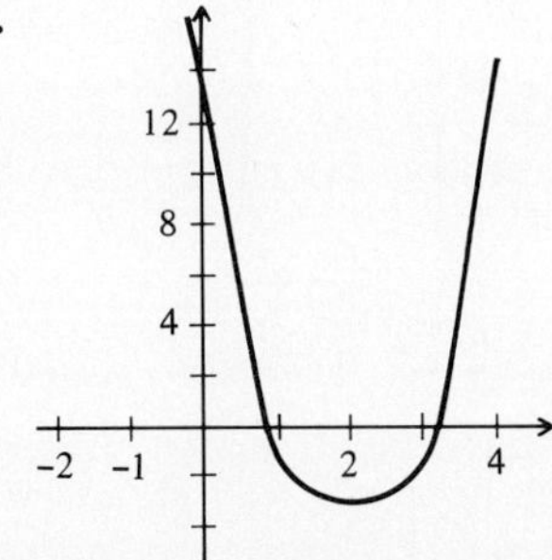

7.

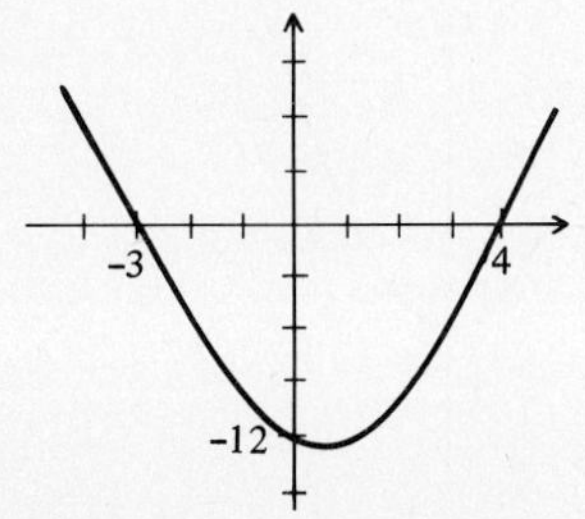

8.

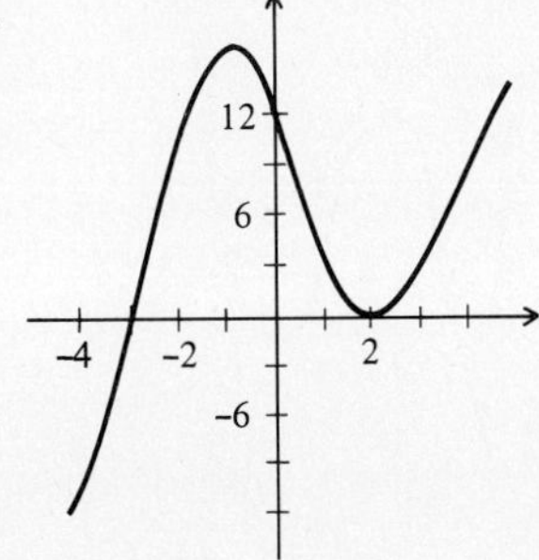

9.

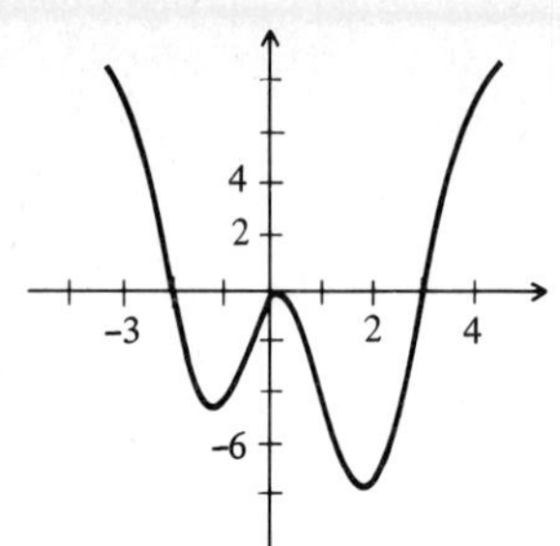

10.

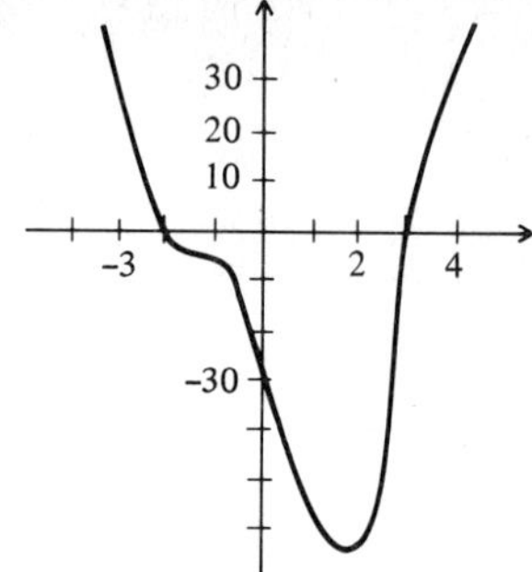

11.

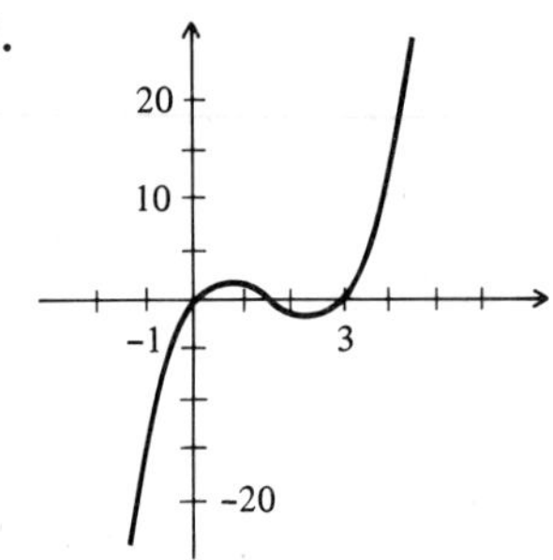

12.

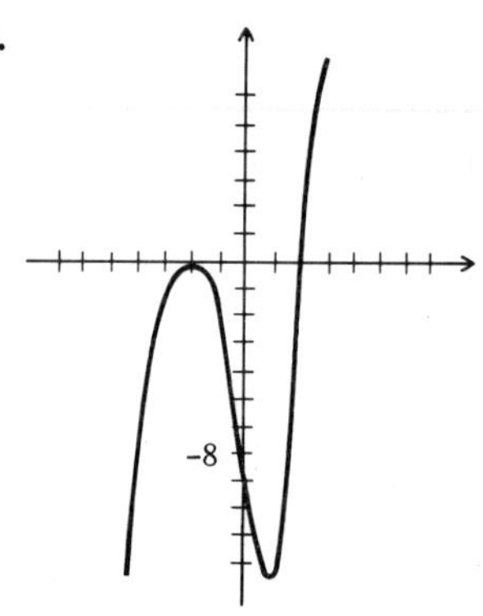

13.

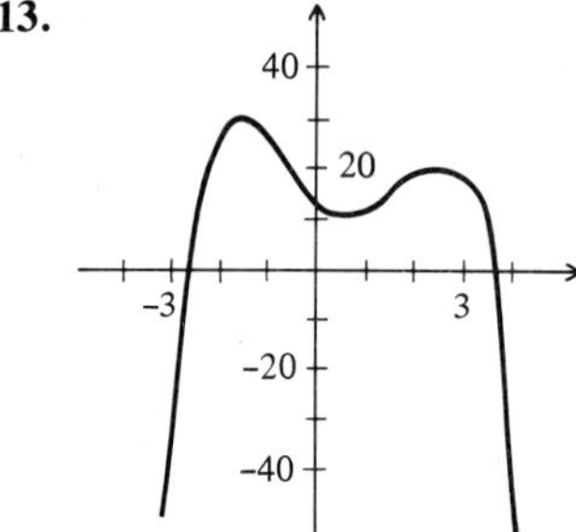

14. $1, -1, \frac{1}{2}, \frac{-2}{3}$

15. $\frac{-2}{5}, \frac{-3+\sqrt{5}}{2}, \frac{-3-\sqrt{5}}{2}$

16.

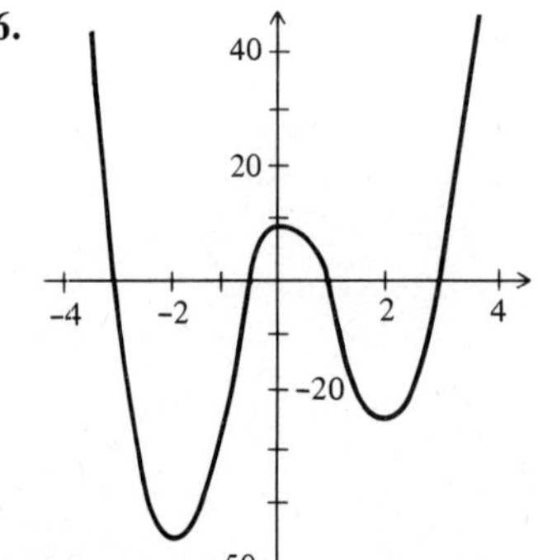

17.

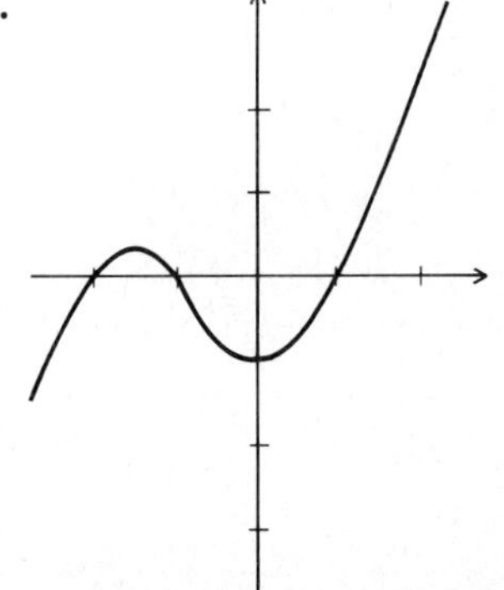

18. (b)

19. (d)

20.

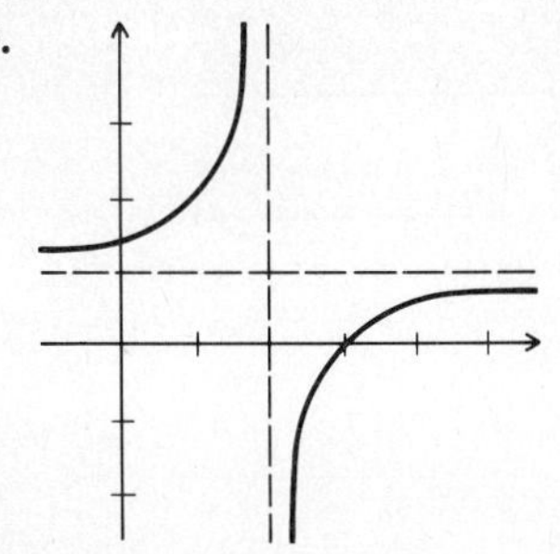

21.

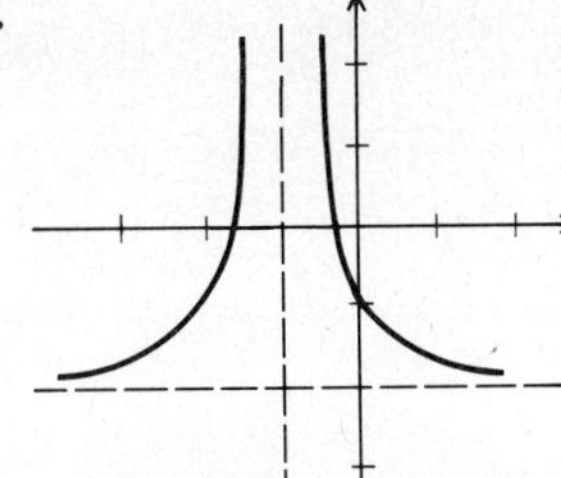

22.

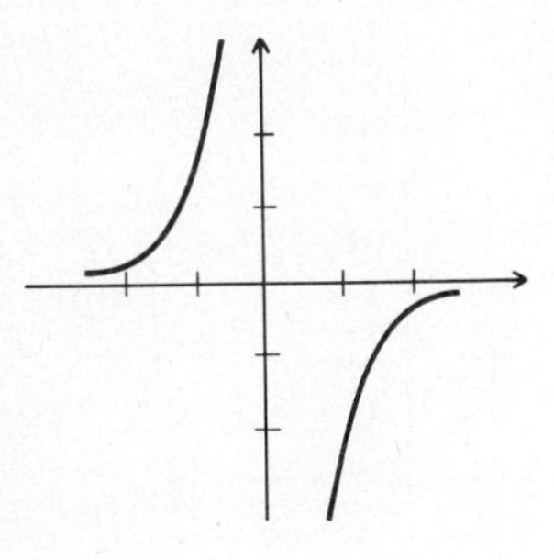

23.

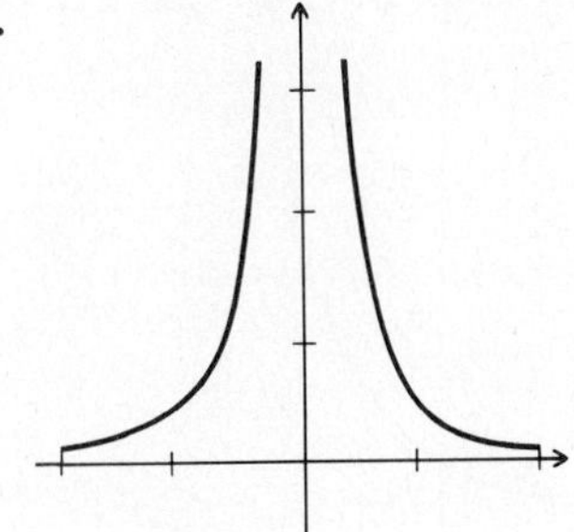

24.

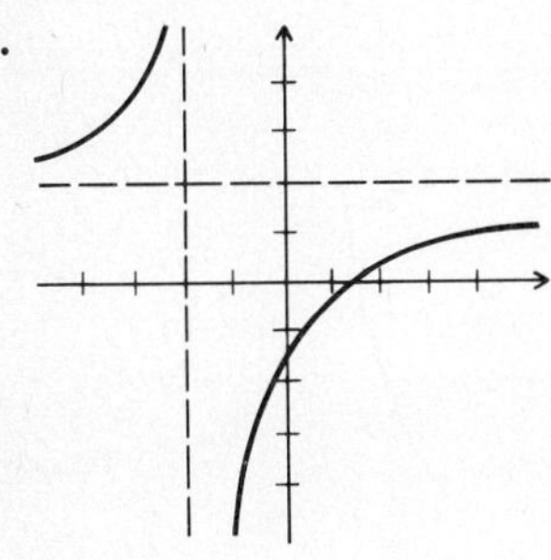

25.

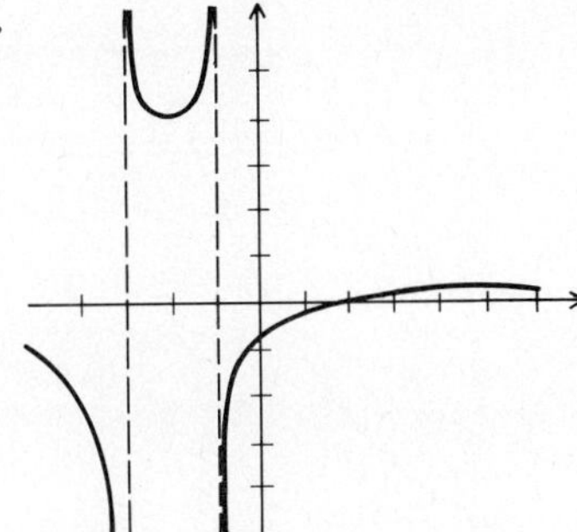

26.

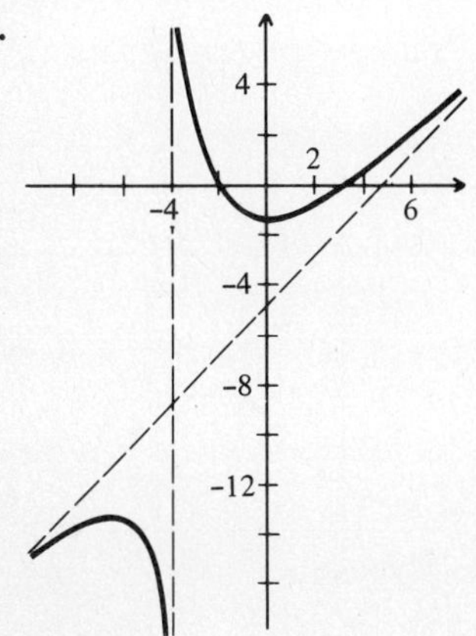

27.

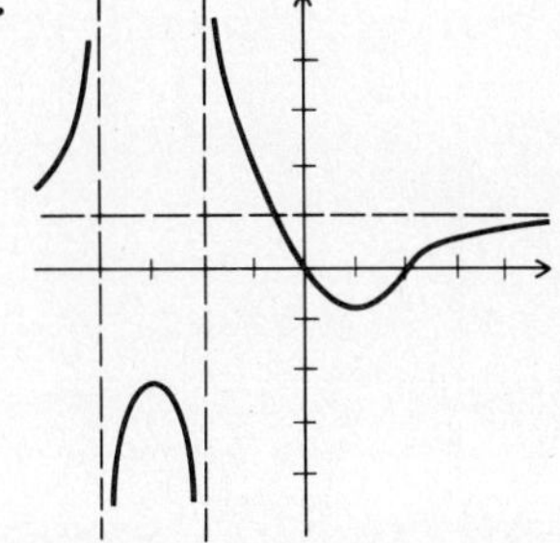

28.

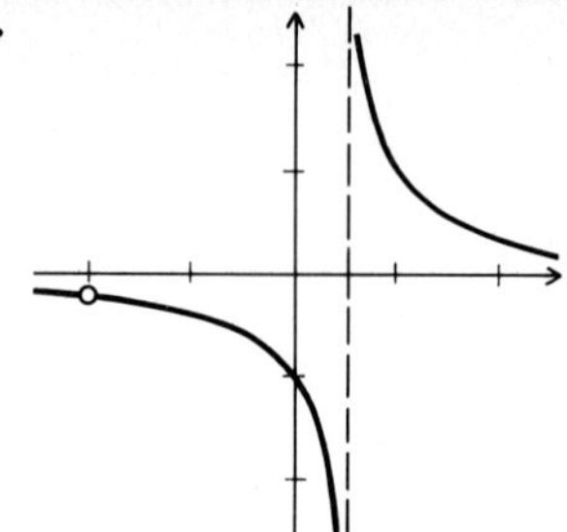

29.

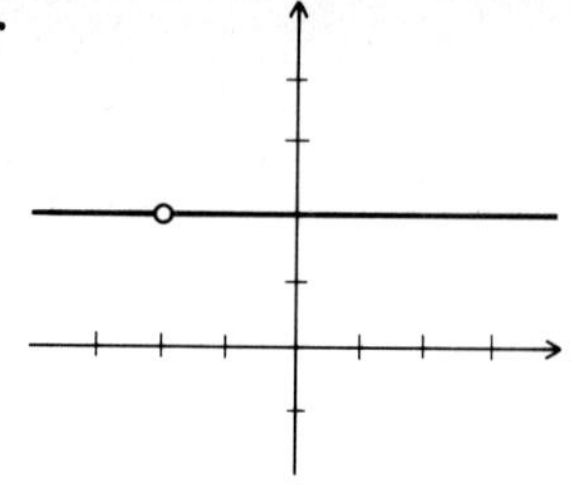

Problem Set 5.1, page 235

1.

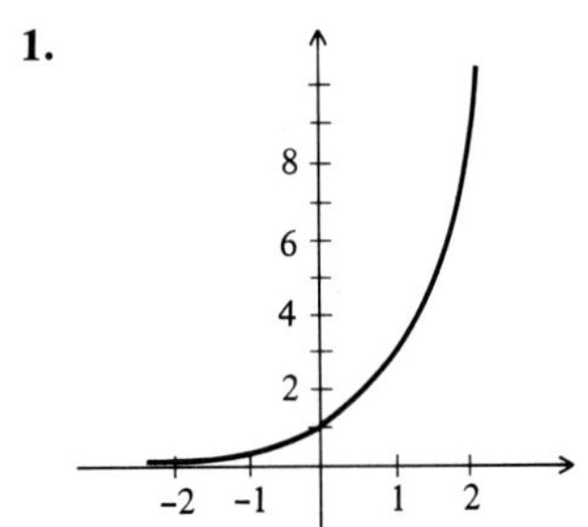

3.

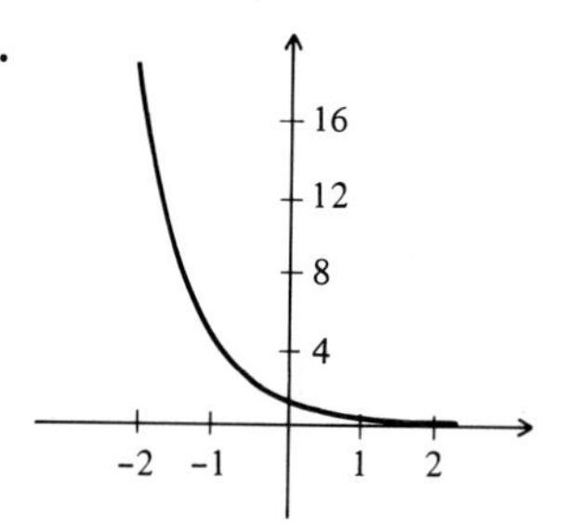

5.

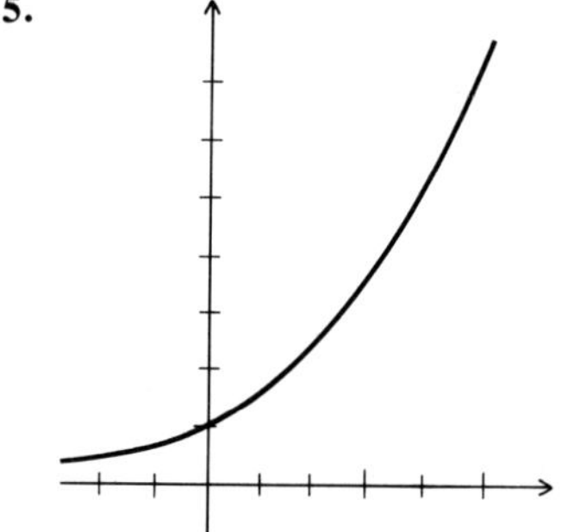

7.

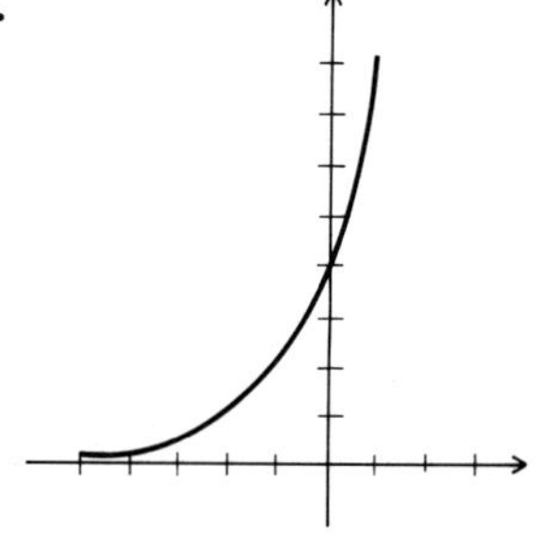

9.

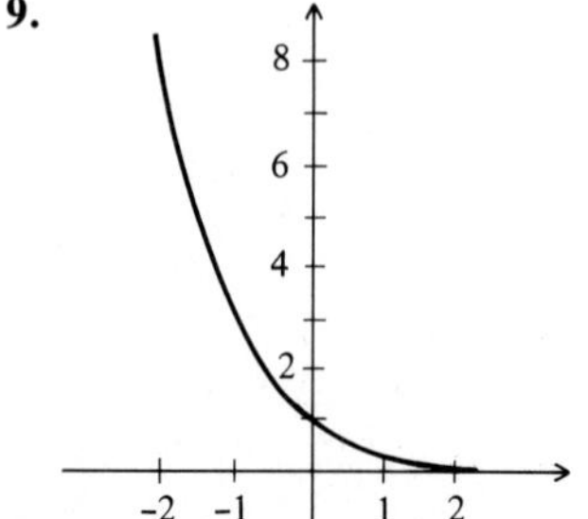

11.

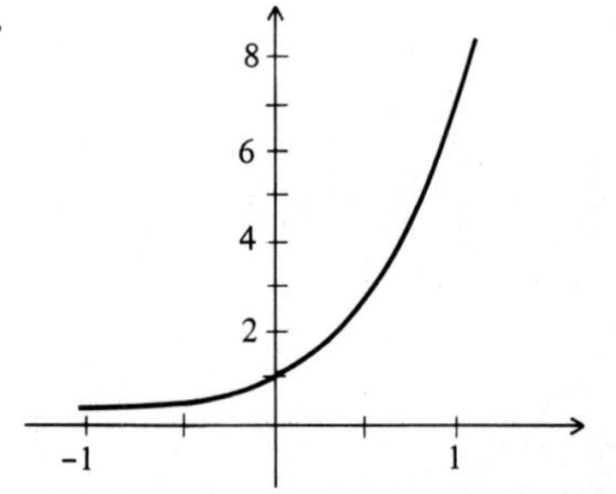

13.

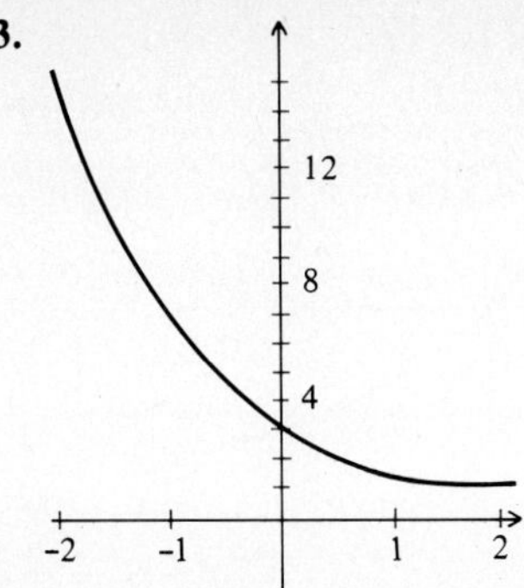

15.

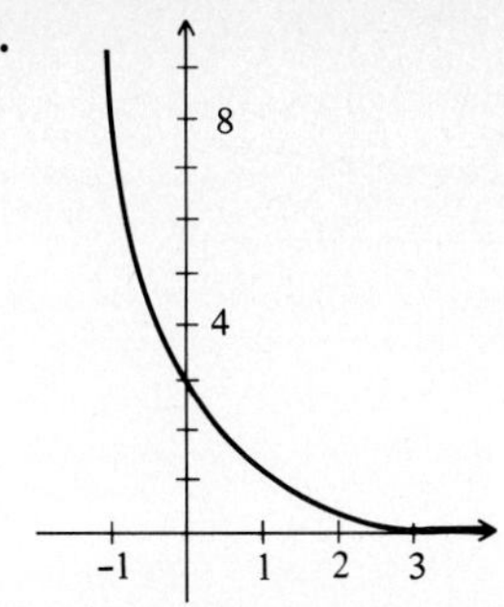

17.

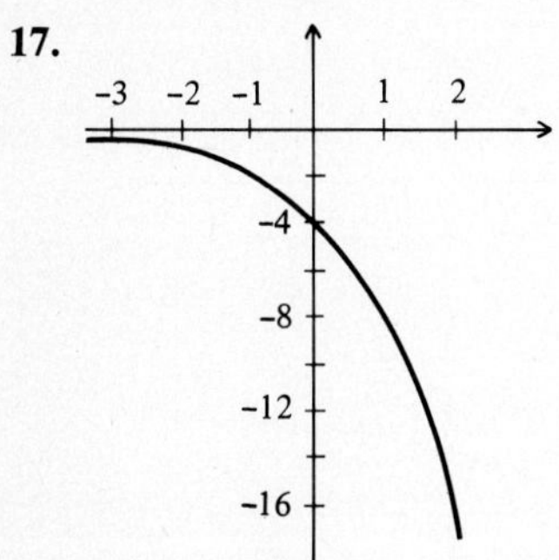

19.

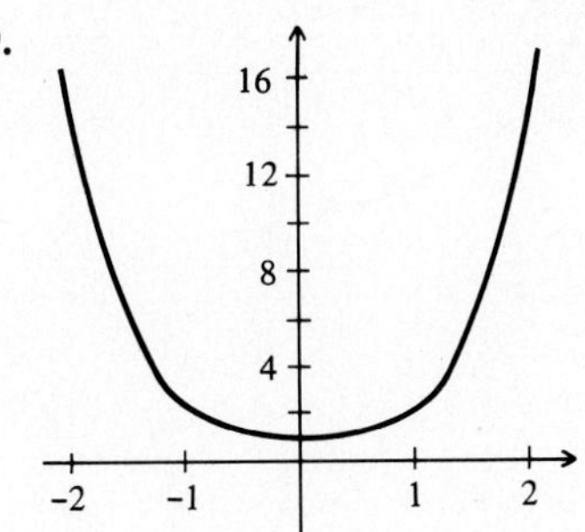

Problem Set 5.2, page 242

1. $4^3 = 64$

3. $\left(\frac{1}{3}\right)^{-2} = 9$

5. $4^{1/2} = 2$

7. $\log \frac{1}{10} = -1$

9. $\log_{1/3} \left(\frac{1}{27}\right) = 3$

11. $\log_{16} \left(\frac{1}{4}\right) = \frac{-1}{2}$

13. 2

15. -1

17. $\frac{1}{2}$

19. k^2

21. x

23. $\sqrt{x}$

25. $\log x + \log y - \frac{1}{2} \log z$

27. $\frac{1}{2} \log x + \frac{1}{2} \log y - 2 \log z$

29. $\log \frac{x\sqrt{z}}{y}$

31. $\log \frac{\sqrt{y}}{x^2 z}$

33.

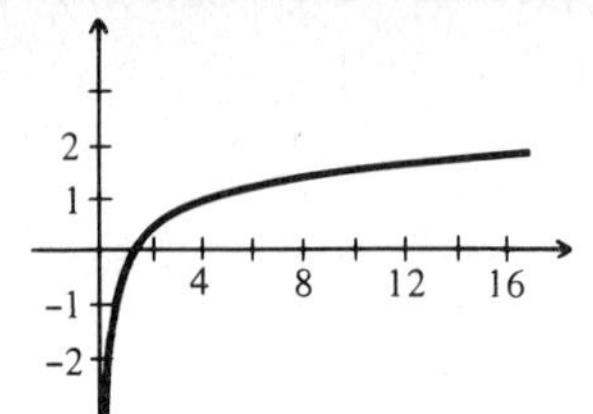

35.

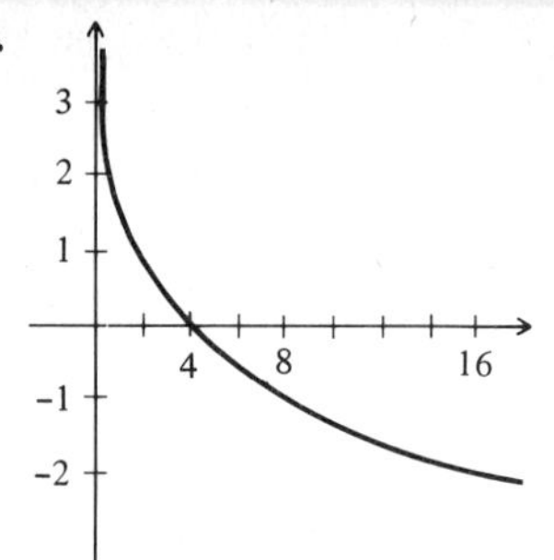

37.

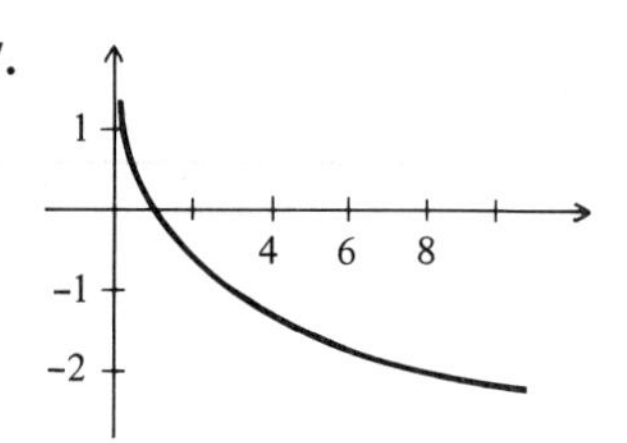

39.

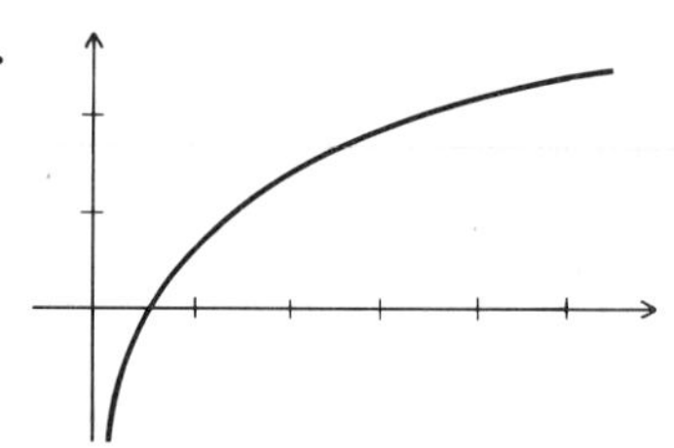

41.

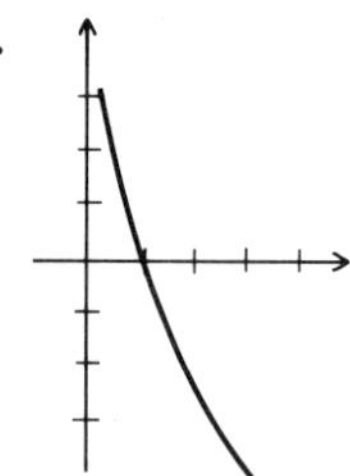

43.

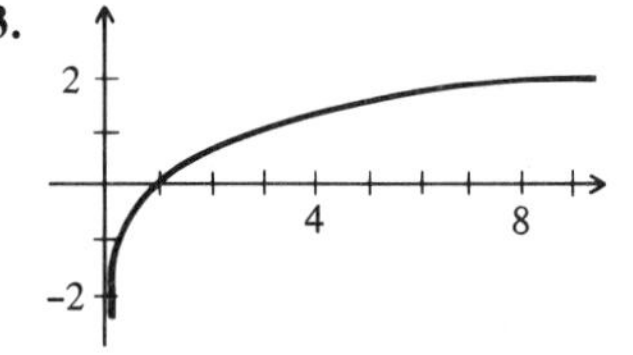

45. $\log (526)^{2/3} = \dfrac{2}{3} \log 526 = \dfrac{2}{3}(2.7210) = 1.8140$

Therefore, $(526)^{2/3} \approx 10(6.52) = 65.2$

Problem Set 5.3, page 249

1. $\left\{\dfrac{\log 14}{\log 6}\right\}$ **3.** $\{\log 2\}$ **5.** $\left\{\dfrac{\log 73}{\log 64}\right\}$

7. $\left\{\dfrac{-\log 2}{\log 5}\right\}$ **9.** $\left\{\dfrac{-1}{2}\right\}$ **11.** $\left\{\dfrac{\log 4}{\log \dfrac{9}{16}}\right\}$

13. $\left\{\dfrac{5 - \log 18}{1 + \log 18}\right\}$ **15.** $\varnothing$ **17.** $\left\{\dfrac{-3 + \ln 15}{2}\right\}$

19. $\{\ln 2, -\ln 2\}$ **21.** $\left\{\dfrac{1}{20}\right\}$ **23.** $\{10\}$

25. $\left\{\dfrac{-\sqrt{10}}{12}\right\}$ **27.** $\left\{\dfrac{25}{2}\right\}$ **29.** $\left\{\dfrac{3}{25}\right\}$

31. $\{1, 100\}$ **33.** $\left\{\dfrac{\sqrt{10}}{10}\right\}$ **35.** $\varnothing$

37. $\{2e\sqrt{e}\}$

39. $\left\{\dfrac{\log 3}{\log 8} - 2\right\}$

41. $\left\{\dfrac{5}{2 - \ln 2}\right\}$

43. $\{1.30\}$

45. $\{-.72\}$

47. $\{2.32\}$

49. $\{2.46\}$

51. $\{1.82\}$

53. $\{1.31\}$

Problem Set 5.4, page 254

1. $\dfrac{100}{7} \ln \dfrac{3}{2}$ (≈ 5.8) yr

3. $\dfrac{5 \ln 4}{3}$ (≈ 2.3) hr

5. $100 \ln 2$ (≈ 69.3) yr

7. $\left(\dfrac{250}{7} \ln \dfrac{6}{5}\right)\%$ ($\approx 6.5\%$)

9. $\dfrac{\ln 3}{6}$ ($\approx 18\%$) per hour

11. $\dfrac{50 \ln 5}{\ln 2}$ (≈ 116) yr

13. $\dfrac{20 \ln \frac{4}{3}}{\ln \frac{3}{2}}$ (≈ 14.2) hr

15. $\dfrac{1100 \ln \frac{14}{15}}{\ln \frac{29}{30}}$ (≈ 2239) ft

17. $10 \ln 4$ (≈ 14) days

Chapter 5 Review, page 255

1.

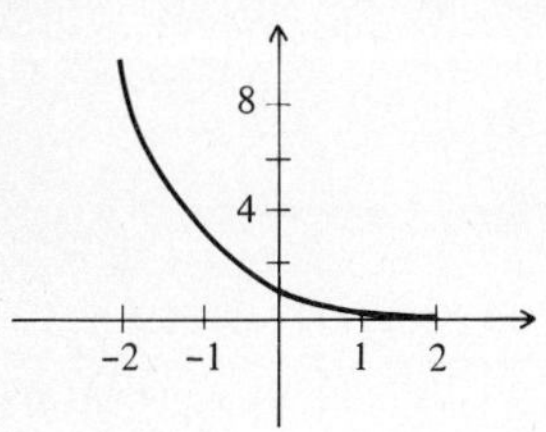

2.

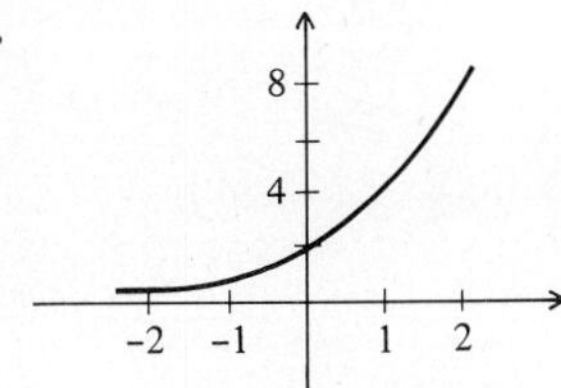

3.

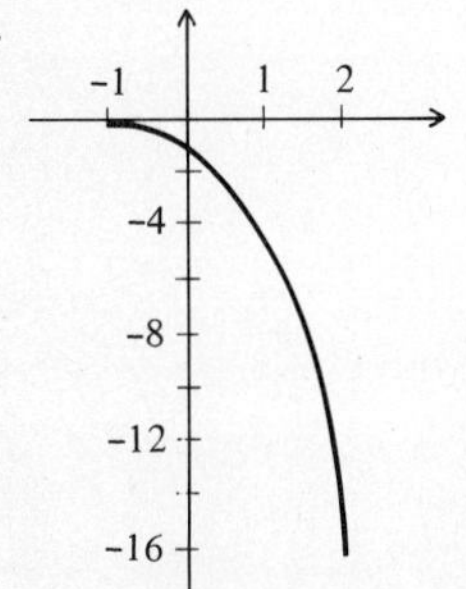

4.

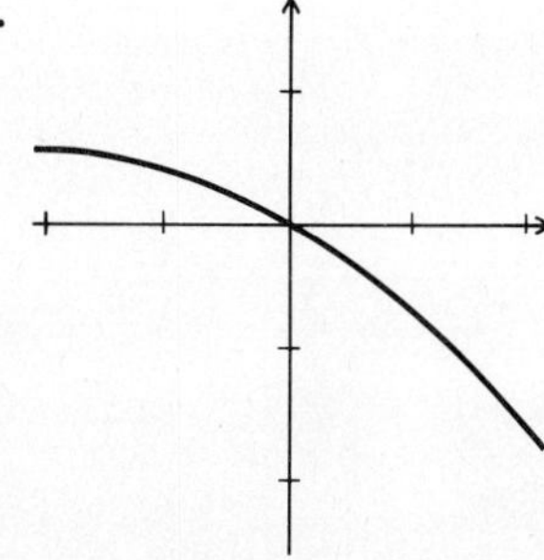

5.

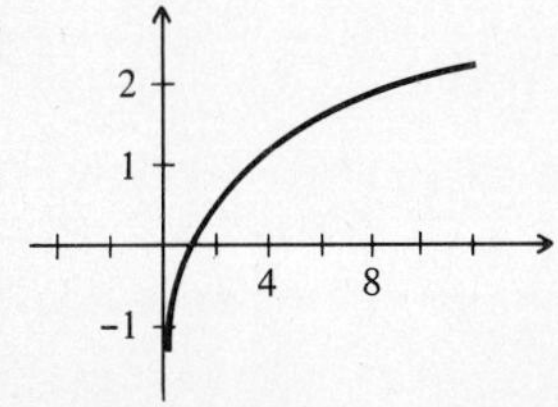

6.

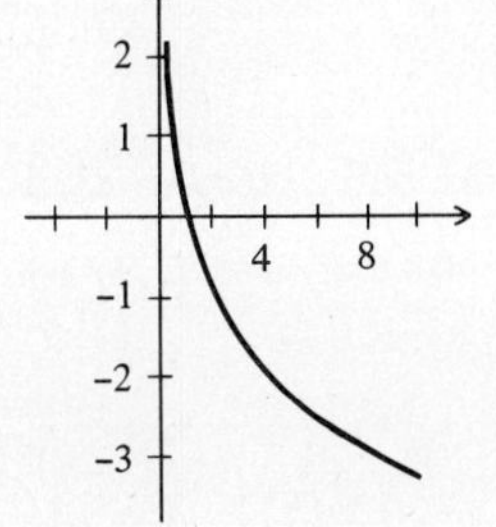

7.

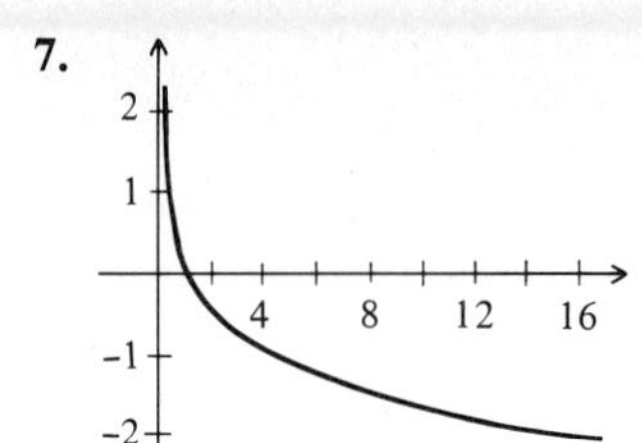

8.

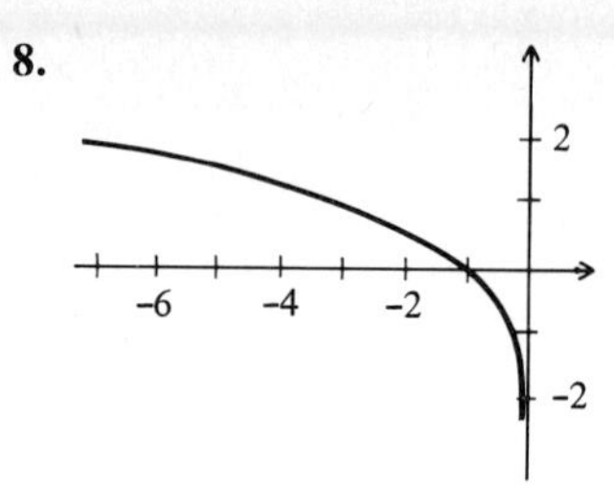

9. $\log_{16} 8 = \frac{3}{4}$

10. $8^{-5/3} = \frac{1}{32}$

11. 1

12. $2 \log x + \frac{1}{2} \log y - 3 \log z$

13. $\log \frac{xz^2}{\sqrt[3]{y}}$

14. $\left\{\frac{\log 20}{\log 5}\right\}$

15. $\left\{\frac{-1}{\log 2}\right\}$

16. $\left\{\frac{-\log 8}{\log 200}\right\}$

17. $\left\{\sqrt{\frac{\log 20}{\log 5}}, -\sqrt{\frac{\log 20}{\log 5}}\right\}$

18. $\varnothing$

19. $\{20\}$

20. $\left\{\frac{\sqrt{10}}{10}\right\}$

21. $\left\{\frac{\sqrt{2}}{10}\right\}$

22. $\{4\}$

23. $\left\{\frac{4}{3}\right\}$

24. $\left\{\frac{3\sqrt{11}}{10}, \frac{-3\sqrt{11}}{10}\right\}$

25. $\{1, 10^{\sqrt{3}}, 10^{-\sqrt{3}}\}$

26. $\left\{\frac{2}{e^5}\right\}$

27. $\{-2e\}$

28. $\{-2.58\}$

29. $\{.83\}$

30. $\{-9.55\}$

31. $\{1.19, 6.20\}$

32. $\frac{\ln 2}{5}$ (≈ 0.139)

33. $\frac{5 \ln 2}{\ln 3 - \ln 2}$ (≈ 8.5) min

34. $\frac{-500 \ln \frac{2}{5}}{\ln 2}$ (≈ 661) yr

35. $\frac{5 \ln 2}{2}$ (≈ 1.7) hr

36. $(10 \ln 2)\%$ $(\approx 6.9\%)$

Problem Set 6.1, page 263

1. $\{(3, 2)\}$

3. $\left\{\left(\frac{-3}{2}, \frac{21}{2}\right)\right\}$

5. $\{(2, -4)\}$

7. $\{(0, 4)\}$

9. $\{(-2, 3)\}$

11. $\{(3, -5)\}$

13. $\left\{\left(\frac{25}{19}, \frac{-9}{19}\right)\right\}$ **15.** $\left\{\left(\frac{-5}{6}, \frac{5}{12}\right)\right\}$ **17.** $\varnothing$

19. $\{(\sqrt{2}, 1)\}$

Problem Set 6.2, page 267

1. $\{(2, 0)\}$

3. $\left\{\left(\frac{5}{3}, 1\right)\right\}$

5. $\{(17, 9)\}$

7. $\{(x, y) \mid 2x - 5y = 5\}$

9. $\{(1, 0)\}$

11. $\left\{\left(\frac{7}{25}, \frac{1}{5}\right), \left(\frac{-1}{5}, 1\right)\right\}$

13. $\varnothing$

Problem Set 6.3, page 272

1. $\left\{\left(-5, \frac{-5}{8}\right)\right\}$

3. $\left\{\left(\frac{1}{5}, \frac{1}{2}\right)\right\}$

5. $\left\{\left(\frac{1}{2}, \frac{1}{3}\right)\right\}$

7. $\{(-3, 2)\}$

9. $\{(-5, 0), (4, 3)\}$

11. $\{(2, \sqrt{3}), (-2, \sqrt{3}), (2, -\sqrt{3}), (-2, -\sqrt{3})\}$

13. $\{(2, 2), (-2, -2)\}$

15. $\{(-1, 2\sqrt{2}), (-1, -2\sqrt{2})\}$

17. $\left\{\left(2, 1\right), \left(\frac{-7}{4}, \frac{-1}{4}\right)\right\}$

Problem Set 6.4, page 275

1. 15 and 12

3. 185 and 186

5. \$2500 at 6%
\$1500 at 8%

7. 11 and 15

9. 55 mph for car, 195 mph for plane

11. 4 liters of 40% solution,
16 liters of 15% solution

13. \$180 and \$205

15. 6.25 by 6.25 cm

17. 5 by 12 in.

19. 2 by 9 or 6 by 3 in.

21. 22 ft

23. current 3 mph, boat 15 mph

25. 20 by 30 in.

27. insufficient information

29. (2, 2)

31. $y = \frac{(x - 1)^2}{8} + 5$

33. $\frac{x^2}{16} + \frac{y^2}{20} = 1$

Problem Set 6.5, page 282

1. $\{(2, 1, 3)\}$

3. $\left\{\left(\frac{1}{2}, -3, \frac{3}{2}\right)\right\}$

5. $\{(1, -1, 2)\}$

7. $\{(0, 2, 1)\}$

9. $\{(-11, -12, 8)\}$

11. $\left\{\left(x, \frac{x+2}{2}, \frac{-x}{2}\right) \middle| x \in \mathsf{R}\right\}$

13. $\varnothing$

15. $\{(2, -3, -1)\}$

17. $\left\{\left(1, \frac{1}{2}, \frac{-1}{3}\right)\right\}$

19. \$10,000 in bonds, \$4000 in savings, \$6000 in stocks.

21. $x^2 + y^2 - 6x + 4y - 12 = 0$

23. 2 hr for pipe A, 4 hr for B, 3 hr to drain

Problem Set 6.6, page 288

1. $\{(2, 3, -4)\}$

3. $\left\{\left(\frac{1}{2}, -3, \frac{3}{2}\right)\right\}$

5. $\{(3, 2, -4)\}$

7. $\{(-1, 5, 2)\}$

9. $\left\{\left(\frac{-1}{2}, 3, \frac{-3}{2}\right\}\right.$

11. $\varnothing$

13. $\{(5, -2, -1, 5)\}$

Problem Set 6.7, page 294

1. 13

3. 7

5. -30

7. $M_{12} = 30$, $M_{23} = 3$

9. 49

11. -168

13. 26

15. -70

Problem Set 6.8, page 300

1. $\{(-1, 6)\}$

3. $\{(1, 1)\}$

5. $\left\{\left(\frac{-9}{19}, \frac{25}{19}\right)\right\}$

7. $\varnothing$

9. $\{(1, 2, -1)\}$

11. $\{(1, 2, 3)\}$

13. $\{(-11, -12, 8)\}$

15. $\left\{\left(\frac{3}{2}, \frac{1}{2}, -3\right)\right\}$

17. $w = 5$

Chapter 6 Review, page 301

1. $\{(-2, 5)\}$

2. $\{(4, -1)\}$

3. $\left\{\left(\frac{-3}{4}, \frac{5}{4}\right)\right\}$

4. $\left\{\left(\frac{-3}{5}, \frac{5}{2}\right)\right\}$

5. $\left\{\left(\frac{17}{3}, -6\right)\right\}$

6. $\left\{(0, 0), \left(\frac{-5}{4}, \frac{15}{8}\right)\right\}$

7. $\left\{\left(\frac{1 + 3\sqrt{41}}{8}, \frac{-3 - \sqrt{41}}{8}\right), \left(\frac{1 - 3\sqrt{41}}{8}, \frac{-3 + \sqrt{41}}{8}\right)\right\}$

8. $\{(-4, 3)\}$

9. $\left\{\left(\frac{-2}{3}, \frac{3}{5}\right)\right\}$

10. $\{(3, -2), (-1, 0)\}$

11. $\{(\sqrt{6}, \sqrt{2}), (\sqrt{6}, -\sqrt{2}), (-\sqrt{6}, -\sqrt{2}), (-\sqrt{6}, \sqrt{2})\}$

12. $\varnothing$

13. 4 in. by 9 in.

14. 6-ft base, 5-ft height

15. 49

16. 80 members, $12 initial cost

17. $\frac{4y^2}{9} - \frac{x^2}{4} = 1$

18. $\left\{\left(\frac{2}{3}, \frac{1}{4}, \frac{1}{2}\right)\right\}$

19. $\{(-4, 5, 1)\}$

20. $\{(0, 2, 1)\}$

21. $\left\{\left(\frac{1}{5}, \frac{-3}{5}, -2\right)\right\}$

22. $\{(-3, 4)\}$

23. $y = \frac{-1}{4}$

Problem Set 7.2, page 311

1. $x^6 + 6x^5y + 15x^4y^2 + 20x^3y^3 + 15x^2y^4 + 6xy^5 + y^6$
3. $x^5 - 5x^4y + 10x^3y^2 - 10x^2y^3 + 5xy^4 - y^5$
5. $x^4 - 8x^3y + 24x^2y^2 - 32xy^3 + 16y^4$
7. $x^4 - 20x^3y + 150x^2y^2 - 500\,xy^3 + 625y^4$
9. $8x^3 + 36x^2y + 54xy^2 + 27y^3$
11. $1 + 8x + 28x^2 + 56x^3 + 70x^4 + 56x^5 + 28x^6 + 8x^7 + x^8$
13. $x^6y^6 + 12x^5y^5 + 60x^4y^4 + 160x^3y^3 + 240x^2y^2 + 192xy + 64$
15. $120x^7y^3$
17. $608{,}256x^{10}y^2$

Problem Set 7.3, page 316

1. $\frac{-1}{3}, \frac{1}{3}, 1$

3. 0, 1, 4

5. $\frac{3}{2}, \frac{5}{6}, \frac{7}{12}$

7. $0, \frac{\log 2}{2}, \frac{\log 3}{3}$

9. $\frac{-1}{2}, \frac{1}{2}, \frac{-3}{8}$

11. $x, -x^2, x^3$

13. 5, 9, 17, 33

15. 1, 0, 1, 0

17. $-1, -1, -2, -6$

19. $x, -x^2, -x^6, -x^{24}$

21. -20

23. 26

25. 14

27. $\frac{-x^2 + x - 1}{x^2}$

29. $\sum_{i=1}^{5} \frac{1}{i^2}$

31. $\sum_{i=1}^{5} (-1)^{i+1} i$

33. $\sum_{i=0}^{2} e^i$

35. $\sum_{i=1}^{5} (-1)^{i+1}x^{2i-1}$

37. 1, 1, 2, 3, 5, 8, 13

39. $1, \frac{1}{2}, 9, \frac{1}{4}$

Problem Set 7.4, page 322

1. **(a)** 1, 6, 11, 16 **(b)** 46 **(c)** 235
3. **(a)** $-2, -8, -14, -20$ **(b)** -56 **(c)** -290
5. 1, 3, 9, 27
7. $-5, 10, -20, 40$
9. $-1, \frac{-1}{3}, \frac{-1}{9}, \frac{-1}{27}$
11. $\sqrt{2}, -\sqrt{6}, 3\sqrt{2}, -3\sqrt{6}$
13. $\frac{200}{27}$
15. -30
17. $a_1 = 4$
19. -1
21. 15 log 3
23. 185 ft
25. 345,600
27. $\frac{235}{8}$ (≈ 29.4) ft

Chapter 7 Review, page 324

6. $x^5 - 5x^4 + 10x^3 - 10x^2 + 5x - 1$
7. $256x^4 + 256x^3y + 96x^2y^2 + 16xy^3 + y^4$
8. $x^6 - 18x^5y + 135x^4y^2 - 540x^3y^3 + 1215x^2y^4 - 1458xy^5 + 729y^6$
9. $180x^8y^2$
10. $1, -5, -15$
11. $-1, \frac{-3}{8}, \frac{-5}{27}$
12. $-3, \frac{9}{2}, -9$
13. $\frac{1}{2}, \frac{-1}{8}, \frac{1}{24}$
14. $x, -2x^2, 3x^3$
15. $x, x, 2x$
16. $-3, 17, -83$
17. $-2, -4, 16$
18. $\frac{7}{6}$
19. 42
20. -14
21. x^2
22. $\sum_{i=1}^{5} (-1)^{i+1}$
23. $\sum_{i=1}^{5} \frac{i}{2i-1}$
24. **(a)** $-3, -8, -13, -18$ **(b)** -255
25. **(a)** 6, 10, 14, 18 **(b)** 240
26. $\frac{-1}{2}, \frac{-3}{2}, \frac{-9}{2}, \frac{-27}{2}$
27. $5, 5\sqrt{2}, 10, 10\sqrt{2}$
28. $\frac{-325}{8}$
29. $12(1 - \sqrt{2})$
30. 54
31. $10,737,418.23

Problem Set 8.1, page 330

1. $y = -30$
3. $y = -2/3$
5. $x = -6$
7. $z = 5/16$
9. 4π
11. 28,000
13. 125 kg
15. $\frac{80}{3}$ ($\approx$26.7) kw
17. 12 ohms
19. 480 lb
21. 5.2

Problem Set 8.2, page 334

1. $A = x(10 - 2x)$
3. $l = \sqrt{36 + x^2} + \sqrt{64 + (20 - x)^2}$
5. $d = \sqrt{(5 - 15t)^2 + (10t)^2}$
7. $C = 3x + \dfrac{120{,}000}{x}$
9. $A = (x - 1)\left(\dfrac{180 - 3x}{2x}\right)$
11. $V = 9d^2$
13. $S = \dfrac{\sqrt{3}\,x^2}{36} + \dfrac{(10 - x)^2}{16}$
15. $S = \dfrac{\pi\sqrt{r^6 + 81^2}}{r}$
17. $A = 30r - 2r^2 - \dfrac{\pi r^2}{2}$
19. $L = 25x\sqrt{36 - x^2} + 225\pi$

Problem Set 8.3, page 342

1. $11 + 6i$
3. $1 - 5i$
5. $2 + 5i$
7. $-4 + 3i$
9. $-2 + 23i$
11. $-24 + 27i$
13. $\dfrac{13}{34} + \dfrac{35}{34}i$
15. $\dfrac{-20}{13} - \dfrac{9}{13}i$
17. 5
19. $-5 + 12i$
21. 1
23. $-i$
29. $\{4i, -4i\}$
31. $\{-3 + 2i, -3 - 2i\}$
33. $\left\{\dfrac{-3 + \sqrt{11}i}{2}, \dfrac{-3 - \sqrt{11}i}{2}\right\}$
35. $\left\{\dfrac{2 + \sqrt{2}i}{2}, \dfrac{2 - \sqrt{2}i}{2}\right\}$
37. $\left\{3, \dfrac{-3 + 3\sqrt{3}i}{2}, \dfrac{-3 - 3\sqrt{3}i}{2}\right\}$
39. $\left\{\dfrac{5 - 4i}{2}\right\}$

Chapter 8 Review, page 343

1. 12-in.
2. $\approx$225 days
3. 0.3 amp
4. $V = \dfrac{4\pi d^3}{75}$
5. $A = \dfrac{1}{4}(x + 10)\sqrt{400 - (x - 10)^2}$
6. $S = \dfrac{\sqrt{3}x^2}{36} + \dfrac{(36 - x)^2}{18}$
7. $V = 24d + 3d^2$
8. $2 + 7i$

9. $5 - 7i$

10. $17 - 7i$

11. $1 - 7i$

12. $\frac{8}{13} + \frac{14}{13}i$

13. $\frac{40}{29} + \frac{13}{29}i$

14. $-5 - 12i$

15. $-i$

16. $\{7i, -7i\}$

17. $\left\{\frac{1 + i}{2}, \frac{1 - i}{2}\right\}$

18. $\left\{\frac{3 + \sqrt{3}i}{2}, \frac{3 - \sqrt{3}i}{2}\right\}$

19. $\left\{\frac{3 + \sqrt{5}i}{2}, \frac{3 - \sqrt{5}i}{2}\right\}$

20. $\{2, -1 + \sqrt{3}i, -1 - \sqrt{3}i\}$

Index

E

F

G

H

I

L

M

N

O

P

Q

R

S

T

U

V

X

Y

Z